cookwise

cookwise

the hows and whys of successful cooking

Shirley O. Corriher

WILLIAM MORROW
75 YEARS OF PUBLISHING
An Imprint of HarperCollins*Publishers*

This book is dedicated to everyone who has ever wondered "Why?"

FOOD PHOTOGRAPHY BY AL CLAYTON

FOOD STYLING BY MARY ANN CLAYTON

BOOK DESIGN BY VERTIGO, NYC

Library of Congress Cataloging-in-Publication Data

Corriher, Shirley O.
    Cookwise: the hows and whys of successful cooking / by Shirley O. Corriher
        p   cm.
    Includes index.
    ISBN 0-688-10229-8
    1. Cookery.   I. Title.
TX714.C6917   1997
641.5—dc21   97-20209
                                                      CIP

03 04 05 WBC/RRD 15

My Gratitude and Thanks

"No man is an island," and there is certainly the love, support, and work of many people in this book.

First of all, thanks to my very huggable husband, Arch, an indefatigable and incredibly creative researcher, who has been through so much on this book—years of working side by side with me into the night. His years as a research engineer and editor served him well in helping me. He can find information that is unfindable. Give him a computer and a good library and he will find the answer to any question, asked or not. Working together is not easy for people who are face-to-face for twenty-four hours a day. But there is no Einstein nor computer whiz I would swap him for in the office, nor any Hollywood Casanova I would swap him for in my life.

To Nathalie Dupree, who has long been the center of cooking in Atlanta, for providing a warm, nurturing, inclusive atmosphere for all Atlanta foodies. Through the years, Nathalie has generously opened her home and her heart. Time and time again she has provided the opportunity for everyone from top food writers to apprentices to spend the evening with cooking celebrities. When a famous culinary person was in town, Nathalie would ask us to call everyone we knew— "Bring a dish and come to Nathalie's to meet . . . "

To Doris Koplin, an outstanding baker and my trainer as a traveling cooking teacher. On our first trip together Doris said, "O.K., Shirley, I know you're a wealth of information, tell me everything you know about yeast and I'll show you how to survive teaching on the road." Doris and I have leaned on each other ever since.

Harold McGee, author of *On Food and Cooking* and *The Curious Cook*—no one could ever ask for a more supportive colleague or a more fun cohort on adventures. Harold is everything from my intellectual hero to my companion in mischief. He has generously given me information and helped me research "mys-teries." It is wonderful to have someone who shares my joy over a really good article on baking powder! Together we have had great times—from co-hosting workshops to drinking hundred-year-old Marsala in Sicily.

Rose Levy Beranbaum, author of *The Cake Bible,* a real cake and pastry expert and a best buddy with whom I have shared and swapped cooking secrets, gossip, and great frolicking times. To try a restaurant we had heard about, we once went to Arboise in France for one evening and stayed a week.

Susan Purdy, author of *Have Your Cake and Eat It, Too,* and Bruce Healy, author of *Mastering the Art of French Pastry*, whose challenging questions and observations have spurred me to greater heights in baking. Swapping information and questions with knowledgeable colleagues like these has been invaluable.

To Judith Weber, agent extraordinaire, who has gone a million miles above and beyond the call of duty to get this book out. For the ten years since Judith got me the contract, she has stood by me through thick and thin. Sometimes it was only her support that bolstered my courage.

To those who have worked directly on bringing this book to press:

Susan Friedland, the editor who first bought this book at Prentice-Hall. Susan was willing to take a chance on a totally unknown writer and got me off to a great start. Not only did Susan buy my book, she also found me an agent, Judith Weber, for which I am exceedingly grateful.

Toula Polygalaktos, whose support kept me on a straight path in the early stages of the book.

Maria Guarnaschelli at Morrow, whose editorial genius went right to the heart of the book's needs. She hired John Willoughby, an experienced editor, to cut and paste to improve cohesion. John was a big help in tightening the organization.

Harriet Bell fearlessly dug into the nitty-gritty of getting a large manuscript into a tight but comprehensive book. Harriet brought in Susan Derecskey, a meticulous line editor who spent many hours doing a detailed edit. This is a better book for Susan's effort.

Ann Bramson, who said, "We are going to get this book out," and meant it. She guided the book forward with a firm, knowledgeable hand, but was ever the diplomat. Ann directed the final edit and guided production through the bound galleys. Chris Benton, an outstanding line editor, fine-tuned the manuscript, and we finally whipped it into shape.

My final editor, Pam Hoenig, who skillfully steered *CookWise* through final production.

Jennifer Kaye, who has patiently and expertly answered my calls and managed to do the impossible.

Deborah Weiss Geline and the excellent copy editors and production staff at Morrow.

Vertigo Design, who did the interior design of the book and made the unusual combination of text, tables, and recipes understandable and attractive.

My talented photographer and food stylist team, Al and Mary Ann Clayton, who translated the points I wanted to make into outstanding photographs.

The typesetters at Westchester Book Composition, who took a heavily edited manuscript and transformed it into exceptionally clean galleys.

Alexandra Nickerson, who understood both the overall purpose and the specific content of this book and did an outstanding job of indexing.

To my ever-faithful recipe testers:

My friend Susan Mack, a former partner at Bear Stearns and an outstanding business executive and organizer, who helped in many ways from editing and advising to literally taking over the recipe testing when she realized that I needed big help.

Week after week, she and her architect husband, Andy Armstrong, held recipe-testing parties in their magnificent professional kitchen, which Andy designed. Susan and I recruited friends and family. Arch bought cases of wine. Every week Susan assigned and dispatched recipes to our heroic testers, who not only prepared the dishes but also filled out lengthy questionnaires on how the cooking went. Then, we all showed up on Wednesday night to taste and critique.

I really appreciate the enormous effort of all these people who came out after a full day's work.

The Wednesday Night Heroes are: Mary Anderson, Andy Armstrong, LaLa Cochran, Hank Corriher, David Darnell, Leslie Fight-Brown, I'Ans Goad, Mary Anne Hill, Jane Holder, Susan Huggins, Fran Johnson, Jean Ellen Jones, Bernadette Leite, Anne Lynde, Barbara Martin, Beth McCool, Bill Meck, George Mende, Avery Morton, Joyce Pillot, Sherry Quarello, Mickey Register, Frances Roberts, Sylvia Robinson, Bill Ross, Doug Roberto, Vicki Smart, Jane Smith, Crit Stuart, Marty Thompson, Trisha Thompson, Mark White, Dick Wilkinson, and John Williams.

Special thanks to Marty Thompson, a talented and experienced food person, and a lifesaver to me. Marty would call on Fridays—"I'm going to have a little time this weekend, just leave recipes you need retested in your mailbox. I'll pick them up and bring you samples when I've finished."

And to my at-a-distance helper, Judith Breen, an excellent experienced cook, who tested recipes on her husband and family in East Lansing, Michigan.

To Dena Dougherty, a knowledgeable friend and cohort, who owned a dessert bakery, DK's Desserts, and let me use her professional equipment or order from her suppliers—an enormous help at times.

To Phyllis Hinton, who computerized some of the original drawings.

To food technologist friends and experts like Dr. Rob Shewfelt, Dr. Frank Bryan, Dr. Carl Hosney, Kay Engelhardt, and others who are named in the book.

To outstanding cooking friends like Margaret Fogleman, who have contributed recipes and encouragement.

To the rest of my family:

My sister Joyce Hutcheson, a home economics major, who was a regional manager for Cuisinarts and is an outstanding cook. Joyce has listened patiently to cooking and ingredient problems and made great suggestions, as well as contributing her recipe for coconut icing.

Hank Corriher and Beth McCool not only tested recipes time after time, they also proofread and typed as well as calmly listened and offered sympathy over various book disasters. Beth contributed the broccoli salad recipe.

My daughter Terry Hecht Infantino, a talented artist, penned many of the original drawings from which the design people worked, and for years has drawn for my handouts. Her husband, Carmelo Infantino, who can design and build anything, figured out a way and made everything from a bubble box for photographs on.

My daughter Sherry Hecht proofread for me early on and heroically helped me cook until three in the morning and drove me all over New York for an important photo shoot. Her fiancé, Joft Saks, patiently carried tons of groceries to the third-floor kitchen and box after box of finished dishes back to the car.

My daughter-in-law, Janice Hecht, an outstanding cook, who contributed the wonderful sweet starter bread.

To Susan Mack's white-china soup tureen, which still holds backup computer disks for multiple versions of the manuscript.

Finally, to many others who are named specifically in the text or recipes and to those who have been inadvertently omitted, I give my thanks.

Contents

Introduction

CookWise is for everyone interested in cooking. If you're a beginning baker making your first pie and need step-by-step directions for a fantastically flaky crust, turn to page 110. If you're a chef having trouble with stringy melted cheese, turn to page 282 (the secret is a little lemon juice). If you're a good home cook looking for the best biscuit recipe or a wonderfully new twist on preparing sweet potatoes for Thanksgiving, turn to pages 77 and 338.

I am thrilled when I learn something that is a major help in improving a dish. When I first learned how important bubbles in the fat are (the creaming step where you beat the butter and sugar to make a cake), all kinds of things became clearer to me. Baking powder and soda do not make a single new bubble; they only enlarge bubbles that are already there. No wonder I made terrible cakes; I barely stirred the butter and sugar together. No wonder Patsy, the pastry chef at a nearby restaurant, who beat the butter and sugar for 7 or 8 minutes, made Queen Mother cakes a mile high while Patrick, the head chef, who followed the same recipe but just threw things in in a hurry as I did, made miserable Queen Mother cakes.

CookWise contains a lot of basics like this. When you learn that too little liquid for the amount of chocolate in a recipe can cause the chocolate to seize and become a solid grainy mass, you can avoid many chocolate disasters. When you know how to estimate the right amount of baking powder, you can tell just by looking at a recipe that a cake is overleavened and may fall. When you know how many eggs you need per cup of liquid for a firm custard or a soft custard, you can see that a quiche doesn't have enough eggs. This kind of know-how puts *you* in control. Being able to write recipes that work is as valuable to home cooks and chefs who want to create their own dishes as it is to cookbook writers, product developers, and test kitchen professionals. Everyone who cooks benefits from knowing how to "fix" problem recipes and how to make ordinary dishes extraordinary.

CookWise offers over two hundred outstanding recipes, carefully chosen to help show the inner workings of an ingredient, technique, or principle. Each recipe lives up to two criteria—it's absolutely delicious and it demonstrates the principles discussed. Most recipes contain a section entitled What This Recipe Shows that summarizes those special points that are being applied. This little bit of science may indicate which steps or ingredients are vital and cannot be left out without consequences.

The recipes in *CookWise* are the ones the cooks I've taught over the years use again and again. In fact, one large bakery chain uses my biscuit and cornbread recipes, worked out for them by their product developer, one of my former students.

Among these recipes you'll find some surprises. Don't be afraid of the unexpected. Just relax and suspend disbelief until you've tried a vinaigrette without vinegar (to keep green veggies bright green) or a crisp, high-egg-white pâte à choux. Many times our age-old traditions of cooking are absolutely scientifically the best way, but not always. You may hesitate at first, but try new approaches and let your taste be the judge. Often, people testing my recipes have said things like "I knew you made a mistake in this dish because the amount of one of the ingredients was all wrong, but when I read your section What This Recipe Shows, I realized you really meant to use that amount. When I tasted it, it was great."

Many concepts shown here are my own inventions, as are many of the techniques. For example, sprinkling croissant or puff pastry dough with ice water before folding to keep it soft and easy to roll, using zest instead of juice to avoid acidic ingredients, using only two whole eggs and the rest whites to create dry, crisp cream puffs. And you will also see many, many people quoted or mentioned—sometimes two within a single recipe. I like to give credit, and if I got concepts that I used in the presentation from one person and ideas for a technique or ingredient from someone else, I tried to credit both.

How to use this book

You can open *CookWise* to any chapter. If your interest is in desserts, you may want to start with the last chapter, Sweet Thoughts and Chocolate Dreams, though there are other dessert recipes elsewhere. They are listed according to category in the front of the book and those in each chapter are listed on the first page of that chapter.

For quick solutions or answers to questions, you can use *CookWise* as a reference book. Forty-two At-a-Glance summary charts (throughout the book) make problem solving quick and easy.

CookWise focuses on how and why things happen in cooking so the roles that ingredients and techniques play are of paramount importance. The nature of the subject dictates the structure of each chapter. As a result, the organization of *CookWise* is different from that of other cookbooks. For example, Chapter 1, The Wonders of Risen Bread, contains extensive flour, yeast, and breadmaking information. Chapter 2, How Rich It Is!, concentrates on dishes where fat is a major component, such as in pie crusts, cookies, and cakes, and also covers the roles of fats in frying, nutrition, and low-fat cooking.

In Chapter 3, Eggs Unscrambled, in addition to basic egg cookery—poached eggs to omelets—custard mysteries are unmasked and soufflé myths are dispelled. Chapter 4, Sauce Sense, does not follow the traditional format based on the five "mother" sauces, but is organized according to the type of sauce—purees, reduction sauces, starch-bound sauces, and emulsion sauces.

You'll gain expertise on fruits and vegetables in Chapter 5, Treasures of the Earth, which opens with a section on how to select the best produce and give it the best care. You'll then learn about the changes that occur in produce during cooking and how to control these changes to your advantage. Chapter 6, Fine Fare from Land, Sea, and Air, begins with how proteins cook, how to tell which meat cuts will be tough or tender, and how to cook to enhance the cut you have.

Chapter 7, Sweet Thoughts and Chocolate Dreams, explores sugar, chocolate, and ice cream, which all involve working with crystals—how to form crystals (tempering chocolate) or how to stop them from forming (making caramel), and how to form baby-fine crystals for smooth creamy textures (ice cream and fudge). You'll also learn about another major dessert ingredient, whipped cream.

CookWise contains knowledge gained from years of cooking curiosity and has been ten years in the writing. From following the grandmother I adored about the kitchen, through my work as a research biochemist for the Vanderbilt University Medical School, through eleven years of feeding over a hundred hungry teenage boys three meals a day, to fun-filled years learning and cooking with home cooks and fellow professionals—bits and pieces of discoveries and know-how from all of this are here. I find great excitement in this knowledge that allows me to make dishes come out as I want. I hope that you too will find pleasure, not only in becoming a more informed and assured cook but in the amazing inner workings of food and cooking—these hows and whys that enable you to make dishes exactly as *you* want. There is great joy in no more failed recipes!

Recipes by Category

Appetizers

Soups and stocks

Eggs

Quick breads and yeast breads

Pies and pie crusts

Cakes

Other desserts

cookwise

the recipes

The Ultimate Brioche I: Light, Airy Breadlike Brioche

The Ultimate Brioche II: Buttery Cakelike Brioche

Incredible Toast Basic Loaf

Crusty French-Type Bread

Great Round Reuben

Brie Filled with Walnuts and Amaretto-Soaked Apricots in Brioche

Sausage in Brioche

Star Spectacular

Grape-Bunch Braid

Buttery Flaky Rolls

Honey Whole Wheat Loaf

Jack, Avocado, and Sprouts on Whole Wheat

Semolina Bread

Rice Bread

Mighty Multigrain—A Hearty Loaf

Firm Sourdough Starter

Wet Sourdough Starter

Sourdough Bread

Starter with Yeast

Yeast Starter Bread

Sweet Yeast Starter

Sweet Starter Bread

Nisu: Rich Finnish Coffee Bread

Focaccia

Light-as-a-Dream Hot Rolls

Crisp-Crusted, Feather-Light Raised Waffles

Touch-of-Grace Biscuits

Portuguese Sweet Bread

Spicy Sugar Lump Loaf

Honey-Walnut Sticky Buns

Savarin

Hot Cross Buns

the wonders
of risen bread

george greenstein in *Secrets of a Jewish Baker* describes perfect rye bread. He stands alone in a silent bakery with two hundred loaves of rye that he has just taken from the ovens. When the cold night air hits the hot crusts, they begin to crack—a two-hundred-participant symphony of precious sounds. He knows that he has made perfect bread.

Peter Reinhart, author of *Brother Juniper's Bread Book,* knows his bread is just right by the crunching sound of the crust when he bites into it. He says it is like the sound a perfectly hit tennis ball makes when it rockets from the sweet spot. When you hear it, you know it's right.

Good bread is no small feat. Ingredients can be as simple as flour, yeast, water, and salt, yet astoundingly different loaves can come from these basic components. When I was visiting master baker Louis Amighetti in St. Louis, he shook his head as he looked at the pale loaves that his inexperienced apprentice had just made—not at all like his richly browned ones. He explained, "You must pay attention. If the timing is not right, the bread is not right."

This chapter will arm you with the knowledge to make your own perfect loaf of bread.

Gluten:
the heart of bread

When you add water to wheat flour and stir, two proteins in the flour, glutenin and gliadin, grab each other *and* water. As you continue to stir, more and more of these proteins connect and cross-connect to form sheets of gluten. These remarkable elastic sheets (below) trap and hold air and the gases made by yeast and enable yeast bread to rise.

Wheat flour is a necessity for these bubble gum–like sheets. Corn, rye, oats, barley, rice, millet, and other grains can be ground into flour to make light baking powder–leavened products like pancakes, muffins, or quick breads. But only wheat flour contains enough of the two proteins glutenin and gliadin to make good sheets of gluten. Rye contains a small amount, and triticale, a rye-wheat crossbreed, has some of these proteins, too, but not enough to make a light bread.

A yeast bread made with any grain that does not contain these two proteins will not rise, no matter how much yeast is used. The yeast can produce millions of bubbles of gas, but without gluten to hold them, the bubbles float off into the air.

This stretchy gluten film traps and holds air bubbles as the dough is mixed and kneaded. Yeast, a one-cell plant that feeds on simple sugars from the dough, oozes out a liquid filled with alcohol and carbon dioxide. When this liquid touches the trapped air bubbles, it releases carbon dioxide gas and enlarges them. The dough becomes lighter and begins to rise. The yeast continues to feed. As long as there is oxygen in the dough, yeast divides and multiplies to produce even more carbon dioxide.

After the dough rises, it is punched down. The clumps of yeast are broken up and spread so that each cell is surrounded by a new food supply. Then the dough is shaped and set aside. Because of all the well-fed new yeast cells, the bread rises faster during this second rise.

Finally, the bread goes into a hot oven and rises even more. Yeast makes carbon dioxide faster as it warms. Also, the alcohol that the yeast has made from the beginning gets hot and changes to a gas, providing more gases to inflate the bubbles, and heat alone makes the gases expand, enlarging the bubbles. This last great rise, called *ovenspring,* continues until the yeast gets so hot that it dies.

Soft starch granules trapped in the film bend and curve themselves around and between the gas bubbles. The dough becomes hotter; the protein gluten film cooks and becomes firm. There are now millions of air cells with delicate thin linings—the incredible texture of bread.

Not all flours are created equal

For hundreds of years bakers have known that flour from one wheat will make a light, well-leavened loaf of bread with good keeping characteristics while flour from another will make a heavy, poor-quality loaf with poor keeping ability. High-protein wheat flour con-

Glutenin

Gliadin

Water

Gluten

taining good-quality glutenin and gliadin, sometimes called *strong flour,* makes good yeast breads. Strong flour is from the hard spring wheats grown in colder climates—in the great northern plains and Canada on our continent during the spring and summer. Soft winter wheats, grown in moderate climates where the ground never freezes to a depth greater than 10 inches, have much less glutenin and gliadin.

Not only can flour from different strains of wheat be different, but flour from the same strain of wheat can vary. Many things—soil, temperature, rainfall, maturity at harvest—influence protein content. Because the two proteins contribute different characteristics to the gluten—glutenin is responsible for elasticity, gliadin for softness—varying ratios will produce subtle differences in bread. Strains of wheat with the same total protein content may have different amounts or different qualities of glutenin. Consequently, one may make a slightly more elastic dough than the other. This may be one of the reasons that French flour, which is lower in protein than American high-protein flours, can make light, airy French bread. Still, the total amount of protein is the best indicator of overall bread-baking quality we have for any plain white flour.

This is where the miller comes in. Wheat kernels, also called *wheat berries,* contain the bran, the germ, and the endosperm, which is starch and protein. In milling, the kernels are cleaned and tempered (soaked in water) for easier removal of the germ and bran. The kernels are crushed and the germ and bran removed. The endosperm goes through one set of rollers and sifters after another, and these grind, sift, and separate the endosperm into fractions called *streams.* It is hard to imagine, but a kernel of wheat may be separated into eighty or more streams. Some streams are high in glutenin and gliadin, while others are high in starch. Every stream is analyzed so the miller knows exactly what is in it. Small millers with simple grinding stones can do some limited separation, too, and they know from experience roughly what is in their different streams.

Just as winemakers blend juices from different vineyards to make fine wines, millers blend flour from different streams of wheat to make different flours. For a bread flour, a miller includes a lot of flour from the high-protein streams. This flour will form good sheets of gluten and make light yeast bread. For a pastry flour, the miller includes very little protein since too much gluten will make a pie crust tough.

Whole wheat flour contains germ and bran. Cup for cup, whole wheat will have less of the gluten-forming proteins than plain flour from the same wheat just because it contains these other parts of the kernel. Whole wheat flour will make a heavier bread and is frequently blended with a high-protein plain white flour for lighter loaves.

What labels will (and won't) tell you

Total amount of protein is an indicator of bread-baking quality for plain white flour alone because rye flour, oat flour, and rice flour contain proteins unconnected with gluten, as does whole wheat flour with the proteins in the wheat germ. That means reading the label on these flours will not reveal much about the bread they will make.

Unfortunately, new Food and Drug Administration (FDA) regulations have made labels less informative even for white flour. The protein content stated on the label of a bag of flour is subject to a round-off rule, so flour labeled as having 9 grams of protein per serving actually can have from 8.50 to 9.49 grams. Under the old regulations (before May 1994), the serving size was 1 cup, and the protein content on the label effectively showed what the flour was best for: a flour labeled as 9 grams was indeed a low-protein flour, ideal for pie crusts and quick breads; a flour labeled as 14 grams (13.50 to 14.49 grams per cup) was a high-protein flour, excellent for yeast breads.

Under the new regulations, however, the serving size is $1/4$ cup, about 30 grams. With rounding, any flour containing 2.50 to 3.49 grams of protein per $1/4$ cup can be labeled as 3 grams. This means both moderately low-protein Southern flour (about 9 grams per cup) and high-protein unbleached flour (about 14 grams) can be labeled as 3 grams per $1/4$ cup. In fact, most flour on the market now says 3 grams of protein, telling you almost nothing about the protein content so important to cooking.

So how *do* you decide which flour to use? Again, as a general rule, high-protein bread flour is best for yeast breads, and low-protein cake flour is best for cakes.

Interestingly, because years ago most consumers buying unbleached flour were using it to make yeast-leavened bread, millers have frequently blended it with a somewhat higher protein content than bleached flour. So many unbleached flours will make light yeast breads. Self-rising flour is low in protein and contains leavening and salt. If you need a low-protein flour without leavening, for a pie crust, for example, you can use a Southern low-protein flour or mix an *instant flour* (such as Wondra), or cake flour, and a national brand of bleached all-purpose. You can call

APPROXIMATE PROTEIN CONTENTS OF DIFFERENT WHITE FLOURS AND THEIR USES

Flour (Examples)*	Protein		Uses
	(grams/cup)**	(percent)	
Cake (Swans Down, Softasilk)	8	7.5 to 8.5	Cakes, quick breads, muffins, pancakes
Instant flours (Shake & Blend, Wondra)	9 to 10	9.5 to 11	Sauce and gravy, blending to lower protein content
Bleached Southern all-purpose (White Lily, Martha White, Gladiola, Red Band)	9	7.5 to 9.5	Pie crusts, biscuits, quick breads, muffins
National brand self-rising (Gold Medal, Pillsbury)	9 to 10	9 to 10	Biscuits, quick breads, muffins
National brand bleached all-purpose (Gold Medal, Pillsbury)	11	9.5 to 12	A little too much protein for best pie crusts, quick breads, muffins, or pancakes; too little protein to make outstanding yeast breads
National brand unbleached all-purpose (Gold Medal, Pillsbury)	12+	10 to 12	Yeast breads, cream puffs
Northern all-purpose (Robin Hood, Hecker's)	11 to 12	11 to 12	Yeast breads, cream puffs, puff pastry
Northern unbleached all-purpose (King Arthur)	13	11.7	Yeast breads, cream puffs, puff pastry, pasta, pizza
Bread flour	13 to 14	11.5 to 12.5	Yeast breads, pasta, pizza
Durum wheat (semolina)	13+	13 to 13.5	Pasta

*Use of brand names as examples does not imply that the values in the table are exact for each brand. There also can be regional variations within some brands.
**Approximate FDA labeling before the label change of May 1994 (see text).

the flour company and ask the exact protein content, but in my experience what you are told by a consumer representative is not always reliable. The best way to find out the approximate protein content of plain flour is to observe how it absorbs water (see page 10).

Wheat is a natural product whose properties vary from region to region and from season to season, so flours vary also. Flour from soft wheats has a greater variability than that from hard wheats.

You can also use the table on page 6 as a general guide, which gives approximate values for protein in both grams per cup and percentages. The measure of flour strength that is used professionally is percentage protein, and this is starting to be used for home cooks. Cookbooks by professional bakers have occasionally used this measure, and *The King Arthur Flour Baker's Catalogue* includes it for King Arthur flours (see Sources—Flour, page 478).

Gluten and the cook

The gluten-forming proteins affect the cook in two different ways. First, a high-protein flour will absorb a lot of water compared to a low-protein flour. This has a major impact on all recipes containing flour: The same amount of flour and water that makes a firm dough with a high-protein flour makes soup with a low-protein flour (see examples on page 8).

Second, as already stated, the amount of these proteins controls the amount of gluten that the flour will form and determines the best use for the flour. Sometimes you need gluten, but sometimes you don't.

When do you need gluten?

The presence of gluten is perfect for yeast breads but disastrous for quick breads, cakes, muffins, and pancakes. Yeast oozes out a liquid that releases carbon dioxide to gently inflate bubbles and stretch gluten with slow, steady pressure. Baking soda and baking powder work in an entirely different way. If you have ever stirred baking powder into hot water to check its potency, you've observed a mass of fine bubbles coming up instantly. Heavy elastic sheets of gluten hold down and interfere with chemical leaveners and make baked goods using them heavy and tough. You want to select a low-protein flour that forms less gluten for light, tender cakes or muffins.

In general, gluten proteins are a blessing when you need strength but a disaster when you need tenderness. Gluten is strong and tough and elastic and holds things together. It is essential for a paper-thin, strong strudel pastry but will toughen a pie crust and make it shrink. Gluten in a cake will toughen it and create tunnels.

Many foods are made light and airy by steam. Water in the dough or batter turns to steam in the hot oven and makes puff pastry, cream puffs, génoise, and other cakes and pastries rise. Some of these steam-leavened baked goods need gluten, and others do not.

Puff pastry, for example, contains no eggs and depends solely on paper-thin layers of dough to hold the steam and puff. The higher-protein flours make much lighter puff pastry in spite of the elasticity as long as the dough is kept moist and soft. (If the dough becomes tight, sprinkle it with ice water just before folding.) Pâte à choux, popovers, and Yorkshire pudding contain eggs, but much of their ability to hold steam comes from gluten, so it is important to have a relatively high-protein flour to make them light. A génoise, on the other hand, is essentially held together and leavened by eggs. A high-protein flour could make this cake tough. So with steam-leavened foods you need to take into account the role played by eggs and whether the food should be tender or it needs strength.

at a _glance_
When you want gluten and when you don't

High Gluten = High-Protein Flour	Low Gluten = Low-Protein Flour
(Strength, Elasticity, Yeast Doughs)	_(Tenderness, Baking Powder– and Baking Soda–Leavened Batters)_
Yeast breads	**Quick breads (baking powder or soda)**
Strudel	**Cakes, muffins, pancakes**
Cream puffs, popovers, Yorkshire pudding	**Pie crusts**
Puff pastry	**Dumplings, Asian soft noodles**
Pasta (durum wheat)	**Génoise**

When gluten—or no gluten—really matters
The right flour for the job can literally mean the success or failure of a dish. Take John Clancy's strudel, for instance. John Clancy was the test kitchen chef for the Time-Life series _Foods of the World,_ a series so outstanding that cooks and chefs rely on it because "the recipes always work."

When John taught in Atlanta several years ago, he prepared the magnificent strudel from Joseph Weschberg's _The Cooking of Vienna's Empire._ For strength and elasticity in a thinly stretched strudel, a high-protein flour makes the needed gluten. I assisted him and bought his groceries. For the strudel, I bought bread flour for strength and elasticity.

John was so thrilled, he hugged me. "Shirley, I did this strudel yesterday in Texas, and the only flour they had was Southern low-protein flour—not much gluten. My poor strudel looked like Swiss cheese!" Using the bread flour, John made a six-foot-long strudel so thin you could read the _Washington Post_ through it.

The right flour makes an incredible difference in the quality of a loaf of bread. A loaf made with a high-protein flour (like bread flour) will rise well in both risings and then bake into a light, airy loaf with good brown crust color. A loaf made with a low-protein flour (like Southern all-purpose or cake flour) will not rise well and will bake into a heavy, dense-textured, pale loaf.

Different kinds of flour can solve many baking problems. A chef from a test kitchen, for example, told me she had a wonderful lemon square recipe, but she could not use it because the lemon squares were so tender that they fell apart. No problem—a little gluten will hold them together. She could change the flour from bleached all-purpose to unbleached all-purpose, which is a higher-protein flour and will form more gluten. Or she could cut the sugar, which interferes with gluten formation. Either way, the lemon squares will be perfect.

Gluten and water absorption
For thousands of years, cooks have known that some flours absorb more water than others, but they usually blamed it on humidity. Humidity really has very little influence. The gluten proteins join with each other _and water_ to form gluten. It's primarily protein content that determines how much water a flour will absorb.

This difference in water absorption can be major. For example, 2 cups of high-protein bread flour absorb 1 cup of water to form a soft, sticky dough. However, 2 cups of low-protein Southern flour or cake flour and 1 cup of water make a thick soup. It takes 1/2 cup more low-protein flour to get the same-consistency dough as with the high-protein flour. This means that even a small recipe with 2 cups of flour can be off by 1/2 cup if the wrong flour is used! This is a difference of 25 percent; commercial recipes with 20 pounds of flour could be off by 5 pounds. Regardless of measure by weight or volume, the type of flour will make a big difference.

Cooks are constantly faced with this problem. The person writing the recipe uses one kind of flour, and the person following the recipe uses another. When driving through Georgia, Mrs. Jones of Connecticut purchases a local church cookbook. When she gets home, she makes Miss Lolly's Cake. The recipe says 2 cups all-purpose flour. Miss Lolly always uses White Lily all-purpose, a low-protein, partially chlorinated flour, and her cakes are superb. Mrs. Jones uses her favorite local all-purpose, high-protein Hecker's, which makes great yeast bread but soaks in the cake liquids and makes a stiff batter and a dry cake. Mrs. Jones thinks this is the worst recipe she has ever followed.

Use the wrong flour, and your quantity could be off by 25 percent—½ cup in a recipe calling for 2 cups flour—because different flours absorb different amounts of water.

As you might suspect, even worse problems can result when trying to translate foreign recipes containing flour. If there can be over ½ cup difference in a 2-cup recipe around the United States, just think about the possible difference in trying to follow a recipe written in another country!

Some time ago, when I was teaching at a chefs' training center in Vancouver, British Columbia, the young chefs were excited about Paul Prudhomme's Cajun cookbook. They loved his recipes, but they were having a real problem with his sweet potato pecan pie. The filling was delicious, but the crust was a disaster. What had they done wrong?

Chef Paul's recipe called for 1 cup all-purpose flour, and he was probably using a low-protein Southern flour. In Canada the flour called *all-purpose* is actually a very-high-protein flour. (Canada produces some of the highest-protein flours in the world. Carol Field, in her book *The Italian Baker,* mentions that a baker who wants to make very light bread will add Manitoba—a Canadian high-protein flour.) So when the young Canadian chefs tried to use it in Paul's recipe, the amount of liquid in the recipe did not even dampen the flour, let alone form a dough. When they added enough liquid to form a dough, the dough was so tough that they could hardly roll it out. "Oh, no!" I said. "What Paul meant was a flour like your cake and pastry flour." This second major type of Canadian flour is low in protein and just right for cakes and pie crusts. So flour labeled "all-purpose" may not suit your purpose at all. All-purpose flour from one place can be totally different from all-purpose flour somewhere else. In fact, the traditional cuisines of the Northern and Southern United States reflect this difference. The South is noted for its pies, biscuits, and cakes, which the low-protein flour from Southern wheat produces beautifully. Cookbooks from New England and the Midwest that date from before the extensive distribution of national brands of flour contain many fine yeast bread recipes, best made with high-protein flour.

Furthermore, flours labeled as all-purpose can differ from each other in the same geographical region. National brands can differ from regional brands in protein content, and unbleached all-purpose is usually different from bleached all-purpose. Two all-purpose flours that you can buy in most markets in the South—Pillsbury unbleached all-purpose, with 12+ grams of protein per cup, and White Lily all-purpose, with about 8.6 grams of protein per cup—vary by almost ½ cup of flour per 2-cup recipe in the amount needed to absorb 1 cup of liquid.

There are no easy solutions. Many recipes using flour are written with an approximate amount, such as "2 to 3 cups flour" or "6 to 8 cups flour." The amount of flour in a recipe is actually a ballpark figure.

The way flour absorbs water tells you a lot about the flour. A flour that absorbs a lot of water is high in protein and good for yeast doughs. They tell of old German bakers who could shove a sweaty arm into the flour barrel and tell what kind of flour it was by how much clung to the arm!

You don't have to master this particular art, but you can perform a similar test by combining an unknown flour and water in the food processor. From the consistency of the dough, you can make a good guess at the protein content by using the following procedure.

HOW TO TELL THE PROTEIN CONTENT OF FLOUR

Measure 2 cups and 1 tablespoon of bread flour and place it in the workbowl of a food processor with the steel knife. If you measure by scooping a dry measuring cup into the flour bag, filling it, and slightly packing the flour as you level it off against the inside of the bag, a little over 2 cups of bread flour will absorb 1 cup of water, producing a sticky dough ball when processed for about 30 seconds. Perform the same test with cake flour and you will find that it takes over 2½ cups to form a similar sticky dough ball.

This gives you a standard. You know that if a little over 2 cups of flour plus 1 cup of water make a sticky dough ball, the flour has about 14 grams of protein per cup. If the dough barely forms a ball and is wet, needing from ¼ to ½ cup more flour to reach the same consistency, it is an all-purpose flour with about 12 grams of protein per cup. If the dough is so wet that it does not form a ball at all and requires over ½ cup more flour to reach the same consistency, you have the equivalent of a Southern low-protein flour or cake flour with 8 to 9 grams of protein per cup, great for pie crusts.

PROTEIN CONTENT AND WATER ABSORPTION CAPACITY OF WHITE FLOURS		
Protein Content	Flour Type	Approximate Volume Needed to Absorb 1 Cup Water*
grams/cup		
14	Bread	2 cups (packed) + 1 tablespoon
13	Unbleached	2 cups (packed) + 2 tablespoons
12	All-purpose	2¼ cups
11	All-purpose	2⅓ cups + 1 tablespoon
10	All-purpose	2½ cups
9	Southern All-purpose	2½ cups + 2 tablespoons
8	Cake	2¾ cups

*To form a sticky dough ball in a food processor. These amounts may vary some with individual measuring techniques.

Breadmaking techniques

Techniques are critical in breadmaking. Give ten bakers the same flour, yeast, salt, and water and have them make bread with just these ingredients and they will make ten different loaves of bread. The sequence in which ingredients are added, the method used (straight dough, sponge, or sourdough), the wetness of the sponge and dough, the kneading time, the temperature and speed of kneading, the length and temperature of rises, the manner of baking—in every step along the way, bakers make decisions about techniques that affect the bread. Breadmaking remains an art, not always completely understood.

The bread recipes in this book

We all have our own ways of doing things. Understanding the way that I have measured, kneaded, and baked will help you get the breads described in this book.

Measuring

Professional bakers weigh ingredients because there can be big differences in volume measure according to the technique used. I realize that many cooks do not have scales and are accustomed to volume measuring. I used to measure as I was trained—by placing the measuring cup on the counter, spooning flour into it, and leveling it off with a straightedge. Through years of teaching, I observed that my students measured by dipping the measuring cup into the flour and leveling it by pressing it against the inside of the bag. This is actually 1 to 2 tablespoons more flour per cup than I was getting.

Since this is the way most home cooks measure, I decided, if you can't lick 'em, join 'em. So, the volume amounts in these recipes were measured by dipping the cup into the bag, filling it with flour, and leveling by scraping against the inside of the bag. If you prefer to work by weight, a cup of bread flour when I measure this way weighs 5.6 ounces.

Substitutions

Any substitutions in types of flour (for example, using unbleached in place of bread flour) will result in a different amount of flour or liquid needed for the correct consistency. Be prepared to adjust the amount of flour or liquid to get the described dough consistency.

Additives

Many of my recipes contain one or more unusual ingredients. I see no reason not to add natural healthful ingredients that improve bread volume or taste. My additives are ingredients that are readily available in most locations at grocery stores, drugstores, or health food stores.

Vitamin C: Because it helps gluten development and produces lighter breads, I frequently add a piece of a crushed vitamin C tablet.

Salt: I like to use sea salt, which contains additional minerals for better gluten development.

Spices: In some breads, spices are there not just for taste but to enhance yeast activity.

Dried beans: In the Crusty French-Type Bread, I included a tablespoon of ground dried beans.

Ground bean flours are a common ingredient in breads in many cultures. Most French flour is fortified with 2 percent fava bean flour, which improves dough performance and enhances flavor, nutrition, and probably yeast growth. Yeast needs minerals and certain amino acids, and beans would certainly contain some of these. Many Mediterranean and Middle Eastern breads contain chickpea (garbanzo bean) flour. I noticed in the grocery stores in Sicily that you can buy small bags of bean flour. Soybean flour is available in most health food stores. Soybean and chickpea flours are available from King Arthur (see Sources—Flour, page 478). To make your own bean flour, you can grind dried beans in a small coffee bean grinder and strain to remove larger pieces.

Barley malt syrup: This syrup helps convert flour to food for yeast. It used to be available in grocery stores, but recently I have had to go to a health food store or a shop with beermaking supplies to find it.

Milk: I scald both milk and nonfat dry milk mixed with water. This may or may not be necessary. There is a protein in whey that can lower loaf volume (see page 80). I think that it is probably a matter of amount—that in small amounts this protein doesn't harm the loaf, but it does in larger amounts and with very fast rises.

Sponge

Most of my bread recipes follow a simplified sponge method. I beat a wet mixture of the liquid, yeast, and a portion of the flour and let it stand in the mixer for a minimum of 30 minutes or up to the ideal 2½ hours, as time permits. Then I add the rest of the ingredients and proceed with the kneading, rising, and shaping as usual. I think that this standing period is so valuable to the texture and flavor that it is well worth the extra time.

Kneading

For detailed information on kneading and overkneading, see pages 27–29. The purpose of kneading is to let the dough proteins connect and cross-connect to form the bubble gum–like sheets of gluten mentioned at the beginning of the chapter.

The dough is stretched out in one direction and allowed to spring back, then stretched again in the same direction and relaxed, time after time. Whether this is done by hand or by machine, the goal is the same—to align protein strands in the same direction for the creation of strong gluten sheets and to incorporate air.

Some feel that the dough's rising, which is a form of molecule-by-molecule stretching, is the key step in dough development. For many years, I made doughs in the food processor and depended heavily on risings for good development. I feel now that I definitely get a better bread in the heavy-duty mixer. I love the satiny texture and strength of a dough kneaded with the dough hook in the mixer for 5 full minutes.

I recently started adding ¼ to ⅓ cup of crushed ice just before kneading. When I read that the gluten proteins absorb more water and make better gluten when cool, I was skeptical. But when I tried it, I found it to be true. In fact, I felt that it made such a difference in my breads that I rewrote many recipes to include this simple addition. Try it for yourself and make your own decision. Recipes have to be adjusted to take into account a little additional liquid. Since you get better absorption, you have to add only a little more flour.

I wrote the bread recipes in this book with kneading directions for a heavy-duty mixer because this is how I feel I get better gluten development and lighter breads. You can certainly knead by hand or in the food processor. You want to start with a wet dough and beat air into the dough either in a mixing bowl with a wooden spoon for several minutes or in the processor for about 30 seconds with the steel knife or for 1 or 2 minutes with the plastic dough blade. Each recipe begins with this step. Then let the sponge stand for 30 minutes to an optimum 2½ hours.

After standing, add the remaining flour as directed in the recipe and knead. My doughs are wetter and softer than most. To knead by hand, oil your hands to prevent their sticking to the dough. Work on a clean counter. A pastry scraper is a great help in kneading sticky doughs. With the heel of your hand or with a pastry scraper, push the dough out in front of you, then pull or scrape it back. Repeat this pushing out and pulling back. The dough gets less sticky as you knead. If you just can't stand the wet dough, add a little more flour, but keep the dough very soft.

The final kneading in the food processor requires just 20 to 30 seconds with the steel knife or 1 or 2 minutes with the plastic dough blade. The dough should be dry enough to form a slightly sticky ball that revolves around on the blade.

Punching down

The operation of punching down is actually turning the dough inside out. Heat has built up in the center of the dough from fermentation. This turning action pulls the cooler outside dough to the center and the warmer dough to the outside.

The yeast also is in clusters. As long as yeast had oxygen from the air bubbles, it was dividing and multiplying, so punching down breaks up clumps of yeast and redistributes them. Some of the carbon dioxide that built up in the dough is released. A small amount of oxygen is incorporated with the manipulation, and a minor amount of working of the gluten takes place.

Oil your hands lightly with vegetable oil. With a closed fist, press down in the center of the dough. Reach to the back of the bowl under the dough and pull the bottom of the dough up and over the center. Repeat the same action at the front of the bowl. You have folded the bottom of the dough over what used to be the dough surface so that this former surface is now in the middle of the dough. Dump the dough ball out on the counter upside down. If the dough is to be divided into several loaves or rolls, divide and round each piece.

Rounding

In good gluten development, protein strands are aligned and stretched in the same direction. Rounding is a step that is emphasized in professional baking books as being very important, yet very few home baking books include it. Rounding the dough pieces aids in aligning gluten and in holding the gas. After punching down, the dough is in an odd shape, and if it has been cut, gases in the dough can easily escape. By tucking the dough into a tight smooth round you create a covering to hold gases better.

Cover this round with plastic wrap and let it stand on the counter for 15 minutes. The relaxed dough is now much easier to shape. I do this with all yeast breads and think it is well worth the few extra minutes.

Shaping

After the rounded dough has rested, it is easy to shape. Each recipe has shaping suggestions and directions. See page 30 for basic shaping directions.

Slashing

Slashing not only produces handsome loaves, it also permits dough expansion and allows steam that could be trapped under the crust to escape. If the top of the oven is hot and the loaf crusts on top before the steam has escaped, you can get a hole just under the crust where the steam was trapped. Loaves can be slashed early or during rising, but most bakers slash just before the bread goes into the oven. With slashing early in the rise, there is less risk of the dough's falling. Such slashes spread into great gashes as the dough rises. When loaves are slashed just before going into the oven, they are well risen and can be deflated easily if not slashed with a sharp razor and a quick, light stroke. Bakers have a special tool, a curved razor with a handle, for slashing. Some artist's cutting tools with replaceable razor-sharp blades are also good for dough slashing. Baguettes are traditionally slashed with the razor held at an angle of 20 to 30 degrees. You are asking for trouble if you try to slash with anything that is not razor sharp.

Baking

In baking bread, a conflict takes place in the oven. Heat gives the yeast a burst of activity until the yeast gets so hot that it dies. The heat also evaporates the alcohol in the dough that the yeast has produced all along, and heat expands the gases, enlarging the bubbles. This gives the dough a large final rise sometimes referred to as *ovenspring* or *oven rise*.

At the same time, the hot, dry air in the oven starts to form a crust on the top of the dough and hold it down. One of the reasons that French bread is so light is that the oven is filled with steam the minute the bread goes in. The air remains moist, the crust stays soft, and the dough rises to a maximum.

For lighter breads, you want to encourage warming the dough fast to help with yeast activity and with alcohol evaporation. This is why placing the dough on hot baking stones or tiles or on a hot brick oven floor makes such nice breads. You get heat from the bottom to warm the dough fast.

BAKING STONES

Baking on hot tiles or a pizza stone gives you excellent light loaves of bread. Large 15 × 16-inch pizza stones are available at cookware shops for around $30. In gas ovens, place a shelf in the lowest position. Place the stone on this shelf and preheat at 450°F (232°C) for 20 to 30 minutes, then turn the oven down to the baking temperature.

In some electric ovens, the heating unit is covered so that you can place the stone on a shelf as low as possible just as in a gas oven. In others, the heating element is exposed. In that case, place the stone on an oven shelf as close above the exposed element as your oven rack holders permit, usually 2 to 3 inches. Preheat for 30 minutes at 450° to 500°F (232° to 260°C).

You can tell from your baked loaf of bread whether the stone is at the correct heat. If the bottom is burned or if the bottom inch of the bread is much denser and tighter in texture than the rest of the loaf, the bottom is cooking before the dough can rise. Your stone is too hot.

Shortly before placing a loaf on the hot stone, I put a small baking pan with 1/2 inch of boiling water on the floor of the oven to provide steam. (For an electric oven, put the pan of water on the lowest possible shelf.) When I place the bread in, I also mist the sides of the oven with a sprayer for more steam. This provides a hot surface to get the dough warm fast plus a moist atmosphere to slow top crusting. Since the loaf is baked in the lower part of the oven, there is a long distance between the bread and the hot top of the oven, which also slows crusting.

A few alternatives to the baking stone

Your goal is to get heat from the bottom and slow crusting. Preheating to a lower temperature or cold-oven starts give heat from the bottom, but you get better loaves with a hot stone. Convection ovens provide excellent heating. However, most ovens that provide both convection and steam are commercial products. With a home convection oven you can place a shallow pan of boiling water on a shelf near the blower for some steam.

Preheating to a lower temperature

If you do not have a baking stone, you can preheat the oven to 25°F (14°C) below the baking temperature and place the bread on a shelf in the lower third of the oven, 5 to 6 inches above the bottom. When you place the bread in, turn the oven up to the baking temperature. Place a shallow pan with ½ inch of boiling water on the floor of the oven. (For an electric oven, put the pan of water on the lowest possible shelf.) If you have a heavy aluminum baking sheet, place this in the oven before preheating, then place the bread on this already hot sheet and turn the oven up to the baking temperature.

Cold-oven starts

Another alternative is a cold-oven start. In most ovens, the bottom unit comes on and stays on until the oven reaches the temperature that you set. Place the bread in the pan in a cold oven on a shelf slightly above the center and turn the oven on to the recommended baking temperature. Place a shallow pan with ½ inch of boiling water on the floor of the oven. (For an electric oven, put the pan of water on the lowest possible shelf.) Black metal or glass pans are a problem with cold-oven starts because they absorb more heat and breads will burn on the bottom. Regular aluminum pans work well. If you have trouble with the bottom getting too brown, place the bread pans on a heavy aluminum baking sheet.

Convection ovens

If you have a convection oven, by all means take advantage of its excellent heating. You can still preheat a stone and place bread pans on it to get a little advantage from it. Convection oven directions recommend setting the temperature 25° to 50°F (14° to 28°C) lower than the recipe directs for a conventional oven. Go by the instructions for your oven. You can achieve a little moisture in the oven by placing a shallow pan of boiling water on a shelf near the blower.

Steam

Steam or moisture does two different jobs when baking bread. Steam in the oven when the bread is placed in it slows down crusting and allows a full rise. Near the end of the baking time, spraying the surface of the bread with a fine mist wets the surface. When this moisture evaporates in the hot oven, it dries the surface of the bread and creates a hard crust. For an even harder crust, misting several times near the end of baking is effective. You can also brush hot loaves with water the minute they come from the oven.

The best way to check that the bread is done is with an instant-read thermometer. If baking in a pan, remove the loaf from the pan and insert the thermometer to the estimated middle of the loaf in the center. If the reading is close to 200°F (93°C), the bread is done. When rapped or thumped on the bottom, the loaf will have a hollow sound.

Sequence of ingredients

The order in which ingredients are added—particularly fat and salt—can make a major difference in bread.

When to add fat

Two brioche recipes given later in this chapter (pages 15 and 17) show the difference that technique alone, particularly when certain ingredients are added, can make. Both loaves are made with the same ingredients and yet are very different.

The first loaf is made in a proper breadlike fashion: You add the yeast and water to the flour, then knead the dough and let the flour proteins join and form good elastic gluten. Then, and only then, is the butter added to the dough. For the second loaf, the butter is added directly to the flour. This greases the proteins so that they cannot grab water or each other and therefore cannot form much gluten. There is no elasticity, and the dough is like a batter.

The loaf put together in the breadlike manner is light with a beautiful golden loose texture. The cakelike brioche of the second recipe is dense and buttery with a fine-grain cake texture and so tender that a knife glides through it. Both are delicious.

Roland Mesnier, the White House pastry chef, inspired my interest in brioche. In a professional pastry class that my friend Rose Beranbaum and I attended, Roland taught magnificent dishes similar to those he prepares for a great White House pastry buffet. He said he had noticed there was usually someone who walked around the buffet and looked at every dessert but never took anything, so he started including a simple loaf of brioche. Sure enough, people who used to just look at the desserts would have a slice of the brioche.

The recipe he gave us in class was 1 stick of butter to 2 cups of flour. But then he threw down the gauntlet: The "ultimate brioche," he said, was 1 stick of butter to 1 cup of flour, but it was very tricky—not something to teach in front of people. Of course Rose and I had to try it, and we thought we had an advantage since we were using bread flour, which is slightly higher in protein than the unbleached flour Roland had used.

We were feeling so smug when we put the bread together and got all the butter incorporated. The dough rose beautifully in the warm room. But when we punched it down, all the butter ran out.

Roland knew that these doughs are so rich that you cannot handle them warm; the butter melts and runs out. You will notice that my recipes emphasize refrigerating the dough. It is imperative that you shape these rich doughs while they are cold.

The brioches are different, but both are wonderful. You may not be able to decide which is your favorite.

The Ultimate Brioche I
Light, Airy Breadlike Brioche

MAKES ONE 4 × 12-INCH LOAF

This is a golden, incredibly buttery bread with a deep brown crust. I considered it a great compliment when a noted professor from Paris and an expert on fine food loved this bread so much that when we met three years later he offered to help me if I would bake another loaf. English food expert Alan Davidson said he had a delightful dinner of several slices of this brioche. Try it with a little Cherry-Chambord Butter (page 78).

what this recipe shows

Mixing and kneading liquid and flour produce elastic sheets of gluten that remain intact when butter is added later, producing a light, well-leavened loaf.

Chilling a very rich dough before punching down makes it possible to work with the dough without losing the butter.

Baking in a long, narrow, shiny pan keeps this rich bread from burning before the center is done.

continued

1 cup warm water (115°F/46°C)

1 teaspoon *and* ¹/₄ cup sugar
 (¹/₄ cup plus 1 teaspoon *total*)

1 tablespoon (1¹/₃ packages) active dry
 yeast

4 large egg whites

2 cups *and* 2¹/₄ cups bread flour
 (4¹/₄ cups *total*)

¹/₄ cup crushed ice

5 large egg yolks

1³/₄ teaspoons salt

Water as needed

1 pound (4 sticks) butter, softened

Nonstick cooking spray

1 egg, beaten, for glaze

Read "The Bread Recipes in This Book" (page 10) before beginning.

1. Stir together warm water, 1 teaspoon sugar, and the yeast in the bowl of a heavy-duty mixer. Let stand 2 minutes until foam appears, indicating yeast is alive and well. Add ¹/₄ cup sugar, the egg whites, and 2 cups flour. With the paddle blade, beat on low-medium speed for 4 minutes to incorporate air. Let stand for 30 minutes to 2¹/₂ hours for improved flavor and texture.

2. Add the crushed ice, egg yolks, remaining 2¹/₄ cups flour, and salt. Remove the paddle blade, insert the dough hook, and knead for 5 minutes on low-medium speed until you have a very elastic, sticky dough. Dough should be wet enough to stick to the bowl. Add more water if necessary and mix in well.

3. Let the dough rise for 10 minutes, then mix in softened butter on low speed. Scrape down the sides and incorporate as much butter as possible. Mix again on low. There is so much butter that the dough will be a stringy mess.

4. Cover and let the dough rise at room temperature for 30 minutes, then refrigerate for 45 minutes. Dough must be cold before you punch it down. Punch down the dough. The dough can be refrigerated overnight at this point if desired. When ready to shape, with both hands gently cup and tuck the dough into a smooth round. Cover with plastic wrap and leave on the counter for 10 minutes.

5. Spray one 4 × 12 × 2³/₄-inch pan (see Notes) with nonstick cooking spray. Divide the dough into 10 pieces and shape into cylinders 3¹/₂ to 4 inches long, about the width of the pan. Arrange the 10 cylinders in a row down the pan, pressing one against the other. Brush the dough with a beaten egg (try not to get egg on the pan) and let bread rise in a warm room until slightly above the pan. This will take about 2 hours, depending on how cold the dough was.

6. About 30 minutes before the dough is fully risen, place a baking stone on the lowest oven shelf and preheat to 450°F (232°C)—see Notes. About 5 minutes before the bread is fully risen, turn the oven down to 375°F (191°C). A few minutes before placing the loaves in the oven, carefully arrange a shallow baking pan with ¹/₂ inch of boiling water on the floor of the oven. (For an electric oven, put the pan of water on the lowest possible shelf.)

7. Brush the loaf with beaten egg again and place the pans directly on the hot stone. Bake until very brown, about 40 minutes. Loaves get a deep brown because of their egg and sugar content. Remove a loaf from the pan and check for doneness by inserting an instant-read thermometer in the bottom center of the loaf to the estimated middle. The loaf is done when the center is about 200°F (93°C). Let the loaves cool in the pan for 5 minutes, then turn out on a rack to finish cooling.

Notes: *If you don't have a stone:*
- Preheat the oven to 25°F (14°C) below the baking temperature, carefully place a shallow pan with ½ inch of boiling water on the oven floor. (For an electric oven, put the pan of water on the lowest possible shelf.) Place the bread on a shelf in the lower third of the oven, 5 to 6 inches above the bottom, and then turn the oven up to the baking temperature.
- Or use the cold-oven procedure described on page 14.

Kaiser Bakeware makes a long, narrow (12 × 4-inch) metal loaf pan with a shiny finish that is ideal for this bread. The center will not get done in a wide pan and the loaf will burn in a gray or dark one. For ordering at a cookware shop, the style number is 610-341.

The best way to see if bread is done? Use an instant-read thermometer.

The Ultimate Brioche II
Buttery Cakelike Brioche

MAKES ONE 4 × 12-INCH LOAF

This loaf is so moist and tender that it almost dissolves in your mouth. The knife falls through it when you slice. The taste is extra buttery.

what this recipe shows

Mixing the butter directly with the flour coats the gluten-forming proteins, and very little gluten is formed. The dough is like a batter, and the bread is very tender.

This dough absorbs much less water than the breadlike brioche since little gluten is being formed.

Chilling a very rich dough before punching down makes it possible to work with the dough without losing the butter.

This dough rises poorly during both rises but amazingly well in the oven because steam is formed from the water that is not tied up in gluten.

This dough is shaped immediately after punching down (there is a minimum of gluten to relax), and the dough must be very cold to handle and shape.

Baking in a long, narrow, shiny pan keeps this rich bread from burning before the center is done.

¾ cup warm water (115°F/46°C)

1 teaspoon *and* ¼ cup sugar (¼ cup plus 1 teaspoon *total*)

1 tablespoon (1⅓ packages) active dry yeast

4¼ cups bread flour

1 pound (4 sticks) butter, softened

5 large egg yolks

4 large egg whites

1¾ teaspoons salt

Water as needed

Nonstick cooking spray

1 egg, beaten, for glaze

continued

Read "The Bread Recipes in This Book" (page 10) before beginning.

1. Stir together the water, 1 teaspoon sugar, and the yeast in a small bowl. Let stand 2 minutes until foam appears, indicating yeast is alive and well.

2. Add 4 tablespoons sugar, the flour, and the softened butter to the bowl of a heavy-duty mixer. Mix with the paddle blade until the soft butter and flour are completely blended. Add the egg yolks and mix for 30 seconds. Add the egg whites, salt, and dissolved yeast and run just to combine. Dough should not form a ball but should be a wet and sticky batter. Add a little water if necessary.

3. Cover with plastic wrap and let the dough rise at room temperature for 30 minutes, then refrigerate dough for 45 minutes. Dough can be refrigerated overnight if desired.

4. Spray one $4 \times 12 \times 2^3/_4$-inch pan (see Notes) with nonstick cooking spray. Dough must be cold before you punch it down. Punch down, divide into 10 pieces, and shape each into a cylinder $3^1/_2$ to 4 inches long, about the width of the pan. Arrange the 10 cylinders in a row down the pan, pressing one against the other. Brush the dough with a beaten egg (try not to get egg on the pan), and let rise in a warm room until slightly above the pan. This will take about 2 hours, depending on how cold the dough was.

5. About 30 minutes before the dough is fully risen, place a baking stone on the lowest shelf of the oven and preheat to 450°F (232°C)—see Notes. About 5 minutes before the bread is fully risen, turn the oven down to 375°F (191°C). A few minutes before placing the loaves in the oven, carefully arrange a shallow baking pan with 1/2 inch of boiling water on the floor of the oven. (For an electric oven, put the pan of water on the lowest possible shelf.)

6. Brush the loaf with beaten egg again and place the pans directly on the hot stone. Bake until very brown, about 50 minutes. Loaves get a deep brown because of their egg and sugar content. Remove a loaf from the pan and check for doneness by inserting an instant-read thermometer in the bottom center of the loaf to the estimated middle. The loaf is done when the center is about 200°F (93°C). Let the loaves cool in the pan for 5 minutes, then turn out on a rack to finish cooling.

Notes: *If you don't have a stone:*
- Preheat the oven to 25°F (14°C) below the baking temperature, carefully place a shallow pan with 1/2 inch of boiling water on the oven floor. (For an electric oven, put the pan of water on the lowest possible shelf.) Place the bread on a shelf in the lower third of the oven, 5 to 6 inches above the bottom, and then turn the oven up to the baking temperature.
- Or use the cold-oven procedure described on page 14.

Kaiser Bakeware makes a long, narrow (12 × 4-inch) metal loaf pan with a shiny finish that is ideal for this bread. The center will not get done in a wide pan and the loaf will burn in a gray or dark one. For ordering at a cookware shop, the style number is 610-341.

The finished baked loaves from the two brioche recipes look more alike than you might think, although there is some difference in volume. The breadlike brioche is lighter and better risen. Going into the oven, the breadlike dough is much higher than the cakelike, so you might expect a major difference in the finished products. Since the cakelike brioche procedure does not form much gluten and therefore does not absorb as much water, the dough is wetter and produces more steam in the hot oven. This results in a higher loaf than you may anticipate.

When to add salt

Salt strengthens and tightens gluten. If you knead a dough without salt, when you add the salt you can lit-erally see the dough tighten. If you stretch the dough before and after the addition of salt, you will feel an amazing difference in the strength of the gluten.

Some bakers like to knead the dough first, then work in the salt. The reasoning is that the dough is easier to knead and requires less work from the mixer and a shorter kneading time without the salt. Then the salt can be mixed in to strengthen the dough just before the rises. Some bakers worry that the salt may not be well distributed throughout the dough when added at the end. You may want to try it and see which method you prefer. The type of salt can make a major difference in how well it blends in. Flaky sea salt and Diamond Crystal kosher salt blend faster and better than granular table salt.

DIFFERENT FORMS OF SALT

Most table salts are dense cubes, made by vacuum pan evaporation and referred to as *granular*. Sea salt and salt made by Akzo Nobel's Alberger process (Diamond Crystal kosher salt) are formed from surface evaporation. A four-sided crystal forms and grows on the sides only, then sinks, and another layer continues to grow on all four sides. A hollow upside-down four-sided flaky pyramid is the resulting shape. With sea salt evaporation, wind usually sprays more water into the hollow space, so the center may fill. With salt made by the Alberger process you can actually see some fragile hollow flaky pyramids with the aid of a magnifying glass.

Thomas Dommer, director of Technical Services at Akzo Nobel Salt, Inc., describes the difference in granular salt and this delicate hollow, flaky pyramid form as the difference between an ice cube and a snowflake. There truly is a dramatic difference. About 90 percent of granular salt dropped onto an inclined surface bounces off, while 95 percent of the flaky pyramid form sticks to the surface. The delicate flaky pyramid form also dissolves in half the time that granular does.

Two other types of salt are compressed granular—granular salt that is pressed into flakes by rollers, like Morton's kosher—and a salt crystallized with a trace amount of prussiate of soda, producing an open porous form called *dendritic,* which is available only commercially.

Not only do these different forms of salt dissolve, mix, and adhere differently, but a given volume, say 1 tablespoon, contains a different weight of salt for each form. To get as much salt as there is in 1 tablespoon of granular salt, you must use $1^1/_2$ tablespoons of Morton's kosher salt or 2 tablespoons of Diamond Crystal kosher salt.

There also are differences in the composition of sea salt. Other salts, regardless of the manufacturer or form, are very pure sodium chloride with a trace amount of anticaking agent or, in the case of iodized salt, some form of iodate added. In addition to sodium chloride, sea salt may contain magnesium chloride, magnesium sulfate, calcium sulfate, potassium sulfate, magnesium bromide, and calcium carbonate. Because of all these different salts, sea salt has a more complex taste with subtle nuances. Some chefs describe it as an almost sweet taste. Different sea salts have different composition, reflecting the minerals in the land that drain into them. Mediterranean sea salt can taste different from New Zealand sea salt and from English sea salt. Sea salt is more expensive than granular salt, so beware of unscrupulous individuals or companies that pass off large-crystal cheap granular salt as sea salt.

I love real sea salt and bring home a suitcase full every time I go to Sicily. One brand from Trapani lists its ingredients as *mare-sole-vento:* "sea-sun-wind." It has delicate light flakes with marvelous complex flavors. For bread, the additional minerals in sea salt can be helpful in gluten development, too.

Breadmaking methods

The three basic methods of breadmaking are the straight or direct-mix method, which is relatively fast; the sponge method, which is only slightly longer; and the sourdough or starter method, which can require long, slow rises. These different methods produce breads with very different textures and flavors and involve different production times—from a few hours for the direct and sponge methods to a few days for sourdough and starter breads. Once you have a good starter going, you can make breads reasonably fast with it. Using it primarily for texture and flavor, you can add a little additional yeast and make breads as fast as with the direct or sponge method.

Straight or direct method

In the straight or direct-mix method, bakers' yeast (the kind you buy, as opposed to what grows in a sourdough starter) is used, all the ingredients are mixed, and the dough is kneaded, allowed to rise, punched down, shaped, allowed to rise again, and baked. The bread has an essentially even texture.

Sponge method

In the sponge method, bakers' yeast is used, and all of the liquid, the yeast, and part of the flour are beaten together. This wet mixture is allowed to sit and ferment, then the rest of the ingredients are added, and the dough is kneaded, allowed to rise, punched down, shaped, allowed to rise again, and baked. This makes loaves with improved taste and more interesting texture than the straight method and allows you to use less yeast. Feeding the yeast and letting it grow a little before the kneading process produces more yeast as well as flavorful fermentation products. If you use a fairly wet sponge (a high ratio of liquid to flour), the bread will also have a greater amount of residual sugar.

The texture of bread depends heavily on rising times and temperatures as well as wetness of the dough. A reasonable amount of yeast and fairly fast rises produce an even texture. For a fine texture, beating air into the sponge is the first step. Gases from yeast do not create any new bubbles; they can only enlarge bubbles that already exist in the dough. Beating millions of tiny air bubbles into a dough is essential for a fine texture, and this is very easy in a soft, wet sponge batter.

The sponge method is not that much trouble. Simply let the fermenting mixture stand for a time in the mixer bowl, then add the rest of the ingredients and knead. Researchers suggest that an ideal standing period is $2\frac{1}{2}$ hours. I find that I get better flavor with even 30 minutes of standing and can get a more open texture. So I believe that whatever time you can allow here is well worth it.

The Incredible Toast Basic Loaf (recipe follows) illustrates how light and well leavened breads can be when made by the sponge method. The texture is fairly even, depending on the wetness of the dough and amount of rise. This dough is strong enough to sustain an overrise and still produce magnificent light loaves with a slightly uneven texture.

Incredible Toast Basic Loaf

MAKES TWO 4½ × 8½-INCH LOAVES

This fast-rising bread makes fabulous toast with a slightly crisp surface crumb. The sponge method produces more flavorful bread than the direct method yet still has a relatively even texture—excellent for sandwiches, toast, or just out of hand.

what this recipe shows

Beating air into the soft sponge dough creates millions of tiny air nuclei for good leavening.

A moderately fast sponge method produces a reasonably even-textured loaf. See Notes.

A wet sponge and potato flakes increase the sweetness of the bread.

Sugars in milk and corn syrup provide plenty of yeast food for fast rises.

Nonfat dry milk improves moisture, tenderness, and flavor of the loaf, contributes minerals that improve gluten development, and contributes proteins and sugar for better crust color and yeast growth.

Vitamin C improves gluten development.

A small amount of fat enhances volume and shelf life.

Corn syrup makes this bread toast beautifully and improves crust color.

Egg yolk adds emulsifiers for better shelf life and better gluten development.

⅓ cup nonfat dry milk	3 tablespoons melted butter
1¼ cups cool water	1 large egg yolk
⅓ cup dried potato flakes	1 tablespoon corn syrup
1 package (2¼ teaspoons) active dry yeast	¼ cup crushed ice
½ cup warm water (115°F/46°C)	1 teaspoon salt, preferably sea salt
1¾ cups *and* 3 cups bread flour (4¾ cups *total*)	1 tablespoon oil for bowl
¼ 500-milligram vitamin C tablet, crushed	Nonstick cooking spray
	1 large egg, beaten, for glaze

Read "The Bread Recipes in This Book" (page 10) before beginning.

1. Stir the dry milk into the cool water in a small saucepan and heat over medium heat until almost simmering. Turn down the heat. Hold below a simmer for 1 minute, then turn the heat off. Let cool for several minutes, then stir in potato flakes.

2. In the bowl of a heavy-duty mixer, stir together the yeast and warm water. Let stand 2 minutes until foam forms, indicating yeast is alive and well. Add 1¾ cups flour. Pour in the milk-potato mixture and beat on low-medium speed with the paddle blade for 4 minutes to beat air into the dough. Let sit for 30 minutes to $2\frac{1}{2}$ hours for improved flavor and texture.

3. Remove the paddle blade and insert the dough hook. Add the vitamin C, melted butter, egg yolk, corn syrup, ice, and salt. Turn the mixer on for a few seconds just to stir in. Add the remaining 3 cups flour. Knead on low-medium speed for 5 full minutes, until the dough is very elastic. The dough should be very soft. Add a little flour or liquid if needed.

4. Place the dough in an oiled bowl and turn to coat well with oil. Cover the bowl with plastic wrap and let the dough rise until doubled in volume, about 1 hour. Punch down the center of the dough with a closed fist. Reach to the back of the bowl down under the dough and pull the bottom of the dough up and over the center. Repeat the same action at the front of the bowl. Turn the dough out on the counter. Divide the dough in half. Using both hands with a gently cupping and tucking action, shape each piece into a smooth round. Now, with both hands, grab the sides of each round and stretch it sideways into an oval. Let it spring back slightly, then pull it out again. Cover both with plastic wrap and leave on the counter for 15 minutes. The dough is now ready to shape.

5. Cup one dough ball with both hands, fingers spread out behind the loaf on either side and thumbs in front of the loaf. Knead by pressing your thumbs into the dough and down against the table. This pulls or tucks in part of the bottom half of the dough. At the same time, pull the top of the dough tight and forward with your fingers. Now move your thumbs down slightly and press down and in again to knead and tuck again. Repeat this motion two to three times until the loaf is stretched taut and well tucked in. The ends go down to a slight taper. Tuck them tightly under; pinch the ends and bottom seam together. Repeat with other dough ball.

6. Spray two $4\frac{1}{2} \times 8\frac{1}{2} \times 2\frac{1}{2}$-inch loaf pans with nonstick cooking spray and place a loaf in each pan, seam side down. Brush with beaten egg. Let the dough rise until slightly more than doubled, about 1 hour.

7. About 30 minutes before the dough is fully risen, place a baking stone on a shelf in the lowest slot and preheat to 450°F (232°C)—see Notes. About 5 minutes before baking, turn the oven down to 375°F (191°C) and carefully place a shallow pan with $\frac{1}{2}$ inch of boiling water on the oven floor. (For an electric oven, put the pan of water on the lowest possible shelf.)

8. Brush the bread again with beaten egg and place the pans directly on the hot stone. Bake until well browned, 45 to 55 minutes. Remove the loaves from the pans. They should sound hollow when thumped on the bottom. You can also check for doneness by inserting an instant-read thermometer in the bottom center of a loaf and pushing it to the center. The loaf is done when the center is about 200°F (93°C). Place the loaf on a rack to cool.

Notes: *For a more even texture substitute $\frac{1}{2}$ cup semolina flour in the last flour added.*
   If you don't have a stone:
· Preheat the oven to 25°F (14°C) below the baking temperature, place the bread on a shelf in the lower third of the oven, 5 to 6 inches above the bottom, and then turn the oven up to the baking temperature. Carefully place a shallow pan with $\frac{1}{2}$ inch of boiling water on the oven floor. (For an electric oven, put the pan of water on the lowest possible shelf.)
· Or use the cold-oven procedure described on page 14.

Sourdough or starter method

In the sourdough method, a starter, made up of strains of yeast and bacteria that have a symbiotic relationship, is used to leaven the dough instead of ordinary bakers' yeast. Since many strains of both yeast and bacteria in doughs live on the same simple sugars (glucose and fructose), in fast, direct breadmaking the yeasts outnumber the bacteria, which don't get to grow and produce their flavorful acid by-products. A sourdough starter has a more balanced population of yeast and bacteria. Also, it may contain a strain of yeast that thrives on some other sugar (say, galactose) in addition to glucose and fructose, and the accompanying bacteria thrive on a still different sugar (say, maltose). Even though the yeast and the bacteria are competing for the glucose and fructose, each has its own additional sugar food source and can grow.

These are acid-forming bacteria. So the strain of yeast must be one that can tolerate high-acid conditions, which bakers' yeast cannot. The yeast in San Francisco sourdough, *Saccharomyces exiguus,* thrives under the acidic conditions that its accompanying bacteria, *Lactobacillus sanfrancisco,* create.

In the sourdough method, there is less yeast available and the dough requires longer rises. With less yeast competing for sugars, the bacteria can grow and produce their flavorful acidic by-products. This gives the characteristic sour taste to these breads. This acidity also breaks gluten strains contributing to the very open, big-holed texture.

Yeast starters A starter made with a small amount of bakers' yeast, flour, and water is sometimes called a *sponge* and sometimes a *yeast starter* or *poolish.* I usually call it a *sponge* if fermentation lasts not much more than 3 hours and all of it is used for one batch of bread. If the fermentation period lasts longer—6 hours to several days—I call it a *yeast starter.*

With a yeast starter you get greatly enhanced flavor—both from products of bacteria and from compounds produced during the lengthy fermentation. Yeast starters are faster and much more reliable than sourdough starters. The resulting breads have a complex, full-flavored taste without the strongly acidic flavors of some sourdoughs. If you want to make sourdoughs and rustic European-type breads, yeast starters are a good way to begin.

Wetness of the dough is always a factor. Very firm doughs make close-textured bread. But, assuming moderately wet soft doughs, the breadmaking method and rising times have a major influence on texture. You can see the difference in flavor and texture that the breadmaking method, amount of yeast, and length of rises make in the Crusty French-Type Bread (recipe follows) and the Yeast Starter Bread (page 65). In the Yeast Starter Bread, you will get a more open texture than in the Crusty French-Type Bread.

Bread made with a sourdough starter will have an even more open texture. Even if you have not succeeded in getting a sourdough starter that leavens well, you can substitute 1/2 cup sourdough starter for an equal amount of yeast starter and follow the Yeast Starter Bread recipe to get an interesting flavor and open texture.

Crusty French-Type Bread

MAKES TWO 18-INCH LOAVES

Long loaves of browned, crusty bread are hard to beat. The center is light and slightly chewy. The crust is crunchy and faintly salted. This is not classic French bread, but I love it with the salted butter glaze.

what this recipe shows

High-protein bread flour makes the loaf light.

A little semolina (durum wheat) flour is a major flavor enhancer.

A small amount of ground dried beans adds to flavor and nutrient content and enhances gluten development.

Vitamin C improves gluten development.

Adding a little crushed ice to chill the dough just before kneading increases absorption of water for better gluten development.

The vinegar simulates some of the flavors of long rises.

1 package (2$^1/_4$ teaspoons) active dry
   yeast

1 tablespoon sugar

1$^1/_2$ cups warm water (115°F/46°C)

1$^3/_4$ cups *and* 2 cups bread flour
   (3$^3/_4$ cups *total*)

$^1/_4$ cup semolina (durum wheat) flour

1 tablespoon finely ground dried fava or
   white lima beans, sifted through a
   coarse strainer, or chickpea or soy
   flour (page 11), optional

$^1/_4$ 500-milligram vitamin C tablet,
   crushed

$^1/_3$ cup crushed ice

1 teaspoon *and* $^1/_4$ teaspoon salt,
   preferably sea salt (1$^1/_4$ teaspoons
   total)

$^1/_2$ teaspoon balsamic, sherry, or cider
   vinegar

1 tablespoon oil for bowl

Nonstick cooking spray

3 tablespoons butter, melted

Read "The Bread Recipes in This Book" (page 10) before beginning.

1. Stir together the yeast, sugar, and water in the bowl of a heavy-duty mixer. Let stand for 2 minutes until a foam forms, indicating yeast is alive and well. Add 1$^3/_4$ cups bread flour, the semolina, and the ground bean flour, if using. With the paddle blade, beat on low-medium speed for 4 minutes to beat air into the dough. Let the sponge sit for 30 minutes to 2$^1/_2$ hours for improved flavor and texture. For a more open texture let the sponge stand a full 2$^1/_2$ hours.

2. Remove the paddle blade and insert the dough hook. Add the vitamin C, crushed ice, 1 teaspoon salt, vinegar, and the remaining 2 cups bread flour. Knead for 5 minutes on low-medium speed, until the dough is very elastic. The dough should be very soft. Add a little more flour or liquid as needed.

3. Place the dough in an oiled bowl and turn to coat well with oil. Cover the bowl with plastic wrap and let the dough rise until slightly more than doubled, 1$^1/_2$ to 2 hours. Punch down the center of the dough with a closed fist. Reach to the back of the bowl under the dough and pull the bottom of the dough up and over the center. Repeat the same action at the front of the bowl.

4. Turn the dough out onto the counter and divide in half. Using both hands with a gently cupping and tucking action, shape each half into a smooth round. With both hands, grab the sides of the round and stretch it sideways into an oval. Let it spring back slightly, then pull it out again. Cover each with plastic wrap and leave on the counter for 15 minutes. The dough is now ready to shape.

5. To shape baguettes, cup the piece of dough with both hands, fingers spread out behind the loaf on either side and thumbs in front of the loaf. Press your thumbs into the dough and down against the table. This pulls or tucks in part of the bottom half of the dough. At the same time, pull the top of the dough tight and forward with your fingers. Now move your thumbs down slightly and press down and in again, pulling the top forward with your fingers to knead and tuck again. Repeat this motion two to three times, until the loaf is stretched taut and well tucked in. The loaf will lengthen as you stretch and tuck and may be long enough if you have pulled your hands outward in the process. If not, lengthen the loaf by placing both hands, spread out, palms down on top of the center of the loaf. Then simultaneously push away against the table with your right hand and pull toward you with your left, pulling the dough out in opposite directions. Repeat this pulling once or twice more if necessary. Now pinch the bottom seam together.

An alternative shaping method simulates machine shaping. Using a rolling pin, roll one piece of the dough into a long rectangle, the approximate length of the pan. Then by hand, roll and tightly tuck and stretch the dough, inch by inch along the long side, to create a tight long loaf. Repeat with the other loaf.

6. Line the underside of a double-trough French bread pan with foil and turn the foil up on each end to form a lip to prevent melted butter glaze from dripping onto the baking stone. Spray the pan with nonstick cooking spray and place the loaves in the pan.

7. Stir together the melted butter and $1/4$ teaspoon salt and brush loaves with it. Let the dough rise until slightly more than doubled, about 2 hours.

8. About 30 minutes before the dough is fully risen, place a baking stone on a shelf in the lowest slot and preheat the oven to 450°F (232°C)—see Note. About 5 minutes before baking, turn the oven down to 425°F (218°C) and carefully place a small baking pan with $1/2$ inch of boiling water on the oven floor. (For an electric oven, put the pan of water on the lowest possible shelf.)

9. Brush the loaves again with butter. Risen bread can be deflated easily. Hold a sharp razor almost parallel to the loaves. With the razor at an angle of about 20 degrees, with a light, quick motion make three diagonal slashes on each loaf and place the pan directly on the hot stone. For a crunchy crust, mist the loaves with water two or three times during the last 10 minutes of baking using a clean spray bottle. Bake until well browned, 30 to 35 minutes.

10. Place bread on a rack to cool. Serve warm or at room temperature.

Note: *If you don't have a stone:*
- Preheat the oven to 25°F (14°C) below the baking temperature, place the bread on a shelf in the lower third of the oven, 5 to 6 inches above the bottom, and then turn the oven up to the baking temperature. Carefully place a shallow pan with $1/2$ inch of boiling water on the oven floor. (For an electric oven, put the pan of water on the lowest possible shelf.)
- Or use the cold-oven procedure described on page 14.

Breadmaking methods and flavor

Even the French complain that bread today does not taste the way it did in the old days. "Bread used to have real flavor!" they say. In the old days, bread was made with starters or with a lengthy sponge method, or part of the dough, with its flavorful fermentation products, was saved to start the next batch. You had a great mix of compounds formed by days of fermentation plus wild yeasts and bacteria with their unique flavorful by-products. Indeed, bread does *not* taste as it did in the old days. Commercially, bread is now made in a similar manner in many parts of the world.

In their article "Influence of the Breadmaking Method on French Bread Flavor," French researchers D. Richard-Molard, M. C. Nago, and R. Drapron tell how they prepared bread by the old lengthy sponge (yeast starter) method, by the sourdough method, and by the modern commercial method. Then they chemically analyzed the breads. There were many changes, as you would expect, but the main change was in the formation of acetic acid: plain old vinegar.

This makes sense. You know that if you leave the cork out of a bottle of wine it will eventually turn to vinegar. Similarly, as dough stands around for a long time, the alcohol made by the yeast in its early fermentation turns into vinegar. Bacteria also is producing acids and flavorful compounds. (See Starters, page 60, for more on bacteria and flavor.)

Breads made with yeast or sourdough starters have many flavorful components. If time does not permit these methods, you can resort to a cheap imitation. I include a tiny bit of vinegar in several bread recipes. The Crusty French-Type Bread recipe (page 24) contains $1/2$ teaspoon of vinegar. This is a poor substitute for all the wonderful flavor elements of long slow rises or starters, but it does contribute to flavor. This is not, however, a case of "If a little is good, a lot will be great." Too much vinegar contributes too much acidity and will kill bakers' yeast and also tear down the gluten.

Flourless sponge or starter Another simple way to enhance flavor is the Flourless Sponge. Joe Ortiz, in his book *The Village Baker,* describes one from the Mugitch Bakery in Vienna. Simply let a mixture of yeast and warm water sit together overnight on the counter and use this as part of the liquid in your recipe the next day (recipe follows). This enhances flavor tremendously and has a slight influence on texture.

The yeast loses its vigorous activity overnight. If you use this as the total leavener, rises will be very slow. I add a little fresh yeast the next day to achieve faster rising times. The Flourless Sponge is still a major flavor enhancer. It is an excellent way to improve the flavor of breads made in bread machines.

Flourless Sponge

The night before you plan to make bread, pour $1/2$ cup warm water (115°F/46°C) into a cup and stir in 1 package ($2^{1}/4$ teaspoons) active dry yeast. Let stand on the counter overnight. Substitute for $1/2$ cup of the liquid in the recipe.

Mixing, kneading, and rising

Mixing, kneading, and rising are major contributors to the texture of the bread. Understanding what each of these procedures does will help you get the texture you want—whether very fine-crumbed loaves with the fast rising times of the direct method or rustic coarse-textured rounds made with starters. Both the length of time and the temperature have an influence.

Mixing

Mixing not only blends ingredients but also traps air bubbles into the dough, which yeast gases will enlarge, so the texture of the bread begins right here.

Kneading

Kneading allows the dough proteins to connect and cross-connect to form sheets of gluten. When water is added to flour, glutenin molecules, which are enormous proteins (500,000 atoms) composed of many subunits, and gliadin molecules (100,000 atoms) are in great disarray. The kneading action pushes the dough out and stretches these long proteins, breaking some of the sulfur-to-sulfur bonds in the long, curly chains. The sulfur bonds pull the sulfurs and the dough back together. This aids in lining up the glutenin molecules parallel to each other and is needed to form good sheets of gluten. The dough is stretched out in one direction and allowed to spring back, then stretched again and allowed to spring back—over and over again. Whether this is done by hand or by machine, the goal is the same—to align protein strands for the creation of strong gluten sheets.

Many different types of bonds—ionic bonding, hydrogen bonding, covalent bonding—and the water-hating ends of molecules (hydrophobic interactions) are involved in joining dough components into these elastic sheets. Even fats in the flour get in on the act. One end of some of the flour lipids bonds itself to glutenin, and the other end bonds to gliadin. But all we need to appreciate is that all sorts of bonds are hooking the flour proteins and water into these elastic sheets that are so effective at holding the gases from yeast.

Kneading is not just the working of the dough that you do by hand or by machine. Yeast backs up your mechanical kneading. Picture the gases from yeast inflating millions of tiny air bubbles throughout the dough, puffing and stretching the gluten proteins so that they can cross-link to their neighbors and form better and better sheets of gluten. The rising of dough is actually kneading on an almost individual cell basis, stretching millions of proteins out at once. It is no wonder that the dough rises faster and better on the second or third rising. It has had an almost cell-by-cell stretching of the proteins to produce gluten.

Some think that yeast is so effective at dough development that it really doesn't matter how well the dough is kneaded mechanically. I have made breads in class in a food processor for years and depended heavily on risings for a good loaf of bread. But I believe that I get much better breads when I knead to have a very elastic, satiny smooth dough before rising.

I love soft doughs, but not so wet that you cannot knead properly. The dough needs to be thick enough for the proteins to be stretched by pulling. If there is too much water, one end of the protein is not held in one place while you pull on the other end.

Temperature is also important. The gluten proteins absorb water better and consequently make more gluten if the dough temperature at the end of kneading is about 80°F (27°C). This is the recommended temperature in bread books in the United States, but Europeans prefer a lower temperature, about 75°F (24°C). By measuring the temperature of doughs before and after kneading, I have found that kneading for 5 full minutes at speed 3 on my heavy-duty Kenwood mixer raises the dough temperature by 6° to 8°F (3° to 4°C). If I start kneading with my dough temperature in the low 70s°F (low 20s°C), I end up with a kneaded dough of ideal temperature.

In my kitchen, which is in the mid-70s°F (mid-20s°C), adding ¼ to ⅓ cup of crushed ice to the sponge just before kneading lowers the starting dough to the ideal temperature. I have written my recipes to include this amount of extra liquid (melted ice). Try my recipes as written before you decide to alter this procedure of chilling just before kneading.I really believe that you get better water absorption and better gluten development with the addition of a little ice.

Is one method of kneading better than another? Although commercial kneading methods are not available in the home, you do have several options. You can knead a dough by hand, with a heavy-duty mixer (KitchenAid, Kenwood, or Braun type), in a food processor, or with bucket-type hand kneaders. You can even knead by folding and running the dough through a pasta machine several times.

An unusual way to knead dough: Fold it and run it through a pasta machine a few times.

You can make an excellent bread by hand. Start by beating together a very wet mixture of all the ingredients, but use only a small part of the flour, as in the sponge method. When you have the ingredients thoroughly mixed with lots of air beaten in, add flour until you get an almost kneadable consistency. Start with the dough sticky enough to adhere to your hands and the counter. With the heel of your hand or a pastry scraper, push and stretch the dough out in front of you, pull it back into a ball, then push out and pull back again. A dough scraper makes it easier to work with a wet dough. Concentrate on stretching the proteins and folding in air as you work.

Heavy-duty mixers do an excellent job of kneading. Their recommended speeds have been worked out by each manufacturer's engineering staff and are usually exactly right. I use speed 3 on the Kenwood or 4 on the KitchenAid. I like to use the sponge method and beat air into a wet mixture with the paddle blade and then, later, add the rest of the dry ingredients and finish kneading with the dough hook.

Food processors alone do not do as good a job of gluten development as mixers, but they do the job in an instant. Good kneading takes about 10 minutes by hand, 4 to 7 minutes in a good mixer, and 30 to 50 seconds in the food processor with the steel knife or 2 minutes with the plastic blade. Since rising aids in kneading, you can make a respectable loaf of bread in the processor, but you never really get the satiny fine gluten that you can get in the mixer.

Many breadmakers like to use the processor in conjunction with some hand kneading. They process a sticky dough for 20 to 30 seconds in the food processor, then turn it out onto a lightly floured counter and knead by hand for 4 or 5 minutes more.

You can make nice breads with the old-fashioned, long-handled, bucket-type kneaders as well. Part of the secret to good kneading is simply being aware of your two goals—incorporating air and stretching proteins—and concentrating on doing just that. One of the greatest faults of beginning bakers is to get the dough too firm. If the dough becomes too tight and stiff, by all means knead in some water. The dough should be soft.

Overkneading and underkneading You can knead a dough too much. The cross-links you worked so hard to create seem eventually to decide that they have had to pull themselves back together one time too many. When their bonds are broken and they are stretched out, the sulfurs start combining with atoms from the water and form thiol groups (page 57). There are no longer sulfurs tugging on sulfurs to pull the dough back together. Alas, there goes your elasticity, and the gluten starts coming apart. The water that was held in the gluten is released. The dough turns into a gooey, inelastic mess. Before your very eyes the satiny smooth dough starts to get sticky. This usually does not happen when kneading by hand because it would take a lot of time and strength to reach that point but it can happen when kneading by machine.

Another type of overkneading occurs when the dough is kneaded at an appropriate speed and then at a much lower speed. The gluten proteins seem suddenly to become disoriented, and a sticky, inelastic dough that looks exactly like an overkneaded dough appears instantly. Scientists have termed this condition *unmixing*. Fortunately, you can redevelop the gluten in this unmixed dough by turning the speed back up. You may get away with this once or twice; however, there's a limit to how many times a dough will tolerate being mixed and unmixed.

With a high-protein flour and appropriate rise times, even underkneaded doughs can produce reasonably good breads. However, I feel strongly that you get superior loaves with adequate kneading. A well-kneaded dough is satiny, extremely elastic, and springs back when you poke it. Doughs that I knead for 5 full minutes with a dough hook in a heavy-duty mixer are like this—satiny and elastic—and produce outstanding breads.

Rise times and temperatures: the final determinants of texture

Again, bread texture begins with mixing. It continues with the breadmaking method and rise times—direct and fast sponge methods with bakers' yeast and fast rises produce more even-textured breads, while starters requiring long rises produce rustic breads with big holes and uneven texture. Wetness of the dough also is always a factor: very firm doughs make close-textured breads. But assuming moderately wet soft doughs, the rising temperature as well as time will influence the texture.

Slowing it down: chilling and retarding Chilling the dough for a significant time, say overnight or longer, will influence the bread in three ways:

Enhances flavor The cold forces the yeast into dormancy so it is no longer gobbling up all the available sugars. The bacteria then can feed and produce some of their flavorful acids, which add subtle complex flavors and enhance keeping quality. Acids produced at cold temperatures (30° to 40°F/−1° to 4°C) are wonderfully flavorful, and even at cool temperatures are good. However, terrible-tasting short-chained acids are produced if the dough is hot for a long period.

To add flavor to bread, chill the dough—bacteria produce wonderfully tasty acids at refrigerator temperatures.

Reduces ovenspring and loaf volume Chilling for a period as long as overnight reduces ovenspring (oven rise) and reduces loaf volume. Alcohol that was made by the yeast from the beginning changes to a gas in the hot oven and is a contributor to ovenspring. During a lengthy cold period, alcohol simply evaporates from the surface of the dough and some goes to vinegar. So, there is a reduced amount of alcohol to contribute to ovenspring. As you see in the section below, chilling reduces the carbon dioxide in the bubbles, which contributes to a slight change in texture and possibly slightly reduced ovenspring and loaf volume.

Opens texture slightly Chilling dough retards the action of yeast and influences texture, though much more subtly than breadmaking method. The air-bubble nuclei worked into the dough during mixing and kneading are enlarged by gases from the yeast. Initially these tiny bubbles contain air, which is mostly nitrogen (78 percent) and oxygen (21 percent). Immediately the yeast uses up the oxygen so that you have mostly nitrogen in these tiny bubbles. The gases from yeast are essentially carbon dioxide, and they enlarge the same bubbles. The longer the dough rises, the more carbon dioxide there is in the enlarged bubbles.

After a long rise, the bubbles contain a very high percentage of carbon dioxide and a very low percentage of nitrogen. The longer the rise, or after multiple punch-downs, the higher the percentage of carbon dioxide. After a very long rise or two or three shorter rises, the bubbles contain essentially 100 percent carbon dioxide.

Carbon dioxide dissolves very well in cold water. So, if the dough is chilled, the carbon dioxide in a great number of bubbles will completely dissolve. The dough now has fewer bubbles for the gases from yeast to enlarge. When the dough is warm again, the yeast puts much more gas into these fewer bubbles, creating bigger bubbles and a coarser bread.

It is possible to have the positive effects of overnight chilling—enhanced flavor and slight opening of texture—without reduced volume. After chilling, you can dissolve a little yeast (1/2 teaspoon) in a few tablespoons of warm water and knead this into the dough before shaping.

Overrise

Overrise of dough in the bowl is not a problem as long as it is not caused by excessive warmth. It's OK for the dough to fall in the bowl since you were going to punch it down anyway. However, if the dough over-rose because it was kept too warm it may have developed off tastes from terrible-tasting short-chained acids made by bacteria when warm.

Overrise of a shaped loaf can be disaster. A beautiful, big overrisen loaf going into the oven must still be able to rise about one third more. If the gluten sheets cannot tolerate this, they will tear, the gases will leak out, and the loaf will fall.

Breads made with 100 percent bread flour have amazing tolerance but even these can fall if pushed too far. If you feel that a loaf has overrisen, it is safer to punch it down, shape it, and let it rise again rather than risk its fall in the oven.

Shaping doughs

The rounding step described in detail earlier (page 12) is a big help in shaping. The dough is tucked and pulled into a smooth round, which minimizes any loss of gases, then covered and allowed to stand on the counter for 15 minutes. This resting period allows the gluten to relax slightly so the dough is much easier to roll or shape.

Rounds (Boules)

Peter Nyberg, one of the country's top sourdough bakers, showed me the amazing difference that shaping techniques make for rounds, or boules. Many bakers pull and tuck the dough under in a cupping motion to form a ball with a tight, stretched surface. Peter shaped one loaf that way. Then he shaped a loaf with his own method, which doesn't deflate the slightly risen round. With the dough ball propped against his left hand, he presses down and forward on the opposite side of the ball with the palm of his right hand to lightly knead the side of the ball. He lets the dough ball rotate slightly forward with his stroke. He continues kneading the sides of the round until he has gone completely around the dough ball. Now, without deflating the dough, he gently pulls the top taut with a light tucking-under motion around the sides. Even immediately after shaping, you can see that the loaf shaped by just tucking under is not

nearly as high as the loaf shaped with Peter's side-kneading method.

Some bakers accomplish this side kneading by using their left hand to hold the dough against their apron at stomach level and then stroking the side of the dough with the palm of the right hand in a right-to-left motion, with little pressure. The dough rolls slightly to the left with this motion. Then the baker repeats the motion to knead a new section of the side, continuing in this manner until he or she has gone around the entire dough ball. The Yeast Starter Bread (page 65), and Mighty Multigrain (page 54) are examples of freestanding loaves with the dough formed this way.

Loaves or pan loaves

To shape a pan loaf, round the dough and then stretch the round with both hands sideways into an oval. Put it on the counter and let it spring back slightly, then pull it out again. Cover with plastic wrap and let stand for 15 minutes. Cup the rested loaf with both hands, your fingers spread out behind the loaf on either side and your thumbs in front of the loaf. Stretch by pressing your thumbs into the dough and down against the table. This pulls or tucks in part of the bottom half of the dough. At the same time, pull the top of the dough tight and forward with your fingers. Now move your thumbs down slightly and press down and in again while at the same time pulling the top forward with your fingers. Repeat this motion two to three times, until the loaf is stretched taut and well tucked in. Now pinch the bottom seam together. The ends go down to a slight taper, and this is the ideal shape for a rustic loaf that is to be placed in a banneton (canvas-lined basket) to rise and then baked freestanding. For a pan loaf, tuck the ends tightly under, pinch the bottom seam and ends together, and place in the pan.

Another method of shaping a loaf for a pan accomplishes the same stretching to create a taut surface by rolling and folding the dough. Roll the rested dough into a large square (approximately 18 × 18 inches) using a rolling pin on a clean counter. Fold the top third down and the bottom third up. Roll over the folded dough with the pin to slightly enlarge the top-to-bottom length and to press the layers together. Rotate the dough 90 degrees so that it is in front of you with the folds on the sides like a book. Now fold the top third down and the bottom third up. Roll over with the pin to press the layers together and get the width of the dough to the length of the pan, about 9 inches. Then, very tightly, stopping every inch or so and pressing the dough together along the crease, roll up along the long side. Press the ends of the dough under well and pinch the bottom seam and ends tightly.

The Honey Whole Wheat Loaf (page 46), Rice Bread (page 52), and Spicy Sugar Lump Loaf (page 84) are all examples of pan loaves.

Baguettes

Shape baguettes as if for a loaf. When doing the tucking in with your thumbs, stretch the dough slightly longer. Tuck in the middle, then farther from the center tuck in again. After several of the tucks the baguette may be long enough. If not, place both hands, spread out, palms down on top of the center of the loaf. Then simultaneously push away against the table with one hand and pull toward you with the other hand, pulling the dough out in opposite directions. Repeat this pulling as necessary to get the length you need.

Bakeries have machines for shaping baguettes that flatten the dough into a long oval, then roll it up tightly. You can roll out dough into an oblong, then roll it up tightly to simulate the machine's process. Using a rolling pin, roll the dough into a long rectangle the approximate length of the pan, then roll up the dough, inch by inch along the long side, tightly tucking and stretching the dough to create a tight long loaf. The Crusty French-Type Bread (page 24) is an example of a baguette.

Shaped bread pans

Many breads are made in unusual-shaped pans. For example, kugelhopf is baked in a Bundt-type pan with decorative ridges or swirls. The tube in the center allows the bread to cook from the center as well as the sides. Savarins (page 90) are always prepared in a ring mold.

Classic brioche à tête, a brioche roll with a knob or head on the top, is prepared in fluted molds that range from roll size to very large. I've tried everything to keep them centered, but still the heads on brioche à tête tip over to one side. Even when I've dug a hole in the roll and placed the narrow end of the pear-shaped piece for the head deep in the hole, they've still leaned to one side or the other.

Here's why: Frequently the dough is slightly warmer on one side than the other, causing the roll to rise more on that side, thus tipping the head. Francois Dionot, head of L'Academie de Cuisine in Bethesda, Maryland, prevents this topple by cutting slits in the roll perpendicular to the side of the pan. These slits act as expansion joints when the roll rises unevenly, absorbing the uneven rise and keeping the heads perfectly straight.

Shaping with water or oil

Is there anything you can do to shape dough easily without flouring the counter? It's always a temptation to add flour during shaping, but adding flour won't keep it soft and moist. Once when I assisted Jacques Pépin, the French chef, in a cooking class, he started to roll out a soft brioche dough and the dough began to stick to the counter. He turned to the sink, wet his hands, and spread a little water on the counter—just enough that the dough slipped slightly as he rolled.

You can use a little oil to do the same thing. A little fat in the dough increases the volume of the bread, so it is good if a little of the oil works into the dough. I have never heard anyone else advocate the use of oil instead of flour, but it keeps the dough soft and moist and produces a lighter bread. It is also a very easy way to work with more complicated dough layering, as in the Great Round Reuben (recipe follows).

If you're making complex breads and need to roll the dough out, oil the rolling pin. If you coated the bowl and the dough with oil, the dough is slightly oily

Use oil when you're rolling out dough—it prevents sticking as well as flour does, keeps the dough soft and moist, and produces a lighter bread.

on the surface, just right not to stick to the counter. Lightly oil the counter if necessary, but keep in mind that too much oil makes the dough slide around so that you cannot roll it. If the dough slides, wipe the work surface lightly with a paper towel to remove excess oil.

Here are three unusual filled bread recipes that use oiled doughs for shaping: a surprising round loaf that is actually a Reuben sandwich and two filled brioches.

Great Round Reuben

MAKES 6 TO 8 SERVINGS

This looks like a great round loaf of bread, but when you slice it each wedge is a fabulous Reuben sandwich with layers of onion-rye bread, corned beef, Swiss cheese, and sauerkraut. It's ideal for tailgate picnics, outings, the Scout troop, or the Tuesday night poker game.

what this recipe shows

High-protein bread flour provides enough gluten to carry the rye.

Vitamin C enhances gluten development.

Milk contributes minerals for gluten development and protein and sugar for better crust color, adds flavor, and helps keep the loaf moist.

Crushed caraway seeds enhance yeast activity.

Baking a filled loaf immediately after shaping produces a thin layer of bread encasing the filling.

Since a thin layer of dough is desired over the filling, the normal procedures for good volume like baking on a stone or a cold-oven start are not used.

$1^1/_3$ cups whole milk

2 packages ($4^1/_2$ teaspoons) active dry yeast

1 tablespoon *and* 3 tablespoons sugar ($^1/_4$ cup *total*)

$^1/_4$ cup warm water ($115°F/46°C$)

$1^1/_3$ cups rye flour

$2^2/_3$ cups bread flour

$^1/_4$ 500-milligram vitamin C tablet, crushed

2 teaspoons salt

$^1/_4$ cup *and* 1 tablespoon oil ($^1/_4$ cup plus 1 tablespoon *total*)

$^2/_3$ cup finely chopped onion

2 tablespoons caraway seeds, crushed

1 tablespoon oil for bowl

Nonstick cooking spray

3 tablespoons Dijon mustard

$1^1/_2$ pounds corned beef, very thinly sliced and cut into $^3/_4$-inch squares

1 can (16 ounces) sauerkraut, drained well, squeezed dry in a towel, and coarsely chopped

$^1/_2$ cup Thousand Island dressing

$^1/_2$ pound Swiss cheese, grated (about 2 cups)

1 large egg, beaten

$^1/_4$ teaspoon coarse salt

Read "The Bread Recipes in This Book" (page 10) before beginning.

1. In a small saucepan, bring milk almost to a boil, then remove from the heat and let cool to warm (115°F/46°C).

2. Dissolve the yeast and 1 tablespoon sugar in warm water in the bowl of a heavy-duty mixer. Let stand for 2 minutes until a foam forms, indicating yeast is alive and well. Add 3 tablespoons sugar, the rye flour, bread flour, vitamin C, salt, and milk. With the paddle blade, mix for 5 minutes on low-medium speed, until dough is elastic. Add ¼ cup oil, onion, and caraway seeds. Mix until well blended. The dough should be soft. Add a little bread flour if too wet or a little water if too dry.

3. Place the dough in an oiled bowl and turn to coat well with oil. Cover with plastic wrap and let rise until doubled. Remove the plastic. Punch down the center of the dough with a closed fist. Reach to the back of the bowl down under the dough and pull the bottom of the dough up and over the center. Repeat the same action at the front of the bowl, then place the dough out on the counter. Using both hands, with a gently cupping and tucking action, shape the dough into a smooth round. Cover with plastic wrap and leave on the counter for 15 minutes. The dough is now ready to shape or, if desired, refrigerate overnight.

4. Spray an 18-inch round pizza pan with nonstick cooking spray.

5. Place a shelf in a slot slightly below the middle of the oven and preheat to 375°F (191°C). Shape and layer the Reuben while the oven is heating.

6. Reserve a third of the dough for the top. Divide the remaining two-thirds of the dough into three pieces—a big piece, a medium-size piece, and a little piece. Oil a rolling pin. The dough should be oily and easy to roll on the counter. Roll out the largest of the last three pieces into a large circle and place on the pizza pan. Spread the dough circle lightly with Dijon mustard. Cover the dough with slightly less than half of the corned beef. Sprinkle slightly less than half of the sauerkraut on top of the corned beef. Spread slightly less than half of the Thousand Island dressing over the sauerkraut, then cover with a thin layer of cheese.

7. Roll out the medium-size piece of dough into a circle about an inch smaller in diameter than the first circle and place on top of the first layer on the pan. Cover the second layer with mustard, corned beef, sauerkraut, dressing, and cheese, reserving enough of each ingredient for the third layer. Roll out the small piece of dough about an inch smaller in diameter than the middle circle and place on top. Cover with layers as before.

8. Roll the reserved third of the dough into a circle large enough to cover the Reuben completely. Wrap the dough across a rolling pin, lift, and transfer to the layered Reuben. Tuck the top dough under the bottom layer of the dough all around the edges. Glaze the loaf with beaten egg and sprinkle lightly with coarse salt. Slash the center in a daisy pattern. Do not let rise.

9. Brush the bread again with beaten egg and place the pan in the oven. Bake until well browned, 50 to 60 minutes. Place the pan on a rack to cool. The Reuben is excellent served warm but quite good at room temperature (see Note).

Note: *The Reuben can be made ahead, wrapped well in foil or a jumbo zip-top plastic bag, and refrigerated or frozen. If frozen, defrost overnight in the refrigerator. To reheat, wrap in foil and heat in a 300°F (149°C) oven for 30 minutes.*

Brie Filled with Walnuts and Amaretto-Soaked Apricots in Brioche

MAKES 18 SERVINGS

Crisp, dark brown brioche cut into a lacy filigree tops brioche-encased creamy Brie that is filled with roasted walnuts, nippy lemon and orange zest, and amaretto-soaked apricots.

what this recipe shows

Less butter than the Ultimate Brioches (pages 15 and 17) makes this dough easier to handle and shape.

Less egg white means it does not puff and pull away from the filling.

Baking a filled loaf immediately after shaping produces a thin layer of bread encasing the filling.

Not beating the dough wet for maximum aeration produces a dense bread.

Having the cheese cold before it goes into the oven limits loss through melting.

Avoiding the normal procedures for good volume, like baking on a stone or using a cold-oven start, produce a thin layer of dough over the filling.

BRIOCHE

1 tablespoon *and* 3 tablespoons sugar
  (¼ cup *total*)

1 package (2¼ teaspoons) active dry
  yeast

1 cup warm water (115°F/46°C)

4 cups bread flour

1 teaspoon salt

2 large eggs

5 large egg yolks

10 ounces (2½ sticks) butter, softened

1 tablespoon oil for bowl

FILLING

10 dried apricots, chopped

¼ cup amaretto (see Notes)

1 1-kilogram (2.2-pound) round Brie

2 cups walnut pieces

3 tablespoons butter, softened

1 teaspoon salt

Finely grated zest of 2 oranges

Finely grated zest of 1 lemon

Nonstick cooking spray

1 large egg, beaten

Read "The Bread Recipes in This Book" (page 10) before beginning.

1. Dissolve 1 tablespoon sugar and the yeast in warm water in the bowl of a heavy-duty mixer with the paddle blade. Let stand 2 minutes until foam appears, indicating yeast is alive and well. Add 3 tablespoons sugar, the flour, 1 teaspoon salt, eggs, and yolks. Knead the dough at low-medium speed for 5 minutes, until very elastic. The dough should be soft and slightly sticky. After 2 minutes of kneading, add water if dough is too dry or flour if dough is too wet.

2. Work the 10 ounces butter into the dough on low speed. Oil a medium mixing bowl, place the dough in the oiled bowl, and turn to coat on all sides. Cover with plastic wrap and let rise until doubled in volume, about 1 hour. Refrigerate until well chilled or overnight if desired.

3. Meanwhile, soak apricots in amaretto in a small bowl.

4. When ready to fill and bake, place the Brie in the freezer long enough to get very cold but not to freeze, about 20 minutes. While the Brie is chilling, prepare the filling.

5. Preheat the oven to 350°F (177°C). Roast the walnut pieces on a large baking sheet until lightly browned, about 12 minutes. While the nuts are hot, stir in 3 tablespoons butter and sprinkle with 1 teaspoon salt. Drain the apricots. Combine the roasted walnuts and apricot pieces in a medium bowl.

6. When Brie is very cold, scrape off as much of the edible white coating as you can easily. With a knife, cut about a $1/2$-inch-deep cut all around the middle of the edge of the Brie so that you can cut it in half horizontally. Insert a piece of fishing line or dental tape in the cut all the way around the Brie. Cross the ends and pull to cut through the Brie. The Brie must be very cold for this to work. Brie that is very soft will glue itself right back together when you cut it, so have it very cold. If you have someone lift the top Brie half as you pull the string through it, this process will be easier. Spread the walnut-apricot filling over the bottom half of the Brie and sprinkle evenly with zest. Place top half back on the Brie and put the Brie in the freezer while you roll out the dough.

7. Punch down the center of the dough with a closed fist. Reach to the back of the bowl down under the dough and pull the bottom of the dough up and over the center. Repeat the same action at the front of the bowl, then turn out onto the counter. Using both hands with a gentle cupping and tucking action, shape dough into a smooth round. Cover with plastic wrap and leave on the counter for 15 minutes.

8. Place a shelf slightly below the middle of the oven and preheat to 375°F (191°C). Reserve a third of the dough for the top covering. Oil the rolling pin. The dough should be oily and roll easily. If it slips too much on the counter, wipe some oil off with a paper towel. Roll out two-thirds of the dough in a circle at least 3 inches larger on all sides than the Brie. Place the filled Brie in the middle of the circle and pull dough together on top to completely cover the Brie. Tear off excess pieces of dough and pinch the dough together. Do not worry about the appearance; just cover and press dough together as well as you can.

9. Roll the reserved third of the dough into a circle large enough to completely cover the top of the wrapped Brie. Using a tiny aspic cutter (tear, diamond, or heart shaped), cut holes in the dough in a pattern covering the whole dough. Start by doing a small circle of cutouts in the center, then a larger circle of cutouts about 1 inch out around the small one, and so on until you have cutout patterns over the whole top. Gently wrap the filigreed dough around a rolling pin, lift it up, and unroll it on top of the wrapped Brie. Glaze the filigree top *only* with beaten egg without getting any around the edges. The top shrinks as it bakes. If you have glued the top around the edges, the top will split across the middle in a great, unsightly gash. For decoration, I usually flatten a small strip of dough, about 1×5 inches, curl it around like a rose, and place it in the center. If you have enough scraps, make two or three leaves. Brush the rose with egg glaze too.

continued

10. Do not let the dough rise. Spray a pizza pan or a baking sheet with sides with nonstick cooking spray. Place the wrapped Brie on it and put it immediately into the oven. Bake until a deep rich brown, about 30 to 40 minutes. Don't worry if some Brie runs out. You can peel that off and eat it later. Let cool on a rack. Serve at room temperature (see Notes).

Notes: *Although soaking the apricots for an hour will do, they are much better if you can soak them overnight.*

The bread can be baked ahead, wrapped in foil or a jumbo plastic zip-top bag, and refrigerated or frozen. Allow the wrapped Brie to come to room temperature while sealed. Unwrap and serve.

Variation: *In Step 6, fill the Brie with a savory mixture of prosciutto cut into small pieces, fresh sage, and lemon zest instead of the walnuts and apricots.*

Soaked Dried Fruits for Breads

Heat 1 cup sugar and 1 cup water just until the sugar is dissolved, then let cool briefly. Stir in $1/4$ to $1/2$ cup good-quality rum such as Myers's or Grand Marnier, amaretto, Chambord, or whatever liquor or liqueur is compatible with the flavors in the bread. Allow the raisins or dried fruit to soak for an hour or, even better, overnight. Drain and knead into the dough.

One day some of my fellow food professionals were hovering over a magazine photograph of sausage in brioche that showed the dough pulled away from the sausage, leaving a great hole between the two. Dough that encases food does, unfortunately, tend to pull away from the filling as it rises. To avoid the fate that one of my colleagues said would make her "jump off a building with embarrassment," use ingredients and techniques like those in the Sausage in Brioche that keep the sausage and dough together.

Sausage in Brioche

MAKES TWO 4$^1/_2$ × 8$^1/_2$-INCH LOAVES

Wine-permeated, herb-scented sausage with buttery brioche and tangy mustard—it's hard to get better than this. These deeply browned loaves make impress-your-guests fare, whether your picnic features simplicity or silver candelabra.

what this recipe shows

Coating the sausage with egg-flour layers glues it securely to the dough and prevents the dough from pulling away as it rises.

Less egg white prevents a brioche dough from puffing and pulling away from the filling.

2 cups Beaujolais or other light red wine

2 pounds sweet or hot Italian sausage

Nonstick cooking spray

1 recipe brioche dough from Brie Filled
with Walnuts and Amaretto-Soaked
Apricots in Brioche (page 34)

2 large egg yolks

Flour for coating

1 large egg, beaten

Read "The Bread Recipes in This Book" (page 10) before beginning.

1. Pour the Beaujolais over the sausages in a saucepan and add enough water to cover them. Cook over medium heat until the pan is almost dry, about 30 minutes. Remove the sausages and let cool. Save the drippings in the pan for Mustard Butter (recipe follows). Remove the skins from the sausage.

2. Spray two $4^{1}/_{2} \times 8^{1}/_{2} \times 2^{1}/_{2}$-inch pans with nonstick cooking spray.

3. Divide the brioche dough in half. Roll each half into a rectangle about 8 inches wide, close to the length of the pan. Brush half of the sausage or sausage pieces with egg yolk, then roll in flour. Now brush 1 rectangle with egg yolk. Lay 1 pound of the sausage—in one piece or in several pieces laid end to end—across the dough. Roll up the dough with the sausage inside and tuck the ends under. Press the seam and ends together. Place the dough, seam side down, in the prepared pan. Repeat with the rest of the sausage and dough. Brush both loaves with a beaten egg. If you have some leftover scraps of dough, make some flowers or decorations and glue them on the top with beaten egg. Glaze the decorations with beaten egg too. If the dough was at room temperature, let rise in a warm room for about 30 minutes. If the dough was cold, let rise for about an hour.

4. About 30 minutes before the dough is fully risen, place a baking stone on a shelf in the lowest slot of the oven and preheat to 450°F (232°C)—see Notes. About 5 minutes before baking, turn the oven down to 375°F (191°C) and carefully place a shallow pan with $^{1}/_{2}$ inch of boiling water on the oven floor. (For an electric oven, put the pan of water on the lowest possible shelf.)

5. Brush bread again with beaten egg and place the pans directly on the hot stone. Bake until deeply browned, about 45 minutes total baking time. Place the loaves on a rack to cool (see Notes). Serve warm with Mustard Butter.

Notes: *If you don't have a stone:*
- Preheat the oven to 25°F (14°C) below the baking temperature, place the bread on a shelf in the lower third of the oven, 5 to 6 inches above the bottom, and then turn the oven up to the baking temperature. Carefully place a shallow pan with $^{1}/_{2}$ inch of boiling water on the floor of the oven. (For an electric oven, put the pan of water on the lowest possible shelf.)
- Or use the cold-oven procedure described on page 14.

Sausage in Brioche can be made ahead, wrapped well in foil or a jumbo zip-top plastic bag, and refrigerated or frozen. If frozen, defrost, still wrapped, overnight in the refrigerator, then reheat wrapped in foil in a 300°F (149°C) oven for 30 minutes.

Mustard Butter

Pan drippings from cooking the sausage
in Sausage in Brioche
(page 36)

2 tablespoons Beaujolais or other light
red wine, optional

$^1/_2$ cup coarse-grain mustard

$^1/_4$ pound (1 stick) butter,
softened

$^1/_4$ teaspoon salt

If there is a little liquid left in the pan, scrape the pan with a spatula to loosen any stuck-on fla-vorful bits. If there is little or no liquid, add a little Beaujolais and scrape. Add the mustard and scrape around to get all the good stuff out of the pan, then scrape the mixture into a small mixing bowl or into a food processor with the steel knife. Add the softened butter and salt and mix to blend well.

Braiding

Braided loaves can be spectacular, and lightly oiled dough makes the weaving a snap.

Here is a recipe that turns standard three-strand braids into a six-pointed star.

Star Spectacular

MAKES 1 LARGE BRAIDED STAR, ABOUT 30 SERVINGS

This delicious, edible centerpiece is a knock-your-socks-off bread star. Dried and varnished, Star Spectacular is a decorator item for kitchen or bakery walls.

what this recipe shows

When dough is oiled lightly, it is very easy to shape into complex braids.

Braiding needs to be kept slightly loose to take rising into account.

3 recipes brioche dough from Brie
Filled with Walnuts and Amaretto-
Soaked Apricots in Brioche
(page 34)

Nonstick cooking spray

1 large egg, beaten

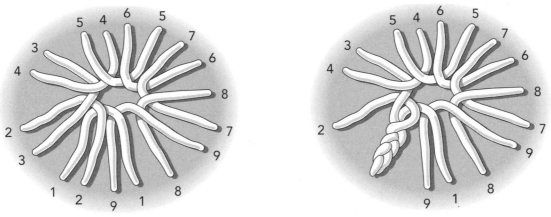

Braiding a Six-Pointed Star

1. Divide the dough into thirds and then each third into thirds again so that you have nine pieces. Between your palms and the counter, roll each piece out into a rope about 24 inches long. Arrange all nine strands, each one overlapping the previous strand as illustrated. Then make six braids of three strands each. You can make a knot of dough for the center or place a bowl of spread in the center after the loaf is baked.

2. Spray a 14 × 17-inch baking sheet or 16-inch pizza pan with nonstick cooking spray. Place braided loaf on sheet and brush well with the beaten egg. Let dough rise until doubled.

3. About 30 minutes before the dough is fully risen, place a baking stone on a shelf in the lowest slot of the oven and preheat to 450°F (232°C)—see Note. About 5 minutes before baking, turn the oven down to 375°F (191°C) and carefully place a shallow pan with ½ inch of boiling water on the oven floor. (For an electric oven, put the pan of water on the lowest possible shelf.)

4. Brush the bread one last time with beaten egg and place the baking sheet directly on the hot stone. Bake until well browned, about 45 minutes. Let cool on the baking sheet on a rack. Serve warm or at room temperature. Place a small bowl of Cherry-Chambord Butter (page 78) or Orange Butter (page 96) in the center of the star.

Note: *If you don't have a stone:*
- Preheat the oven to 25°F (14°C) below the baking temperature, place the bread on a shelf in the lower third of the oven, 5 to 6 inches above the bottom, and then turn the oven up to the baking temperature. Carefully place a shallow pan with ½ inch of boiling water on the oven floor. (For an electric oven, put the pan of water on the lowest possible shelf.)
- Or use the cold-oven procedure described on page 14.

Grape-Bunch Braid

MAKES 1 BRAIDED LOAF, ABOUT 5 × 9 INCHES

This unusual loaf is a little knot braid that resembles a bunch of grapes. The drawings show the steps for braiding. You can use many different doughs for a braided loaf like this. The Nisu is excellent.

1 recipe Nisu dough (page 70), prepared through Step 3

Nonstick cooking spray

3 tablespoons instant coffee or instant cappuccino powder dissolved in $^1/_3$ cup hot water

3 tablespoons dark corn syrup

1 large egg, beaten

$^1/_2$ teaspoon cardamom seeds, freshly crushed

$^1/_4$ cup coarse sugar

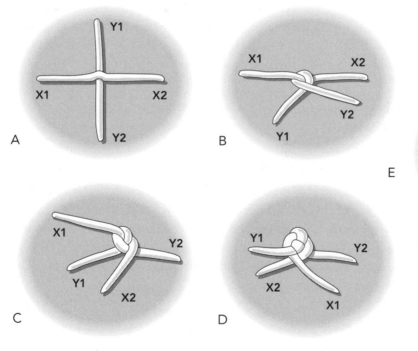

Grape-Bunch Braid

1. Divide the dough in half. Between your palms and the counter, roll each piece out into a rope about 18 inches long. Look at the drawings and arrange the ropes crossed in the center with the rope that is perpendicular to the counter edge on the bottom and the one that is parallel to the counter edge on top. Bring the upper tip (Y1) of the perpendicular rope down across the other rope and over to the left. Then bring the lower tip (Y2) of the perpendicular rope over it to the right. Now bring the right tip (X2) of the parallel rope down across the one just moved (Y2). Work the left tip (X1) of the parallel rope down under the upper tip (Y1) and across the right tip (X2). You now have four strands. Finish with a regular four-strand braid.

2. Spray a 12 × 15-inch baking sheet or a 14-inch pizza pan with nonstick cooking spray and place the braided loaf on it. Brush the dough with some of the coffee mixture. Let dough rise until doubled, about 1 hour, brushing the dough with the coffee mixture two or three times.

3. About 30 minutes before the dough is fully risen, place a baking stone on a shelf in the lowest slot of the oven and preheat to 450°F (232°C)—see Note. About 5 minutes before baking, turn the oven down to 375°F (191°C) and carefully place a shallow pan with ½ inch of boiling water on the oven floor. (For an electric oven, put the pan of water on the lowest possible shelf.)

4. Stir the corn syrup, egg, and cardamom together in a small bowl. Brush the loaf with the syrup-egg mixture and place baking sheet directly on the hot stone. After 30 minutes, quickly pull the loaf out of the oven and brush again with syrup-egg mixture. Sprinkle heavily with coarse sugar. Bake until well browned, about 15 minutes more. Let cool on the baking sheet on a rack. Serve warm or at room temperature. Excellent with Orange Butter (page 96).

Note: *If you don't have a stone:*
- Preheat the oven to 25°F (14°C) below the baking temperature, place the bread on a shelf in the lower third of the oven, 5 to 6 inches above the bottom, and then turn the oven up to the baking temperature. Carefully place a shallow pan with ½ inch of boiling water on the oven floor. (For an electric oven, put the pan of water on the lowest possible shelf.)
- Or use the cold-oven procedure described on page 14.

Layering dough for flakiness

You saw what technique can do in two Ultimate Brioches that were very different simply because of when the butter was added. The form of the butter can also completely change a bread. The secret of flakiness, for example, is layers or patches of firm, unmelted fat between layers of dough (page 107). The fat acts as a spacer just long enough to allow the dough to start to set; then the fat melts and liquid in the dough turns to steam and puffs the dough layers apart. If you are using a fat like butter that melts at a moderately low temperature, you must keep the dough and the fat cold. If the fat melts, and soaks into the dough, there goes your flakiness.

Croissants—classic or innovative

There are several ways to produce flakiness. In the classic method for making croissants, Danish, or puff pastry, a layer of cold fat is placed on top of a layer of dough. The two layers are folded two to four times, flattened, refolded, and flattened several times, producing a dough with over a thousand thin layers (*mille-feuilles*) of unmelted fat between layers of dough. It is vital that the initial dough is underkneaded before folding in the butter, wet to the point of being sticky, and very soft. Placing it in a well-oiled bowl makes it easier to handle. This wet dough will provide steam for extraordinarily light croissants and, more important, a soft wet dough allows you to make the many turns (folds) with ease.

Rolling the dough out is essentially kneading it. Each time the dough is rolled out it becomes more elastic and firm. It is important to start with a soft, underkneaded dough or you will not be able to roll out and make all the turns.

If you thought you had the dough wet enough to begin with, but during the turns it becomes firm and difficult to roll, all is not lost. Roll the dough out as large as you can, brush off all the flour, brush moderately heavily with ice water, fold, brush newly exposed areas with ice water, finish the turn, and refrigerate as usual. Continue brushing with ice water as necessary to keep the dough soft so that you can finish all the turns.

It is important that the dough be kept cold during the entire rolling process. If it gets warm and the butter melts into the dough you will lose flakiness.

In shaping croissants, a frequently made error is to roll them too tightly. This limits the rise and lightness and the tops will split. The split is hidden by the flakiness in the final baked product but, nevertheless, the croissant is not as high and light as it would be if rolled loosely.

Just as in puff pastry, a high oven temperature (425°F/218°C) is needed to create steam for a good puff before the dough starts crusting and setting. After they are puffing nicely (10 to 15 minutes), turn the oven down to prevent overbrowning.

Another method is to knead chips of frozen butter into cold dough, shape, let rise at a cool temperature of 65° to 70°F (18° to 21°C), and bake. Helen Fletcher wrote wonderful recipes for flaky baked goods using grated chips of frozen butter that appeared both in magazine articles and in her book *The New Pastry Cook*.

A third way to create flaky doughs is to flatten and coat big pieces of frozen butter with flour by rolling and rerolling a rolling pin over a mixture of flour and big lumps of frozen butter as described for making flaky pie crust (page 110), then folding this flour-coated frozen butter into an already lightly kneaded dough. I have not seen a recipe that uses exactly this technique for croissantlike rolls, so that's the type demonstrated by my Buttery Flaky Rolls.

You can convert the Buttery Flaky Rolls to a classic croissant recipe simply by reducing to 1/2 cup the flour that is tossed with the butter lumps, then rolling and pressing this butter-flour mixture into a 6 × 6-inch square. You may need 1/4 cup more bread flour in the dough itself, but keep it sticky. After its rise, roll the dough into a 9 × 14-inch rectangle. Place the butter-flour square on the upper half, fold the dough up and over, and seal the edges well. Flatten and proceed with turns as directed in classic croissant recipes.

Buttery Flaky Rolls

MAKES ABOUT 2 DOZEN ROLLS

These flaky, buttery puffs are wonderful and show clearly how cold flattened pieces of fat create flakiness. They are a cross between puff pastry and croissants. This dough works best as a loosely folded-over roll like a Parker House. If you want to cut and roll as a croissant, keep them small and roll loosely or the center does not get done.

what this recipe shows

Flattening pieces of cold butter creates flakiness.

Milk supplies minerals for yeast food and gluten development and protein and sugar for better crust color and adds flavor and moisture.

Using barley malt syrup provides enzymes to convert starch into sugars for yeast food.

Keeping the dough wet—quite sticky—and soft provides steam for well-puffed layers, and the soft dough is easy to roll.

¾ pound (3 sticks) unsalted butter, cut into ½-inch cubes

1½ cups *and* 1¾ cups bread flour (3¼ cups *total*)

1 package (2¼ teaspoons) active dry yeast

3 tablespoons sugar

1 cup warm whole milk (115°F/46°C)

¼ cup all-purpose flour

1 teaspoon salt

1 teaspoon barley malt syrup (see Notes) or 1 tablespoon light brown sugar

2 tablespoons heavy or whipping cream

1 tablespoon oil for bowl

Nonstick cooking spray

1 large egg, beaten

1. Toss butter cubes and 1½ cups bread flour together in a medium bowl and place in the freezer for 30 minutes. Dump flour–frozen butter mixture onto a clean counter, roll over the mixture with a rolling pin, scrape butter off the pin, pile the mixture together, and roll over it again. Repeat this process three times, scrape back into the bowl, and place in the freezer for 10 minutes. Then dump out and flatten three more times until you have flattened, flour-coated flakes of butter that look like paint peeling off a wall. Scrape the mixture back into the bowl and place in the freezer again.

2. Stir the yeast and sugar into the milk in the bowl of a heavy-duty mixer. Let stand 2 minutes, until foam appears, indicating yeast is alive and well. Add 1¾ cups bread flour, all-purpose flour, salt, barley malt syrup, and cream. Knead with the paddle blade on medium speed for 2 minutes. The dough should be soft and quite sticky. Place the dough in an oiled bowl and, with an oiled spatula, scrape the wet dough in. Turn the dough to coat with oil. Cover with plastic wrap and let rise at room temperature for about 45 minutes. Remove the plastic. Punch down the center of the dough with a closed fist. Reach to the back of the bowl under the dough and pull the bottom of the dough up and over the center. Repeat the same action at the front of the bowl, cover with plastic wrap, and refrigerate until quite cold, at least 1 hour or overnight if desired.

3. Lightly flour the counter and roll dough out as large as easily possible. Cover with plastic wrap, let stand 10 minutes, then roll larger. Spread three-quarters of the frozen butter–flour mixture over the middle of the dough, leaving the right and left quarters uncovered. Fold left quarter to middle, fold right quarter to middle, spread the remaining quarter of the butter on top, then fold one double side over the other. Roll as flat as easily possible. Fold left quarter to center, right quarter to center, then fold bottom half over top half. Roll out as flat as possible. Place on a baking sheet, cover with plastic wrap, and place in the refrigerator for 10 minutes, then in the freezer for 10 minutes.

4. When the dough is quite cold, roll out on a lightly floured counter to less than ½ inch thick. Cut into long strips about 4 inches wide with a very sharp knife or pizza cutter. Cut every 2 inches so that you have 4 × 2-inch rectangles. Fold each in half so that you have 2-inch-square double-layered rolls. Spray a baking sheet with nonstick cooking spray and place the puffs on the baking sheet. Cover with plastic wrap and let rise in a cool place—cool room temperature is fine—for 2 to 3 hours. Do not let the puffs get warm, or the butter will melt and they will lose flakiness.

continued

5. About 30 minutes before the dough is fully risen, place a baking stone on a shelf in the lowest slot of the oven and preheat to 450°F (232°C)—see Note. About 5 minutes before baking, turn the oven down to 400°F (204°C) and carefully place a shallow pan with ¹/₂ inch of boiling water on the oven floor. (For an electric oven, put the pan of water on the lowest possible shelf.)

6. When the puffs are fully risen, brush the tops *only* with the beaten egg and place the puffs in the freezer for 15 minutes. Brush the puffs again with egg and place the pan directly on the hot stone. Bake until well browned, about 25 minutes. Cool briefly on a rack. Serve hot.

Notes: *Barley malt syrup is available at health food stores and beer- and winemaking supply shops. If you don't have a stone:*
- Preheat the oven to 25°F (14°C) below the baking temperature, place the bread on a shelf in the lower third of the oven, 5 to 6 inches above the bottom, and then turn the oven up to the baking temperature. Carefully place a shallow pan with ¹/₂ inch of boiling water on the oven floor. (For an electric oven, put the pan of water on the lowest possible shelf.)
- Or use the cold-oven procedure described on page 14.

Slashing

Have you ever had a slice of bread with good texture but a great hole just under the crust? When bread goes into a hot, dry oven, its top begins to crust immediately. As the dough gets hot, liquid in the dough turns to steam. The steam rises but cannot escape because the top has already formed a crust. This trapped steam forms a hole right under the crust. Slashing the bread on top not only has a beautiful effect but also vents steam. Loaves are normally slashed just before going into the oven. Because they are well risen at that point, they can be deflated easily if not slashed with a sharp razor and a quick, light stroke.

A slash is not the only way to vent bread. I have a friend who makes beautiful big loaves of Jewish rye bread by very carefully poking several holes in the top with a thin sharp ice pick when the dough is half risen.

Slashes can be impressive decorations, especially when heat from the bottom as from a baking stone makes the slash spread dramatically. The slashes on baguettes are traditionally made holding the blade at about a 20-degree angle. The Yeast Starter Bread (page 65) can be slashed in a number of dramatic ways.

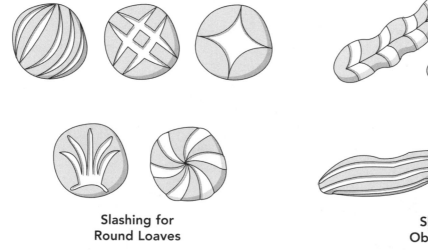

Slashing for Round Loaves

Slashing for Oblong Loaves

at a *glance*
Breadmaking techniques

What to Do	Why
Sequence of ingredients	
Liquid and flour combined first	Produces light, breadlike loaf.
Fat and flour combined first	Produces dense, tender cakelike loaf.
Salt added before kneading	Strengthens and tightens gluten and is incorporated evenly and thoroughly.
Salt added after kneading but before rising	Dough is easier to knead and takes a shorter kneading time, then gluten is strengthened and tightened for good rises. Use flaky sea salt or kosher salt for faster and better blending.
Breadmaking methods	
Straight Dough	Produces medium-fine-textured loaf.
Sponge	Increases flavor, with very fine to open texture, depending on rise.
Sourdough	Strong flavor, open texture.
Kneading	
Methods—food processor, hand, or heavy-duty mixer	Heavy-duty mixer is personal preference.
Overkneading	Gluten will come apart if dough is overworked or kneaded at a slow speed after medium-speed kneading.
Rise time and temperature	Determine bread texture.
Punching Down	Redistributes yeast.
Rounding	Allows dough to relax, making shaping easier, and aids in aligning gluten.
Layering	Creates flakiness.
Shaping	Tight shaping of loaves produces even loaves.
Slashing	Decorative, vents steam, and allows expansion of loaf.
Baking	Must get heat from the bottom to have a good rise before crust forms to hold loaf down.
Steam or Moisture	Steam early in baking helps prevent crusting for maximum rise. Water misted directly on loaf at end of baking creates crunchy crust.
When shaped loaf overrises	Punch down and let rise again rather than risk fall in the oven.

Ingredients

Picking bread ingredients involves more than meets the eye. Many ingredients play multiple roles, and any change in an ingredient can make a totally different bread. For example, honey instead of corn syrup in a bread can mean different volume, texture, shelf life, or color. Milk in place of water will produce a loaf of different color and texture.

Flour: the essential ingredient

As already explained, flour can be made from any grain, but only flour from wheat contains enough gluten-forming proteins to make a light loaf of yeast bread. Flours from grains other than wheat can be mixed with wheat flour to make excellent flavorful loaves, but wheat is a necessity.

Wheat flours

Plain white flour contains essentially the two gluten-forming proteins and starch and was discussed in the gluten section (page 4).

Other wheat flours include whole wheat, white wheat, vital wheat gluten, gluten flour, and semolina (durum wheat). Whole wheat contains germ and bran as well as starch and gluten-forming proteins. Per cup, whole wheat from the same grain will have less of the two gluten-forming proteins than white flour. This lower amount of gluten results in heavier, denser loaves of bread. Also if the whole wheat is not finely ground, sharp edges on the bran can cut gluten strands during kneading.

Whole wheat flour and high-protein bread flour are frequently combined to produce lighter loaves, as in my Honey Whole Wheat Loaf (recipe follows). Adding vital wheat gluten or gluten flour is another way to increase gluten formation and lighten whole wheat and multigrain loaves.

Vital wheat gluten, which is available in health food stores, is gluten that has been extracted from flour. It contains 45 to 60 grams of protein per cup or has approximately three to five times the gluten-forming potential of bread flour. Gluten flour technically is flour blended by a miller (not extracted) that contains high concentrations of the gluten-forming proteins; however, some flour labelers do not acknowledge this distinction. You will find that some products labeled gluten flour are extracted vital wheat gluten.

Whatever the label, vital wheat gluten and gluten flour are both strong stuff. Since either forms a lot of gluten, it absorbs a lot of water. Too much can make a loaf tough and dry. Cautious amounts of vital wheat gluten are ideal for lightening heavy loaves of multigrain like Mighty Multigrain (page 54).

White whole wheat is a high-protein strain of hard white wheat grown in Kansas. It contains more gluten-forming proteins than regular whole wheat and as a result rises better than regular whole wheat. You may still want to blend it with some high-protein bread or unbleached flour for a very light loaf. It has fewer tannins than regular whole wheat, a sweeter taste, and a pale color. It makes bread similar to that made with unbleached flour. Information on suppliers is available through the White Wheat Board (see Sources—White wheat information, page 479).

Honey Whole Wheat Loaf

MAKES ONE 5 × 9-INCH LOAF

My favorite whole wheat bread in all of Atlanta is a long French-style loaf from the Georgia Grille. I love it so much that owner Karen Hilliard gives me a loaf every Christmas. It also prompted me to use nonfat dry milk and honey as they do at the Georgia Grille and add my favorite ingredients and techniques to arrive at this big, handsome loaf—plain and simple but simply delicious.

what this recipe shows

Dry milk contributes minerals for gluten development and protein and sugar for better crust color, adds flavor, and helps keep the loaf moist.

Good gluten-forming bread flour is combined with the whole wheat for a lighter loaf.

Vitamin C is included for better gluten development.

Sugars from the milk and honey provide lots of yeast food for good rises.

Honey adds flavor, feeds the yeast, and absorbs moisture from the air for good keeping qualities.

Working wheat germ in carefully by hand after kneading prevents its sharp edges from cutting gluten strands and reducing volume.

1/3 cup nonfat dry milk

1 1/2 cups cool water

1 package (2 1/4 teaspoons) active dry
   yeast

2 tablespoons warm water
   (115°F/46°C)

2 1/2 cups bread flour

1/4 500-milligram vitamin C tablet,
   crushed

1/4 cup crushed ice

1 teaspoon salt, preferably sea salt

2 tablespoons honey

1 large egg yolk

1 1/4 cups whole wheat flour

1/4 cup wheat germ

1 tablespoon oil for bowl

Nonstick cooking spray

1 large egg, beaten

Read "The Bread Recipes in This Book" (page 10) before beginning.

1. Stir the dry milk into the cool water in a small saucepan and heat over medium heat almost to a simmer. Turn down the heat. Hold below a simmer for 1 minute, then turn heat off and let cool at least 5 minutes.

2. In the bowl of a heavy-duty mixer, stir together the yeast and warm water. Let stand 2 minutes, until foam appears, indicating yeast is alive and well. Add the bread flour. Then pour in the milk and beat on low-medium speed with the paddle blade for 4 minutes to beat air into the dough. Let the sponge sit for 30 minutes to 2 1/2 hours for improved flavor and texture.

3. Remove the paddle blade and insert the dough hook. Add the vitamin C, crushed ice, salt, honey, egg yolk, and whole wheat flour. Knead on low-medium speed for 5 minutes, until dough is very elastic. The dough should be very soft and slightly sticky. Add a little flour or liquid if needed. Sprinkle wheat germ over dough and work in by hand.

4. Place the dough in an oiled bowl and turn to coat well with oil. Cover the bowl with plastic wrap and let the dough rise until slightly over doubled in volume, about 1 1/2 hours. Punch down the center of the dough with a closed fist. Reach to the back of the bowl under the dough and pull the bottom of the dough up and over the center. Repeat the same action at the front of the bowl and turn it out on the counter. Using both hands with a gently cupping and tucking action, shape dough into a smooth round. Grab the round with both hands and stretch it sideways into an oval.

continued

Put it on the counter and let it spring back slightly, then pull it out again. Cover with plastic wrap and leave on the counter for 15 minutes. The dough is now ready to shape.

5. Cup the dough with both hands, fingers spread out behind the loaf on either side and thumbs in front of the loaf. Press your thumbs into the dough and down against the table. This pulls or tucks in part of the bottom half of the dough. At the same time, pull the top of the dough tight and forward with your fingers. Now move your thumbs down slightly and press down and in again to knead and tuck again. Repeat this motion two to three times, until the loaf is stretched taut and well tucked in. Pinch the bottom seam tightly together. The ends go down to a slight taper. Tuck them tightly under, then pinch the ends and bottom seam together.

6. Spray a 5 × 9 × 3-inch loaf pan with nonstick cooking spray and place the loaf in, seam side down. Brush with beaten egg. Let the dough rise until more than doubled, about 1 hour.

7. About 30 minutes before the dough is fully risen, place a baking stone on a shelf in the lowest slot of the oven and preheat to 450°F (232°C)—see Note. About 5 minutes before baking, turn the oven down to 375°F (191°C) and carefully place a shallow pan with ½ inch of boiling water on the floor of the oven. (For an electric oven, put the pan of water on the lowest possible shelf.)

8. Brush the bread again with beaten egg and place the pan directly on the hot stone. Bake until well browned, 45 to 55 minutes. Remove the loaf from the pan. The loaf should sound hollow when thumped on the bottom. You can also check for doneness by inserting an instant-read thermometer in the bottom center and pushing it to the middle. The loaf is done when the center is at least 200°F (93°C). Place the loaf on a rack to cool.

Note: *If you don't have a stone:*
- Preheat the oven to 25°F (14°C) below the baking temperature, place the bread on a shelf in the lower third of the oven, 5 to 6 inches above the bottom, and then turn the oven up to the baking temperature. Carefully place a shallow pan with ½ inch of boiling water on the floor of the oven. (For an electric oven, put the pan of water on the lowest possible shelf.)
- Or use the cold-oven procedure described on page 14.

Jack, Avocado, and Sprouts on Whole Wheat

MAKES 4 SERVINGS

This is my favorite open-face sandwich. The Monterey Jack and avocado mixture puffs to a rich golden brown. Sweetness from the apple, crisp sprouts, and seeds creates delicious taste and texture contrasts.

what this recipe shows

The egg in the mayonnaise makes the topping puff.

The protein in the mayonnaise and cheese and the sugar make the topping brown beautifully.

$^1/_2$ small carrot, peeled and quartered

2 ounces Monterey Jack, cut into 6 pieces

$^1/_2$ cup mayonnaise, homemade (page 289) or store-bought

$^1/_2$ teaspoon light brown sugar

$^1/_4$ teaspoon salt

$^1/_8$ teaspoon white pepper

$^1/_4$ Granny Smith apple, peeled and cut into $^1/_2$- to $^3/_4$-inch chunks

$^1/_2$ ripe avocado, cut into $^1/_2$-inch cubes

4 slices whole wheat bread

2 tablespoons roasted sunflower seeds

$^1/_2$ cup alfalfa sprouts

Turn the broiler on. Process the carrot to a fine chop in a food processor with the steel knife. Add the cheese and process to a fine chop. Add the mayonnaise, brown sugar, salt, and pepper and process to mix well. Add apple chunks and process with three or four quick on/off pulses only. Spoon into a bowl and stir in avocado cubes. Toast the whole wheat slices very lightly. Spoon a quarter of the avocado-cheese mixture onto each slice and spread evenly. Broil until puffed and lightly browned. Sprinkle each with seeds and sprouts and serve hot.

Semolina: the pasta flour

Some food reference books describe semolina as a coarse grind of cereal and say that white semolina is ground from rice—kasha is a semolina from buckwheat. Restaurant critic, Elliott Mackle, who has toured durum wheat fields, explained to me that in the United States anything called *semolina* must by law be made of durum wheat with no more than 3 percent other flour. So, in the United States semolina is exclusively durum wheat flour that has been chipped or sliced (not ground) into a coarse texture—about the texture of coarse cornmeal.

Durum wheat is a high-protein, extremely hard wheat with a large kernel. Durum varieties belong to an entirely separate botanical species from ordinary wheats and are completely different. In durum wheat the starch is locked inside a hard protein shell. This means that when you overcook pasta made with semolina it does not become a gooey, starchy mess. In fact that's why pasta makers use semolina: its protein content, about 13 grams per cup, is high enough to provide good strength and elasticity, and the cooked pasta has little starch coating.

Pasta manufacturers order from the durum

wheat miller a specific texture of semolina. Their pasta is pressed at high pressure through Teflon-coated brass rollers to produce a hard surface on the pasta that will not get soft and overcooked before the center is done.

Be aware, though, that semolina in Italian recipes and semolina in the United States grocery stores are not necessarily the same. Semolina machines—the machines for processing durum wheat—are of Swiss or Italian manufacture and can be set to produce different textures. Most of the semolina in grocery stores in the United States has a coarse texture similar to cornmeal, but it can be produced in finer textures like flour, too. I have recently been able to find semolina flour in all the health food stores. Susan Derecskey, cookbook writer and editor, tells me that she has had good success making Sicilian pastries and breads with the coarse semolina by processing it in the food processor until it is silky fine.

Most important of all, in spite of its high protein content, durum wheat does not form as high-quality gluten as other varieties. A mixture of bread flour and semolina flour, however, produces a light loaf with a golden tint and an excellent flavor, as in the next recipe.

Semolina Bread

MAKES ONE 10-INCH ROUND LOAF

Golden semolina flour gives this round loaf a pale yellow crumb and distinctive taste. With full flavor, a crunchy browned crust, and a medium texture, this bread is excellent as a hearty accompaniment to salad and one-dish dinners.

what this recipe shows

Semolina forms some gluten but not as much as some other hard varieties of wheat.

High-protein bread flour, vitamin C, and crushed ice added before kneading contribute to better gluten development.

Olive oil in the loaf imparts flavor and slightly enhances volume.

1 package (2 1/4 teaspoons) active dry yeast

1 tablespoon sugar

1 1/2 cups warm water (115°F/46°C)

2 cups bread flour

1/4 500-milligram vitamin C tablet, crushed

1/4 cup crushed ice

2 cups semolina flour

1 1/4 teaspoons *and* 1/4 teaspoon salt, preferably sea salt (1 1/2 teaspoons *total*)

2 tablespoons *and* 1 tablespoon *and* 3 tablespoons good-quality olive oil (6 tablespoons *total*)

3 to 4 tablespoons coarse cornmeal as needed

Read "The Bread Recipes in This Book" (page 10) before beginning.

1. Stir the yeast, sugar, and water together in the bowl of a heavy-duty mixer. Let stand 2 minutes, until foam appears, indicating yeast is alive and well. Add the bread flour. With the paddle blade, beat on low-medium speed for 4 minutes to beat air into the dough. Let the sponge sit for 30 minutes to $2^{1}/_{2}$ hours for improved flavor and texture.

2. Remove the paddle blade and insert the dough hook into the mixer. Add the vitamin C, crushed ice, semolina flour, and $1^{1}/_{4}$ teaspoons salt and knead 5 minutes on low-medium speed, until the dough is very elastic. The dough should be soft but all together and pulled away from the bowl. Add a little more bread flour or liquid if needed. Pull the dough into several pieces and drizzle 2 tablespoons olive oil over it. Knead on medium speed to incorporate oil well, about 30 seconds.

3. Oil the bowl with 1 tablespoon olive oil and place the dough in the oiled bowl. Turn to coat well with oil. Cover the bowl with plastic wrap and let the dough rise until slightly more than doubled, $1^{1}/_{2}$ to 2 hours. Punch down the center of the dough with a closed fist. Reach to the back of the bowl under the dough and pull the bottom of the dough up and over the center. Repeat the same action at the front of the bowl.

4. When ready to shape and bake, turn the dough out onto the counter and, using both hands with a gentle cupping and tucking action, shape into a smooth round. Cover with plastic wrap and leave on the counter for 15 minutes. The dough is now ready to shape.

5. Sprinkle a medium baking sheet or 16-inch round pizza pan heavily with cornmeal. If you are right-handed, prop the dough ball against your left hand and press down and forward on the opposite side of the ball with your right palm to lightly knead the side of the ball. Let the dough ball rotate slightly forward with the stroke and continue to knead the sides of the round until you have gone completely around. Pinch the bottom to seal and place on the cornmeal on the baking sheet. Stir together 3 tablespoons olive oil and $^{1}/_{4}$ teaspoon salt and brush some of it on the loaf. Let the dough rise until doubled, about 1 hour.

6. About 30 minutes before the dough is fully risen, place a baking stone on a shelf in the lowest slot of the oven and preheat to 450°F (232°C)—see Note. About 5 minutes before baking, turn the oven down to 375°F (191°C) and carefully place a shallow pan with $^{1}/_{2}$ inch of boiling water on the oven floor. (For an electric oven, put the pan of water on the lowest possible shelf.)

7. Brush the bread again with oil-salt mixture and slash decoratively (page 44). Place the baking sheet directly on the hot stone. Mist the loaf with water two or three times during the last 10 minutes of the baking time, using a clean spray bottle. Bake until well browned, about 50 minutes. Remove the loaf from the pan. The loaf should sound hollow when thumped on the bottom. You can also check for doneness by inserting an instant-read thermometer in the bottom center of loaf and pushing it to the middle. The loaf is done when the center is over 200°F (93°C).

8. Place the loaf on a rack to cool. Serve warm or at room temperature.

Note: *If you don't have a stone:*
- Preheat the oven to 25°F (14°C) below the baking temperature, place the bread on a shelf in the lower third of the oven, 5 to 6 inches above the bottom, and then turn the oven up to the baking temperature. Carefully place a shallow pan with $^{1}/_{2}$ inch of boiling water on the oven floor. (For an electric oven, put the pan of water on the lowest possible shelf.)
- Or use the cold-oven procedure described on page 14.

GRAINS AND DRIED INGREDIENTS IN BREADS

Grains other than wheat form little or no gluten, so they are essentially dead weight in doughs. If the grains are coarsely ground, they can cut gluten strands during kneading. Any seeds or coarsely ground grain should be worked in by hand after kneading.

In addition to forming less gluten and being heavy, breads containing whole grains such as rice, millet, or wheat berries can be dry. Some recipes call for presoaking grains and some for cooking grains before adding them.

Fully cooked grains can contribute moisture and good keeping characteristics. The cooked rice in the following Rice Bread does exactly this.

Soaking dried fruits or any other dried ingredient before you add it to the dough solves the same problem and can add rather than take away moisture. Raisins, dried apples, and dried apricots are delicious soaked in a solution of sugar, water, and liquor or liqueur.

Rice Bread

MAKES ONE 5 × 9-INCH LOAF

This big golden-brown loaf has a fascinating honeycomb texture. It is a savory loaf with the Parmesan and cayenne, but it can be converted to a sweet loaf by substituting spices, orange and lemon zest, roasted nuts, and chopped dried fruit.

what this recipe shows

Cooked rice adds moisture and good keeping characteristics and also imparts an unusual texture.

Grated Parmesan is a slightly salty taste enhancer that does not interfere with yeast growth.

1 package (2¼ teaspoons) active dry yeast	1 teaspoon salt
1 tablespoon light brown sugar	2 tablespoons olive oil
1¼ cups warm water (115°F/46°C)	⅓ cup freshly grated Parmesan
1½ cups *and* 1¾ cups bread flour (3¼ cups *total*)	½ teaspoon recently opened cayenne
¼ 500-milligram vitamin C tablet, crushed	1 cup cooked long-grain white or brown rice
¼ cup crushed ice	1 tablespoon oil for bowl
	Nonstick cooking spray
	1 large egg, beaten

Read "The Bread Recipes in This Book" (page 10) before beginning.

1. Stir the yeast and brown sugar into the warm water in the bowl of a heavy-duty mixer. Let stand 2 minutes, until foam appears, indicating yeast is alive and well. Add 1½ cups bread flour. With the paddle blade, beat on low-medium speed for 4 minutes to beat air into the dough. Let the sponge sit for 30 minutes to 2½ hours for improved flavor and texture.

2. Remove the paddle blade and insert the dough hook into the mixer. Add vitamin C, crushed ice, salt, oil, and remaining 1³/₄ cups flour. Knead on low-medium speed for 5 minutes, until the dough is very elastic. Add Parmesan, cayenne, and rice. Work into dough by hand or in the machine. The dough should be very soft and slightly sticky. Add a little flour or liquid as needed.

3. Place the dough in an oiled bowl and turn to coat well with oil. Cover the bowl with plastic wrap and let the dough rise until slightly more than doubled in volume, about 1¹/₂ hours. Punch down the center of the dough with a closed fist. Reach to the back of the bowl under the dough and pull the bottom of the dough up and over the center. Repeat the same action at the front of the bowl, then turn the dough out onto the counter. Using both hands with a gentle cupping and tucking action, shape the dough into a smooth, tight round. Grab the sides of the round and stretch it sideways into an oval. Let it spring back slightly, then pull it out again. Cover with plastic wrap and leave on the counter for 15 minutes. The dough is now ready to shape.

4. Cup the dough with both hands, fingers spread out behind the loaf on either side and thumbs in front of the loaf. Press your thumbs into the dough and down against the table. This pulls or tucks in part of the bottom half of the dough. At the same time, pull the top of the dough tight and forward with your fingers. Now move your thumbs down slightly and press down and in again to knead and tuck again. Repeat this motion two to three times, until the loaf is stretched taut and well tucked in. Pinch the bottom seam together tightly. The ends go down to a slight taper. Tuck them under, then pinch the ends and bottom seam together.

5. Spray a 5 × 9 × 3-inch loaf pan with nonstick cooking spray and place the loaf in, seam side down. Brush with beaten egg. Let the dough rise until slightly more than doubled, about 1 hour.

6. About 30 minutes before the dough is fully risen, place a baking stone on a shelf in the lowest slot of the oven and preheat to 450°F (232°C) for 30 minutes (see Note). About 5 minutes before baking, turn the oven down to 375°F (191°C) and carefully place a shallow pan with ¹/₂ inch of boiling water on the oven floor. (For an electric oven, put the pan of water on the lowest possible shelf.)

7. Brush the bread again with beaten egg and place the pan directly on the hot stone. Bake until well browned, 45 to 55 minutes. Remove the loaf from the pan. The loaf should sound hollow when thumped on the bottom. You can also check for doneness by inserting an instant-read thermometer in the bottom center of loaf and pushing it to the middle. The loaf is done when the center is at least 200°F (93°C). Place the loaf on a rack to cool.

Note: *If you don't have a stone:*
- Preheat the oven to 25°F (14°C) below the baking temperature, place the bread on a shelf in the lower third of the oven, 5 to 6 inches above the bottom, and then turn the oven up to the baking temperature. Carefully place a shallow pan with ¹/₂ inch of boiling water on the oven floor. (For an electric oven, put the pan of water on the lowest possible shelf.)
- Or use the cold-oven procedure described on page 14.

Flours other than wheat

Rye and triticale, a wheat-rye crossbreed, are the only nonwheat flours that contain glutenin and gliadin, but neither has enough for risen bread. Therefore both must be combined with high-protein wheat flour to produce good breads.

The same is true of most grains or flour from other grains—they are simply dead weight in a loaf of bread and must be mixed with high-protein bread flour or vital wheat gluten. Rye, barley, cornmeal, and millet, to name a few, are added to bread for their unique flavors. The more of these other dead-weight grains you have in a loaf, the heavier the loaf will be. Still, as Mighty Multigrain illustrates, the vital wheat gluten provides a much needed gluten lift.

Also, if you knead coarse grains, seeds, or even whole wheat, which contains bran with sharp edges, with the dough, these particles can tear gluten strands and wreck leavening. Any textured ingredient should be worked in by hand after kneading.

Mighty Multigrain—A Hearty Loaf

MAKES ONE 9-INCH ROUND

Barley flour and brown sugar give this loaf deep, rich flavors. Little millet balls and flecks of flax seeds add texture and taste. This bread is incredibly good toasted. It also makes memorable grilled cheese sandwiches.

what this recipe shows

Vital wheat gluten and bread flour provide enough gluten to lift the barley and seeds.

Barley malt syrup provides enzymes to make food for yeast for a good rise.

Ground ginger enhances yeast activity.

Vitamin C helps gluten development.

Coarse ingredients that can tear gluten like wheat germ, millet, and flax seeds are worked in by hand after kneading.

1 package (2 1/4 teaspoons) active dry yeast

1 teaspoon barley malt syrup (see Notes) or 1 tablespoon dark brown sugar

1 cup warm water (115°F/46°C)

1 3/4 cups and 1 cup bread flour (2 3/4 cups total)

1/4 cup semolina flour

1/2 teaspoon ground ginger

3/4 cup buttermilk

1/4 500-milligram vitamin C tablet, crushed

1/3 cup crushed ice

1 1/2 teaspoons salt, preferably sea salt

1 large egg yolk

1 tablespoon oil

1/4 cup gluten flour or vital wheat gluten

2 tablespoons dark brown sugar

1/4 cup barley flour

1/4 cup cornmeal

1/2 cup wheat germ

2 tablespoons millet

1/4 cup flax seeds

1 tablespoon oil for bowl

Coarse cornmeal as needed

1 large egg, beaten

Read "The Bread Recipes in This Book" (page 10) before beginning.

1. In the bowl of a heavy-duty mixer, stir the yeast and barley malt syrup into warm water. Let stand 2 minutes, until foam appears, indicating yeast is alive and well. Add 1¾ cups bread flour, semolina, ginger, and buttermilk. With the paddle blade, beat on low-medium speed for 4 minutes to beat air into the dough. Let the sponge sit for 30 minutes to 2½ hours for improved flavor and texture.

2. Remove the paddle blade and insert the dough hook. Add the vitamin C, crushed ice, salt, egg yolk, oil, gluten flour, 1 cup bread flour, brown sugar, barley flour, and cornmeal. Knead on low-medium speed for 5 minutes, until dough is very elastic. Add the wheat germ, millet, and flax seeds. Work into dough by hand. The dough should be fairly firm. Add a little flour or liquid as needed.

3. Place the dough in an oiled bowl and turn to coat well with oil. Cover the bowl with plastic wrap and let the dough rise until doubled in volume, about 1½ hours. Punch down the center of the dough with a closed fist. Reach to the back of the bowl down under the dough and pull the bottom of the dough up and over the center. Repeat the same action at the front of the bowl, then turn the dough out onto the counter. Using both hands with a gentle cupping and tucking action, shape dough into a smooth tight round. Cover with plastic wrap and leave on the counter for 15 minutes. The dough is now ready to shape.

4. If you are right-handed, prop the dough ball against your left hand and press down and forward on the opposite side of the ball with the palm of your right hand to lightly knead the side of the ball. Let the dough ball rotate slightly forward with the stroke and continue to knead the sides of the round until you have gone completely around. Pinch the bottom to seal.

5. Sprinkle a baking sheet or small pizza pan heavily with coarse cornmeal and place the loaf on it. Brush the loaf with beaten egg. Let the dough rise until more than doubled, about 1 hour.

6. About 30 minutes before the dough is fully risen, place a baking stone on a shelf in the lowest slot of the oven and preheat to 450°F (232°C) for 30 minutes (see Notes). About 5 minutes before baking, turn the oven down to 375°F (191°C) and carefully place a shallow pan with ½ inch of boiling water on the oven floor. (For an electric oven, put the pan of water on the lowest possible shelf.)

7. Brush the bread again with beaten egg. Slash in desired pattern and place the pan directly on the hot stone. Bake until well browned, 45 to 55 minutes. Remove the loaf from the pan. The loaf should sound hollow when thumped on the bottom. You can also check for doneness by inserting an instant-read thermometer in the bottom center of loaf and pushing it to the middle. The loaf is done when the center is at least 200°F (93°C). Place the loaf on a rack to cool.

Notes: *Barley malt syrup is available at health food stores and beer- and winemaking supply shops.*

> *If you don't have a stone:*
> - Preheat the oven to 25°F (14°C) below the baking temperature, place the bread on a shelf in the lower third of the oven, 5 to 6 inches above the bottom, and then turn the oven up to the baking temperature. Carefully place a shallow pan with ½ inch of boiling water on the oven floor. (For an electric oven, put the pan of water on the lowest possible shelf.)
> - Or use the cold-oven procedure described on page 14.

HOW THE MILLER HELPS: BLEACHES AND DOUGH IMPROVERS

Just-milled flour, which has a yellowed look like old lace, makes gummy doughs and poor-quality bread. As flour stands exposed to the air, however, oxygen combines with the yellow carotenoid pigments and converts them to a colorless form, thus naturally bleaching the flour. Oxygen also reacts with the thiol groups in dough (page 57) and prevents their interfering with elasticity. Thus oxygen improves baking qualities in several ways.

Millers use bleaches and oxidants to speed up this natural oxidizing process, which takes from eight to twelve weeks. Some of these additives or processes bleach only or improve gluten development only, while some, like chlorination, do a number of things at once. Bleaching, a natural occurrence, has been given a bum rap. Some claim that it removes vitamin E from the flour. The fact is that flour's vitamin E content is so minute that nutrition databases list flour as containing no vitamin E.

Bleach only

Benzoyl peroxide is a commonly used bleach-only ingredient that millers add. Bleach-only ingredients convert the yellow carotenoid and xanthophyll pigments to a colorless form. Although health advocates may prefer unbleached flour, protein content is essentially unaffected by the added bleaching process. A high-protein flour may be bleached or unbleached according to the needs of the miller. It just happens that millers frequently blend a higher-protein flour for the unbleached label than for bleached.

Bread flour is never chlorinated, but cake flour and some all-purpose flours are.

Chlorination

Chlorination is a method of bleaching in which flour is exposed to chlorine dioxide gas. Chlorination leaves the flour slightly acidic, aids in distribution of the fat, and enhances water absorption. (See page 137 for more about chlorination and cake flour.)

Dough improvers

Dough improvers or dough conditioners are umbrella terms that include many different additives that improve product quality. They include oxidizing and reducing

Why Dough Springs Back

Bonds Hold Gluten Tight   Stretch Dough, Bonds Pull Apart   Bonds Pull Back Together

Sulfur Compounds in Just-Milled Flour Wreck Elasticity

Bonds Hold Gluten   Stretch Dough, Bonds Pull Apart   Sulfur Compounds Join Bonds

Leaveners: lightening the dough

Whether it's an airy baguette or a feather-light biscuit, bread relies on leaveners to lighten the dough. Gases from yeast or chemical leaveners, air, and steam from moisture in the dough are all leaveners. Usually, several leavening agents are involved at once. Incorporation of air into the dough is vital. Yeast and chemical leaveners do not create new bubbles; they only enlarge air bubbles already existing in the dough. With both types of leavener, steam is also a major contributor to a well-leavened loaf.

As with other groups of bread ingredients, leaveners are not always interchangeable. For example, baking soda is four times as powerful as baking powder and requires some acidic ingredient in the dough to work. And certainly you can get a disastrous result if you substitute yeast for a chemical leavener. Yeast breads and quick breads—those made with chemical leaveners and so called because they require no rising before baking—are quite dissimilar. But whether a biscuit or a baguette, knowing how leaveners work can help you make better bread.

agents that improve gluten formation, enzymes that help in providing food for yeast, and emulsifiers that improve both gluten formation and antistaling properties.

Oxidizing and reducing agents

When gluten-forming proteins are stretched out, the sulfur of a neighboring protein is not the only thing to which a protein's sulfur can link. In freshly milled flour, there are villainous little things called *thiol groups*. These little fellows are all over the place in the new flour, both on the proteins themselves and on other molecules in the flour. These thiol groups hook onto a sulfur when it is pulled loose from another sulfur. Now that there is this outsider, the thiol group, hooked onto each sulfur, the two sulfurs cannot pull back together. There goes your elasticity.

Oxidizers like iodate of calcium or potassium or potassium bromate are added to flour not to bleach but to oxidize these villainous little sulfur groups in freshly milled flour. Oxidizing these little rascals (S-H groups) permits better cross-linking of the gluten and gives dough better air-holding ability. The miller can add these oxidants for better gluten formation. Frequently, commercial bakers add them as a dough conditioner mix.

These oxidants vary in how fast they work. The iodates work right away in the dough to produce better gluten. Potassium bromate does not improve the dough until it becomes more acidic from yeast activity and is heated to 140°F (60°C). Therefore, you do not see the results of bromate addition until the dough is baking. The process of adding an oxidizer to flour is called *bromation* when potassium bromate is the oxidizing agent. What is important to the bread baker is that these oxidizers can produce significantly better loaves of bread.

Vitamin C works in conjunction with oxidizers—whether oxygen, iodates, or potassium bromate—to improve gluten development. Since tiny amounts of vitamin C can improve gluten quality and is a healthful nutrient, I frequently add it to recipes.

Bromates and iodates—good for you?

Conditioned flour contains bromates more frequently than iodates. Some nutrition books assume that the flour contains iodates and list flour with dough conditioners as being a source of iodine to prevent goiter. Either a deficiency or an excess of iodine can cause thyroid problems. In areas where iodized salt is not available, it could well be that conditioners in the flour are providing a vital nutrient to prevent goiter.

The bromates or the iodates used as oxidizers in flour convert in baking to plain bromide salt or iodide salt (potassium bromide, calcium or potassium iodide). These salts occur naturally, in sea salt, for example, and are close cousins to sodium chloride—plain old table salt.

There is a major anti-bromate movement, and the public is either ignorant of or unwilling to believe the fact that in bread, potassium bromate is converted to a harmless salt. Legislation has passed in California banning bromate. Nearly all bread flour previously contained bromate, but it is now being replaced with ascorbic acid, a poor substitute.

Yeast

Yeast is a chlorophyll-free, one-cell plant that feeds only on simple sugars. Alpha amylase, an enzyme, breaks damaged starch molecules in the flour into these simple sugars that yeast needs. Trace amounts of this enzyme are in flour. In commercial breadmaking, bakers add this and other enzymes in the form of malted flour, malt, or barley malt syrup to provide a good food supply for the yeast. Semolina flour has more damaged starch than that from other strains of wheat and is excellent food for yeast.

How yeast grows As the dough rises, yeast eats sugars around it, gives off carbon dioxide and ethyl alcohol, and grows. As long as yeast can get oxygen from air in the dough, it reproduces by dividing. One yeast cell divides into two, then both of them divide, then all four divide to make eight, and so on. Soon a clump of cells has replaced each single yeast cell.

The cells in the center of the clump may not be able to get food from the flour and therefore may not be able to give off gases after a while. When you punch down the dough and reshape it, you break up these clumps of cells and spread them out in the dough. Now each yeast cell has fresh flour around it, an abundant food supply, and can produce much more carbon dioxide. This is why the second rise goes so much faster than the first. You have many, many more yeast cells, and each has a fresh food supply. Without oxygen—they have probably used all the oxygen that was in air bubbles in the dough in the first rise—they cannot continue to divide, but they can produce carbon dioxide and alcohol.

Different types of yeast I used to think that yeast is yeast. But yeast is yeast only in the way that a rose is a rose. There are giant yellow roses, pink primroses, and classic American Beauty roses, and there are many different kinds of yeast. Each strain or each different brand can have different characteristics. Some strains absorb water faster than others; others produce more carbon dioxide; some strains like one kind of food, and other strains like another.

At the Red Star Yeast research facility in Milwaukee, a microbiologist showed me a large culture dish holding several different strains of yeast stained bright orange against a green background. Pointing to a big orange patch on the dish, the microbiologist said, "This fellow loves the food we are feeding in this dish." Then he pointed to two tiny orange spots in the same dish and said, "But these two do not like it at all."

The strains of yeast in sourdough starters are quite different from bakers' yeasts. They grow well in acidic surroundings; bakers' yeasts do not. Bacteria in doughs that produce flavorful by-products live on the same simple sugars that bakers' yeasts do. In ideal growth conditions for yeast, it takes all the sugars, and the bacteria do not get to produce these flavors. Sourdough yeasts, in addition to consuming simple sugars, usually can utilize another sugar, one not sought by their accompanying acid-forming bacteria. This enables both of them to grow well at the same time.

For many years, microbiologists have been collecting, growing, and studying yeast varieties from all over the world. Now, with modern biotechnology, scientists are creating exciting new varieties of yeast.

Three forms of bakers' yeast Bakers' yeast is now available in three different forms: active dry yeast, cake or compressed yeast, and fast-rise (with trade names such as RapidRise and Quick Rise) yeast.

The most widely available yeast is active dry yeast. This is yeast that has been oven-dried at controlled temperatures. It is sold in standard packages, two or three packages to a sleeve (American packages are 7 grams—2¼ teaspoons per package), and in jars. Fleischmann's, Red Star, and Hodgson's are some of the more familiar brand names in the United States. If you do much baking, you can save a lot of money by buying yeast in 2- to 3-pound bags from bakery supply companies or large grocery discount stores. Keep it in the freezer in sealed plastic freezer bags.

Cake (compressed) yeast is the form in which yeast was available in our grandmothers' time. It is yeast whose cells are not dried and comes in little squares sold in the refrigerated case. The strains of yeast sold in cakes are very active, producing great amounts of carbon dioxide for wonderful breads, better than most dried strains make. Unfortunately, these strains cannot be dried without damaging or killing them.

Also unfortunately, cake yeast is a poor keeper. You must keep it refrigerated, and even then it has a limited shelf life. When I was teaching in Boston, Jane Doerfer, who wrote the cookbook for the Legal Seafoods restaurants in that city, took me to buy ingredients. When I saw cake yeast in the store, I said, "Oh, Jane, I want to get this. It's great yeast, and you can never find it."

"I wouldn't," Jane replied. "Every time I've bought it, it was bad."

I insisted. "There are three different boxes here, two with different dates. Surely, one of the three will be OK."

I took two squares from each box. Jane was right. When we proofed them, to my sorrow, all six squares were dead yeast. Because of the short life of cake yeast, it is vital to proof it.

If you want to try cake yeast, go to a bakery supply house, where it

Don't skip the proofin[g] step with cake yeast— its shelf life is short, and it's often dead.

is sold in 1-pound blocks like butter, very fresh and active. I have never had a bad one. Or ask your local baker. One in my area will sell an occasional block to me.

Fast-rise yeast is a product of genetic engineering. Biologists took the best characteristics from many different strains—the ability to produce great quantities of carbon dioxide from one strain, the ability to soak in water fast and become alive and active quickly from another, resistance to damage from small changes in temperature from yet another—and created new beings, what I call "super babies." They are air dried instead of oven dried.

Each company has its own line of super babies, and you can expect differences among the brands. They do, however, share certain qualities. All the super babies are fine, open, porous small particles that soak in water and become alive and active instantly.

Ironically, the marvelous qualities of these super babies caused problems when they first came out. Yeast companies were so excited about their wonderful new creations that they forgot that they were dealing with the same old public. They reduced the yeast's food supply and made what they called a *high-protein yeast*. These super babies soaked in liquid so fast that they could be mixed right with the flour (their food supply). The directions on the package said, "Mix yeast directly with flour and then add warm water." Unfortunately, home bakers who had been proofing their yeast for twenty years were not going to stop now, directions or no directions. The yeast companies had once run an unsuccessful campaign to get consumers to add yeast directly to flour, so they probably should have known better. Their directions called for adding unusually hot water (130°F/54°C) to the yeast-flour mixture to produce a warm dough that rose fast. However, consumers who used the traditional cooler water ended up with a dough filled with tiny undissolved lumps of yeast because the large amount of flour instantly cooled it to a temperature too low to dissolve the yeast.

Consumers who had tried this method unsuccess-fully were reluctant to try a similar method with the new yeast, so they continued to proof the new super babies, just as they had the older yeast strains. However, when warm water and sugar were added to the early super babies to proof them, they multiplied so fast that they instantly ate up all the available sugar and, so to speak, ate themselves up. By the time they were added to the flour, they were dead and produced failed loaves of bread. To add to the confusion, some brands had retained a reasonable food supply, so these did work if you proofed them.

Fortunately, processors now take into consideration human frailty. The "new improved" fast-rise yeast is now packaged with a food supply large enough to proof it for short periods even though the producers tell you not to. If you tried the fast-rise yeast earlier and were unhappy with it, you may want to try it again. It does work beautifully if mixed directly with flour.

The new fast yeast was popular in Europe some time before it hit the American market, and you will see European brands of this product available here too in cookware shops and at bakery supply companies. Recently, a French instant yeast was put on the U.S. market in 11-gram (1 tablespoon) foil small packages. It has been available for several years in 1-pound vacuum-sealed packages. Do not accept bags if the vacuum seal has been broken, because this type of yeast starts to deteriorate in air. It remains in excellent condition and can be opened and resealed, time after time, if it is kept in the freezer tightly sealed in a plastic zip-top freezer bag.

Yeast and cold water Water or dough that is too hot (over 140° to 145°F/60° to 63°C) kills yeast. But heat is not the only way it can be damaged. Dry yeast has to hydrate (soak in water) to become operative; it needs warm water to plump up its primitive organs and get going. Cold water will kill or damage dry yeast.

Cake yeast cells are already wet, alive, and active; cold water does not harm them. It is only during the initial step of pro-

Cold water kills or damages dry yeast.

viding water to dried cells to change them to active, alive plants that the yeast is so vulnerable to cold water.

I heard a good example of this difference several years ago. Jack Lirio, an excellent teacher and baking specialist from San Francisco, asked at a meeting of cooking professionals, "Who tried the Thirty-Minute Croissants? I tried them six times, and they never did work!" This was a recipe that had appeared in a national magazine.

The concept behind the recipe was good. The secret of flakiness is thin layers of dough separated by thin layers of firm, unmelted fat. Cold is the big secret: keep the fat cold so that it will not melt. In an effort to keep things cold from the beginning, the writer called for dissolving the yeast in cold water.

The person who developed the recipe probably used cake yeast, but the magazine's policy was to use active dry yeast, so it had substituted dry for cake yeast in the recipe. Dissolving in cold water damages the dry yeast but not cake yeast.

You will notice that the Cuisinart company is aware that cold water can harm dry yeast. Some of its recipes read, "Dissolve yeast in 2 tablespoons warm water, then add ice water to make 1 cup."

Starters Most strains of bakers' yeast thrive on fructose and glucose, the sugars that table sugar, sucrose, breaks down into. Bacteria in doughs that produce flavorful acidic by-products also feed on these two sugars and, unfortunately, get a limited supply because yeast consumes most of the sugars under straightforward mixing and rising conditions. You can, however, set up conditions that give the bacteria an opportunity to be active.

If there is originally no bakers' yeast (a sourdough starter) or a minimum of yeast (a yeast starter) or a piece of dough from previously made bread and a little yeast (a mixed starter) and the dough is allowed to stand at cool room temperature (68° to 72°F/20° to 22°C) for one or more days, bacteria in the dough can thrive and produce their marvelously flavorful products. These are the three main ways to produce starters and the wonderful complex

flavored, open-textured breads that they make. By-products of this lengthy yeast fermentation also contribute to these remarkable flavor changes in the bread. Even the sponge method of making bread—allowing flour, water, and yeast to stand for up to 2½ hours—improves bread flavor.

Sourdough starters Sourdough starters illustrate the symbiotic relationship through which some strains of yeast and bacteria grow well at the same time because they thrive on different sugars. The strains of yeast and bacteria in San Francisco sourdough are examples (page 23). A similar situation exists with a strain of yeast patented as Barm, *Saccharomyces dairensis,* and a lactic acid bacteria, *Lactobacillus brevis.* Both use the sugars glucose and fructose. In addition, the yeast uses galactose, and the bacteria use maltose, providing each with a private food supply. For an ideal starter, both yeast and bacteria need to grow in balance. Large sourdough bakeries analyze the dough periodically for lactic acid (produced by the bacteria) and acetic acid (from alcohol produced by the yeast) to make sure that they are balanced. If the dough has too much lactic acid, they know that bacteria are getting the upper hand and so provide conditions (like warming the dough slightly) to favor yeast growth.

Because of their acidity, breads made with sourdough starters have a good shelf life. Acidic doughs inhibit mold growth and slow staling.

A good sourdough starter is not always easy to achieve. You must get both the yeast and the bacteria growing. Starters contain water and a substance that supplies sugars to feed both yeast and bacteria. Wheat flour contains 1.7 percent sugar, over a third of which is glucose, the simple sugar that yeast and bacteria both feed on. A starter can be simply flour and water. I have had success with the Wet Sourdough Starter (page 62)—whether because of the mix of flours or my dusty house, I'm not sure. Food science writer Harold McGee has had great success with the Firm Sourdough Starter (page 62) using organically grown wheat.

Many starters contain flour, water, and an extra substance that supplies sugars like cooked or raw potatoes, grapes, cherries, figs, or even an onion. I haven't tried apples or pears yet, but I think they would make good flavorful starters too. Some like to crush the fruit and put it in a cheesecloth bag in flour and water. Nancy Silverton has an excellent recipe for such starters in *Breads from the La Brea Bakery.*

Some like to crush the fruit, let it ferment for several days, and strain juice from this ferment to add to flour and water. You can make your own starter using recipes such as those that follow or purchase one. An active Russian starter and others are available from Sourdoughs International, and King Arthur also sells starters (see Sources—Sourdough starters, page 479).

When the starter is at its peak and ready to use, I always freeze a portion of it. Who knows? You may get something that tastes wonderful and then be unable to grow more of it if you have used it all. You lose a percentage of the yeast when you freeze it, but there is plenty left to grow more.

Sourdough baker Peter Nyberg and I decided the best time to freeze a starter would be at its peak except that the yeast would have consumed its food supply, leaving it nothing to eat if you defrosted it for a day in the refrigerator. So we took some starter at its peak, added a little flour and water (giving it a little more food), and froze it immediately. This worked fine.

As another method of storage, I have dried starter by pouring a thin layer on parchment-covered jelly-roll pans and leaving it at room temperature until dry. I break up pieces of the caked dry starter and store them at room temperature in small freezer-type zip-top bags. To restart, I crumble the pieces as fine as possible, soak them for a day in slightly warm water (105°F/40°C), then begin feeding regularly three times a day (see page 63).

SOURDOUGH STARTER TIPS

There are no guarantees in making sourdough starter, but some things that improve your chances are:

- Use unchlorinated water. I get a gallon or two of supermarket bottled spring water, not distilled.
- Use the least-processed flour you can get. Good choices are health-food organic unbleached and whole wheat, King Arthur, or wheat berries hand-ground in a coffee mill or spice grinder.
- Keep at room temperature (low 70s°F/20s°C) (if possible part of the time outdoors with the wet starter) for 5 or 6 days to get it going.
- After the initial start, whenever the starter is at room temperature feed it every 4 to 6 hours. If an active starter sits at room temperature over 6 hours without feeding, acid-producing bacteria get the upper hand and the starter becomes very acidic. Such a starter can impart a good sour taste, but it will not leaven well.
- Refrigerate the starter every night and whenever you are not using it.

- Feed the starter a volume equal to its volume at each feeding. For example, if there are 2 cups of starter, feed it a total of 2 cups of flour and water.
- Feed the starter slightly more flour than water at each feeding. For example, for a 2-cup feeding, use a little under 1 cup water and a little over 1 cup flour.
- Get used to discarding some starter down the drain with running water. Bite the bullet. It doubles with every feeding, so unless you have a walk-in cooler and are trying to make starter for a bakery, you will soon have a refrigerator, a freezer, and a counter full of starter.
- Keep the starter about the consistency of thick pancake batter.
- If the starter discolors or becomes foul smelling, discard immediately and begin again.
- When making bread with the starter, mix, shape, and bake after one rise. Some starters are not strong enough to rise again without a feeding.

Firm Sourdough Starter

Starters are prepared by two basic methods, a firm dough starter or a wet starter. For the firm dough starter, knead ½ to ¾ cup minimally processed flour (such as King Arthur unbleached) with enough water to make a very firm dough ball. Bury the ball in the center of a 5-pound bag of flour.

After 3 days, dig up the ball and discard the hard crust and any additional dough necessary to keep only about half the original ball. Knead in enough fresh flour and water to return the dough to its former size, then bury it again. Repeat discarding half, kneading in a fresh half, and reburying for 2 more days.

On the sixth day, discard half and knead in a fresh half, but add enough water to make the dough very soft. Place in a plastic or glass container, cover with plastic wrap touching the surface to prevent drying, and leave overnight. On the seventh day, discard half, knead in a fresh half, and add enough water to bring the starter to the consistency of a heavy wet batter. Let stand at cool room temperature, uncovered. In 4 to 6 hours, feed an amount equal to the volume of the batter—about 1½ cups, so this would be a little less than ¾ cup water and a little more than ¾ cup flour. In 4 to 6 hours, feed again, this time a little less than 1½ cups water and a little over 1½ cups flour. After 2 to 4 hours, cover with plastic wrap and refrigerate for the night.

Now start feeding regularly three times a day for 5 days. When you remove the starter from the refrigerator, you have about 3 cups. Stir well and discard half, then feed a scant ¾ cup water and a generous ¾ cup flour. In 4 to 6 hours, stir well, then feed a scant 1½ cups water and a generous 1½ cups flour. Feed again after 4 to 6 hours. You may want to discard half the starter so you are feeding only 1½ cups each of flour and water at the last feeding. After 2 to 4 hours, refrigerate for the night.

Repeat this feeding three times a day for 4 more days. You should have a good strong starter that bubbles happily after every feeding. It can now be kept refrigerated and fed only once or twice a week during periods that you are not using it.

To make bread, pull out your starter, feed it, let it warm up a couple hours, then use 1½ to 2 cups to make a loaf of bread. Feed the remaining starter once or twice more and store again in the refrigerator.

Wet Sourdough Starter

I like to make starters with a mixture of flours to get a variety of flavor components and ingredients for yeast and bacteria growth. The durum (semolina) flour is one of them because I had noticed starters fed this flour grew faster—an observation confirmed by Peter Nyberg. Why? Because of its hardness, food science writer Harold McGee explained, semolina has more damaged starch, which supplies more food for yeast, than many other wheat flours.

5½ ounces (about 1 cup) unbleached
   flour

2 ounces (about packed ⅓ cup)
   semolina (durum) flour

2 ounces (about packed ⅓ cup) whole
   wheat flour (see Note)

2 cups water (or more if needed) at
   room temperature

Combine flours and water in a mixer bowl. Knead on low speed for 5 minutes. Place in a plastic or glass container and cover with cheesecloth. Let stand at room temperature, ideally in the low 70s°F (20s°C), for 3 days. Ed Wood, sourdough expert, recommends leaving the starter outdoors if possible to catch more wild yeasts.

FIRST FEEDING

5½ ounces (about 1 cup) unbleached flour

2½ ounces (about ½ cup) semolina (durum) flour

1 cup water at room temperature

On the fourth day, place starter and all feeding ingredients in the mixer bowl and knead on low speed for 5 minutes. Place in a plastic or glass container and cover with cheesecloth. Let stand for 3 more days. On the eighth day begin regular three feedings a day.

SCHEDULE FOR REGULAR THREE FEEDINGS PER DAY

1. Discard, freeze, or dry all but 1½ cups starter. Feed a scant ¾ cup water and a generous ¾ cup flour. In 4 to 6 hours, stir well, then feed a scant 1½ cups water and a generous 1½ cups flour. Feed again after 4 to 6 hours. You may want to discard half the starter so you are feeding only 1½ cups each flour and water at the last feeding. After 2 to 4 hours, refrigerate for the night.

2. Repeat this feeding three times a day for 4 more days. You should then have a good strong starter that bubbles happily after every feeding. It can now be kept refrigerated and fed only once or twice a week during periods that you are not using it.

3. To make bread, pull out the starter, feed it, let it warm up a couple hours, then use 1½ to 2 cups to make a loaf of bread. Feed the remaining starter once or twice more and refrigerate again.

Note: *Harold McGee buys organically grown wheat kernels from a health food store and grinds them himself in a hand-cranked coffee grinder (so it will not overheat as it would in an electric grinder). He finds this makes excellent starters. I have started doing this too.*

Sourdough Bread

MAKES ONE 7-INCH ROUND LOAF

Your own sourdough loaf is a true achievement—a beautiful crusty, rugged-looking marvel. Once you have a good starter going, the bread is simple to make. Shaping after mixing, and baking after only one rise, ensures a better leavened loaf.

what this recipe shows

Homegrown yeast lives together with thriving bacteria to produce a leavened slightly acidic, extremely flavorful loaf.

continued

2 cups starter, homemade (pages 61–63), or store-bought

1 teaspoon barley malt syrup (see Notes)

¼ cup crushed ice

2 cups *and* 1 cup bread flour (3 cups *total*)

1 teaspoon salt

1 tablespoon oil for bowl

Coarse cornmeal as needed

Read "The Bread Recipes in This Book" (page 10) before beginning.

1. Place the starter, barley malt syrup, and ice in the bowl of a heavy-duty mixer with the dough hook. Add 2 cups bread flour and salt and beat on low speed about 1 minute. Add the last 1 cup of flour as needed to get a firm dough. Knead on low-medium speed for 5 minutes, until dough is very elastic. The dough should be firm. Add a little flour if needed.

2. Oil your hands. Using both hands with a gentle cupping and tucking action, shape the dough into a smooth tight round. Cover with plastic wrap and leave on the counter for 15 minutes. The dough is now ready to shape.

3. If you are right-handed, prop dough ball against your left hand and press down and forward on the opposite side of the ball with the palm of your right hand, to lightly knead the side of the ball. Let the dough ball rotate slightly forward with the stroke and continue to knead the sides of the round until you have gone completely around. Pinch the bottom to seal.

4. Sprinkle a baking sheet or large pizza pan heavily with cornmeal and place the loaf on it. Sprinkle the loaf lightly with a tablespoon of bread flour and cover loosely with plastic wrap. Let the dough rise until doubled, about 3 to 4 hours.

5. About 30 minutes before dough is fully risen, place a baking stone on a shelf in the lowest slot of the oven and preheat at 450°F (232°C) for 30 minutes (see Notes). About 5 minutes before baking, turn the oven down to 375°F (191°C) and carefully place a shallow pan with ½ inch of boiling water on the oven floor. (For an electric oven, put the pan of water on the lowest possible shelf.)

6. Slash the loaf in the desired pattern and place the pan directly on the hot stone. Bake until well browned, 45 to 55 minutes total baking time. Loaf should sound hollow when thumped on the bottom. You can also check for doneness by inserting an instant-read thermometer in the bottom center of loaf and pushing it to the center. The loaf is done when the center is over 200°F (93°C). Place the loaf on a rack to cool.

Notes: *Barley malt syrup is available at health food stores and beer- and winemaking supply shops.*

If you don't have a stone:
· Preheat the oven to 25°F (14°C) below the baking temperature, place the bread on a shelf in the lower third of the oven, 5 to 6 inches above the bottom, and then turn the oven up to baking temperature. Carefully place a shallow pan with ½ inch of boiling water on the oven floor. (For an electric oven, put the pan of water on the lowest possible shelf.)
· Use cold-oven procedure described on page 14.

Starter with Yeast

MAKES ABOUT 1 POUND

2½ cups unbleached flour

½ cup semolina (durum) flour

½ teaspoon active dry yeast

¼ cup warm water (110°F/43°C)

1¾ cups cool water (65° to 70°F/18° to 21°C)

Combine flours in a mixer bowl. Stir yeast into warm water in a small cup. Let stand 2 minutes, until foam appears, indicating yeast is alive and well. Add dissolved yeast and cool water to flour. Blend on low speed about 1 minute. Knead on low-medium speed for 5 minutes. Place in a plastic or glass container and cover with plastic wrap touching the surface of the starter to prevent crusting. Let stand in a draft-free area at a temperature about 71°F (22°C) for 10 to 12 hours. The starter is now ready to make bread. This is enough to combine with 1 pound of flour (about 3 cups) or enough for two loaves, or one large round.

Yeast Starter Bread

MAKES TWO 7-INCH ROUNDS OR ONE 10-INCH ROUND

Made as one round, this is a great rustic crusty bread—handsome and flavorful.

what this recipe shows

A small initial amount of yeast multiplies steadily as the starter stands for 10 hours.

Acids and other fermentation products in the starter enhance taste.

1 recipe Starter with Yeast (above)

¼ 500-milligram vitamin C tablet, crushed

1 teaspoon barley malt syrup (see Notes)

¼ cup crushed ice

3 cups bread flour

1½ teaspoons salt

1 tablespoon oil for bowl

Coarse cornmeal as needed

Read "The Bread Recipes in This Book" (page 10) before beginning.

continued

1. Place the starter, vitamin C, barley malt syrup, and ice in the bowl of a heavy-duty mixer with the dough hook. Mix on low speed just to blend. Add the bread flour and salt and knead on low-medium speed for 5 minutes, until dough is very elastic. The dough should be firm. Add a little flour or liquid if needed.

2. Place the dough in an oiled bowl and turn to coat well with oil. Cover the bowl with plastic wrap and let the dough rise until doubled in volume, about 1$^{1}/_{2}$ hours. Punch down the center of the dough with a closed fist. Reach to the back of the bowl down under the dough and pull the bottom of the dough up and over the center. Repeat the same action at the front of the bowl, then turn the dough out onto the counter. For two loaves, divide the dough in half. Using both hands with a gentle cupping and tucking action, shape each half into a smooth tight round. Cover with plastic wrap and leave on the counter for 15 minutes. The dough is now ready to shape.

3. If you are right-handed, prop one dough ball against your left hand and press down and forward on the opposite side of the ball with the palm of your right hand to lightly knead the side of the ball. Let the dough ball rotate slightly forward with the stroke and continue to knead the sides of the round until you have gone completely around. Pinch the bottom to seal. Repeat this procedure with the other round.

4. Sprinkle a baking sheet or large pizza pan heavily with cornmeal and place the loaves on it. Sprinkle each loaf lightly with a tablespoon of bread flour and cover loaves loosely with plastic wrap. Let the dough rise until doubled, about 2 to 3 hours.

5. About 30 minutes before dough is fully risen, place a baking stone on a shelf in the lowest slot of the oven and preheat to 450°F (232°C)—see Notes. About 5 minutes before baking, turn the oven down to 375°F (191°C) and carefully place a shallow pan with $^{1}/_{2}$ inch of boiling water on the oven floor. (For an electric oven, put the pan of water on the lowest possible shelf.)

6. Slash the loaves as desired and place the pan directly on the hot stone. Bake until well browned, 45 to 55 minutes. Remove the loaf from the pan. Loaves should sound hollow when thumped on the bottom. You can also check for doneness by inserting an instant-read thermometer in the bottom center of loaf and pushing it to the middle. The loaf is done when the center is over 200°F (93°C). Place the loaf on a rack to cool.

Notes: *Barley malt syrup is available at health food stores and beer- and winemaking supply shops.*

> *If you don't have a stone:*
> · Preheat the oven to 25°F (14°C) below the baking temperature, place the bread on a shelf in the lower third of the oven, 5 to 6 inches above the bottom, and then turn the oven up to the baking temperature. Carefully place a shallow pan with $^{1}/_{2}$ inch of boiling water on the oven floor. (For an electric oven, put the pan of water on the lowest possible shelf.)
> · Or use the cold-oven procedure described on page 14.

Sweet Yeast Starter

1 package (2¼ teaspoons) active dry yeast	**FEEDING**
¼ cup warm water (115°F/46°C)	¾ cup sugar
¾ cup whole milk at room temperature	3 tablespoons instant potatoes
1 cup bread flour	1 cup warm water (115°F/46°C)

In a glass quart or half-gallon jar, stir together yeast, ¼ cup warm water, milk, and bread flour. Let stand uncovered at room temperature for 24 hours. Stir and cover tightly with plastic wrap. Refrigerate for 2 to 3 days. Then feed by stirring together sugar, instant potatoes, and 1 cup warm water and adding to the starter. Let stand at room temperature for 8 to 12 hours, then refrigerate. After you feed the starter and let it stand for 8 hours, remove 1 cup for bread or discard 1 cup. Feed every 3 to 5 days.

Sweet Starter Bread

MAKES THREE 4½ × 8½-INCH LOAVES

You would think that if your mother-in-law were a professional cook, you would hesitate to take her your own personal baking as a gift. My daughter-in-law, Janice, is fearless, however, and comes from a long line of good Moravian cooks. She never hesitates to bring outstanding goodies for Christmas. Her Sweet Starter Bread is always a big hit. Even though she brings three loaves, they are all gone in a day and a half.

what this recipe shows

A wet sponge and potatoes in the starter add sweetness to the loaves.

The long refrigerator stays of the starter give bacteria a chance to grow, producing very pleasant sweet by-products at these cold temperatures, yet still imparting good keeping characteristics.

¼ teaspoon active dry yeast (see Notes)	1 tablespoon *and* ½ teaspoon salt (1 tablespoon plus ½ teaspoon *total*)
⅓ cup sugar (see Notes)	¼ cup oil *and* 1 tablespoon (⅓ cup *total*)
1¾ cups warm water (115°F/46°C)	Nonstick cooking spray
1 cup Sweet Yeast Starter (above)	5 tablespoons lightly salted butter, melted
3 cups *and* 3 cups bread flour (6 cups *total*)	

Read "The Bread Recipes in This Book" (page 10) before beginning.

continued

1. Stir the yeast, sugar, and warm water together in the bowl of a heavy-duty mixer with the paddle blade (see Notes). Add the starter and 3 cups bread flour and beat on low-medium speed for 4 minutes to beat air into the dough. Let the sponge sit for 30 minutes to 2½ hours for improved flavor and texture.

2. Remove the paddle blade and insert the dough hook in mixer. Add 3 cups bread flour and 1 tablespoon salt and knead 6 minutes on low-medium speed, until dough is very elastic. Divide dough into several pieces and add ¼ cup oil. Knead 30 seconds on low, then a minute on low-medium to incorporate oil. Dough should be very soft. Add a little flour or water if needed. Place dough in a very large mixing bowl with 1 tablespoon oil. Turn dough to coat well with oil. Cover bowl with plastic wrap and let stand overnight at room temperature.

3. Punch down the center of the dough with a closed fist. Reach to the back of the bowl under the dough and pull the bottom of the dough up and over the center. Repeat the same action at the front of the bowl. Turn out onto the counter and divide into three equal pieces. Using both hands with a gentle cupping and tucking action, shape each piece into a smooth round. Cover each with plastic wrap and leave on the counter for 15 minutes. The dough is now ready to shape.

4. Spray three 4½ × 8½ × 2½-inch pans with nonstick cooking spray. Roll one of the rested doughs into a rectangle approximately 18 inches wide and 10 to 12 inches from top to bottom using a rolling pin on a clean counter. Fold the right third of the dough to the center, then fold the left third of the dough over it. Press the three layers together with your fingertips. Fold the top third down and the bottom third up. Grab each side of the dough and pull until it is the length of the pan. Tuck the ends of the loaf firmly under. Repeat with the other two loaves and place each in a prepared pan. Stir ½ teaspoon salt into melted butter in a small bowl and brush loaves with some of the mixture. Cover loosely with plastic wrap and let rise in a warm place for 5 to 8 hours.

5. When dough is 30 minutes from being fully risen, place a baking stone on a shelf in the lowest slot of the oven and preheat to 450°F (232°C)—see Notes. About 5 minutes before baking, turn the oven down to 375°F (191°C) and carefully place a shallow pan with ½ inch of boiling water on the oven floor. (For an electric oven, put the pan of water on the lowest possible shelf.)

6. Brush the loaves one last time with butter and place the loaf pans directly on the hot stone. Bake until well browned, 45 to 55 minutes. Remove the loaf from the pan. The loaf should sound hollow when thumped on the bottom. You can also check for doneness by inserting an instant-read thermometer in the bottom center of loaf and pushing it to the middle. The loaf is done when the center is at least 200°F (93°C). Place the loaves on a rack to cool.

Notes: *If you want much faster rising time on this bread, increase the active dry yeast in the dough recipe to ½ teaspoon or even 1 teaspoon. It will still have excellent flavor.*

If you want a less sweet bread, simply leave out the sugar in the dough only (not in the starter).

If you are using a food processor, this large amount of dough is too much for even a large processor. You can divide the yeast starter mixture in half and prepare the bread in two batches.

If you don't have a stone:

· Preheat the oven to 25°F (14°C) below the baking temperature, place the bread on a shelf in the lower third of the oven, 5 to 6 inches above the bottom, and then turn the oven up to baking temperature. Carefully place a shallow pan with ½ inch of boiling water on the oven floor. (For an electric oven, put the pan of water on the lowest possible shelf.)

· Or use the cold-oven procedure described on page 14.

Spices to enhance—or fight—yeast I asked Elmer Cooper, one of Red Star Yeast's top chemists, about the old German bakers' saying that a pinch of ginger will make your yeast work better. (Often old chefs' tricks have scientific explanations.) Elmer chuckled, "If you had asked me several years ago, Shirley, I would have laughed at you. But I'll be glad to send you a copy of an article in *Cereal Chemistry,* 'The Effect of Spices on Yeast Fermentation.' "

The old German bakers' tale about a pinch of ginger was true—certain spices do enhance yeast's activity. Small amounts of ginger, ground caraway, cardamom, cinnamon, mace, nutmeg, and thyme all improve yeast activity. Dry mustard, like salt, strongly inhibits yeast growth. The accompanying table shows the effects of some of these spices.

Dry mustard and salt strongly inhibit yeast growth.

EFFECT OF SPICES ON YEAST ACTIVITY		
Spice	Amount	Change in Yeast Activity
	(grams of spice with 2 grams sugar and 1 gram yeast in 30 ml water)	*(ml of gas increase or decrease in 3 hours)*
Cardamom	0.1	+ 85
	0.5	+140
Cinnamon	0.05	+102
	0.1	+103
	0.5	+ 46
	1.0	− 30
Ginger	0.1	+ 87
	0.75	+172
	1.0	+136
	2.0	+ 72
Dry mustard	0.25	−120
Nutmeg	0.1	+ 40
	0.5	+111
Thyme	0.1	+ 92
	0.5	+157
	1.0	+154

Wilma J. Wright, C. W. Bice, and J. M. Fogelberg. "The Effect of Spices on Yeast Fermentation." *Cereal Chemistry,* Vol. 31 (March 1954), pp. 100–112.

The more you add of some spices (ground caraway, for example), the more they improve yeast activity. With most of the spices, however, the old German "good pinch" is appropriate. Some spices improve yeast activity when added at low levels but retard it at high levels. This is especially true of cinnamon, which improves activity when used in amounts from 5 to 10 percent of the weight of the yeast but much less at 50 percent of the yeast weight. At a weight equal to the yeast, it dramatically reduces activity.

Cinnamon and cloves contain chemical compounds known as *cinnamic aldehyde* and *eugenol*, respectively, which act as preservatives by inhibiting mold growth and aflatoxin production. But they are also detrimental to yeast growth. Use of these spices, as well as nutmeg and allspice, in the dough itself should be limited to amounts below about ¼ teaspoon per cup of flour.

Fenugreek and rosemary have a very interesting characteristic—they also act as preservatives. Since they are antioxidants, they prevent oxygen in the air from combining with compounds in the dough, thus acting as a preservative by preventing chemical changes in the dough. The effect of fenugreek on yeast activity was not reported in the Wright article, but other references indicate that it also improves yeast activity.

Want your bread to stay fresh longer? Add fenugreek or rosemary.

In addition to possibly improving yeast activity, spices in doughs make breads taste wonderful. What would Hot Cross Buns (page 94) be without the aroma and taste of their spices? Nisu, the rich Finnish coffee bread, is irresistible with freshly crushed cardamom seeds and glazed with strong coffee and sugar crystals. Without the distinctive scent of rosemary, Focaccia (page 72) would be just a flat bread.

Nisu

Rich Finnish Coffee Bread

MAKES ONE 12-INCH BRAIDED LOAF

This unusual bread is stained dark with strong coffee, glazed with an egg-honey coating, and sprinkled with coarse sugar. It is buttery with a rich, exotic taste from the cardamom seeds and coffee.

what this recipe shows

A very wet sponge helps impart sweetness.

Cardamom enhances yeast activity as well as imparts flavor.

Evaporated milk supplies lactose, a sugar that yeast does not like, leaving more residual sugar for a sweeter loaf.

Egg yolk adds color and taste and, as an emulsifier, enhances keeping quality.

1 package (2¼ teaspoons) active dry yeast

1 teaspoon barley malt syrup (see Notes), light molasses, or light brown sugar

1½ cups warm water (115°F/46°C)

2½ cups *and* 2 cups bread flour (4½ cups *total*)

1 small can (5 ounces) evaporated milk

¼ 500-milligrams vitamin C tablet, crushed

¼ cup crushed ice

1 teaspoon salt

3 tablespoons light brown sugar

1 large egg yolk

2 teaspoons *and* ¹/₂ teaspoon cardamom seeds, freshly crushed (2¹/₂ teaspoons *total*)

¹/₄ pound (1 stick) butter, softened

1¹/₂ tablespoons oil for bowl

Coarse cornmeal as needed

3 tablespoons instant coffee or instant cappuccino, dissolved in ¹/₃ cup hot water

3 tablespoons dark corn syrup

1 large egg

¹/₄ cup coarse sugar

Read "The Bread Recipes in This Book" (page 10) before beginning.

1. Stir yeast and barley malt syrup into warm water in the bowl of a heavy-duty mixer with the paddle blade. Let stand 2 minutes, until foam appears, indicating yeast is alive and well. Add 2¹/₂ cups bread flour and evaporated milk. Beat on low-medium speed for 4 minutes to beat air into the dough. Let the sponge sit for 30 minutes to 2¹/₂ hours for improved flavor and texture.

2. Remove the paddle blade and insert the dough hook. Add the vitamin C, crushed ice, salt, brown sugar, and egg yolk. Run on low to mix, then add 2 cups flour and knead on low-medium speed for 5 minutes, until the dough is very elastic. Pull the dough off the dough hook and divide into four pieces in the bowl. Distribute 2 teaspoons cardamom seeds and the butter over the dough. Mix for 1 to 2 minutes on low speed to incorporate well. Dough should be very soft. Add a little flour or liquid if needed.

3. Place the dough in an oiled bowl and turn to coat well with oil. Cover the bowl with plastic wrap and let rise until doubled in volume, about 1¹/₂ hours. Punch down the center of the dough with a closed fist. Reach to the back of the bowl under the dough and pull the bottom of the dough up and over the center. Repeat the same action at the front of the bowl. Then place the dough on the counter and cut it into three equal pieces. Using both hands with a gentle cupping and tucking action, shape each piece into a smooth round. Now, with both hands, grab the sides of each round and stretch it sideways into an oval. Let it spring back slightly, then pull it out again. Cover all three with plastic wrap and leave on the counter for 15 minutes. The dough is now ready to shape.

4. Place your palms in the center of the dough and, rolling and stretching each piece, roll out into a rope about 18 inches long. Press the strands together at one end and loosely braid like a pig-tail, pressing the strands tightly together at the end.

5. Sprinkle a baking sheet or medium pizza pan with cornmeal and place the braided loaf on it. Brush heavily with the coffee-water mixture. Loosely cover with plastic wrap. Let rise until at least doubled, over an hour. Brush again heavily with coffee. Then, in a small bowl, whisk together corn syrup, egg, and ¹/₂ teaspoon cardamom seeds. Brush the loaf with this syrup-egg mixture.

6. About 30 minutes before the dough is fully risen, place a baking stone on a shelf in the lowest slot of the oven and preheat to 450°F (232°C)—see Notes. About 5 minutes before placing the loaves in the oven, turn the oven down to 375°F (191°C) and carefully place a shallow pan with ¹/₂ inch of boiling water on the oven floor. (For an electric oven, put the pan of water on the lowest possible shelf.)

7. Brush the bread again with the egg mixture and place the baking sheet directly on the hot stone. Bake until well browned, 45 to 55 minutes. Remove the loaf from the pan. The loaf should sound hollow when thumped on the bottom. You can also check for doneness by inserting an instant-read thermometer in the bottom center of loaf and pushing it to the middle. The loaf is done when the center is at least 200°F (93°C). Place the loaf on a rack to cool.

Notes: *Barley malt syrup is available at health food stores and beer- or winemaking supply shops.*

If you don't have a stone:
- Preheat the oven to 25°F (14°C) below the baking temperature, place the bread on a shelf in the lower third of the oven, 5 to 6 inches above the bottom, and then turn the oven up to the baking temperature. Carefully place a shallow pan with ¹/₂ inch of boiling water on the oven floor. (For an electric oven, put the pan of water on the lowest possible shelf.)
- Or use the cold-oven procedure described on page 14.

Focaccia

MAKES 18 APPETIZER SERVINGS

This garlicky herbed bread is a rustic wonder that makes incredible sandwiches.

what this recipe shows

Rosemary enhances yeast activity and keeping quality in addition to imparting flavor.

1 medium-large red onion, chopped

¹/₂ teaspoon dried red pepper flakes

2 teaspoons chopped fresh rosemary

2 tablespoons *and* 3 tablespoons *and* 3 tablespoons olive oil (¹/₂ cup *total*)

3 large garlic cloves, minced

20 oil-cured olives, pitted and chopped

1 recipe Crusty French-Type Bread (page 24), prepared through Step 3 but not divided

Bread flour as needed

4 garlic cloves, sliced

¹/₂ teaspoon fresh rosemary needles

Coarse sea salt

Read "The Bread Recipes in This Book" (page 10) before beginning.

1. Sauté the onion, dried pepper flakes, and chopped rosemary in 2 tablespoons olive oil in a large skillet over medium heat until soft, about 2 minutes. Stir in the minced garlic and olives and sauté 1 minute. Transfer to a strainer set over a bowl and let stand until lukewarm. Dab with a paper towel to remove excess liquid.

2. Roll the dough out into a thin rectangle with a rolling pin and spread onion mixture on it. Sprinkle with about ¹/₂ cup flour, then roll dough up tightly like a jelly roll. Press the roll out with a rolling pin and roll it into a 10 × 12-inch rectangle. If the onion mixture makes the dough too wet, knead in more flour, then roll back to a 10 × 12-inch rectangle.

3. Oil a 10 × 15-inch jelly-roll pan and stretch the dough out with your hands into the pan. Heat on low 3 tablespoons olive oil, garlic slices, and rosemary needles for several minutes in a small saucepan. Let cool for a couple of minutes. Poke finger holes all over dough. Drizzle with the garlic and rosemary olive oil. Let focaccia rise until almost doubled, 30 to 45 minutes.

4. About 30 minutes before the dough is fully risen, place a baking stone on a shelf in the lowest slot of the oven and preheat to 450°F (232°C)—see Note. About 5 minutes before placing the loaves in the oven, turn the oven down to 425°F (218°C) and carefully place a shallow pan with ½ inch of boiling water on the oven floor. (For an electric oven, put the pan of water on the lowest possible shelf.)

5. Place the pan of focaccia on the hot stone. Bake until browned, about 30 minutes. Brush the focaccia with oil and sprinkle with coarse salt. Slide focaccia onto a large rack to cool. Cut in 2- to 3-inch squares or other desired size and serve warm.

Note: *If you don't have a stone:*
- Preheat the oven to 25°F (14°C) below the baking temperature, place the bread on a shelf in the lower third of the oven, 5 to 6 inches above the bottom, and then turn the oven up to the baking temperature. Carefully place a shallow pan with ½ inch of boiling water on the oven floor. (For an electric oven, put the pan of water on the lowest possible shelf.)
- Or use the cold-oven procedure described on page 14.

Chemical leaveners: baking powder and baking soda

Both baking soda and baking powder produce carbon dioxide gas, which helps leaven cakes, quick breads, muffins, pancakes, and so on.

Baking soda (sodium bicarbonate) alone breaks down when heated to form carbon dioxide gas and sodium carbonate (washing soda), which has an unpleasant soapy taste and is moderately alkaline. However, if baking soda is combined with an acid, carbon dioxide gas comes off much faster and a small amount of a milder-tasting salt is left behind. Baking soda works beautifully with foods that contain mild acids like chocolate, honey, molasses, citrus juice, sour cream, buttermilk, and brown sugar, to name a few.

Baking powder contains both baking soda (1 teaspoon baking powder contains ¼ teaspoon soda) and the exact amount of acid to use up all the soda. Baking powder also contains cornstarch to separate these two ingredients and to absorb moisture to keep both ingredients dry. Baking powders can be fast-acting, slow-acting, or double-acting, depending on the acid or acids they contain. Fast-acting baking powders contain an acidic ingredient that dissolves fairly rapidly in cold water (like cream of tartar, tartaric acid, or monocalcium phosphate monohydrate). If the baking powder has an acidic ingredient that does not dissolve easily (like sodium aluminum sulfate or anhydrous monocalcium phosphate), it does not start producing gas until the batter is hot; it is a slow-acting baking powder.

Double-acting baking powders contain both a fast-dissolving and a slow-dissolving acidic ingredient to produce a small amount of gas during mixing, then a maximum amount in the hot oven. Encapsulation of a single acidic ingredient is also used to make some baking powders double-acting. Most widely distributed grocery store brands of baking powder, such as Rumford, Calumet, and Clabber Girl, are double-acting.

You can make your own fast-acting baking powder from baking soda by mixing it with cream of tartar and cornstarch. A good formula is 1 tablespoon baking soda with 2 tablespoons cream of tartar and 1½ tablespoons cornstarch.

Some recipes call for both baking powder and baking soda. This may seem redundant since they are both essentially baking soda, but there are good reasons to use both. Baking powder is very reliable since it has just the right amount of acid for the amount of soda. But sometimes a recipe also contains an acidic ingredient, and adding a little soda will neutralize the extra acidity. Half a teaspoon of soda will neutralize 1 cup of a mildly acidic ingredient like buttermilk or sour cream. If you have a recipe that normally requires 2 teaspoons baking powder, and it contains ½ cup buttermilk, you may want to substitute ¼ teaspoon soda (to neutralize the buttermilk) plus 1 teaspoon baking powder. You will still have the equivalent leavening power of 2 teaspoons baking powder.

An important fact to remember about baking soda and baking powder is that baking soda is four times as strong as baking powder—a little soda goes a long way. You should use about 1 teaspoon baking powder (no more than 1¼ teaspoons) per cup of flour, or ¼ teaspoon baking soda per cup of flour in

> **Less is more with baking powder—too much makes baked goods fall.**

a recipe. Although it sounds strange, *too much baking powder makes baked goods fall.* The bubbles get big, float to the top, and pop. The baked goods get heavy and fall. It's a common problem.

There are also specialty commercial baking powders like ammonium carbonate and ammonium bicarbonate, used in cookies and crackers. These baking powders totally break down into gases when heated, leaving no solid residue at all. However, since ammonia is a smelly gas, these can be used only in flat products like cookies so that all the gas can escape.

Some of the very best roll recipes use both yeast and baking powder. The addition of baking powder will give you a very light roll even with a little more sugar than you can normally use in yeast breads. The amount of sugar in the Light-as-a-Dream Hot Rolls would make a plain yeast roll heavy. With baking powder and sugar, they are wonderfully light, have a sweet flavor, and brown beautifully. Marion Cunningham's incredibly light waffles (page 76) also take advantage of leavening from both yeast and baking soda.

Light-as-a-Dream Hot Rolls

MAKES ABOUT 20 MEDIUM ROLLS

These rolls have real homemade hot roll taste. They are extraordinarily light and tender and very hard to stop eating. This recipe is from Marion, Arkansas, and is my version of Margaret Fogleman's version of her neighbor's recipe. (We don't know where her neighbor got it!)

> **what this recipe shows**
>
> **The addition of chemical leaveners makes the high sugar content possible and produces a very light roll.**

1¾ cups whole milk

4 tablespoons shortening

½ cup sugar

2 cups *and* 2 cups *and* 1 cup bread flour (5 cups *total*)

1½ packages (scant 3½ teaspoons) active dry yeast

2 large eggs

1 large egg yolk

4 tablespoons (½ stick) butter, softened

2 teaspoons *and* ¼ teaspoon salt (2¼ teaspoons *total*)

1 teaspoon baking powder

½ teaspoon baking soda

Nonstick cooking spray

6 tablespoons butter, melted

Read "The Bread Recipes in This Book" (page 10) before beginning.

1. Scald the milk in a medium-size heavy saucepan. Remove from the heat, stir in the shortening and sugar, and let stand 5 minutes to cool. Blend 2 cups flour and the yeast in a heavy-duty mixer with the paddle blade for a few seconds on low, then stir in the warm milk mixture. Beat on low-medium speed for 4 minutes to incorporate air. Let stand for 30 minutes to $2^1/2$ hours for improved flavor.

2. Add 2 cups bread flour, $1/2$ cup at a time, running the mixer on low speed for 1 minute after each addition. Let the dough rise in the mixer bowl for about 1 hour. This dough rises fast, so keep an eye on it. You may need to punch it down after 30 minutes so that it won't come out of the bowl and then let it rise again for 30 minutes. Add 1 egg and beat in, then add the remaining egg, egg yolk, and softened butter and beat in. Cover the dough with plastic wrap and refrigerate several hours or overnight if desired.

3. Sift together 1 cup bread flour, 2 teaspoons salt, the baking powder, and baking soda into a medium mixing bowl. Beat this mixture into the dough. The dough should be very soft.

4. Divide the dough in half, then divide each half in half again. Now divide each fourth into five rolls so that you have twenty in all. Shape each roll into a smooth round. You can flatten and fold each round into Parker House rolls if desired or simply spray muffin tins with nonstick cooking spray and place a roll in each cup. With sharp scissors or a razor, cut a cross in the top of each dough ball. Stir together the melted butter and $1/4$ teaspoon salt in a small bowl. Brush rolls well. Let rise until doubled, about $1^1/2$ hours.

5. When the rolls are about 30 minutes from being fully risen, place a baking stone on a shelf in the lowest slot of the oven and preheat to 450°F (232°C)—see Note. About 5 minutes before placing the rolls in the oven, turn the oven down to 375°F (191°C) and carefully place a shallow pan with $1/2$ inch of boiling water on the oven floor. (For an electric oven, put the pan of water on the lowest possible shelf.)

6. Brush the rolls again with the butter mixture and place the muffin tins directly on the hot stone. Bake until lightly browned, 30 to 40 minutes. Serve hot.

Note: *If you don't have a stone:*
- Preheat the oven to 25°F (14°C) below the baking temperature, place the bread on a shelf in the lower third of the oven, 5 to 6 inches above the bottom, and then turn the oven up to the baking temperature. Carefully place a shallow pan with $1/2$ inch of boiling water on the oven floor. (For an electric oven, put the pan of water on the lowest possible shelf.)
- Or use the cold-oven procedure described on page 14.

Crisp-Crusted, Feather-Light Raised Waffles

MAKES 8 WAFFLES

Marion Cunningham graciously shared this waffle recipe, which appears in her little book of delicious breakfast dishes, *The Breakfast Book*. These are the best waffles I have ever had. They are now a traditional part of the holidays in my household. I make double batches and keep the batter in the refrigerator (see Notes). Two waffle irons are out on the counter, and those who get up before I do know that they can make themselves a waffle (see Notes).

> **what this recipe shows**
>
> Yeast and overnight fermentation products impart excellent flavors. Baking soda and eggs make the waffles extraordinarily light.
>
> Eggs and sugar give good color and a crisp surface. For an even crisper, browner surface, add 2 tablespoons corn syrup.

1 package (2¼ teaspoons) active dry yeast	1 teaspoon sugar
½ cup warm water (115°F/46°C)	1 teaspoon salt
2 cups warm whole milk (115°F/46°C)	2 cups bleached all-purpose flour
¼ pound (1 stick) or less butter, melted	2 large eggs
	¼ teaspoon baking soda

1. Sprinkle the yeast on warm water in a very large mixing bowl and let stand for 5 minutes. Add the milk, butter, sugar, salt, and flour and beat until smooth. (A hand beater does this well.) Cover the bowl with plastic wrap and let stand overnight. In warmer climates, refrigerate; in an air-conditioned house around 70°F or cool climate, leave at room temperature.

2. When ready to cook the waffles, beat in the eggs and baking soda. The batter is very thin, and most waffle irons will need ½ to ¾ cup batter. Bake in a hot waffle iron. Serve immediately.

Notes: *This batter keeps for several days in the refrigerator. Use a large container since it will expand.*

A good waffle iron (one that browns both top and bottom evenly) is a treasure. I love my Vitantonio waffle iron for both its even browning and its speedy 90-second cooking time.

Leavening by steam

Steam is a powerful leavener. The secret of many light breads is a wet dough whose liquid turns to steam in the hot oven.

I learned at an early age that the amount of liquid in doughs is crucial. As a little girl, I followed my grandmother around the kitchen. For breakfast, lunch, and dinner she made the lightest, most wonderful biscuits in the world. She was a very busy woman with all my uncles and grandfather to feed three meals a day—and she did not get her chickens at the grocery store.

I used her bread bowl, her flour, her buttermilk—I did everything the same, and I shaped the biscuits just as she did. But mine always turned out a dry, mealy mess. I would cry, "Nannie, what did I do wrong?" She would lean down and give me a big hug and say, "Honey, I guess you forgot to add a touch of grace."

It took me twenty years to figure out what I was missing. I had watched my grandmother shape those biscuits time after time and thought that the dough had to be dry enough to shape by hand.

Actually, Nannie made a very wet dough in the back of her big wooden bread bowl, then dried her hands, took a handful of flour from the front of the bowl, and sprinkled it on the dough. She floured her hands well, pinched off a biscuit-size piece of wet dough, and dipped it in the flour. Only then was she ready to shape it.

This was the secret. Her dough was not dry enough to shape—it was so wet that she had to flour it to handle it at all. Then she baked them in a very hot oven, where the wet dough created plenty of steam to puff and make feather-light biscuits.

Now I make biscuits almost as good as my grandmother's, and so can you, with a good wet dough and a touch of grace. My grandmother never made Cherry-Chambord Butter, but it's a wonderful spread for her biscuits.

Touch-of-Grace Biscuits

MAKES ABOUT 10 BISCUITS

These are my grandmother's feather-light, real Georgia biscuits. For company, or just for yourself, you need to make these biscuits with part cream at least once—I got a standing ovation for them at an international food conference in Sicily.

what this recipe shows

Low-protein flour helps make tender, moist biscuits.

A very wet dough makes more steam in a hot oven and creates lighter biscuits.

Nonstick cooking spray

2 cups Southern self-rising flour (see Notes)

1/8 teaspoon baking soda

1/2 teaspoon salt

1/4 cup sugar

4 tablespoons shortening

2/3 cup heavy or whipping cream

1 cup buttermilk, or until dough resembles cottage cheese

1 cup bleached all-purpose flour for shaping (see Notes)

2 tablespoons butter, melted

continued

1. Preheat the oven to 425°F (218°C) and spray an 8- or 9-inch round cake pan with nonstick cooking spray.

2. Combine the self-rising flour, salt, and sugar in a medium mixing bowl. With your fingers or a pastry cutter, work the shortening into the flour mixture until there are no shortening lumps larger than a big pea.

3. Stir in the cream and buttermilk and let stand for 2 or 3 minutes. This dough is so wet that you cannot shape it in the usual manner. It will look like cottage cheese.

4. Pour the cup of all-purpose flour onto a plate or pie tin. Flour your hands well. Spoon a biscuit-size lump of wet dough into the flour and sprinkle some flour over the wet dough to coat the outside. Pick up the biscuit and shape it roughly into a soft round. At the same time, shake off the excess flour. The dough is so soft that it will not hold its shape. As you shape each biscuit, place it in the pan. Push the biscuits tightly against each other so that they will rise up and not spread out. Continue shaping biscuits in this manner until all of the dough is used. To make a large batch of biscuits in a hurry, spray a medium-small (about 2-inch) ice cream scoop with nonstick cooking spray. Cover a jelly-roll pan with all-purpose flour. Quickly scoop biscuits onto the flour, sprinkle with flour, shape, and place in small pans.

5. Bake just above the center of the oven until lightly browned, 15 to 20 minutes, then brush with melted butter. Cool for 1 or 2 minutes in the pan, then dump out and cut the biscuits apart. "Butter 'em while they're hot!" Split the biscuits in half, butter or spread with Cherry-Chambord Butter (recipe follows), and eat immediately.

Notes: *If low-protein Southern self-rising flour is not available, use 1 cup national-brand self-rising all-purpose and 1/2 cup instant flour (such as Shake & Blend or Wondra) or cake flour, plus 1/2 teaspoon baking powder. If self-rising flour is not available, use a total of 1 1/2 teaspoons baking powder.*

Do not use self-rising flour for shaping since the leavener will give a bitter taste to the outside of the biscuits.

Cherry-Chambord Butter

1/4 pound (1 stick) butter

1 package (8 ounces) cream cheese

2 tablespoons Chambord or other raspberry liqueur

1/3 cup confectioners' sugar

1 small jar (5 ounces) good-quality cherry preserves

Finely grated zest of 1 orange

Process the butter, cream cheese, liqueur, and confectioners' sugar in a food processor with the steel knife to blend well. Stir in the preserves and zest by hand. Chill well before serving. Keeps well in a closed jar for several weeks in the refrigerator.

A wet dough is not just the secret of light biscuits; it will give you a lighter yeast bread, too. If you have never made very soft, slightly sticky doughs, you are in for a treat. The lightness of the breads they make is lovely. Throughout this chapter, the recipes have cautioned you to keep the doughs soft and some slightly sticky. Admittedly, you need the dough to have a certain firmness to knead it properly—if the dough is too wet, you won't be able to stretch the proteins out as they should be stretched. Still, probably the greatest fault of beginning breadmakers is making an overly dry dough.

Wet-dough breads need to be baked in a pan with sides. When simply placed on a baking sheet, they spread out and flatten like a thick pizza. Even in a pan, their shape is not as perfect as bread from a firmer dough, but the lightness of the bread may be worth sacrificing the perfect shape. Although this is a matter of personal choice, you should try a dough much wetter than you normally use at least once to experience the changes that varying the moisture makes.

For the best of all worlds, you can beat together very wet dough ingredients with part of the flour. Beat this soupy mess to blend all ingredients and to incorporate air, let it stand for 30 minutes to 2½ hours for the yeast to get a good start, and then add flour to get the dough firm enough to pull well but remain sticky. Knead to stretch the proteins and develop gluten. If necessary, add a little water near the end of kneading to finish up with a soft, slightly sticky dough.

A choice of liquids Yeast dissolves and gets going best in warm water, but you can always dissolve it in a small amount of warm water and use another liquid for a major part of the moisture in the recipe. Milk, yogurt, sour cream, and cream, which make doughs soft and tender, are frequently used. Fruit juices, which add a little sugar and acidity and a subtle hint of fruit flavor, can be used as well. Some wines or liquors/liqueurs can be used, but remember that alcohol evaporates at a lower temperature than water and you won't have as much steam for leavening in the oven. You will need to combine an alcoholic beverage with a reasonable percentage of water, a dairy product, or juice—something that is mostly water—for a good-rising bread.

Dairy products in breadmaking

Milk and other dairy products have several effects on breads. The proteins in milk improve the nutritional profile of the bread, and the minerals enhance yeast growth and activity. The sugar in milk, lactose, is not used by yeast, so it is left in the loaf as residual sugar to give a sweeter bread. Lactose also tenderizes and moistens for a softer, moister crumb and adds flavor to the loaf and color to the crust. The powdered milk in the Honey Whole Wheat Loaf recipe (page 46), for example, produces a soft, moist, rich crumb.

Milk is also an excellent buffer, a substance that prevents something from changing acidity easily. A buffer in bread made with bakers' yeast, for example, prevents the dough from becoming so acidic that the bakers' yeast is damaged. (There are other strains of yeast that can grow in more acidic conditions, like the strains produced in a sourdough starter.)

TO SCALD OR NOT TO SCALD?

Bread recipes in old cookbooks advise you to scald the milk before using it in a bread. I thought this heating was taken to kill some enzyme in the milk and that modern pasteurization had made scalding unnecessary. I had also heard that after running tests on scalded and unscalded milk a food magazine staff concluded that it made no difference.

I did notice when I was developing the Honey Whole Wheat Loaf recipe, however, that an increase in the nonfat dry milk to 1/2 cup significantly reduced the bread volume. Attributing this to the sugar in the milk, I was then amazed to read in a professional baking book, "It has long been known in the art that nonfat dry milk seriously affects the volume, symmetry, cellular structure, and texture of bread when used as such. Research by prior investigators has found that such nonfat milk, when heat treated by holding it at about 180°F to 190°F over a period of time, overcomes this undesirable property to a certain degree when it is used in the conventional system of bread-making."

Keep nonfat dry milk to a minimum: It can reduce bread volume.

When I delved into the literature on the subject, I found that in 1975 two Michigan State University researchers, T. Volpe and M. E. Zabik, isolated a protein in the whey that was responsible for this reduced volume and poor texture. Unfortunately it is present in nonfat dry milk as well as fresh. I believe that the *amount* of this protein is the key to whether it will harm the bread—that would explain why less than 1/2 cup nonfat dry milk had no ill effects on my bread. I imagine you will find that it does not matter in some recipes, but it may in others.

But, as they say, "a little knowledge is a dangerous thing." Personally I am going to scald both milk and reconstituted nonfat dry milk just for my own peace of mind.

How sweet it is: sugar, syrups, and honey in bread

Sugar and syrups

Have you ever seen a yeast bread recipe that calls for a lot of sugar? There are sweet rolls with sweet, gooey fillings, but there is not much sugar in the dough itself. The really sweet breads are carrot bread and zucchini bread—quick breads made with baking powder or soda. Why is this?

When dough contains a lot of sugar, the two gluten proteins link with the sugar instead of with each other. Glutenin combines with sugar, gliadin combines with sugar, and very little gluten is formed. You can see this dramatically for yourself. Prepare a small gluten ball by processing 1 cup bread flour with 1/2 cup water for about 40 seconds in a food processor. Thoroughly rinse the dough in a strainer to remove the starch by squeezing the dough with your fingers under running water. You will have insoluble, springy gluten left—usually a gray blob about the size of a small lemon.

Now do the same thing again, but this time add 1/3 cup of sugar to the flour. You can tell even in the food processor that gluten is not forming. Instead of a dough ball that bounces around, you now have a thick soup. Not enough water was taken up in the gluten formation to make a ball. This time, when you rinse the dough, the glutenin-sugar and the gliadin-sugar compounds that were formed dissolve in water and wash down the drain. You will end up with nothing left in the strainer!

Too much sugar is also damaging to yeast. It draws water from the yeast and inhibits its growth. Bakers sometimes add extra yeast to sweet breads just to compensate for some loss.

Two tablespoons of sugar per cup of flour is the maximum that you can add without major gluten damage. But, you may ask, what about Hawaiian bread and Portuguese Sweet Bread, which taste so sweet? How about some sweet starter breads with their sweet taste?

Hawaiian bread and Portuguese Sweet Bread do contain close to the maximum sugar that a yeast dough can handle. Sweet breads also often contain barley malt, barley malt syrup, or malted flour, ingredients containing enzymes that help break down big molecules of starch into the simple sugars that yeast can use. It could be that not all sugars combine equally well with glutenin and gliadin. Perhaps some of these sugars formed by barley malt and barley malt syrup are not as detrimental to gluten formation as table sugar (sucrose). Or maybe it is because sugar is being produced in the dough right up to the time that the dough gets hot enough to kill the enzymes.

Breads made with a sponge that has a high ratio of liquid to flour have a higher residual sugar content. So, using a very wet sponge as I have in the Portuguese Sweet Bread is a good idea.

Commercially, the sweeter doughs are kneaded almost twice as long as regular doughs. You can do the same at home and get better gluten development as well. Kneading time can be increased easily enough with a heavy-duty mixer. If you want a dough of maximum sweetness, you can also use a tablespoon of barley malt syrup (available at health food stores). Use a very high-protein flour and gluten flour or vital wheat gluten, knowing that you will need all the gluten you can get, and increase the kneading time.

Different sugars, different characteristics All sugars interfere with gluten development, but sugars are not identical in other ways. Glucose, corn syrup, browns at lower temperatures than many sugars. Most sugars are hygroscopic and absorb water from the atmosphere, but fructose, fruit sugar, which is one of the two sugars in table sugar, is more hygroscopic than most. We have seen that some strains of yeast can thrive on sugars that others can't thrive on, and some strains of bacteria like sugars that yeast doesn't like. Again, yeast doesn't like lactose, the sugar in milk, so breads with milk and dairy products will have more residual sugars than those containing only sucrose.

Portuguese Sweet Bread

MAKES ONE 8-INCH ROUND LOAF

Southerners are accused of having a sweet tooth. I know I do, and I love this bread. The spicy, lemony loaf is glazed with sugar and spice and cream. It keeps well and is excellent toasted.

what this recipe shows

Potato is used for extra sweetness and moisture.

Extra yeast is used to compensate for the yeast damaged by sugar.

Barley malt syrup is used to contribute enzymes to supply yeast with food.

Bread flour, gluten flour, and vitamin C are used for optimum gluten development.

Nutmeg enhances yeast activity.

Sweetened condensed milk adds sucrose and lactose for more residual sugar, good crust color, tenderness, moisture, and flavor.

The dough is kneaded slightly longer for better gluten development.

continued

1 small potato, peeled and chopped

1^1/$_2$ cups water

1 tablespoon (1^1/$_3$ packages) active dry yeast

2 tablespoons barley malt syrup (see Notes)

1 cup *and* 2 cups bread flour (3 cups *total*)

^{1}/$_2$ teaspoon *and* ^{1}/$_4$ teaspoon freshly grated nutmeg (^{3}/$_4$ teaspoon *total*)

2 tablespoons vital wheat gluten or gluten flour

^{1}/$_4$ 500-milligram vitamin C tablet, crushed

^{1}/$_4$ cup canned sweetened condensed milk

1 large egg

1^1/$_4$ teaspoons salt

Finely grated zest of 1 lemon

Finely grated zest of 1 orange

3 tablespoons butter, softened

1 tablespoon oil for bowl

Nonstick cooking spray

3 tablespoons heavy or whipping cream

2 tablespoons sugar

Read "The Bread Recipes in This Book" (page 10) before beginning.

1. Boil the potato in the water in a small saucepan until the potato is quite soft. Strain and catch the liquid in a measuring cup. Add or subtract water to make ^{3}/$_4$ cup liquid. Place in the bowl of a heavy-duty mixer with the paddle blade. Using a fork, mash at least ^{1}/$_4$ cup potato in a small bowl and add back to the liquid. When it has cooled to 115°F (46°C), add the yeast, barley malt syrup, 1 cup flour, and ^{1}/$_2$ teaspoon nutmeg. Beat on low-medium speed for 4 minutes to beat air into the dough. Let the sponge sit preferably for 30 minutes to 2^1/$_2$ hours for improved flavor and texture.

2. Remove the paddle blade and insert the dough hook in mixer. Add 2 cups flour, vital wheat gluten, vitamin C, condensed milk, egg, and salt. Knead for 7 minutes on low-medium speed, until dough is very elastic. The dough should be soft and slightly sticky. Add a little flour or water if necessary. Pull the dough into four pieces and add lemon and orange zest and softened butter and mix to blend in, about 1 minute.

3. Place the dough in an oiled bowl and turn to coat well with oil. Let rise until doubled, about 1^1/$_2$ hours. Punch down the center of the dough with a closed fist. Reach to the back of the bowl under the dough and pull the bottom of the dough up and over the center. Repeat the same action at the front of the bowl, then turn the dough out onto the counter. Using both hands with a gentle cupping and tucking action, shape the dough into a tight, smooth round. Cover with plastic wrap and leave on the counter for 15 minutes. The dough is now ready to shape.

4. Spray an 8-inch springform pan with nonstick cooking spray. If you are right-handed, prop the dough ball against your left hand and press down and forward on the opposite side of the ball with the palm of your right hand to lightly knead the side of the ball. Let the dough ball rotate slightly forward with the stroke and continue to knead the sides of the round until you have gone completely around. Pinch the bottom to seal. Place the dough in the pan and brush with some of the cream. Let rise until doubled. Brush again with cream.

5. About 30 minutes before the dough is fully risen, place a baking stone on a shelf in the lowest slot of the oven and preheat to 450°F (232°C)—see Notes. About 5 minutes before placing the loaves in the oven, turn the oven down to 375°F (191°C) and carefully place a shallow pan with ½ inch of boiling water on the oven floor. (For an electric oven, put the pan of water on the lowest possible shelf.)

6. Brush the bread again with cream and sprinkle with sugar and ¼ teaspoon nutmeg. Place the springform pan directly on the hot stone. Bake until well browned, about 45 minutes. Remove the loaf from the pan. The loaf should sound hollow when thumped on the bottom. You can also check for doneness by inserting an instant-read thermometer in the bottom center of loaf and pushing it to the middle. The loaf is done when the center is at least 200°F (93°C). Place the loaf on a rack to cool. Serve warm or at room temperature.

Notes: *Barley malt syrup is available at health food stores and beer- and winemaking supply shops.*

 If you don't have a stone:
 · Preheat the oven to 25°F (14°C) below the baking temperature, place the bread on a shelf in the lower third of the oven, 5 to 6 inches above the bottom, and then turn the oven up to the baking temperature. Carefully place a shallow pan with ½ inch of boiling water on the oven floor. (For an electric oven, put the pan of water on the lowest possible shelf.)
 · Or use the cold-oven procedure described on page 14.

I used a few tricks to get the Portuguese Sweet Bread as sweet as possible and still keep it light. Here is another good trick that I learned from Bernard Clayton, a great bread baker and author of *The Complete Book of Breads,* for incorporating sugar into a loaf of bread: Use sugar lumps. The small, hard lumps of sugar will stay together during brief mixing, so they don't interfere with gluten formation. In the hot oven, the sugar finally melts and leaves a puddle of syrup in the slice of bread. Large, rectangular lumps are too big, but you can break them up into medium-size pieces. The lumps called *sugar dots* are ½-inch cubes, the ideal size.

Spicy Sugar Lump Loaf

MAKES ONE 5 × 9-INCH LOAF

Lumps of sugar coated with spices melt during baking and produce wonderful little surprises of syrupy, spicy spots in the loaf. Orange zest and tiny bits of candied ginger impart a zippy freshness. A slice of this bread toasted and spread with Orange Butter (page 96) is wonderful.

> **what this recipe shows**
>
> A wet sponge helps produce a sweeter bread.
>
> Buttermilk adds flavor, color, and lactose, which yeast does not consume, so there is more residual sugar and a sweeter loaf.
>
> Crushed ice added before kneading cools dough for better liquid absorption and gluten development.

1 package (2¹/₄ teaspoons) active dry yeast	1¹/₂ teaspoons salt
1 teaspoon barley malt syrup (see Notes)	Finely grated zest of 1 orange
1 cup warm water (115°F/46°C)	3 tablespoons butter, softened
2 cups *and* 1¹/₂ cups bread flour (3¹/₂ cups *total*)	1 tablespoon oil for bowl
¹/₂ cup warm buttermilk (115°F/46°C)	1 cup sugar dots or rectangular sugar lumps, broken into small pieces (see Notes)
2 tablespoons dark brown sugar	Nonstick cooking spray
¹/₄ cup crushed ice	2 teaspoons ground cinnamon
¹/₄ 500-milligram vitamin C tablet, crushed	1 teaspoon ground allspice
1 large egg yolk	6 slices candied ginger, finely chopped
1 teaspoon pure vanilla extract	1 large egg, beaten

Read "The Bread Recipes in This Book" (page 10) before beginning.

1. Stir the yeast, barley malt syrup, and warm water together in the bowl of a heavy-duty mixer with the paddle blade. Let stand 2 minutes, until foam appears, indicating yeast is alive and well. Add 2 cups bread flour, buttermilk, and brown sugar. Beat on low-medium speed for 4 minutes to beat air into the dough. Let the sponge sit for 30 minutes to 2¹/₂ hours for improved flavor and texture.

2. Remove the paddle blade and insert the dough hook. Add the crushed ice, vitamin C, egg yolk, vanilla, 1¹/₂ cups bread flour, and salt and knead for 6 minutes on low-medium speed, until the dough is very elastic. The dough should be soft and slightly sticky. Pull the dough into four pieces. Add orange zest and softened butter and mix for 1 to 2 minutes to incorporate well.

3. Place the dough in an oiled bowl and turn to coat well with oil. Cover the bowl with plastic wrap and let rise until slightly more than doubled in volume, about 1¹/₂ hours.

4. Punch down the center of the dough with a closed fist. Reach to the back of the bowl under the dough and pull the bottom of the dough up and over the center. Repeat the same action at the front of the bowl, then turn the dough out onto the counter. With both hands, with a gentle cupping and tucking action, shape the dough into a smooth tight round. Now grab the sides of the round and stretch it sideways into an oval. Let it spring back slightly, then pull it out again. Cover with plastic wrap and leave on the counter for 15 minutes. The dough is now ready to shape.

5. Place the sugar dots in a plastic bag and spray them lightly with nonstick cooking spray (so spices will stick to sugar), add cinnamon and allspice, and toss to coat sugar with spice.

6. Roll the dough into a rectangle approximately 18 inches wide and 10 to 12 inches from top to bottom, using a rolling pin on a clean counter. Sprinkle two-thirds of the candied ginger and sugar lumps over the left two-thirds of the dough. Roll a rolling pin across the top to press ginger and sugar slightly into the dough. Fold the right third of the dough to the center, then fold the left third of the dough over it. Press the three layers together with your fingertips. Flatten slightly with the rolling pin. Sprinkle the remaining third of the ginger and sugar over the bottom two-thirds of the dough and roll over them with the rolling pin. Fold the top third down and the bottom third up. Grab each side of the dough and pull until it is the length of the pan. Tuck the ends of the loaf firmly under.

7. Spray a 5 × 9 × 3-inch loaf pan with nonstick cooking spray. Place the loaf in the pan and brush with egg. Let rise until slightly more than doubled, about 45 minutes. Brush again with egg.

8. About 30 minutes before the dough is fully risen, place a baking stone on a shelf in the lowest slot of the oven and preheat to 450°F (232°C)—see Notes. About 5 minutes before placing the loaf in the oven, turn the oven down to 375°F (191°C) and carefully place a shallow pan with ½ inch of boiling water on the oven floor. (For an electric oven, put the pan of water on the lowest possible shelf.)

9. Brush the bread again with egg and place the pan directly on the hot stone. Bake until well browned and done, about 50 minutes. Remove the loaf from the pan. It should sound hollow when thumped on the bottom. You can also check for doneness by inserting an instant-read thermometer in the bottom center of loaf and pushing it to the middle. The loaf is done when the center is at least 200°F (93°C). Place the loaf on a rack to cool. Serve warm or at room temperature.

Notes: *Barley malt syrup is available at health food stores and beer- and winemaking supply shops.*

If you're using large sugar lumps, place them in a heavy-duty plastic bag and break with a hammer or meat pounder. Try to hit each lump to break it into two or three pieces, but avoid making too much powder.

If you don't have a stone:
- Preheat the oven to 25°F (14°C) below the baking temperature, place the bread on a shelf in the lower third of the oven, 5 to 6 inches above the bottom, and then turn the oven up to baking temperature. Carefully place a shallow pan with ½ inch of boiling water on the oven floor. (For an electric oven, put the pan of water on the lowest possible shelf.)
- Or use the cold-oven procedure described on page 14.

Another way around the sugar limitation is to use a sweet filling. The dough can be mildly sweet, flavored, and spiced, with a sweet, rich filling. You may have noticed that in a sweet roll ring the dough rises and pulls away from its filling, leaving a space between the filling and the dough. This problem is solved in cinnamon rolls and in the Honey-Walnut Sticky Buns by having the dough and filling press against each other in the pan as they rise.

Rose Levy Beranbaum solved this problem with a filled brioche by cutting the filled dough into sections and placing them in a pan vertically to create a cake of tall cinnamon rolls. As the dough expanded, it pressed against the filling.

Another solution to prevent dough from pulling away from a sweet filling is to twist the dough and filling together.

Honey-Walnut Sticky Buns

MAKES 12 BUNS

Gooey, nutty, and incredibly delicious.

> **what this recipe shows**
>
> When a filled dough is sliced and the slices placed beside each other in a pan, as the dough rises it is pushed against the filling, preventing gaps around it.
>
> The secret of the filling is the coconut. Flaked coconut holds moisture and sweetness.

1/4 pound (1 stick) *and* 4 tablespoons (1/2 stick) *and* 4 tablespoons (1/2 stick) butter (1 cup—2 sticks— *total*)

1/4 teaspoon *and* 1/2 teaspoon salt (3/4 teaspoon *total*)

1 1/2 cups light brown sugar

1/4 cup dark corn syrup

2 teaspoons pure vanilla extract

1 1/2 cups walnut pieces

1 1/2 cups almond slivers

1 recipe Nisu dough (page 70), prepared through Step 3

2 teaspoons ground cinnamon

1/2 cup flaked coconut, frozen or canned

Nonstick cooking spray

1/4 cup whipping cream

3/4 cup honey

Read "The Bread Recipes in This Book" (page 10) before beginning.

1. Melt 1/4 pound butter in a saucepan and stir in 1/4 teaspoon salt, the brown sugar, and corn syrup. Heat until the brown sugar is dissolved. Remove from the heat. Let cool 2 minutes, then stir in the vanilla.

2. Preheat the oven to 350°F (177°C). Spread the walnuts on one 12 × 15-inch baking sheet and the almonds on another. Toast in the oven until lightly browned, 8 to 10 minutes for the almonds,

12 to 14 minutes for the walnuts. Add 2 tablespoons butter to each sheet of nuts and sprinkle 1/4 teaspoon salt over each. Toss nuts, butter, and salt while the nuts are warm.

3. Roll the Nisu dough into an 18 × 10-inch rectangle with a rolling pin on the counter. Leave the top inch of dough along the long side uncovered. Sprinkle the rest of the dough evenly with cinnamon, then with half the walnuts and half the almonds. Roll across the top of the nuts to press them slightly into the dough. Sprinkle with coconut. Drizzle with half of the cooled butter–brown sugar mixture. Begin at the front along the long side and roll the dough and filling into a long, tight log.

4. Spray two 9 × 2-inch round cake pans with nonstick cooking spray. Divide the other half of the cooled butter–brown sugar mixture and the remaining walnuts and almonds and place in the bottom of the two pans. Cut off 1 1/2-inch slices of the rolled dough with a very sharp serrated knife or by wrapping a piece of dental tape or fishing line around the log, crossing the string, and pulling. Arrange six buns in each pan, allowing room for them to rise. Brush with cream and let rise until doubled, about 50 minutes.

5. About 30 minutes before the dough is fully risen, place a baking stone on a shelf in the lowest slot of the oven and preheat to 450°F (232°C)—see Note. About 5 minutes before placing the buns in the oven, turn the oven down to 375°F (191°C) and carefully place a shallow pan with 1/2 inch of boiling water on the oven floor. (For an electric oven, put the pan of water on the lowest possible shelf.)

6. Brush the buns one last time with cream and place the cake pans directly on the hot stone. Bake until lightly browned, about 25 minutes. Remove from the oven.

7. Turn two stovetop burners on to very low heat. On a third, heat the honey and 4 tablespoons butter in a small saucepan. Drizzle the mixture over the buns, dividing between the two pans. Shake and let stand 1 minute for the syrup to drain to the bottom. Then heat each pan on a warm burner for about 2 minutes. Run a spatula around the edge of one pan and invert onto a serving dish with a rim. Repeat with the other pan. Cool 10 to 15 minutes. Serve warm.

Note: *If you don't have a stone:*
- Preheat the oven to 25°F (14°C) below the baking temperature, place the bread on a shelf in the lower third of the oven, 5 to 6 inches above the bottom, and then turn the oven up to baking temperature. Carefully place a shallow pan with 1/2 inch of boiling water on the oven floor. (For an electric oven, put the pan of water on the lowest possible shelf.)
- Or use the cold-oven procedure described on page 14.

Honey

Honey not only adds a pleasant flavor and sweetness to bread but keeps baked products moist as well. One of the main sugars in honey, fructose, is more hygroscopic than either table sugar (sucrose) or glucose; thus honey takes in and retains moisture from the atmosphere.

Make your cakes last longer with honey, which keeps them moist.

Because they stay moist, cakes and breads made with honey are good keepers.

Honey does, unfortunately, have a couple of drawbacks. First, it is composed of sugars that can destroy gluten just as table sugar does, so it must be used in limited amounts: 1/4 cup per 2 cups of high-protein flour is the maximum. Second, some honeys kill yeast. I found this out when I used yeast from the same large bag to make various breads in a class, and all rose nicely except my favorite honey whole wheat recipe. But why didn't this happen every time?

To find out more about honey's sometime yeast-killing tendencies, I called Elmer Cooper at the Red Star Yeast laboratory. A little digging and a day later, he reported that some honey processors use an additive to prevent mold from growing on honey. If something could prevent mold from growing, I thought, it could probably kill yeast, too.

I didn't have time to experiment to see which honeys consistently killed yeast, so I simply avoided the issue by replacing most of the honey in my recipe with brown sugar and some spices for flavor. But that wasn't the end of the story.

A couple of years later, Harold McGee, author of *On Food and Cooking,* asked what I thought of a quote from a historical cookbook that went something like this: "Thou knowest if thou addeth honey to thy bread which killeth yeast thou must addeth potash [baking soda] to leaven."

Obviously, we agreed, it was not just a commercial additive that could make honey a yeast killer.

It turns out, the literature on honey revealed, that it has been used as a pharmaceutical for hundreds of years. In times before refrigeration and good food supply, honey killed many things that grew on spoiling food. Some honeys have an excellent ability to kill bacteria; some will even kill staph. Part of honey's antibacterial ability comes from hydrogen peroxide, which is destroyed when honey is heated. However, even after heating, some honey still has very good antibacterial activity.

I had to find out why, so I contacted bee expert Dr. Thomas Rinderer, who suggested honey's hygroscopicity might kill yeast if the honey were concentrated enough. But Harold McGee noted that when sugar, also hygroscopic, is used in the same concentrations, it doesn't affect yeast. Our quest continued. . . .

. . . And took us all the way back to 401 B.C., when, Xenophon reported, soldiers on one of Cyrus's expeditions behaved like madmen after eating honey, then lay unconscious for twenty-four hours. A little later in history, three squadrons of Pompey's troops were killed while under the influence of honey fed to them by the natives.

These stories are confirmed today: Honey produced from the nectar of rhododendrons is known to be poisonous, though fortunately the intoxication is rarely fatal and generally lasts only twenty-four hours. Honey from certain plants of New Zealand is also poisonous.

Our conclusion: If poisonous honey had been gathered from plants whose nectar was poisonous, then honey from plants whose nectar possessed good antibacterial activity, as some nectars do, could have exceptional antibacterial activity.

At any rate, honey that kills yeast is the exception, not the rule. One yeast expert told me that he had not encountered a honey that killed yeast in many years in the baking field. So, after all of this research, my advice is: If you make a bread with honey that does not rise, perhaps you should use a different source of honey. Use the honey that killed the yeast to soothe a sore throat instead.

Salt

If the water you're using in a dough is soft (with few natural mineral salts), you'll want to be sure to add salt, which plays three roles in dough: It enhances flavor, controls bacteria, and strengthens dough by tightening gluten. That last quality is a mixed blessing: adding salt to the dough increases dough strength and prevents weakness and stickiness but also increases the mixing time needed to reach maximum dough development. This raises the question of when to add the salt—after kneading to reduce kneading time or before for better incorporation. The form of salt used—granular, Morton kosher, sea salt, or Diamond Crystal kosher—holds part of the answer because form influences how easily it blends. See "Different Forms of Salt," page 19, for more details.

Despite all of its good qualities, however, salt can be harmful to yeast activity. Even a small amount, such as $1/3$ teaspoon per cup of flour, slows yeast growth noticeably. If there is a reason to slow down fermentation—say, until the next shift comes on in a bakery—bakers can do just this by adding a little more salt to the dough.

This is a challenge not only because salt tightens and improves gluten but even more important because salt is a major flavor component of bread. If you have ever tasted Tuscan bread, or any bread without salt, you know how strange it tastes. A little salt makes such a difference! The very small amount of salt in bread does not provide a salty taste but does bring out the flavor of the fermentation products and the delicious components of freshly baked bread. Ernst John Pyler, in *Baking Science and Technology*, says, "Bread without salt is insipid." Whenever you get a group to taste bread samples, they favor those with the higher salt content.

Another tightrope to walk! Bread will taste better with more salt. The small amounts that we are dealing with ($1/2$ to $3/4$ teaspoon of salt per loaf of bread) are not enough to add significantly to dietary salt intake. But salt significantly slows down yeast fermentation. What can you do? If you cannot use much more than $1/3$ teaspoon of salt per cup of flour without real interference, why not put a little salt in the glaze? One of the winning ways of the Crusty French-Type Bread (page 24) is the salt in its glaze.

In breads containing cooked rice or other grains, you can cook the rice with an excess of salt to contribute to overall loaf flavor. Adding salty ingredients like cheese, olives, or sun-dried tomatoes to bread is another way to enhance flavor. The Parmesan in the Rice Bread (page 52) is a major flavor enhancer. As just mentioned, a salty glaze is an effective taste enhancer, too.

Fats: less is more

Fats, too, play several roles in dough. If the fat is added directly to the flour before any liquid, it coats gluten proteins and reduces the gluten formed or, in other words, tenderizes. The Ultimate Brioche II (page 17) is a good example of this. On the other hand, if small amounts of fat are added after mixing and just before kneading, fats increase the gas-holding ability of doughs. Up to 3 percent of flour weight, fats increase bread volume. Even the amount incorporated into the dough from a well-oiled bowl during rising can increase volume slightly.

In recipes like the two brioches and the Nisu (page 70), fats produce a moister crumb and better keeping characteristics.

So, a small amount of fat has major advantages in a loaf of bread.

> A little oil added after mixing and before kneading will give your yeast bread more volume. Even the amount incorporated from a well-oiled rising bowl will make your bread a little lighter.

Eggs

What exactly is the role of eggs in bread? In cooking, the whole egg is one thing and the white and the yolk are two different things altogether. Whole eggs leaven and also hold things together. Egg whites leaven, hold things together, and are the most incredible drying agent that cooks have at their disposal. Yolks, on the other hand, do not leaven as well as whites or whole eggs. They do hold things together, but their real contribution is their emulsifying ability—they hold fat and moisture together and produce smooth, creamy textures. In breads, emulsifiers slow staling and help breads stay soft longer.

Extra egg white is perfect in pâte à choux (page 229), for instance, because you want your cream puffs light and dry and crisp. You may have noticed how a brioche containing a lot of whole eggs dries out. Extra yolk is better in a loaf of bread that you do not want to be dry.

In addition to drying, whole eggs lighten and strengthen breads. The savarin (page 90) is an example of the perfect use of eggs in bread. Savarin is a dessert bread that you soak in a rum-sugar syrup. Most breads would fall apart if you tried to soak them, but the eggs set and give good strength to the savarin so that it does not disintegrate. They also leaven, so the savarin is very porous—almost spongelike and perfect to hold a lot of rum-sugar syrup. It does not matter in the least that eggs dry the savarin because it is soaked in syrup.

at a *glance*
Whole eggs, yolks, or whites—which to use in bread and when

Portion of Egg	Uses in Baking	Dish
Whole eggs	Lighten, strengthen, add nutrients or color	Savarin
Whites	Lighten, dry and crisp, bind	Pâte à choux
Yolks	Emulsify, bind ingredients, leaven	Brioche, custards, truffles

DOUBLE-GREASING FOR STICKY SITUATIONS

Remember how eggs stick? Savarins are no exception. If you don't have nonstick cooking spray that you can apply liberally to your pan, try this trick I learned from my French friend Christiane Le Guen.

I was in cooking school in London preparing savarins, with no nonstick cooking spray in sight. I was already in the doghouse in that class and had to think of something. I buttered the savarin ring heavily and placed it outside the window for a minute in the icy London air.

Then I oiled the pan on top of the firm, cold butter and poured in my batter. The savarins of all the other students stuck and tore apart when they tried to remove them from the pans. Mine fell right out and was gorgeous. You can use this technique whenever oil will not interfere with foam structure (as it will, for example, in a sponge cake). Grease the pan with butter, chill in the freezer, then oil. I also use this trick in soufflés that I dump out.

Savarin

MAKES 1 LARGE SAVARIN, 8 LARGE SERVINGS

This porous, spongelike bread is ideal to soak up rum-sugar syrup. A rum-filled slice served with fresh, ripe strawberries or peaches and a generous blob of sweetened whipped cream is a magnificent dessert. Ever since I made Madeleine Kamman's savarin from *The Making of a Cook* (I learned so much from that book!), I have been hooked on cream in savarins. As Madeleine says, it makes all the difference in the world.

what this recipe shows

Eggs make this bread very light and porous.

Cream and sugar slightly tenderize the dough.

The very wet batter puffs well to create a spongelike texture.

The protein network of cooked eggs makes the bread strong enough to hold up when soaked with the rum-sugar syrup.

Blending the rum with the sugar syrup gives a better taste than sprinkling the rum separately.

1 teaspoon *and* 3 tablespoons sugar
(3 tablespoons plus 1 teaspoon *total*)

1 package (2¼ teaspoons) active dry
yeast

¼ cup warm whole milk (115°F/46°C)

2 cups bread flour

½ teaspoon salt

¼ cup heavy cream at room
temperature

4 large eggs at room temperature

6 ounces (1½ sticks) butter, softened

Nonstick cooking spray

SYRUP AND GLAZE

1 cup sugar

2 cups water

Finely grated zest of 1 lemon

Finely grated zest of 1 orange

½ cup Myers's dark rum

1 jar (8 ounces) apricot preserves

FILLING

2 cups heavy or whipping cream

¼ cup confectioners' sugar

½ teaspoon Myers's dark rum

8 large strawberries for garnish

Fresh ripe fruit (strawberries,
raspberries, plums, or peaches),
peeled and sliced as necessary

1. Preheat the oven to 400°F (204°C).

2. Stir together 1 teaspoon sugar, yeast, and warm milk in the bowl of a heavy-duty mixer with the paddle blade. Let stand 2 minutes, until foam appears, indicating yeast is alive and well. Add the bread flour, 3 tablespoons sugar, salt, cream, and eggs. Mix on low-medium speed for 4 minutes to incorporate air into dough. Cover with plastic wrap and let the dough rise for 45 minutes.

3. Uncover and place back on mixer. Spoon the softened butter into the dough in tablespoon-size pieces. Mix on low until the butter is well blended in. Spray an 8-cup ring mold well with nonstick cooking spray. Make sure that the mold is well greased. If you do not have nonstick cooking spray, butter the mold well, place in the freezer for 3 minutes, then oil it. Pour the batter into the mold, cover with plastic wrap, and let rise in a warm spot for about 35 minutes.

4. About 30 minutes before the dough is fully risen, place a baking stone on a shelf about 2 inches from the floor of the oven and preheat to 450°F (232°C)—see Note. About 5 minutes before placing the loaf in the oven, turn the oven down to 375°F (191°C) and carefully place a shallow pan with ½ inch of boiling water on the oven floor. (For an electric oven, put the pan of water on the lowest possible shelf.)

5. Remove the plastic wrap and place the savarin pan directly on the hot stone. Bake until lightly browned, about 30 minutes. Remove from the oven.

6. To make the syrup, stir the sugar, water, and lemon and orange zest together in a medium bowl. When the sugar is dissolved, add rum. Let the savarin cool in the mold for several minutes. If the dough has risen above pan, trim it off level with pan. Unmold the savarin onto a round cake cooling rack. Place the savarin and rack over a bowl. Pour the rum syrup over the savarin, catching the drippings in the bowl. Spoon the drippings back over the savarin. Continue spooning syrup over savarin until it is thoroughly soaked.

7. With large spatulas, place the savarin on a pedestal cake dish. Heat the apricot preserves in a small saucepan and strain. Brush the glaze on the savarin. Coat several times, until very shiny.

continued

8. To make the filling, place a bowl, beaters, and the cream in the freezer for 5 minutes to chill well. Whip the cream in a cool room until stiff peaks form. Gently stir in the sugar and rum by hand. Pipe eight swirls of cream evenly around the top of the savarin. Make strawberry flowers by slitting strawberries into fourths. Place a strawberry flower in each whipped cream swirl. Pipe cream swirls around the base of the savarin. Put the remaining whipped cream in the center of the savarin and fill with fresh fruit. Serve cold.

Note: *If you don't have a stone:*
- Preheat the oven to 25°F (14°C) below the baking temperature, place the bread on a shelf in the lower third of the oven, 5 to 6 inches above the bottom, and then turn the oven up to the baking temperature. Carefully place a shallow pan with ½ inch of boiling water on the oven floor. (For an electric oven, put the pan of water on the lowest possible shelf.)
- Or use the cold-oven procedure described on page 14.

Flavoring ingredients

In plain bread, the major flavoring components are yeast fermentation products and acids made by bacteria. Salt accents and enhances these intriguing flavors. You can add many flavoring ingredients to doughs, depending on your individual taste—fruit, citrus zest, nuts, cheeses, vanilla, lemon, almond, any of the extracts, liquors, or liqueurs. I like fruit zest, nuts, or a little Parmesan. The saltiness of Parmesan enhances taste without interfering with the action of the yeast. It brings out the wonderful flavors from fermentation and bacteria, letting the real bread flavors predominate.

When choosing a flavoring ingredient, keep in mind that bread dough will end up some three to four times its original size. Since flavoring ingredients do not enlarge, they tend to get lost. Even if you knead in as much of an ingredient as you can get into a dough, the final baked loaf of bread may appear to have very little of the ingredient's flavor.

> Bakers who want to avoid chemical preservatives sometimes use raisin juice concentrate and raisin pulp for longer bread shelf life.

Ingredients that are intense in flavor are best, because a little must go a long way. Grated lemon or orange zest is good. For savory breads, Parmesan cheese, olives, sun-dried tomatoes, and intense-flavored herbs like rosemary or sage are winners. The Focaccia (page 72) makes use of garlic and rosemary for flavor.

Dried fruits are more intense in flavor than fresh fruits, but dried fruits can rob doughs of valuable moisture. Soak raisins or dried fruits in a warm sugar-water solution flavored with rum, brandy, or a good-quality liquor/liqueur for better flavor and moisture in doughs (page 94).

Better and better: additives for improved breads

Mold inhibitors

You may think that, as a home baker, you do not have access to the preservatives that bakers use. Maybe not, but you do have both techniques and ingredients to make your breads keep well. Doughs made with starters or doughs that stay overnight in the refrigerator have wonderful flavors and good keeping quality. Cinnamon in some concentrations helps prevent mold growth (page 70).

Bakers who make breads without preservatives sometimes use raisin juice concentrate and raisin pulp to extend the breads' shelf life. There are small amounts of propionic acid, which is a mold inhibitor, in the concentrate and pulp. For this, or some other unknown reason, raisin products do work. A simple home remedy to slow mold growth is to add cinnamon and some softened, ground raisins to your dough.

Antistaling agents

Staling is complex and involves changes in starch and probably many other changes. It is not just drying out, because a tightly wrapped piece of bread can stale.

Bread keeps best frozen or at room temperature—not in the refrigerator. The long, straight starch in bread, called *amylose,* changes to a hard crystalline form near refrigerator temperatures. Bread will actually stay softer longer at room temperature than refrigerated, but it is more likely to mold.

This change in the starch amylose is also the reason that bread softens so nicely when reheated in a regular or toaster oven (not in the microwave). With heat, the hard crystallike amylose melts back to its soft form.

We know that emulsifiers slow staling. For this reason, I frequently add an egg yolk, nature's great emulsifier, to the dough. Emulsifiers also help in joining proteins, starch, and other components in gluten. We also know that bread from doughs that have spent a night in the refrigerator do not stale as fast as others. Acidic doughs stale more slowly.

Actually, there are still many unanswered questions about staling. The famous starch expert at Kansas State University, Dr. Carl Hoseney, points out that the changing form of the starch is only part of the answer to why bread hardens upon standing. Cooks know that many breads are good keepers, such as those containing honey. Rice bread with its moist, cooked grain also keeps well.

Cooks also know of the amazing shelf life of breads like panettone, breads made with starters, and breads prepared with cold slow rises. There is something in these processes that delays the staling. It is not actually known whether it is the acids produced by bacteria in cold rises, the strain of yeast used, or some other factor that slows staling.

Homemade dough improvers—a combination of effects

Dough developers or dough conditioners used by commercial bakers and millers may contain vitamin C (ascorbic acid), inorganic salts, some form of sugar, and possibly potassium bromate for better gluten cross-linking. One of the dough developers that I have seen frequently in French bakeries in the United States is from Belgium and contains dextrose, vitamin C, and other oxidizing agents.

You can simulate a professional baker's dough improvers for yeast breads in your own home baking by proofing yeast with a little sugar and adding a crushed vitamin C tablet just before kneading.

For even more comprehensive dough improvement, add vitamin C for better gluten development; an egg yolk for antistaling and better gluten development; $1/4$ teaspoon ground fenugreek, rosemary, or cinnamon for enhanced yeast activity and better keeping qualities; 1 tablespoon sugar to feed yeast; and 1 teaspoon barley malt syrup, which contains enzymes to convert flour to yeast food.

A newspaper wire service writer asked me for an Easter bread, preferably one in which she could include a minilesson about an interesting point in breadmaking. I am fascinated by the fact that leaving dough in the refrigerator overnight improves keeping as well as flavor, so I designed these Hot Cross Buns (page 94). To my amazement, when I wrapped a bun in plastic and kept it at room temperature, the bread was still in good shape after more than two weeks!

Hot Cross Buns

MAKES 12 BUNS

Because these buns contain a fair amount of sugar and protein, they bake up dark brown and are dramatic with drizzles of white icing. These strongly spiced and orange zest–flavored buns are a real treat served with Orange Butter (page 96).

what this recipe shows

Soaking dried raisins adds moisture and flavor to the loaf as well as inhibiting mold.

Barley malt syrup adds enzymes to supply sugars for better yeast growth.

An egg yolk adds flavor and acts as an emulsifier to improve keeping quality.

Cinnamon enhances yeast activity and cinnamon and raisins improve keeping quality.

Evaporated milk adds flavor, nutrients, and moisture, enhances crust color, and supplies lactose, which yeast doesn't like, leaving more residual sugar and a sweeter loaf.

$^1/_2$ cup sugar

1 cup boiling water

$^1/_2$ cup raisins

3 tablespoons good-quality dark rum, such as Myers's

$1^1/_2$ teaspoons active dry yeast

1 teaspoon barley malt syrup (see Notes) or unsulfured molasses

$^1/_2$ cup warm water (115°F/46°C)

$1^1/_2$ cups *and* 1 cup *and* 2 tablespoons bread flour (a scant $2^3/_4$ cups *total*)

$^1/_4$ cup dark brown sugar

Finely grated zest of 2 oranges

2 teaspoons Spice Mixture (page 96)

1 small can (5 ounces) evaporated milk

1 teaspoon salt

1 large egg yolk

4 tablespoons ($^1/_2$ stick) butter, softened

1 tablespoon oil for bowl

$^1/_2$ cup finely chopped candied fruit

2 tablespoons finely minced candied ginger

Nonstick cooking spray

1 large egg, beaten

ICING

$^1/_2$ cup confectioners' sugar

$^1/_2$ teaspoon dark rum

$1^1/_2$ to 2 teaspoons heavy or whipping cream

Read "The Bread Recipes in This Book" (page 10) before beginning.

1. Stir the sugar into boiling water in a small bowl. Add the raisins, then stir in rum. Cover loosely with plastic wrap and let stand at least an hour or overnight.

2. Stir together the yeast, barley malt syrup, and warm water in the bowl of a heavy-duty mixer with the paddle blade. Let stand 2 minutes, until foam appears, indicating yeast is alive and well. Add $1^1/_2$ cups bread flour, the brown sugar, orange zest, Spice Mixture, and evaporated milk. Beat on low-medium speed for 4 minutes to beat air into the dough. Let the sponge sit for 30 minutes to 2½ hours for improved flavor and texture.

3. Remove the paddle blade and insert the dough hook in the mixer. Add 1 cup bread flour, salt, and the egg yolk. Knead on low-medium speed for 5 minutes. Pull the dough into four or five pieces. Spread butter over the dough and knead for another 1 to 2 minutes to incorporate butter. Place the dough in an oiled bowl and turn to coat well with oil. Cover with plastic wrap and refrigerate overnight.

4. When ready to shape and bake, let the dough warm at room temperature for about 30 minutes, then place the dough on the counter and, using both hands with a gentle cupping and tucking action, shape into a smooth round. Now grab the sides of the round and stretch sideways into an oval. Let it spring back slightly, then pull it out again. Cover with plastic wrap and leave on the counter for 15 minutes. The dough is now ready to shape.

5. Drain the raisins. Toss the candied fruit, candied ginger, and raisins with 2 tablespoons flour in a small mixing bowl. Roll the dough into a rectangle approximately 18 inches wide and 10 to 12 inches from top to bottom, using a rolling pin on a clean counter. Sprinkle two thirds of the candied fruit mixture over the left two-thirds of the dough. Roll the rolling pin across the fruit to press it partially into the dough. Fold the right third of the dough to the center, sprinkle with the remaining candied fruit, then fold the left third of the dough over it. Press the three layers together with your fingertips. Flatten slightly with the rolling pin. Divide the dough into twelve equal pieces. With a cupping, pulling, and tucking action, shape each into a roll.

6. Spray two 7 × 11 × 1-inch baking pans with nonstick cooking spray. Place six buns in each. Brush with beaten egg and let rise until doubled, about an hour.

7. About 30 minutes before the buns are fully risen, place a baking stone on a shelf in the lowest slot of the oven and preheat to 450°F (232°C)—see Notes. About 5 minutes before placing the buns in the oven, turn the oven down to 375°F (191°C) and carefully place a small baking pan with ½ inch of boiling water on the oven floor. (For an electric oven, put the pan of water on the lowest possible shelf.)

8. Brush the buns again with egg and place the baking pans directly on the hot stone. Bake until deeply browned, about 18 minutes. Because of the high protein and sugar content, these buns get quite brown. Remove from the pans and place on a rack to cool.

9. When the buns have cooled for about 20 minutes but are still slightly warm, stir the confectioners' sugar, rum, and cream together in a small bowl. Spoon an icing cross on top of each bun. Serve at room temperature.

Notes: *Barley malt syrup is available at health food stores and beer- and winemaking supply shops.*
   If you don't have a stone:
   · Preheat the oven to 25°F (14°C) below the baking temperature, place the bread on a shelf in the lower third of the oven, 5 to 6 inches above the bottom, and then turn the oven up to the baking temperature. Carefully place a shallow pan with ½ inch of boiling water on the oven floor. (For an electric oven, put the pan of water on the lowest possible shelf.)
   · Or use the cold-oven procedure described on page 14.

Spice Mixture

MAKES ABOUT 3½ TEASPOONS

2 teaspoons ground cinnamon

½ teaspoon freshly grated nutmeg

½ teaspoon ground allspice

⅛ teaspoon ground cloves

¼ teaspoon ground ginger

¼ teaspoon ground cumin

Stir all of the spices together in a small bowl. Any mixture not used can be kept tightly sealed in a small jar or zip-top plastic bag.

Orange Butter

MAKES ABOUT 1½ CUPS

¼ pound (1 stick) butter, softened

2 tablespoons confectioners' sugar

Finely grated zest of 1 orange

2 tablespoons orange marmalade

1 tablespoon Grand Marnier or other orange liqueur

Process the butter, sugar, and orange zest in a food processor with the steel knife. Spoon into a medium mixing bowl and stir in orange marmalade and Grand Marnier. Keeps well in a tightly sealed jar in the refrigerator.

at a glance

The roles of ingredients in bread

Ingredient	Role in Bread	See Page
White flour (from wheat)	Contains gluten-forming proteins that make gluten and allow a bread to rise.	4
Gluten flour	Very high percentage of gluten proteins lightens whole wheat and multigrain breads.	46
Whole wheat and multigrain flours	Contain few or no gluten-forming proteins and make heavier breads, but contain good nutrition ingredients and fiber.	46, 54
Leaveners		
Bakers' yeast	Leavens and imparts flavor.	58
Natural yeast	Leavens, imparts excellent flavors and good keeping characteristics.	60
Baking soda and baking powder	Leaven and impart flavor.	73

Ingredient	Role in Bread	See Page
Liquids	Combine with glutenin and gliadin to form gluten. Turn to steam in the hot oven, make doughs rise more, and produce lighter breads.	77
Sweeteners (sugar, syrups, honeys)	Provide food for yeast, flavor, tenderize, make crumb moister, make crust darker, but also make bread heavier.	
Sugar	All listed under "Sweeteners."	80
Syrups	Malt and barley malt syrup do all listed under "Sweeteners" and provide enzymes to break starch into sugar for yeast action.	81
Honey	All listed under "Sweeteners," keep baked goods moist, but a few honeys kill yeast.	87
Salt	Tightens gluten, enhances flavor, slows yeast activity, controls bacterial growth.	88
Dairy products	Add protein, tenderize, enhance crust color.	79
Eggs		
Whole	Lighten, strengthen, but dry.	89
Whites	Dry, lighten, and strengthen.	89
Yolks	Strengthen, moisten, and provide emulsifiers.	89
Flavoring ingredients	Enhance flavor.	
Spices that enhance yeast: ginger, nutmeg, cinnamon, thyme, allspice, cardamom, ground caraway	Enhance flavor and in small amounts improve yeast activity.	69
Spices that are antioxidants: rosemary, fenugreek	Enhance flavor and improve keeping quality.	70
Spices that deter yeast: dry mustard, salt	Enhance flavor but slow yeast activity.	69
Dough improvers	Improve gluten formation, sometimes feed yeast.	
Bleaches	Convert color pigments to colorless forms.	56
Chlorination	Converts color pigments to colorless forms, makes dough or batter more acidic, which gives cakes a better texture, distributes fat more evenly in cake batters (produces better texture), and enhances water absorption by starch.	56
Oxidizers	Improve gluten formation, improve dough elasticity.	57
Bromination	Improves gluten formation, improves dough elasticity.	57

Problem solving

How to tell why the bread did not rise

There are two reasons why bread does not rise: One is that the yeast is dead, and the other is that you do not have good gluten development. This is about all that can go wrong with a yeast dough. If the yeast is producing carbon dioxide and the satiny sheets of bubble gum–like gluten are there to catch and hold the carbon dioxide, the bread is going to rise.

You can tell instantly whether you have a yeast problem or a gluten problem. Poke at the uncooked dough and try to pull a piece off. If the dough is elastic and springy, your gluten is fine, but the yeast is dead. Dissolve another package of yeast (this time hopefully alive!) in a few tablespoons of warm water and knead the live yeast into the dough. It should rise nicely now.

If the dough appears to have fine bubbles in it, tears easily when you pull at it, and has little elasticity, you have a gluten problem. You can add bread flour or gluten flour and more liquid to moisten the additional flour. Then the dough must be kneaded again.

How to spot bad yeast bread recipes

Unfortunately, there are a lot of bad yeast dough recipes out there—even in books by famous food writers. For example, there is a rye bread recipe by a prominent author that starts out with 2 packages of yeast. There is nothing wrong with that; this must be a big batch of bread. Then you get to the flour—3 cups rye flour. Uh-oh!

One package (2¼ teaspoons) of yeast will leaven 6 to 8 cups of flour slowly and 2 to 3 cups fast. When you see that there are 2 packages of yeast and 3 cups of flour, you know that the recipe has big problems. If a recipe does not work and the bread does not rise, most people think that all you have to do is to add more yeast—but that is not the problem here.

As you now know, rye flour does not have enough glutenin and gliadin to form elastic sheets of gluten. Unless the rye is blended with a high-protein wheat flour, the bread is not going to rise. You can put 10 packages of yeast in that dough, and the yeast can produce great quantities of carbon dioxide, but if there is no gluten to hold it, the carbon dioxide will drift away in the air. You will get a good understanding of how important gluten is when you try to make a loaf of bread with 100 percent rye flour. The author of the recipe does say that this is a "hearty bread"! You can add ⅓ cup gluten flour and a little liquid to the recipe to get a risen bread.

One of my students in New Orleans brought in a recipe from a newspaper; both times she had tried the recipe it had failed. Again, the recipe called for 2 packages of yeast and 3 cups of flour. Whenever you see a recipe like that, you know that there are problems.

This was a recipe for King's Bread, a traditional bread at Mardi Gras. It is shaped like a wreath, covered with coarse sugar, and has a crown or a doll hidden in the loaf. Whoever gets the crown becomes king or queen. This recipe also called for 1 full cup of sugar. No wonder the bread would not rise! With that much sugar in a loaf containing 3 cups of flour, very little gluten was formed. The dough had no air-holding ability at all. My student tried the recipe with 3 tablespoons of sugar instead of 1 cup, and it worked beautifully.

at a *glance*
Yeast bread problems

Problem	Probable Causes
Bread does not rise	Dead yeast, or poor gluten formation
Poor volume and dense	Low-protein flour
Dense cakelike texture	Inadequate rising time Fat added at wrong time Too much sugar Too much salt Oven or stone too hot
Too chewy—tough	Too much gluten—from high-protein flour Not enough fat and/or fat added at wrong time
Lack of flavor or off flavors	No salt or too little salt Too much yeast Rising too warm and too fast
Bubble under crust	No slash or vent Oven preheated too hot Bread too near top of oven
Crust too pale	Not enough protein and sugar in the dough
Crust too dark	Too much protein and sugar in the dough
Poor keeping characteristics	Not enough sugar or honey Rising too warm and too fast

Storing bread

Freezing is an excellent way to store bread. In a small household, where you do not consume large quantities of bread daily, it is a good solution. Bread can be warmed briefly in a toaster oven just to defrost and soften. I do not like to microwave bread because of the drying and toughening. Bread is porous and heats in a very few seconds in the microwave. Unfortunately, most people overheat it. It is fine to eat instantly out of the microwave, but the moment it cools it becomes very dry and tough.

To freeze bread, seal it tightly in freezer plastic wrap, bags, or foil. Not all plastic is impermeable, and freezer odors can penetrate many plastics. Make sure that you use a freezer-type plastic.

Uncooked dough can be frozen, too. There is some loss of subsequent yeast activity in freezing, but not enough to do much harm to a well-made loaf. If you know you are going to freeze unbaked dough, you can add a little extra yeast when you are making it. The ideal time to freeze dough is after it has risen once, been punched down, and shaped. If you are baking in a pan, you can freeze the dough in the same pan that you plan to bake it in. After it is frozen, remove it from the pan, wrap well, and return to the freezer. When you are ready to let this dough have its final rise, unwrap it, place it back in the pan in which you froze it, and let it rise. This rise will take at least 2 to 3 hours longer than usual because the bread must both defrost and rise as it warms.

Make sure you freeze breads in wrap and bags made specifically for the freezer to prevent penetration of freezer odors. Not all plastic is impermeable.

the recipes

how rich it is!

The amazing roles of fats

fats are a naturally occurring component of all foods. Even watermelon has fat. Chemically speaking, fats are compounds composed of three connecting carbon atoms with a fatty acid hooked onto each carbon. But to cooks, fats are far more.

Fats are vital ingredients responsible for the success or failure of many dishes. Fats leaven—bubbles in the fat help cakes rise. Fats tenderize—they coat flour proteins and prevent their joining to form gluten. Fats act as a spacer to hold thin layers of dough apart for flaky pastry. Fats influence the amount that cookies spread. Fats thicken sauces. Fats can be heated to high temperatures, providing immediate and completely surrounding high heat for fast cooking. Fats not only carry flavors but coat and hold flavors in the mouth for a different taste impact than water-soluble flavors. Fats are an essential nutrient.

As never before, we are aware of the type and amount of fat in foods and of the role of fats in health. The man on the street knows his cholesterol level. He may even know his HDL and LDL levels.

The many forms of fat

Numerous fats are available for use in cooking. The challenge is to decide which is best for the job at hand. Fats at room temperature may be solids like shortening, lard, and coconut oil or liquids like corn, soy, or canola oil. In some instances, a solid fat creates a better dish because solid fat traps air bubbles and aids in aeration. In other instances, liquid fat is preferable because liquid coats flour proteins and acts as a tenderizer.

Some solid fats, like butter, have a sharp melting point. They are firm and solid at one temperature and then, at a slightly higher temperature, they become soft and melt. Cookies made with a fat like this spread in a hot oven. Other fats, like shortening, stay the same consistency over a wide range of temperatures. Cookies made with shortening do not spread so much.

In selecting a fat, deciding between a solid and a liquid fat is not the cook's only concern. Some processed fats have features for specific uses, features that may improve a dish in one cooking use but be detrimental in another. For example, you may have read *mono- and diglycerides* in the ingredient list on shortening. Mono- and diglycerides, which are emulsifiers, help produce moister cakes, but they also lower the smoke point and make the fat less desirable for high-temperature frying. Silicone in oil keeps the oil from becoming contaminated by oxygen in the air, cuts the risk of fire, and reduces foaming in frying, but in baking silicone can damage the foam structure of a cake.

This chapter details fat's many roles in pie crusts, in cookies, in cakes, in frying, and in nutrition. When you know what fat does in a dish, you can better select the right fat or even a fat substitute to do the job.

Pie crusts

I was taught to make tender, flaky pastry as if it were one word, *tenderflaky.* Nothing could be further from the truth—tenderness is one characteristic, flakiness a totally different characteristic. The techniques and ingredients needed for tenderness are completely different from those needed for flakiness. The recipe, the ingredients, and the techniques all influence the kind of pie crust that you make.

Basic Formula

The basic formula for a pie crust is one to three, one part fat to three parts flour (by volume) with a little salt and liquid. When you vary too far from this ratio, you get an unmanageable crust. Many crust recipes call for 1 1/2 cups flour and 1/2 cup (8 tablespoons) fat. For a slightly richer and tenderer crust, go heavier on the fat, 9 to 10 tablespoons for lard or shortening, 10 to 11 for butter. With butter, which is only 81 percent fat, you need about 10 tablespoons to have 8 tablespoons of fat. The recipe that I start with is:

1 cup national-brand bleached all-purpose flour (see Note)

1/2 cup instant flour (such as Wondra or Shake & Blend)

1/4 teaspoon salt

8 tablespoons unsalted butter, cut into 1/2-inch cubes

2 tablespoons shortening or lard

The amount of liquid depends on how many gluten-forming proteins in the flour are left uncoated with fat after the fat is worked into the flour. The type of flour, the technique used to work the fat into the flour, and the type of liquid will all influence this.

This recipe is just a good ballpark starting point. To meet specific needs and conditions you may have to make numerous changes. In "Pie Crusts in the Real World," page 112, I go through the step-by-step evolution of a pie crust recipe for a small bakery.

Note: *If pastry flour or a low-protein flour like White Lily is available, use $1^1/_2$ cups in place of the bleached all-purpose and the instant flour.*

Ingredients

Flour

The perfect tenderness of a pie crust is a delicate matter. Very little gluten is needed to hold the crust together, and more gluten than this makes the crust tough. One way to minimize the amount of gluten-forming proteins is to use a flour that has less protein in the first place. As explained in Chapter 1, pastry flour and a few Southern all-purpose flours are low in protein (8 to 9 grams of protein per cup). These low-protein flours produce much less gluten and much more tender pie crusts, biscuits, and cakes.

Commercial buyers have access to pastry flour. In many locations these low-protein flours are available to home cooks only in health food stores or by mail order (see Sources—Flour, page 478). To get a lower-protein mixture, combine two parts national-brand bleached all-purpose flour and one part cake flour or Gold Medal instant flour (Wondra) or Pillsbury's Shake & Blend flour. Wondra and Shake & Blend flours are available nearly everywhere in cylinders like salt and in some locations in 5-pound bags or 2-pound boxes. These are low-protein, pregelatinized wheat flour with malted barley flour. Some bakery pastry chefs use 100 percent Wondra or Shake & Blend for pie crusts. Cake flour is slightly acidic, has a very fine texture, and has an advantage for some applications. It does, however, have a definite taste.

Fat

Fats have two major roles in pie crusts. They tenderize by coating flour proteins, which prevents their joining to form gluten, and they create flakiness by serving as a spacer between layers of dough. Pieces of cold, firm fat keep layers of dough apart in the hot oven just long enough for the dough to begin to set. The fat melts, then steam from the dough forms and puffs the layers apart. Different types of fat—butter, margarine, shortening, lard, and oil—have different tastes, different protein-coating abilities, and different abilities to serve as a spacer. Here are some characteristics and how they affect fat's roles in a pie crust.

Fat content of a fat Solid fats that are 100 percent fat, like shortening and lard, make flakier pastry than butter or margarine, which are only about 80 percent fat. The same amount of butter or margarine will produce a less tender and/or less flaky crust than other fats under the same conditions. Spreads can vary all over the place in the amount of fat and water that they contain. Some of the low-fat spreads are 50 percent water or even more.

Texture, temperature, and melting characteristics At the same temperature, butter, shortening, lard, and oil can have very different textures. At, say, 50°F (10°C), which is the temperature shortly after removal from the refrigerator, butter is quite firm, shortening and lard are soft and pliable, and oil is a liquid. So, at the same temperature, they have very different protein-coating abilities and different ease of use. Using the same techniques, oil and softer fats will coat many more proteins and make a more tender crust.

If you have ever made one pie crust with butter and another with shortening, you know how much easier it is to work with the shortening crust. When the butter crust is cold, the butter is hard and the crust is difficult to roll. If you let the crust get too warm, the butter gets soft and may soak into the dough. You lose all the flakiness. Shortening, on the other hand, is designed to remain about the same consistency over a wide range of temperatures. There is not as much risk that shortening will get so hot that it will soak into the dough, and it is considerably easier to handle.

The texture and the melting characteristics of the fat are very important in creating flakiness—a fat must remain solid in the hot oven long enough for the dough on either side of it to begin to set. Most of the fats that are solid at room temperature can do this. Butter must be very cold going into the oven to hold up long enough. Fats that hold their shape over a wide range of temperatures, such as shortening, lard, and some margarines, hold up well and create flakiness more easily.

Structure of the fat The size and type of fat crystals can influence the kind of job that a fat does in cooking. The structure—the type of crystals—of some types of lard such as leaf lard, which comes from the fat around the hog's kidneys, produces the flakiest pastry of all.

I do not share the common paranoia concerning animal fat. Certain saturated fats and trans fatty acids (in hydrogenated fats like margarines) raise everybody's serum cholesterol. A moderate amount of dietary cholesterol does not elevate serum cholesterol in most people. I feel that small amounts of chicken fat and lard, which contain large amounts of mono- and polyunsaturated fats, are better for my serum cholesterol level than butter or many margarines. (See pages 178–180 for details.)

The problem I have found with lard is that, with the exception of a few brands that are consistently good, the products in the grocery stores may be off-tasting. Lard is 56 percent mono- and polyunsaturated fat, which reacts more readily than saturated fat and can change taste more easily. Since there is such limited use of lard in the United States, the shelf turnover is probably slow.

Taste of fat Nothing tastes like real butter, and it is frequently the fat of choice. It is, however, difficult to handle, does not make as flaky or as tender a crust with the same amounts and conditions as 100 percent fats, and is probably nutritionally less desirable than many other animal fats. Lard, which produces marvelously flaky pastry and is very easy to work with, unfortunately may have off flavors.

Sugar

Sugar wrecks gluten formation (page 80), which makes it undesirable in a yeast dough but ideal for tenderizing a pie crust.

The three types of basic French pastry doughs are pâte brisée, pâte sucrée (sugar pastry), and pâte sablée. This classification used to make no sense to me. Pâte brisée is a regular crust with little or no sugar. Pâte sucrée has some sugar—several tablespoons—but pâte sablée has much more sugar. So why didn't they call pâte sablée *sugar pastry*, since it is the one with the most sugar? *Sablée* means "sandy," and indeed it is crumbly. Sometimes pâte sablée crusts do not even have enough gluten to roll them out. The crust has to be sprinkled in the pan and then pressed in with the fingers. This is exactly what a large amount of sugar will do: prevent gluten formation to the point that the crust is crumbly, like sand.

A crust can be made crumbly (almost totally lacking in gluten) in a number of ways, not just through high sugar content. For example, the crust can contain such a high percentage of fat and so little water that little or no gluten is formed. Such a crumbly crust is desirable when you want an easy-to-cut crust with a filling that is quite firm. The Savory Herbed Cheesecake (page 225), with a very tender press-in crust, is an example.

Acidic ingredients

Many pastry recipes contain lemon juice, vinegar, buttermilk, sour cream, or some other acidic ingredient. When you marinate meat, you soak it in wine or a vinaigrette—something acidic. In both cases, acids tenderize by breaking down long, stringy protein molecules into smaller pieces. Gluten is protein, so acidic ingredients are ideal for tender crusts. In Flaky Crisp Crust (page 108) and in other crusts, sour cream is an excellent tenderizer.

We have a good old Southern chef in Atlanta who makes great pie crust using a tablespoon of frozen orange juice concentrate. It is partially frozen, so it helps chill the crust for flakiness. It is acidic, so it tenderizes. It is sweet, so it tenderizes. It adds a nice hint of flavor and just a little color. What a good idea!

Liquids

The two gluten-forming proteins in wheat flour *must* have water to join together and form gluten. An easy way to prevent their joining is to cut their water supply. You may have read an Italian pastry recipe that called for a tablespoon or two of oil or a pastry recipe that called for an egg yolk (31 percent fat) as part of the liquid. Many pastry recipes use cream, sour cream, or cream cheese. In all of these cases, some of the water in the recipe is replaced with liquid fat—a good idea for minimizing gluten.

Cream and sour cream are excellent examples of liquids that are part fat for making pastry tender. Just think about all the nice things that sour cream does. It is cold, so it maintains flakiness. It is acidic, so it tenderizes. It contains sugar (lactose), so it tenderizes. It is part fat, so it tenderizes by replacing some of the water. It also adds flavor, and the additional protein and sugar of dairy products produce a browner crust.

Cream cheese is another favorite ingredient. Regular (not low-fat) cream cheese has a high fat content, so you are adding a soft, cold, malleable, part-fat ingredient instead of water. Like cream and sour cream, cream cheese makes tender, easy-to-work crusts that have good color.

Ingredients for more or less color

Many times you want more color in a crust. A pale crust looks anemic and sad. A rich-looking, nicely browned flaky crust is much more enticing. There are also times when you need less color. Prebaking the empty crust makes it crisper, but with a pie that requires a long time to set, like a high-sugar pecan pie, the crust can get too brown or burn on the edges.

Color is controlled primarily by three ingredients: the amount of protein, the amount and type of sugar, and the acidity. I like to think of browning as low-temperature caramelization. Sugar does not caramelize below 325°F (163°C). To get the wonderfully sweet flavors of caramel and rich brown color at lower temperatures, you must have protein and sugar and a nonacidic or mildly acidic environment. The more protein and sugar, the browner the crust will be. A crust made with milk, which contains both protein and sugar, will get much browner in the same baking time than a crust made with water. Some sugars, like glucose (corn syrup), brown at lower temperatures than others. Even a teaspoon of corn syrup will make a noticeably browner crust.

Acidic ingredients tenderize, but they make the crust paler. If you decide that you can tenderize in another way, you can simply add a pinch ($1/8$ teaspoon) of baking soda to neutralize the acidity and make the crust browner. Many cookie recipes have soda in them not for leavening but for color.

What can you do in the case of the pecan pie, where you want to avoid browning? You will have to limit the sugar and protein and increase the acidity of the crust. Using part cake flour is a help since it is both low in protein and acidic. Butter, which contains about 20 percent water and dairy products, browns better than shortening or lard. Making the fat part butter and part shortening or lard will help. I used both cake flour and shortening to make the Pie Crust for Longer Cooking Times for the Bourbon Pecan Pie (page 123).

at a *glance*
Pie crust ingredients

What to Do	Why
For tenderness:	
Use low-protein flour.	The flour contains less gluten-forming proteins.
In sweet pies, or other appropriate pies, add sugar.	Gluten-forming proteins bond with sugar instead of each other.
Use some acidic ingredients.	Acid breaks down gluten proteins.
Use some fat in liquid form to moisten, like oil or egg yolk.	Gluten-forming proteins cannot join together without water.
Instead of using water, use an ingredient that is part fat, like sour cream or cream cheese.	Gluten-forming proteins cannot join without water.
For flakiness:	
Use fats that are solid at room temperature, like lard or shortening.	These fats will remain firm in the oven long enough for the dough to set.
Use cold ingredients.	Cold ingredients keep fat firm longer.
Roll large ($1/2$-inch-square) nearly frozen butter lumps with flour to flatten.	Large flat pieces of fat create flakes.
For more color:	
Use ingredients that contain protein or sugar, like higher-protein flour, dairy products, or sugars.	Sugar and protein are needed for the browning reaction.
Use corn syrup.	Glucose (corn syrup) browns at lower temperatures than other sugars.
Add a little baking soda (alkali).	Baking soda neutralizes acidity and enhances browning.
For less color:	
Avoid proteins and sugars (use lower-protein flour).	Proteins and sugars are needed for the browning reaction.
Use part cake flour.	It is acidic and low-protein.
Use acidic ingredients like vinegar.	These ingredients increase acidity and limit browning.

Pie crust techniques

Much depends on technique. Your perfect pie crust is very much up to you. For some, perfection is layers of delicate flakes. For others, it is an almost disintegrating tenderness. What is important is that you know how to achieve the crust that you want. The nature of the crust depends on the incorporation of the fat with the flour.

The goal for flakiness is to achieve firm, cold, flat pieces of fat coated with flour. Experienced pastry chefs do this just by flattening cold lumps of fat in a bowl of flour with their fingertips. Some cooks like to slice lumps of fat in a bowl of flour with a pastry cutter or two knives. Others prefer to let the blades of a beater cut cold lumps of fat into pieces in flour. For me, the best way is to dump cold flour and butter lumps out on a clean counter and roll with a rolling pin, scrape together, and reroll over the butter several times. This gives me wonderfully flat, flour-coated flakes of cold fat.

The goal for tenderness is to coat or grease flour proteins so that they cannot join together with water to form tough gluten. The softer the fat and the more it is worked into the flour, the better the flour proteins are greased. A cook may rub the fat and flour together or use a pastry cutter or knives to cut the pieces very fine or a mixer or food processor to incorporate the fat thoroughly into the flour.

Getting perfect flakiness

What is flaky? Flaky pastry is made up of delicately crisp, thin layers, as in puff pastry, croissants, and Danish pastry. Layers of cold, firm fat serve as spacers that keep the thin layers of dough apart in the hot oven just long enough for the dough to begin to set. The fat melts, and steam from the liquid in the dough forms and puffs the layers apart beautifully.

The fat must remain unmelted during the whole folding operation. If the butter gets warm and starts to melt and soak into the dough, there goes the flakiness! For flakiness, you must have cold, firm pieces of fat, big enough not to melt immediately—($\frac{1}{3}$- to $\frac{1}{2}$-inch cubes). Recipes that say to cut the fat into the flour until it resembles coarse meal will not make flaky pastry. The fine pieces of fat melt too fast.

The pieces of cold fat must not only be large but also flattened out. Large lumps that are not flattened can be so big that they go all the way through the dough. When the fat melts, you end up with a hole in the crust.

The butter can be flattened with the flour dry or wet. Jim Dodge, pastry chef and author of *Baking with Jim Dodge,* flattens the butter in a dry mixture first, then does a turn (folds and layers the dough) on the finished pie crust dough. To do this, place a bowl containing the flour and butter lumps in the freezer for 10 minutes, then dump the mixture out on the counter and roll over it with a large rolling pin. Some butter will stick to the pin. Scrape it off, scrape the mixture back into a pile, and roll over it again. Rapidly repeat this rolling and scraping at least six times, until big flat flakes of butter are coated with the flour. The mixture should look like paint flaking off a wall. If at any time during the rolling the butter starts to melt, scrape the mixture back into the bowl and place in the freezer for 5 minutes.

Flattening the butter and coating it with dry flour like this produce a more tender crust than folding and rolling out a finished dough containing liquid. When you flatten the butter after the liquid is mixed in, every time you work the dough by rolling it out, flour proteins join together to form gluten and make the crust tougher. This is why Jim Dodge's method produces excellent pastry. When making hand-held pastries or pastries that need to hold together, you want more gluten. You may want to do a few turns (folding and reflattening) on the finished dough. For a more tender crust, do all the flattening with the dry mixture and incorporate one or more of the ingredients that aid tenderness.

You can use both methods. Jim Dodge does a fold or so on his flaky pastry doughs even after flattening the fat in the dry flour.

> Recipes that tell you to cut fat into flour until the fat resembles coarse meal will not make flaky pastry.

The Flaky Crisp Crust recipe that follows includes both the dry flattening technique and a technique for excellent crispness—coating the bottom of the crust with fine dark cracker crumbs. The Simple Flaky Crust (page 110) is a simplified, quick recipe with the dry flattening technique. It is still very tender and a great basic crust.

Flaky Crisp Crust

MAKES TWO SINGLE OR ONE DOUBLE 9-INCH CRUST

This is an easy-to-roll, marvelously flaky crisp crust. Years ago Judy Brady, an outstanding cook and restaurateur in Fort Walton Beach, Florida, introduced me to the joy of sour cream in pie crusts. It has been one of my favorite crust ingredients ever since. This crust is quite tender, too. The simplified version on page 110 is my favorite crust.

what this recipe shows

Low-protein instant flour mixed with all-purpose flour makes a more tender crust.

Rolling near-frozen 1/2-inch cubes of butter with flour flattens and coats the butter for a very flaky crust.

Lard enhances flakiness.

Sour cream, which is acidic, also contains milk sugars and fat to make a more tender crust.

Chilling the mixed dough allows the moisture to distribute evenly.

A dark crumb coating on the bottom of the crust speeds up baking and dries for crispness.

Chilling the shaped crust in the freezer firms the fat for flakiness and holds the crust shape.

Baking blind (empty—without the filling) produces a crisp crust.

---

1 3/4 cups bleached all-purpose flour (see Note)

1/2 cup Wondra or Shake & Blend flour

1/2 teaspoon salt

12 tablespoons (1 1/2 sticks) very cold unsalted butter in 1/2-inch cubes

2 tablespoons very cold lard or shortening in tablespoon-size pieces

1 carton (8 ounces) sour cream

1 to 2 tablespoons cold whole milk if needed

4 double graham crackers, crushed to fine crumbs

1 large egg, beaten

1. Stir together both flours, the salt, and the butter lumps in a medium mixing bowl and place in the freezer for 10 minutes.

2. When the flour-butter mixture is cold, dump it out onto a clean counter. Roll over the mixture with a rolling pin to flatten the butter and coat it with flour. Some butter will stick to the pin. Scrape it off and scrape the mixture together. Roll over the mixture again. Rapidly continue rolling and scraping together three times. Scrape back into the bowl and place in the freezer for 5 minutes. Dump back out, roll and scrape together one more time. The mixture should look like paint flakes that have fallen off a wall. Add the pieces of lard to the mixture and roll in, scraping and rerolling twice. Scrape the mixture back in the bowl and return to the freezer for 10 minutes. If the butter becomes very soft at any time during the rolling, immediately return the mixture to the freezer for 5 minutes.

3. Gently stir the sour cream into the flour-fat mixture. Add milk only if needed to get the mixture wet enough to hold together. Pull the dough together into a round, wrap well in plastic wrap, and refrigerate for 30 minutes to several hours or overnight.

4. Preheat the oven to 400°F (204°C).

5. Divide the dough in half. Shape each half into a 6- to 8-inch disk about ³⁄₄ inch thick. For a one-crust pie, freeze or refrigerate one disk for later use. Lightly flour the counter, place the other disk on the counter. Sprinkle lightly with flour and roll out evenly, placing the rolling pin in the center of the dough disk, rolling forward and back (taking care not to roll off the dough and thin the edges), rotating the dough 45 degrees, and rolling again. Keep a little flour on the counter to one side. If the dough tends to stick when rotating, drag it through the flour. When the dough is nearing the desired thickness, place spacers (page 119) or rulers on either side of the dough, rest the rolling pin on the two spacers or rulers, and roll it across the dough.

6. Spread the crumbs evenly over a counter area as large as the crust. Lightly flour the crust top, brushing off any excess. Fold the dough in half, then in half again. Unfold the dough on top of the crumbs and lightly roll over the top with the rolling pin twice to press a light crumb coating into the bottom side of the crust. Refold the dough as before, place in the pie pan, and unfold. Trim and shape the edges as desired (see "Shaping the Edges," page 119).

7. Bake blind right side up as described on page 119.

Note: *If pastry flour or a low-protein flour like White Lily is available, use 2¹⁄₄ cups in place of the bleached all-purpose and the instant flour.*

Variations:
- *Flakier*—After Step 3, roll the dough out in a large rectangle (18 × 12 inches). Fold a quarter of the dough along the long side to the center, then fold a quarter of the dough on the other end to the center. Now fold one double layer over the other to form four layers. Roll over the layers to press together and proceed to Step 4.
- *Cream Crust*—Substitute ¹⁄₂ to ³⁄₄ cup heavy or whipping cream for the sour cream.
- *Cream Cheese Crust*—Substitute 8 ounces softened cream cheese for the sour cream.
- *Cheese Crust*—Stir ¹⁄₃ cup finely grated Parmesan, ¹⁄₈ teaspoon cayenne, and ¹⁄₄ teaspoon freshly grated nutmeg into the crust along with the sour cream and other ingredients in Step 3.
- *Nut Crust*—Grind ¹⁄₂ cup roasted walnuts, pecans, almonds, or hazelnuts. Stir the nuts, ¹⁄₂ teaspoon grated orange zest, and 2 tablespoons sugar into the crust after Step 2.
- *Lemon or Orange Crust*—Stir 2 teaspoons finely grated lemon or orange zest and 2 tablespoons sugar into the crust after Step 2.
- *Sturdier Pastry for Hand-Held Pies*—Follow the Flakier Variation and use all-purpose flour in place of the flour mixture. Roll the butter into the flour only four times in Step 2 and omit the cracker crumb coating.

Simple Flaky Crust

Mix 2 cups bleached all-purpose flour, $1/2$ cup instant flour (Wondra or Shake & Blend), and $1/2$ teaspoon salt. Cut $1/2$ pound (2 sticks) butter into $1/2$-inch cubes, add to flour mixture, and toss to coat. Place in the freezer for 10 minutes.

Dump the flour-butter mixture on the counter and roll over it with a large rolling pin to flatten butter lumps. Scrape together and roll over again. Repeat one more time, then scrape back into the bowl and place in the freezer for 5 minutes. Dump onto the counter and roll and scrape together three more times. Place in the freezer for 10 minutes, then gently fold in 1 carton (8 ounces) sour cream. The dough should be moist enough to hold together in a ball. Add 1 to 2 tablespoons milk if needed.

Shape into a ball, cover with plastic wrap, and refrigerate for at least 30 minutes before dividing in half and rolling out. This is enough for two single or one double 9-inch crust.

Jim Stacy, an excellent professional baker in San Francisco, produced hundreds of tarts a day in the bakery that he started, Tarts. Jim developed a method with a large mixer to cut the butter into the flour until it resembled oatmeal flakes—not big enough to be lumps, but just big enough to give flakiness. The Flaky Butter Crust is a home version.

Flaky Butter Crust

MAKES THREE SINGLE 9-INCH CRUSTS

Jim Stacy's combination of a slightly high ratio of butter to flour and perfect-size fat produces a lovely, buttery, flaky crust. His result does not have the extreme flakiness of very large lumps of fat that are flattened by hand, but it is a fast, flaky crust that is prized by caterers and bakers. The fat and the dough must be kept cold throughout the entire operation of blending, rolling, and shaping.

what this recipe shows

Dissolving the salt in water evenly distributes its flavor.

Chilling ingredients, bowl, and mixer blade keeps the butter cold and in larger pieces.

The paddle blade can be used to cut the butter into the flour in oatmeal-flake-size pieces.

$1/2$ teaspoon salt

$1/3$ cup cold water

$1/2$ pound (2 sticks) cold unsalted butter, cut into $1/2$-inch cubes

11 ounces (about 2 cups) all-purpose flour

$2 1/2$ cups raw rice, dried beans, or pie weights

1. Dissolve the salt in the water and refrigerate. Place the butter and flour in a mixer bowl. Chill bowl, ingredients, and paddle blade for mixer in freezer until well chilled, at least 30 minutes.

2. Using the paddle blade and the slowest mixer speed, cut the butter into the flour until the butter resembles flakes of oatmeal. With the mixer running, slowly pour in the cold salted water and mix until the dough forms a ball. Divide the dough into three equal amounts. Shape each into a disk 6 to 8 inches in diameter and ³/₄ inch thick. Wrap the disks in plastic wrap and refrigerate for 30 minutes.

3. Preheat the oven to 375°F (191°C).

4. Remove one dough disk from the refrigerator and roll it out into an 11-inch circle. Place the dough in a 9-inch pie pan and press well into the pan. Lift the edges, fold under slightly, and press down so that the edges are slightly thicker. Place a piece of wax paper or parchment over the crust and fill the crust with rice to hold it down in the pan. Press the rice down and out to sides. Return the crust to the refrigerator or freezer while rolling out the next disk. Repeat rolling, filling, and refrigerating with the other two crusts. The crusts must be cold when they go into the hot oven, so do not leave one out while working on another.

5. Bake until the edges are lightly browned, about 20 minutes. Remove the rice and press the paper liner to force the crust down into the pan well, then remove the liner. Prick each crust several times with a fork. Bake 5 to 10 minutes longer.

Puff pastry

Puff pastry is one of the ultimate examples of flakiness. Classic puff pastry is made by wrapping a flat slab of butter in dough, rolling it out, folding, rolling, and refolding several times to produce a dough with over a thousand layers of unmelted fat between layers of dough. "Rough puff" pastry is a faster, less formal pastry, made by rolling, folding, and rerolling large frozen flakes of butter and dough.

In the rolling and folding process, when you fold the bottom third up and the top third down over it and then roll this out, it is called making one turn. If, instead, you fold the bottom fourth to the center, the top fourth to the center, then fold the top two layers down over the bottom two layers, it is a double turn.

The rolling procedure in the Simple Flaky Crust (page 110) makes an excellent rough puff pastry. Toss together 3 cups bread flour, 4 sticks butter cut into ½-inch cubes, and ½ teaspoon salt. Place in the freezer for 10 minutes, then roll and scrape together 5 times as described in the recipe directions. Place in the freezer for 10 minutes.

Toss this flour-butter mixture gently with enough ice water to form a soft dough that holds together. Shape the dough into an 8 × 10 × 1-inch rectangle. Wrap well in plastic wrap and refrigerate for 1 hour. On a lightly floured surface, roll into a large rectangle—about 18 × 20 inches. Do two double turns on the dough, then wrap and refrigerate for 30 minutes.

This is a very flaky puff pastry at this point. If even greater puff is desired, you can do one more single or double turn on the dough after refrigeration. Remember that it is imperative to keep the dough cold throughout the rolling and folding. If the dough gets warm and the butter soaks into the dough, the flakiness is lost.

Every time you roll the dough, which is like kneading it, it becomes more elastic and firm. If the dough is difficult to roll, roll it out as large as you can, brush off any flour, and brush it heavily with ice water; then do the folds and refrigerate. Brush with ice water as needed to keep the dough soft so that you can finish all the turns.

Puff pastry needs to be baked at a high initial temperature (425°F/218°C) to get steam for a good puff. After it has puffed (10 to 15 minutes), turn the oven down to 375°F (191°C) to finish baking.

Pie crusts in the real world

Whether you are a home cook or a pastry chef in charge of a large bakery, you frequently have to face nonideal situations. For some insight into problem solving, here is the way that I developed a basic pie crust recipe for a small bakery. I started by multiplying Jim Stacy's recipe by 10, which makes 5 pounds fat and about 7 pounds flour.

The bakery was just getting into pies, and its personnel were completely inexperienced at rolling or working with dough. The dough had to be extremely easy to work with, and an all-butter crust might be too difficult to handle. I decided to use part shortening (about 2 pounds—4 cups) with 3 pounds (6 cups) of butter. This would make the dough easier to handle but still give it a real buttery flavor.

For tenderness I used White Lily, essentially a pastry flour. I placed the mixing bowl with the flour and butter, cut into $1/2$-inch cubes, along with the paddle blade in the walk-in freezer for 30 minutes. Then I mixed just until the butter was in big flakes, added the shortening, mixed for a few seconds, then added the water with the salt dissolved in it. The crust was far too tender and an unappealing snow white.

These crusts had to survive delivery and definitely had to be darker. I knew that I could get the crusts a beautiful color with milk or sour cream, but the bakery did not use milk in any products. They did use sour cream, but it is relatively expensive, so I tried to work with other products on hand. I cut the shortening down to 1 pound (2 cups) to make the crusts tougher and switched to unbleached flour to toughen and give better color.

Alas, even with these drastic changes the crust was still too tender and too pale. I baked a pecan pie for 50 minutes, and the crust was still snow white. I decided I would really toughen them up. After I cut the butter in, leaving it in big flakes, I added half of the water and ran the mixer for a few seconds to develop some gluten. Then I added the shortening. Finally I had the texture perfect—very flaky, but strong enough to hold together and to cut without crumbling.

The color, however, was still not good. I couldn't understand this. I was using high-protein nonacidic flour, and I had developed some gluten, but still no color. Browning is determined by protein, sugar, and acidity. I had hesitated to add sugar because I didn't want the crust any more tender. I decided to try everything at once. I replaced some of the liquid with protein by using 3 large eggs and added a little sugar ($1/2$ cup). Mainly out of curiosity, I checked the acidity of the tap water at the bakery and, to my amazement, it was not a neutral pH 7 but down to a pH 6, slightly acidic. I added a little baking soda (1 tablespoon) to the water. At last the crust had good color; however, it was thicker than it should be—slightly puffed. I should have thought of that. The egg whites were giving it a puff. So I dropped the whole eggs and went with 5 large egg yolks. With the acidity cut by the soda, we finally had our perfect crust—rich brown, wonderfully flaky, and tough enough to cut beautifully and stand up to delivery.

Our final recipe ran something like this: Place 7 pounds unbleached flour with 3 pounds butter, cut into $1/2$-inch cubes, in the large mixer bowl. Place the bowl containing the flour-butter mixture and the paddle blade in the freezer for 30 minutes. Also, place 2 cups water with 1 tablespoon salt stirred into it and 2 cups shortening in the freezer. Remove the bowl and beater and mix until the butter is in very large flakes. With the mixer on low, drizzle in the cold salt water from the freezer. Mix for 30 seconds. Add the cold 2 cups of shortening, spooned in in large tablespoon-size chunks, and mix for 1 minute. In a large measuring cup, stir together 5 large egg yolks, $1/2$ cup sugar, 1 cup cold water, and 1 tablespoon baking soda. With the mixer running on low, drizzle in this mixture. Mix for 1 minute and add more cold water if needed to get the dough to hold together when mashed into a ball. Cover and refrigerate the dough for an hour or overnight. Divide into individual crust amounts, shape into 6×1-inch disks, and wrap each in plastic. Refrigerate those needed in the next 2 days and freeze the rest.

Techniques for tenderness

For tender pastry, you need to keep the gluten-forming proteins apart. Greasing the proteins is a very effective way to prevent their joining with water and with each other. This is exactly what happens when you work the fat into the flour. The more you work the fat into the flour, the better you coat the proteins. For a tender crust, you need to work the fat in well—to the consistency of coarse meal. This is just the opposite of the big chunks of fat that you need for a flaky crust. Also for tenderness, warm fat will spread out and grease the proteins better than cold. In fact, the most tender crust of all is the hot-oil crust that was popular in the 1930s, which is tender to the point of falling apart and not the least bit flaky.

Very Tender Flaky Crust

MAKES TWO SINGLE OR ONE DOUBLE 9-INCH CRUST

This is a more tender version of the Flaky Crisp Crust (page 108). It is almost fall-apart tender.

what this recipe shows

Less protein in the flour mixture means more tenderness.

Sugar increases tenderness.

Additional shortening increases tenderness.

Sour cream, which is acidic, also contains milk sugars and fat and makes a tender crust.

Frozen orange juice concentrate adds sugar and acid for more tenderness.

Glucose (corn syrup) browns at a low temperature for good browning.

Chilling the mixed dough allows the moisture to distribute evenly.

A dark crumb coating on the bottom of the crust speeds up baking and dries for crispness.

Chilling the shaped crust in the freezer firms the fat and holds the crust shape.

Baking blind produces a crisp crust.

1 1/4 cups bleached all-purpose flour (see Note)

1 cup instant flour (Wondra or Shake & Blend)

1/2 teaspoon salt

2 tablespoons sugar

1/4 pound (1 stick) very cold unsalted butter, cut into 1/2-inch cubes

1/3 cup very cold shortening, cut into tablespoon-size pieces

1 carton (8 ounces) sour cream

1 tablespoon dark corn syrup

1 tablespoon frozen orange juice concentrate

1 to 2 tablespoons cold whole milk if needed

4 double graham crackers, crushed to fine crumbs

1 large egg, beaten

1. Stir together both flours, the salt, sugar, and butter cubes in a medium mixing bowl and place in the freezer for 10 minutes.

continued

2. When the flour-butter mixture is cold, dump it out onto a clean counter. Roll over the mixture with a rolling pin to flatten the butter and coat it with flour. Some butter will stick to the pin. Scrape it off and scrape the mixture together. Roll over the mixture again. Rapidly continue rolling and scraping together three times. Scrape back into the bowl and place in the freezer for 5 minutes. Dump onto the counter again; roll and scrape together three more times. The mixture should look like paint flakes that have fallen off a wall. Add the pieces of shortening to the mixture and roll them in, scraping and rerolling two to three times. Scrap the mixture back into the bowl and return to the freezer for 10 minutes. If the butter becomes very soft at any time during the rolling, immediately return the mixture to the freezer for 5 minutes.

3. Stir together the sour cream, corn syrup, and frozen juice concentrate. Stir the sour cream mixture into the flour-fat mixture. Add milk only if needed to get the mixture to hold together. Pull the dough together into a ball, wrap well in plastic wrap, and refrigerate for 30 minutes to several hours or overnight.

4. Preheat the oven to 400°F (204°C).

5. Divide dough in half. Shape each half into a 6- to 8-inch disk about 1 inch thick. For a one-crust pie, freeze or refrigerate one disk for later use. Lightly flour the counter, place the other disk on the counter. Sprinkle lightly with flour and roll out evenly, placing the rolling pin in the center of the dough disk, rolling forward and back (taking care not to roll off the dough and thin the edges), rotating the dough 45 degrees and rolling again. Keep a little flour on the counter to one side. If the dough tends to stick when rotating, drag it through the flour. When the dough is nearing the desired thickness, place spacers (page 119) or rulers on either side of the dough, rest the rolling pin on the two spacers or rulers, and roll it across the dough.

6. Spread the crumbs evenly over a counter area as large as the crust. Lightly flour the top of the crust, brushing off any excess. Fold the dough in half, then in half again. Unfold the dough on top of the crumbs and lightly roll over the top with the rolling pin twice to press a light crumb coating on the bottom side of the crust. Refold the dough as before, place in the pie pan, and unfold. Trim and shape edges as desired (see "Shaping the Edges," page 119).

7. Bake blind as described on page 119.

Note: *If pastry flour or a low-protein flour like White Lily is available, use 2¼ cups in place of the bleached all-purpose and the instant flour.*

Variations:

- *Cream Crust*—Substitute ½ cup heavy or whipping cream for the sour cream.
- *Cream Cheese Crust*—Substitute 8 ounces cream cheese for the sour cream.
- *Cheese Crust*—Stir ⅓ cup finely grated Parmesan, ⅛ teaspoon cayenne, and ¼ teaspoon freshly grated nutmeg into the crust after Step 2.
- *Nut Crust*—Grind ½ cup roasted walnuts, pecans, almonds, or hazelnuts. Stir the nuts, ½ teaspoon finely grated orange zest, and 2 tablespoons sugar into the crust after Step 2.
- *Lemon Crust*—Stir 2 teaspoons finely grated lemon zest and 1 tablespoon sugar into the crust after Step 2.
- *Orange Crust*—Stir 2 teaspoons finely grated orange zest and 2 tablespoons sugar into the crust after Step 2.

at a glance
Pie crust techniques

What to Do	Why
For tenderness:	
Have the fat warm—soft.	Better coats gluten proteins and prevents gluten formation.
Cut fat into flour well to coat proteins.	Greased proteins cannot join together to form gluten.
Add sugar.	To reduce gluten formation.
Use acidic ingredients.	To break apart some gluten.
Limit working the dough after the liquid is added.	To limit gluten formation.
Use liquid fat as part of the liquid.	To limit gluten formation.
For flakiness:	
Cut cold fat into ¼- to ½-inch cubes.	The fat must be large enough that it will not melt easily.
Flatten large pieces of cold fat by rolling out with the flour dry or by folding and rolling out dough a couple of times after the liquid is added or, best of all, use both together.	If fat is not flattened, large lumps of fat will go all the way through the crust, melt, and leave holes.
Keep the dough cold during the entire rolling and shaping process.	For flakiness, the fat must remain firm and unmelted.

WHEN TO ADD SALT

Different cooks have different preferences for adding salt to a pastry dough. Some like the ease of adding it to the flour in the beginning and feel that it gets distributed well enough. Others like to add it to the liquid for more even distribution.

Salt enhances gluten formation. If it is mixed with the flour, some of it will be greased and probably will not affect gluten formation, yet it will still be there for taste. Adding it to the liquid puts it in a perfect position to enhance gluten formation and toughness.

Also, unless you are using consistent techniques and ingredients and a recipe that you have done many times, you may not know exactly how much liquid the crust will require. To make sure that you get all the salt in, you have to put all of the salt in part of the liquid, add that, then add the rest of the liquid. The choice of method is essentially a matter of personal preference.

How to make it crisp

A marvelous book called *Pie Marches On,* originally published in 1939 when pie was the king of desserts, shows the tall, handsome author, Monroe Boston Strause, "The Pie King" himself, on the frontispiece. Standing beside him, and coming just a little above his waist, is Mary Pickford on tippy toes in 4-inch heels, looking up adoringly at him.

For a very crisp crust, Monroe crushes graham cracker crumbs, places his crust bottom side down on these fine dark crumbs, and rolls across the top of the crust with a rolling pin to press a thin layer of dark crumbs into the outside of the crust. This gives the crust a darker exterior so that it will bake faster, enhancing flakiness, and the dry crumbs pull moisture out of the crust, producing a drier, crisper crust. I have used this technique in the preceding two crust recipes.

Prebaking the crust empty, then baking on a glaze is another outstanding way to get a crisp crust. The crust is baked blind until it is dry and crisp. It is then brushed with a glaze to seal it and reheated to set the glaze. Frequently used glazes are: egg white or whole egg; for fruit pies, red currant jelly; and for quiche, Dijon mustard, which not only seals but also adds flavor. One of my students, who runs a hunting lodge on a river in Oregon, says that when you do not have time to bake blind, brushing the uncooked dough for a quiche with Dijon before filling makes the crust less soggy.

Another technique for a crisp crust is to partially or totally cook the filling on the stove, as in the Big-Chunk Fresh Apple Pie in Flaky Cheese Crust (page 121), then pour it into a precooked crust.

French chefs use a layer of caramel over the crust in classic fruit tarts. You make the caramel and pour a very thin layer over the crust before you spoon in the pastry cream. The problem is that even the thinnest layer of rock-hard caramel that you can pour is still thick enough to make slicing difficult.

White House pastry chef Roland Mesnier sprinkles the crust heavily with confectioners' sugar, then melts the sugar to a paper-thin layer of caramel either with a propane torch (à la Julia Child) or by covering the edge with foil and slipping the crust under the broiler for a few seconds. My Renaissance Classic Fruit Tart makes use of this technique.

Renaissance Classic Fruit Tart

MAKES 8 SERVINGS

Dark cherries are piled two layers high for a bounteous, luxurious look. Contrasting light-colored apple wedges are arranged to resemble blossoms against the dark background. The glaze makes the tart look like a Renaissance painting. It is almost too beautiful to eat, but once you taste the rich custard with fresh, cold fruit, crisp, flaky pastry, and a hint of caramel, the tart disappears like magic

what this recipe shows

A removable-bottom tart pan makes a minimal crust so the fruit is accented.

Vitamin C in orange juice prevents cut apples from discoloring, and orange juice is not as acidic as lemon juice.

Zip-top plastic bags allow easy rearranging for marinating.

Plastic wrap placed on the surface of the pastry cream prevents formation of a skin (dried casein).

Hot pastry cream will not thin if stirred as cold, firmly set pastry cream does.

2 to 3 tablespoons confectioners' sugar

½ Orange Crust variation of Very Tender Flaky Crust (page 113), Orange Crust variation of Flaky Crisp Crust (page 108), or Simple Flaky Crust (page 110), with 2 teaspoons finely grated orange zest, prebaked and glazed in a 9-inch removable-bottom tart pan

1 batch warm Orange Pastry Cream (below)

1 recipe Macerated Fruit for Tarts (page 118)

⅓ cup apricot preserves

1 tablespoon dark corn syrup

1. Sprinkle the sugar evenly over the bottom of the pie crust. Wrap the edges with aluminum foil. Watching carefully every minute, slip the crust under the broiler and broil just until the sugar melts. If preferred, melt the sugar with a quick blast from a propane torch with a flame spreader. Let cool thoroughly. Spoon the hot pastry cream into the crust without excessive stirring and smooth with a minimum of strokes.

2. Drain the macerated strawberries and cherries well and lay out on a paper towel for a minute. Arrange a row of strawberries around the outer edge of the cream with the berries lying on their side, pointing outward. Fill the center with a layer of cherries. Arrange another layer of strawberries on top of the first layer, but slightly more to the center so that you can still see the bottom layer. Fill the cherries in the center to level with the second layer of strawberries.

3. Drain the macerated apple wedges and place on a paper towel for a minute to drain further. In the upper right of the pie, fan the wedges from one apple in a circle to resemble a blossom. Have the outer edge standing up a little. Arrange the wedges from another apple in a similar manner slightly below and to the right of the first apple. Then arrange the last apple in a similar manner, slightly below and to the left of the second apple.

4. Melt the apricot preserves and corn syrup in a small saucepan. Strain into a small bowl. Brush the strained liquid over the fruit. Brush several times to cover well. Refrigerate to chill. Serve cold.

Orange Pastry Cream

2 cups half-and-half

Zest of 1 orange in strips

1 cup sugar

¼ cup bleached all-purpose flour

⅛ teaspoon salt

6 large egg yolks

1. Heat the half-and-half with the orange zest in a saucepan over medium heat just to a simmer. Remove from heat.

2. Stir together the sugar, flour, and salt in a medium mixing bowl. Whisk in the egg yolks. Remove the strips of zest from the half-and-half and whisk about 1 cup of the hot half-and-half into the egg yolk mixture. Then pour all of the egg mixture back into the saucepan with the remaining half-and-half. Cook over medium heat, stirring constantly, until thick. Remove from the heat and cover with a piece of plastic wrap touching the custard surface.

Macerated Fruit for Tarts

1 quart fresh strawberries, stems
  removed

3 tablespoons dark brown sugar

2 tablespoons *and* 2 tablespoons *and*
  1 tablespoon Grand Marnier or other
  orange liqueur (5 tablespoons *total*)

2 pounds dark ripe cherries, pitted, or
  4 cans (16.5 ounces each) dark pitted
  cherries in syrup, drained

3 tablespoons sugar if using fresh
  cherries

3 ripe (yellowish, not green) Golden
  Delicious apples, peeled, cored, and
  cut into wedges

1/2 cup orange juice

1. Gently toss the strawberries with the brown sugar and 2 tablespoons Grand Marnier in a heavy-duty zip-top plastic bag. Refrigerate until ready to use, turning the bag once or twice.

2. Toss the cherries with sugar and 2 tablespoons Grand Marnier in another zip-top plastic bag. Refrigerate until ready to use, turning the bag once or twice.

3. Soak the apple wedges in orange juice with 1 tablespoon Grand Marnier in a third zip-top plastic bag. Refrigerate until ready to use, turning the bag once or twice.

Giving it a rest

Many recipes say, "Shape dough into a ball and place in the refrigerator for 30 minutes for the gluten to relax." In a properly made crust, there should be very little gluten, but distribution of moisture and firming of the fat are other reasons to let the dough stand. Even in a properly made dough some spots may be slightly wet and sticky, others dry and crumbly. If you roll out the dough right after you put it together, some spots may stick to the counter while other spots may crack. The time in the refrigerator allows the moisture in the dough to become more evenly distributed.

Storing the dough

The dough should stay in the refrigerator at least long enough to distribute the moisture, but there's no harm in giving it a longer visit to suit your baking schedule. Commercial bakers do just that: They refrigerate a large amount of dough for even moisture distribution—which takes several hours or overnight because of the volume—and then weigh out portions for individual crusts, shape them into disks 6 to 8 inches in diameter and 3/4 inch thick, and wrap the disks in plastic wrap. Those that are needed for the next day are refrigerated, the rest frozen. Storing the dough in flat disks makes it very easy to thaw what you need and roll out with just a few strokes.

Rolling it out

Many kitchen counters are not large enough to roll a large rolling pin at all angles. In limited space, rotating the dough works very well. This technique also prevents the dough from sticking to the counter or pastry cloth. Flour the rolling pin, place it in the center of the dough disk, and roll forward and back, taking care not to roll off the dough and thin the edges. Rotate the dough 45 degrees, roll forward and back again, rotate, and so on. Keep a little flour on the counter to one side. If the dough tends to stick when rotating, drag it through the flour.

Pastry cloths Pastry cloths, the nonstick sheets designed for pastry rolling, not only prevent sticking but frequently also have circles printed on them in the sizes needed for different pan sizes—helpful in rolling an even circle of correct size.

Dough is also easy to roll between two sheets of wax paper or plastic wrap or a layer of foil on the bottom and plastic wrap or wax paper on top. You will need to peel the paper or plastic off once or twice and relocate it during rolling. Regardless of the surface, flour lightly to prevent the dough from sticking.

Rolling pins Rolling pins come in many sizes, shapes, and types. If the edges of your pastry are often thin while the center is thick, a French rolling pin may correct this. It is thin on the ends and fat in the center, so it presses the dough down more in the center.

I like to use a fairly large rolling pin and spacers, which hold your rolling pin at the same height above the work surface on both sides to give you an even thickness. The best way to find your personal best rolling pin is to try different sizes and types.

Controlling the thickness Rolling the dough to a uniform thickness is vital to prevent burning in thinner spots. Spacers, which you can purchase at cookware shops, create a dough of even thickness. As a substitute, you can use two rulers that are the thickness you want the dough to be. When you have partially rolled the dough and it is not far from the desired thickness, lay one spacer or ruler outside the edge of the dough and the other on the opposite side, rest the rolling pin on the two, and roll it across the dough.

Spacers like fat rubber bands that fit on each end of the rolling pin do the same job perfectly, and they come in sets of different thicknesses. These are available in Europe, but unfortunately I have not seen them in the United States. You may be able to improvise by layering several wide, thick rubber bands on each end of the rolling pin.

Moving the dough from counter to pan Some cooks like to lightly flour the top of the rolled dough, brush off any excess flour, and fold the dough in half, then in half again. This makes a quarter-circle that is easy to move. Place the dough in the pan with the point of the folded circle in the center and unfold it.

Others like to roll the dough around the rolling pin, carry it to the pan, and unroll it across the top of the pan. If you use this technique, be sure to roll most of the dough around the pin—do not leave half of it dangling down. If the dough stretches and thins out, it may shrink during baking.

Shaping the edges

Roll the crust several inches larger than the pan so that $1/2$ to $3/4$ inch hangs over the edge after trimming. Fold under $1/2$ inch of crust all around the edges, making an even double-thick edge to flute. If the crust on the edge is not of even thickness, the thin spots will brown excessively or burn. With good shaping techniques, you can be bold with your designs and ideas.

After shaping, place the dough in the refrigerator to relax any gluten that you developed in pushing and pulling the crust into shape—30 minutes will minimize shrinkage. To chill the fat well for maximum flakiness, place the crust in the freezer for 10 minutes just before baking.

Baking blind

Prebaking a crust without the filling—called *baking blind*—makes a crisp crust. You can bake blind right side up or upside down if the edge is flat, not fluted. To bake the crust right side up, cover it with parchment or foil and fill the covered crust with weights to hold it down. For weights you can use raw rice, beans, pie weights, or—my favorite—pennies. It is important to weight the crust well, filling completely to the top of the edges to hold them in place.

Bake 15 minutes in the center of a preheated 400°F (204°C) oven. Remove the parchment and weights. Cover the edges with foil if they are browning fast. Bake 5 more minutes. Glaze with your choice of glazes and bake 4 to 5 minutes to cook glaze.

For a plain-edged crust, not fluted, baking blind upside down is a little easier and may give you a better crust. This procedure stretches the sides and gives you a slightly higher crust. When you bake the crust right side up, the sides can sag when the crust gets hot before the dough proteins set well to hold the shape.

To bake upside down, after you place the crust in the pan and chill it, cover the crust with a sheet of crumpled and smoothed wax paper and place a pie pan the same size as the one the crust is in on top of the wax paper. Turn both pans upside down, with the crust in the middle. Bake upside down for about 12 minutes in the center of a preheated 400°F (204°C) oven. Remove the top pan so that the pie crust bottom is exposed. Return the crust to the oven upside down and bake for 5 to 10 minutes longer. Replace the pan on top of the crust, turn the crust right side up, and remove the inside pan and wax paper. Bake about 5 minutes more to dry the inside well. Glaze the inside of the crust with your choice of glazes and bake for 4 or 5 minutes to cook the glaze.

Baking blind produces a superior crust. For a demonstration, I bake two pies made with the same dough and the same filling—one baked with the filling in, the other baked blind, glazed, filled, and baked. The crust baked blind is always much nicer—definitely crisper. A crisp, thoroughly cooked, and glazed crust has a good chance of staying crisp when the filling is added.

In many cases, like custard pies, baking the crust and the filling at the same time creates a dilemma. The crust needs to be baked at a high temperature to set the proteins before the fat melts (for flakiness and shape), but a delicate filling like a custard should be baked at a lower temperature. One solution is to bake for a short period at a high temperature (for a decent crust) and then turn the oven down to a temperature that will not overcook the filling. This is a compromise. Baking the crust blind at a high temperature and then baking the filling at an appropriate lower temperature in the prebaked crust produces a better filling and crust.

Pie pans When it comes to pie pans, there are many options—black pan, dull pan, shiny pan, ovenproof glass (Pyrex), removable-bottom quiche pan, or flan ring. Which is best? The number-one consideration: the crust needs to cook fast. The proteins need to start to set before the fat melts. If not, both shape and flakiness are lost. A glass pie plate will cook very fast because it is cooking by both conduction and radiant energy that goes through the clear glass directly to the crust. All the other pans cook by heat conduction only. The drawback: even Pyrex is hard-pressed to go directly from the freezer into a hot oven without breaking.

Color and type of metal also affect cooking speed. Dark or dull metal pans absorb heat faster and cook faster than shiny pans. A heavy pan made of a good conductor like aluminum is certainly going to give you a more evenly baked crust than thinner, less conductive, tin-plated steel. Removable-bottom tart pans produce attractive dishes; I have some heavy, dark ones that I love. For a traditional pie pan, my favorite is heavy, dull aluminum. For a crust that has to cook for a long time, as in a pecan pie, I like a shiny (to slow browning) removable-bottom quiche pan.

There is another pan consideration: holes in the crust or holes in the pan? When you bake the crust, air that was trapped between the crust and the pan gets hot and expands. The crust bubbles up. You can prick holes in the crust and weigh it down with pie weights, beans, or rice. If you prick holes in the crust, even though you later glaze the crust to seal it, some fillings will leak.

Monroe Boston Strause in *Pie Marches On* has the solution: Don't prick holes in the crust—punch them in the pan. Then you don't have to worry about leaks in the crust. In the thirties these pans were quite popular. I have found them in antique shops and junk stores. They are still available in some cookware shops and by mail order (see Sources—Pie pans, page 477). You can also drill holes in a pan yourself.

Problems with double-crust pies

Because double-crust pies (top and bottom crusts) are not baked blind, it is easy to get an improperly baked, soggy crust. You can solve this problem by baking a large decorative pattern of crust separately for a top crust. A large cutout pattern that covers most of the pie gives the illusion of the traditional double-crust pie. The recipe for the incredibly good Big-Chunk Fresh Apple Pie in Flaky Cheese Crust does just this.

Bake the bottom crust blind and glaze it for perfect crispness. Bake the top crust on the bottom of an inverted metal bowl and cook the filling on the stove. When you are ready to serve, simply assemble the three completely cooked components.

Big-Chunk Fresh Apple Pie in Flaky Cheese Crust

MAKES 8 SERVINGS

I love apple pie with huge chunks of sweet, ripe apples that are not cooked too long. They are still slightly firm and glossy with a brown sugar–butter–cinnamon glaze. When you add mellow roasted walnuts and a flaky cheese crust, this is a memorable apple pie. I got raves from fellow professionals at a bring-a-dish meeting. The fancy top crust makes it spectacular.

This recipe does take time. I recommend preparing the crust dough and filling a day ahead—then you are just left with the baking.

what this recipe shows

Finely grated Parmesan in the crust adds rich cheese flavor and creates a well-browned smooth crust. Larger lumps of grated cheese make a coarse-textured crust.

Baking the bottom crust, top crust, and filling separately and assembling just before serving gives crisp, nonsoggy crusts.

Roasting walnuts enhances flavor.

Soda makes cells mushy and softens dates.

Briefly sautéing and poaching apples softens them slightly before the sugar, which prevents cells from falling apart and preserves texture, is added.

Vanilla, cinnamon, and nutmeg lightly spice apples and dates.

Cake crumbs and walnuts sprinkled on the crust absorb excess liquid and keep the crust crisp.

CRUST

1 recipe Cheese Crust variation of Flaky Crisp Crust (page 108), prepared through Step 5, or Simple Flaky Crust (page 110), with cheese added

1 large egg, beaten

Nonstick cooking spray

3 to 4 tablespoons granulated sugar

1. Preheat the oven to 400°F (204°C).

2. Leave one dough disk in the refrigerator for the leaf pattern top crust while you follow rolling directions in Step 5 of the crust recipe for the bottom crust with the other disk. You can coat the bottom of the crust with crumbs or leave it plain as desired. Place the crust in a 9-inch pie pan. Flute the edges, chill, and bake according to right-side-up baking blind directions (page 119). Glaze with a beaten egg and rebake 3 minutes to set and dry.

3. Roll out the other pastry disk according to directions in Step 5 for the crust. With a pizza wheel or sharp knife, cut out a 12-inch circle. Turn an 8- to 9-inch round-bottom metal bowl upside down and spray the bottom with nonstick cooking spray. Arrange the dough circle on it. From scraps, cut out leaves and press vein patterns into each leaf with the end of an ice pick or toothpick. Make three or four branches about 10 inches long and several small branches by rolling strips of dough between your palm and the counter and then twisting them. Stick the leaves and branches on the dough circle by dampening them with a few drops of water and pressing. Arrange branches curved about in artistic patterns. Twist or curl up the edges of the leaves so that they stand up and have a three-dimensional effect. Have some stuck-on branches curl over themselves and cross on top of or under stuck-on leaves. Carefully place the bowl and decorated dough in the freezer for 15 minutes.

4. To bake, brush with beaten egg and sprinkle with granulated sugar. Bake on the upside-down bowl for 15 to 20 minutes until deep brown.

FILLING

2 cups *and* $^1/_2$ cup chopped walnuts, pecans, or almonds ($2^1/_2$ cups *total*)

3 tablespoons *and* 6 tablespoons lightly salted butter (9 tablespoons *total*)

$^1/_8$ teaspoon *and* $^3/_4$ teaspoon salt (scant 1 teaspoon *total*)

1 cup chopped dates

$^1/_2$ cup boiling water

$^1/_2$ teaspoon baking soda

2 tablespoons cornstarch

$^1/_4$ cup *and* $^1/_2$ cup water ($^3/_4$ cup *total*)

12 to 14 medium (about 6 pounds) Golden Delicious apples (select yellowish, ripe apples), peeled, each cut into 8 to 12 wedges

1 tablespoon pure vanilla extract

1 cup granulated sugar

1 cup light brown sugar, packed

1 teaspoon ground cinnamon

$^1/_4$ teaspoon freshly grated nutmeg

$^1/_3$ cup fine cake or bread crumbs

1. Preheat the oven to 350°F (177°C).

2. Roast the nuts on a baking sheet for 10 to 12 minutes. While they are hot, stir in 3 tablespoons butter and $^1/_8$ teaspoon salt.

3. Place dates in boiling water and stir in soda.

4. Stir cornstarch into $^1/_4$ cup cool water.

5. Sauté the apple chunks in 6 tablespoons of butter in a very large, heavy skillet or large casserole (or 2 skillets if necessary) for about 2 minutes, turning gently with a large spatula. Stir vanilla into $^1/_2$ cup water in a cup, then add to apples. Simmer, uncovered, 1 minute. Restir cornstarch and stir in. Add the sugars, $^3/_4$ teaspoon salt, the cinnamon, and nutmeg, and bring just to a simmer over low to medium heat. Simmer until the liquid in the skillet is thick and bubbly. Add $1^1/_2$

cups roasted nuts. Drain dates and add. Toss gently until the apples, dates, and nuts are well coated with the sugar mixture.

6. Sprinkle the remaining $^1/_2$ cup nuts and the cake crumbs over the prebaked cheese crust. Spoon in the apple mixture with a slotted spoon. If any liquid left in the skillet is thick, pour it over the apples. If thin, boil to thicken to a syrup, then spoon it over the apples. Cover the filling with the leaf pattern top crust. Serve hot or at room temperature.

How to avoid overbrowning

To slow browning, you can cover the edges of the crust with foil or the pie tape available from Maid of Scandinavia (see Sources—Pie tape, page 477) when baking. Food stylists who must have picture-perfect crusts swear by the pie tape. It is like heavy crepe paper and shapes easily, so it is fast and easy to press over the edge of a pie crust.

Another way to avoid overbrowning is to use a pan that minimizes the edge—the removable-bottom tart pan. I like to use a tart pan for pies that require a long cooking time, like pecan pie.

Heating or even partially or completely cooking the filling in a pan on top of the stove helps with the problem of overbrowning. In the Bourbon Pecan Pie (page 125), I heat the syrup for filling in a saucepan, bake it in a tart pan, and use ingredients and techniques to reduce browning as in the recipe for Pie Crust for Longer Cooking Times.

Pie Crust for Longer Cooking Times

MAKES ONE SINGLE 9-INCH CRUST

This crust is formulated for fillings that take a long time to bake and pose the potential problem of an overbrowned crust. If this crust is too pale even after 25 to 30 minutes of blind baking and 45 minutes to an hour of baking time, go to 25 percent cake flour and 75 percent all-purpose and/or 10 percent shortening by volume and 90 percent butter.

what this recipe shows

Using half cake flour, with its acidity and low protein content, reduces browning.

Using half shortening limits browning.

Omitting sugar limits browning.

Using water instead of a dairy product limits browning.

Acids like cider vinegar tenderize and limit browning.

Chilling the dough allows the moisture to distribute evenly.

Chilling the shaped crust in the freezer firms the fat for flakiness and to hold crust shape.

Baking blind produces a crisp crust.

continued

³/₄ cup bleached all-purpose flour

³/₄ cup cake flour

¹/₄ teaspoon salt

4 tablespoons (¹/₂ stick) very cold
    butter, cut into tablespoon-size pieces

4 tablespoons shortening, very cold,
    cut into tablespoon-size pieces

3 tablespoons ice water

¹/₂ teaspoon cider vinegar

¹/₂ teaspoon pure vanilla extract

1 large egg white, beaten

1. Stir together both flours, the salt, and the butter lumps in a medium mixing bowl and place in the freezer for 10 minutes.

2. When the flour-butter mixture is cold, dump it out onto a clean counter. Roll over the mixture with a rolling pin to flatten the butter and coat it with flour. Some butter will stick to the pin. Scrape it off and scrape the mixture together. Roll over the mixture again. Rapidly continue rolling and scraping together three times. Scrape back into the bowl and place in the freezer for 5 minutes. Add the pieces of shortening to the mixture and roll in, scraping and rerolling two or three times. Scrape the mixture back into the bowl and return to the freezer for 10 minutes. If the butter becomes very soft at any time during the rolling immediately return the mixture to the freezer for 5 minutes.

3. Stir together the water, vinegar, and vanilla in a small bowl. Then gently stir the water mixture into the flour-fat mixture. Pull the dough together into a round, wrap well in plastic wrap, and refrigerate for 30 minutes to several hours or overnight.

4. Preheat the oven to 400°F (204°C).

5. Shape the dough into a 6- to 8-inch disk about ³/₄ inch thick. Lightly flour the counter, place the disk on counter, sprinkle lightly with flour, and roll out evenly. Place the rolling pin in the center of the dough disk, rolling forward and back (taking care not to roll off the dough and thin the edges), rotate the dough 45 degrees, and roll again. Keep a little flour on the counter to one side. If the dough tends to stick when rotating, drag it through the flour. When the dough is nearing the desired thickness, place one spacer or ruler at each side of the dough, rest the rolling pin on the two spacers or rulers, and roll it across the dough.

6. Place the dough in a shiny metal (not black) removable-bottom 9 × 1¹/₂-inch tart pan and bake blind (page 119). Glaze with egg white and bake 3 to 5 minutes to set.

Variation: *Orange Crust—Stir 2 teaspoons finely grated orange zest into the crust with the water in Step 3.*

Bourbon Pecan Pie

MAKES 8 TO 10 SERVINGS

As a child in Georgia, I spent exciting times searching for fallen pecans hidden in the leaves. On cold winter evenings, my grandfather and my uncle and I played rummy on a card table near the stove. We all cheated, and when we started to fuss with each other, my grandmother would make us put away the cards and shell pecans. Finding it just a fun new game, I would get involved in trying to shell a perfect pecan half.

Pecans and bourbon are a natural match. Good bourbon has a sweet, rich, mellow flavor that is excellent in many dishes, especially good with sweet potatoes and pecans.

what this recipe shows

A removable-bottom tart pan, producing minimal edges, reduces overbrowning.

Boiling the corn syrup, sugar, and butter for 3 minutes reduces the liquid and heats the filling before it goes into the oven, shortening the oven cooking time and helping to spare the crust.

High sugar concentration and starch (arrowroot) prevent eggs from curdling.

1 cup pecan pieces

2 tablespoons *and* 4 tablespoons *and* 3 tablespoons lightly salted butter (9 tablespoons *total*)

$^1/_8$ teaspoon *and* $^1/_8$ teaspoon *and* $^1/_8$ teaspoon salt (scant $^1/_2$ teaspoon *total*)

1 cup *and* 2 tablespoons light corn syrup ($1^1/_8$ cups *total*)

$^3/_4$ cup *and* $^1/_2$ cup light brown sugar, packed ($1^1/_4$ cups *total*)

2 tablespoons arrowroot

2 tablespoons water at room temperature

2 tablespoons bourbon

2 teaspoons *and* $^1/_2$ teaspoon pure vanilla extract ($2^1/_2$ teaspoons *total*)

3 large eggs at room temperature

3 large egg yolks at room temperature

1 recipe Orange Crust variation of Pie Crust for Longer Cooking Times (page 123), prebaked in a 9-inch shiny or light-colored removable-bottom tart pan (see Note)

$1^1/_3$ cups pecan halves

1. Preheat the oven to 350°F (177°C).

2. Roast the pecan pieces on a baking sheet in the center of the oven for 8 minutes only. While the pecans are hot, stir in 2 tablespoons butter and $^1/_8$ teaspoon salt.

3. Raise the oven temperature to 400°F (204°C).

4. Boil 1 cup corn syrup, $^3/_4$ cup brown sugar, 4 tablespoons butter, and $^1/_8$ teaspoon salt in a medium-size, heavy saucepan on medium heat for 3 minutes. Remove from the heat and let cool for 2 minutes. Stir together the arrowroot and water in a small bowl. Stir in the bourbon and 2 teaspoons vanilla. Stir this mixture well into the hot syrup. Stir the eggs and egg yolks together in a large bowl, then stir the hot syrup in a little at a time. Sprinkle pecan pieces over prebaked crust. Pour the filling over the pecans and bake in the lower third of the oven for 35 minutes. If the crust edge is getting too brown, cover just the edge with foil.

5. While the pie is baking, stir together and boil 3 tablespoons butter, $1/8$ teaspoon salt, 2 tablespoons corn syrup, and $1/2$ cup brown sugar in a medium-size, heavy saucepan for 1 minute. Remove from the heat and stir in pecan halves and $1/2$ teaspoon vanilla.

6. Pile or arrange coated pecan halves on top of pie and return to the oven for 10 minutes.

Note: *If the tart pan is $1^1/2$ inches deep, use a 9-inch pan. If the pan is 1 inch deep, use a 10-inch pan.*

Variation: *For a plain pecan pie, omit the bourbon.*

at a glance
Techniques for more or less crust browning

What to Do	Why
For more browning:	
Use heatproof clear glass or dark pan.	Both cook the crust faster.
Coat outside of crust with a dark crumb coating.	Dark coating absorbs heat and produces a crisper crust.
Use ingredients for browning.	See page 112.
For less browning:	
Precook or heat filling.	Precooking or heating will permit a shorter cooking time.
Use shiny or light-colored metal pans.	Light-colored metal reflects heat and bakes more slowly.
Cover edges with foil or pie tape.	Covering reduces browning.

at a *glance*
Pastry faults

Fault	Cause
Crust not flaky	Fat pieces were too small or cut into flour too finely. Fat was not cold enough. Type of fat used melts fast.
Crust tough	Too much gluten has been developed for one or more reasons: Flour had too much protein. Some fat was not worked well into flour. Too much water was added. Dough was worked too much.
Crust soggy	Crust was not cooked long enough. Crust was cooked with wet filling. Crust was cooked with filling at too low a temperature. Crust was not glazed. There was not enough heat from the bottom.
Crust loses shape	Crust was not cold enough when going into oven. Oven was not hot enough.
Crust burned in spots	Crust was not evenly rolled out—had thin spots.
Crust too brown	There was too much protein and sugar in recipe. Crust was baked too long or too hot.
Crust too tender	Fat was too warm and/or was worked into flour too well.
Crust shrank	All causes under "Crust tough." Dough stretched in moving.
Crust bubbles during baking	Air was trapped between crust and pan (use pan with holes and pie weights).

Cookies

Cookies are a microcosm of cooking. The role of each ingredient is magnified in this low-liquid situation. Not just a change in the amount of liquid (many cookies have no liquid as such), but any change in ingredients that changes the available liquid in the batter changes the cookie.

Ingredients

Fat

Cooks have the choice of a number of fats for cookies—butter, shortening, margarines, spreads, or oils. Your selection depends on the type of cookie that you want.

Water in fat The first difference the choice of fat makes is in the amount of water in the dough. Different fats contain different amounts of water. Lard and shortening are 100 percent fat. Butter is only about 81 percent fat. Margarines by law are a minimum of 80 percent fat. Low-fat and fat-free spreads vary widely in their ingredients, so expect very different results from different brands. Some of the "fake-fat" spreads contain as much as 58 percent water! This makes a staggering difference in the cookies. Cookies using fats with high water content are going to be soft and puffy. They will steam up to a fair height.

Melting pattern of fat How the fat melts makes a big difference in the cookie. Butter is as hard as a rock at one temperature. Then, only slightly warmer, it is soft and just a little warmer it melts. If you use a fat like butter that melts over a narrow temperature range, shortly after the cookies go into a hot oven the butter will melt and the cookies will spread.

On the other hand, shortening stays the same texture over a wide temperature range. Cookies made with shortening or part butter and part shortening do not spread as much as all-butter cookies. Some of the fake fats and spreads that contain gums and starch remain the same consistency regardless of temperature. The cookie stays exactly as you placed it on the baking sheet—no spread.

A restaurant in Atlanta that specializes in big chocolate chip cookies took advantage of these differences when it tried to mass-produce the cookies. Finding that for their size requirements, the all-butter cookies spread too much—the bakers replaced just a little of the butter with shortening, making them producible without losing their buttery flavor.

Flour

Some cookies are simply butter, sugar, flour, and flavorings. Butter is about 81 percent fat and 18 percent water (the remaining 1 percent being milk solids). The only liquid in the cookies comes from the butter. The smallest change in free liquid (liquid in the fat, eggs, or liquid like milk) in a cookie recipe can affect the spread and thus the shape of a cookie. Since different flours absorb different amounts of water, the choice of flour can change the free liquid in a cookie and accordingly change its shape. Flour also can influence a cookie's tenderness and color.

A high-protein flour that soaks in and ties up water makes cookies dry and crisp and hold together better. Cake flour or a low-protein Southern flour that does not soak up much water leaves water free in the dough to turn to steam in the hot oven. Being chlorinated and slightly acidic, cake flour also makes cookies set faster and spread less. On both counts cake flour is ideal for tender, puffed cookies.

The protein content of the flour also influences the cookie's color: the more protein, the darker the cookie. Both effects of flour were illustrated in a demonstration in which I made the same cookie recipe with five different kinds of flour—cake flour, Southern all-purpose, national-brand bleached all-purpose, unbleached all-purpose, and bread flour. The cookies with the cake flour were pale, soft, and

To make flatter cookies, use butter; shortening will not melt as quickly in the oven, so the cookie dough won't spread as fast.

puffed. The cookies with the Southern all-purpose were similar, but slightly darker with a little less puff. The bleached all-purpose cookies were considerably darker and flatter. The unbleached and the bread flour cookies were much darker, flatter, and crisper—the bread flour being the darkest and crispest. There was an incredible difference between the cookies made with cake flour and those made with bread flour—from pale, soft, and puffed to dark and crisp.

"This explains my cookie mystery!" exclaimed the owner of a large catering company who watched the demonstration. Her shop can never bake enough of a certain excellent cookie recipe to meet holiday demands, so some of her employees—experienced cooks all—bake cookies elsewhere to keep up. The cookies baked at the shop are always the same, but frequently the other cookies are different. How could that be? It was the flour, of course—the cookies baked away from the shop often were made with a different kind.

Flour and technique also influence a cookie's tenderness. Cookies can be so tender that they crumble and fall apart, or they can be chewy. If some water gets to the flour before the flour proteins are thoroughly coated with fat, these proteins and water will join together, and some gluten will be formed. If your cookies are too crumbly, sprinkling a little water on the flour before mixing it with the other ingredients may solve the problem. Under the same conditions, flour that has less protein will make tender cookies while flour with more protein holds dough together better and can make chewy cookies.

Cookies too crumbly? Sprinkle a little water on the flour before mixing with other ingredients.

Sugar

Sugar makes cookies tender, contributes to browning, and can make cookies crisp. The type of sugar affects browning and crispness.

Glucose (corn syrup) browns at a lower temperature than table sugar. Substituting as little as 1 tablespoon of corn syrup for sugar will make a cookie browner. Corn syrup also makes the surface of the cookie crisper.

In a cookie with a high sucrose (table sugar) content, a low moisture content, and no acidic ingredients, the sugar crystallizes when the cookie cools, turning the cookie hard and crisp.

Brown sugar contains molasses and sugars other than plain table sugar. Some of these sugars absorb moisture from the atmosphere to make baked products soft. Cookies made with brown sugar can soften upon standing.

Honey contains 42 percent fructose, the sugar in fruits. This sugar is very hygroscopic (absorbs moisture from the atmosphere). Cookies made with honey soften fast upon standing.

Liquid and egg

The perfect amount of liquid in a cookie is a delicate balance. Less than a tablespoon of milk, cream, or water in the usual recipe provides steam for a little puff in a cookie. A little more, however, and the batter thins, causing the cookies to spread.

Eggs in cookies set in the hot oven and hold the cookies together. You can have a batter with extra egg so wet that it is just short of runny, yet it will just puff, not spread. Egg whites will dry baked goods out. A good cookie recipe needs enough sugar to make up for the drying effect of the eggs.

Baking powder and baking soda

A small amount of baking powder (1 teaspoon per cup of flour) or baking soda (1/4 teaspoon per cup of flour) contributes to leavening. Baking powder contains its own acids and does not neutralize acids in the dough. This leaves the dough acidic and makes cookies bake faster and spread less.

Sometimes you will see a cookie recipe with 3/4 teaspoon or more of baking soda. These larger amounts of baking soda do not contribute to leavening but are there for better browning. Baked goods made with an acidic batter set faster but do not brown well. Baking soda neutralizes the acidity so the cookies will brown better.

Cookie ingredients

Ingredient	Influence on Cookie
High-protein flour	Makes cookies darker in color and flatter.
Low-protein flour	Makes cookies pale, soft, and puffy.
Fat with sharp melting point, like butter	Makes cookies spread.
Fat that maintains same consistency over a wide temperature range	Makes cookies that do not spread as much.
Reduced-fat spreads	Makes cookies soft and puffy.
Corn syrup	Makes cookies browner.
Brown sugar and honey	Makes cookies that soften on standing.
Baking soda	Makes cookies browner.

Cookie-making techniques and equipment

Shaping cookies through heat control

You gain a little control over the shape of the cookies through the temperature of the dough and the temperature of the oven. If the dough is very cold going into a hot oven, the fat will stay cold and firm long enough to help the cookie retain its shape. You may have noticed when you do not refrigerate the remaining cookie dough while baking a first batch that the second batch spreads much more. Some cookie makers like to mix the dough and then refrigerate it for a while so that it is very cold going into the oven.

- If you want the cookie to retain its shape and not spread, have the dough very cold and the oven temperature slightly higher than usual.
- If you want the cookies to spread and be flat, put soft room-temperature dough into a slightly cooler oven.

Baking time and temperature

Cookies bake in a short time, usually at 350° to 375°F (177° to 191°C), and they overcook easily. In fact a few degrees and a few minutes matter a lot. What should have been a wonderful moist cookie can become a dried-out disaster in minutes. So check the oven temperature with an oven thermometer shortly before baking and use a timer.

Cookies are also deceptive. They may look undercooked when in fact they are perfect. Watch for browning on the very edges. It is better to slightly undercook cookies than to overcook them.

Sticking

Many of us were taught not to grease baking sheets because a slick pan encourages spreading. There are, however, more effective ways to prevent spreading, as already discussed. Cookies can really stick to baking sheets, especially when the dough contains egg. The perfect time to remove cookies from the baking sheet is a narrow window about 2 or 3 minutes after they come out of the oven for most recipes. Before then they are soft and tear easily; after that they can act as if welded to an ungreased sheet.

I personally believe in using nonstick cooking spray or greasing even nonstick pans. Baking on greased foil can also solve the sticking problem. Parchment may be the best of both worlds. It limits spread, and the cookies are easy to remove.

Pans

A heavy baking sheet is a must. Thin pans that buckle and do not evenly distribute the heat pro-duce unevenly cooked, sometimes burned, disap-pointing cookies. The insulated cookie sheets are good as are heavy nonstick baking sheets. My per-sonal favorites are simply thick ($3/32$ inch—about $1/10$ inch) pieces of aluminum from a sheet metal com-pany. I go through the company's bin of scrap pieces, pick out the sizes I need, and have the com-pany smooth the edges for me. These extraordinarily heavy sheets distribute the heat evenly and bake cookies beautifully.

Three different cookies from the same recipe

You can alter any recipe to tailor-make cookies exactly as you want. As proof, here is a basic choco-late chip cookie formula (page 132) that can be made into thin, crisp cookies; soft, puffy cookies; and something in between simply by changing the type of flour and fat.

For a thin cookie, use butter, which will melt and spread the cookies. Use a little corn syrup for crisp-ness and good color and increase the baking soda for good color. More liquid in the batter will enhance spread, but with the corn syrup only a little milk is needed. Stay with bleached all-purpose flour, which won't absorb a lot of liquid.

For a puffed cookie, use shortening and cake flour to limit spread. Switch from baking soda to bak-ing powder to keep the dough acidic, which limits spread. Use an egg for liquid to puff well. Use brown sugar, which contributes some moisture, to balance the drying effect of the egg. To limit spread further, cut fat and sugar just a little.

The thin and puffed cookies are dramatically different. The thin ones are about 3 inches in diame-ter, slightly crisp, and nicely browned. The puffed ones are about 2 inches in diameter, puffed quite tall, and pale.

For my favorite cookie, the in-betweener, I use half butter and half butter-flavored shortening, and I like the crispness from corn syrup. Cake flour, baking powder, and an egg limit the spread a little. Reducing the sugar by 2 tablespoons from the basic formula also limits spread.

Chocolate Chip Cookies

MAKES ABOUT 2½ DOZEN COOKIES

Basic	Thin	Puffed	In Between
1 cup coarsely chopped pecans	1 cup	1 cup	1 cup
2 tablespoons butter	2 tablespoons	2 tablespoons	2 tablespoons
1½ cups flour	1½ cups bleached all-purpose flour	1½ cups cake flour	1½ cups cake flour
¾ teaspoon salt	¾ teaspoon	¾ teaspoon	¾ teaspoon
¼ teaspoon baking soda	¾ teaspoon baking soda	1½ teaspoons baking powder	1½ teaspoons baking powder
10 tablespoons fat	10 tablespoons butter	9 tablespoons butter-flavored shortening	5 tablespoons butter and 5 tablespoons butter-flavored shortening
1 cup sugar	½ cup sugar and ⅓ cup light brown sugar and 3 tablespoons light corn syrup	1 cup minus 1 tablespoon brown sugar	¾ cup light brown sugar and 2 tablespoons light corn syrup
3 tablespoons liquid or 1 large egg	2 tablespoons milk	1 large egg	1 large egg
1 tablespoon pure vanilla extract	1 tablespoon	1 tablespoon	1 tablespoon
1 cup semisweet chocolate chips (6 ounces)	1 cup	1 cup	1 cup
Nonstick cooking spray			

1. Preheat the oven to 350°F (177°C).

2. On a large baking sheet, roast the pecans for 10 to 12 minutes. While the nuts are still hot, stir in 2 tablespoons butter.

3. Turn the oven up to 375°C (191°C).

4. Sift together the flour, salt, and baking soda or baking powder in a medium mixing bowl.

5. Using an electric mixer, cream fat and sugar in a large bowl until light and fluffy. Add the corn syrup, if using. Add the liquid or egg and beat thoroughly. Beat in the vanilla. On low speed, gradually add the dry ingredients until thoroughly combined. Scrape down the sides once with a rubber spatula. Add the pecans and chocolate chips. Beat 5 seconds on low. Use the rubber spatula to finish mixing in well.

6. Spray cookie sheets lightly with nonstick cooking spray. With a tablespoon or small ice cream/food scoop (I use a No. 40, about 1½ inches in diameter), drop slightly heaped tablespoons of batter about 2 inches apart onto the greased sheets. Bake the cookies for about 12 minutes or until the edges just begin to brown. Remove from the oven and let the cookies cool on the sheet on a cooling rack for 3 minutes, then remove the cookies to a rack to cool completely.

Variations: *Replace the pecans with walnuts or replace semisweet chocolate chips with white chocolate and the pecans with macadamia nuts.*

For "super cookies," in addition to pecans, add 1 cup roasted chopped walnuts.

at a *glance*
Fine-tuning cookies

	What to Do for More	What to Do for Less
Spread	Use all butter. Add 1 to 2 tablespoons liquid (water, milk, or cream—not egg). Use a low-protein flour like bleached all-purpose (but not one that is chlorinated). Add 1 to 2 tablespoons sugar. Use room-temperature ingredients or let dough stand at room temperature.	Use shortening or reduced-fat spread. Use an egg for liquid. Use cake flour. Cut sugar by a few tablespoons. Switch from baking soda to baking powder (which contains mild acids). Use cold ingredients or chill dough before it goes into the oven.
Puff	Everything under "What to Do for Less Spread."	Everything under "What to Do for More Spread."
Tenderness	Use cake flour (low-protein flour). Add a few tablespoons of sugar. Add a few tablespoons of fat.	Use unbleached or bread flour (high-protein flour). Cut sugar by a few tablespoons. Cut fat by a few tablespoons. Add a tablespoon or more of water to the flour before combining with other ingredients.
Color	Substitute 1 to 2 tablespoons of corn syrup for sugar. Use an egg for liquid. Use unbleached or bread flour.	Use water for liquid. Use cake flour or bleached all-purpose.

Crackers

The same principles that apply to cookies apply to crackers. Thin, crisp crackers, such as the following recipe, are a perfect example of what happens when you have a recipe with a high liquid content. As with cookies, the type of flour helps determine the consistency of the batter, which must be just right for these Crisp Corn Wafers. If the batter is too thick, they do not dry out. I tried using cake flour since it has the same protein content everywhere in the country, but I didn't like the taste. So for consistency of protein content I went with instant flour like Wondra or Shake & Blend. You'll be surprised by how easy it is to make these professional-looking gourmet crackers.

Crisp Corn Wafers

MAKES 3 PANS, ABOUT 75 SMALL CRACKERS

Chuck Allen, a former national network reporter, has been on the scene of many of the major events of our time. He tells spellbinding stories of hiding out all night with a camera crew in a van to be at just the right spot for a story the next day. In addition to being a storyteller extraordinaire, Chuck is a master cook and his food is superb. This is my version of Chuck's wafers. They look like elegant, expensive, store-bought wafers, and they are addictive—you can't eat just five.

what this recipe shows

Liquid creates steam to make very wet batters into light crackers.

The batter consistency is thin enough to coat the pan.

Sugar and egg contribute to browning.

4 tablespoons (1/2 stick) butter at room temperature

1 tablespoon sugar

1 large egg

1 cup instant flour (Wondra or Shake & Blend)

1 teaspoon baking powder

1/4 teaspoon salt

2/3 cup whole milk at room temperature

1/2 cup water at room temperature

1/4 cup cornmeal, white or yellow

Butter to grease pans

1. Preheat the oven to 375°F (191°C).

2. Beat together the butter and sugar in a mixer on medium speed until light and fluffy. Add the egg and beat in well.

3. Sift together the flour, baking powder, and salt in a small bowl.

4. Sift half of the flour mixture into the butter-egg mixture and beat in well on low speed. Stir in the milk, then sift in the remaining flour mixture and stir in. Stir in the water. Sprinkle the cornmeal on top and stir just to mix in well. Strain to remove lumps. You should have about 2 cups of batter, enough for 2/3 cup per pan for three pans. Butter well three 10 × 15-inch jelly-roll pans by spreading the butter with a piece of crumpled wax paper. Pour 2/3 cup of batter along the long side of the pan and tilt as needed to spread the batter completely over the pan as evenly as possible. Repeat with the second pan. You can bake two pans at once, but alternate the shelves once during the baking for even cooking. Bake for about 8 minutes or until just the edges are lightly browned.

5. Remove the pans from the oven and use a pizza cutter or sharp knife to cut the crackers into squares. Rotate the crackers around the edge of each pan so that the brown edge is inside. Place the crackers back in the oven and continue baking another 8 minutes. Remove all lightly browned crackers to paper towels to cool. Return pale crackers to the oven to brown lightly, several minutes. Butter each pan before reuse. Store in an airtight can.

Cakes

Cakes, cakes, cakes—pound cakes, layer cakes, angel food cakes, sponge cakes, génoises, chiffon cakes, tortes, gâteaux, roulades, biscuits (beesQWE), meringues, dacquoises, vacherins, et cetera. The number and varieties seem infinite. Fortunately, cakes fall into two major groups: foam cakes and shortened (butter) cakes:

- Foam cakes have a high ratio of eggs to flour and fall into three categories: those containing no fat in any form (angel cakes, dacquoises, and meringues), those whose only fat is egg yolk (sponge cakes, some biscuits, some roulades, etc.), and those that contain oil or fat in addition to egg yolks (chiffon cakes, génoises, etc.). Since foam cakes do not depend so much on fat, but rather on eggs, they are not discussed here but are in Chapter 3.
- Shortened cakes contain fat, frequently in a solid form such as butter or shortening. Examples of cakes in this category are pound cakes and all types of creamed cakes (yellow cakes, butter cakes, etc.).

The fascinating roles of fats in shortened cakes

Years ago, when *The Cook's Magazine* once asked me to do an article on the roles of fats in cooking, I wasn't sure that I could be enthusiastic about the subject. I asked for a few days to research fats.

I went to my local university library and made copies of the articles about fats in leading technical journals. Armed with a stack of articles, I settled down for an evening of fat reading.

I was fascinated by articles using modern research equipment and video cameras to take pictures inside a cake batter while the cake is baking. The fat in the batter is stained one color, and the milk portion another, so you can see exactly where the bubbles are and what is going on. It is as if you are standing right in the middle of the batter.

I had always assumed, since the baking powder dissolves in the liquid portion of a cake batter, that was where the baking powder made the bubbles. This is not the case at all. Baking powder does not form a single new bubble. It only enlarges the bubbles that already exist in the fat. This adds new significance to the creaming step in shortened cakes. Since the only bubbles in the cake are those in the fat and those beaten into the batter, the creaming of the butter and sugar to incorporate bubbles into the fat is crucial.

All at once, all kinds of things began to make sense to me. I was a terrible cake baker. I never paid any attention to the creaming step. I barely blended the sugar and butter together and then added the eggs and other ingredients. No wonder my cakes were heavy.

Then I remembered the cakes at a nearby restaurant where I sometimes worked for an evening when I thought I wanted to start a restaurant (one frantic evening of heavy-duty cooking was always enough to get me over it!). Those made by head chef Patrick, an ex-Dominican monk who happily and casually threw in one ingredient after the other as I did, were about as flat as mine as well. Those contrasted with cakes made by assistant Patsy, who tended to throw the butter and sugar into the mixer, turn it on, and go on to something else until 5 or 6 minutes later, when she returned to add a few eggs. Her chocolate cakes done by the same recipe were a mile high. No wonder—she had aerated the fat like crazy!

It was now midnight, but I was so excited over how important bubbles in the fat were, that I had to share this vital information. I realized that my friend Doris Koplin, an outstanding professional baker, got up at 4:00 A.M.—but this was so thrilling that I knew she would want to know immediately. I called Doris and described the enormous importance of bubbles in the fat, even though it was hard to do over the phone, without the use of both hands. Doris mumbled, "Shirley, I'm so happy for you!" and hung up.

Ingredients

To cake lovers, food literally means cake. If you have ever had chocolate cake with swirls of gooey icing on cake so moist it looked wet, you too may visualize a luscious slice of cake when someone says "food."

What most cake bakers—home cooks and professionals alike—know well is that cake baking is an exacting art. To get the cake you want, you need to understand how the ingredients work in balance with each other and in concert with cake-making techniques.

Versatile fats

In the delicate balance of ingredients, fats and sugar are the heart and soul of shortened (butter) cakes. As tenderizers, they soften and moderate the firm structure made by proteins in flour, eggs, and milk. Fats perform three crucial roles. They make the cake light and delicate by holding tiny air bubbles that are expanded by gases from baking powder or baking soda. They help make the cake melt-in-your-mouth tender by coating the flour proteins so they cannot form tough gluten. And they carry rich flavors and essential nutrients.

Fats in leavening Solid fats that hold air bubbles well include butter, shortening, margarines, spreads, and solid animal fats. Animal fats, which can have a slightly meaty flavor, are used primarily in Christmas steamed puddings and mincemeat baked goods.

With most solid fats, most of the vital bubbles in the batter that must be enlarged by baking powder and baking soda need to be beaten in during the creaming or mixing of the cake. Shortening has an advantage over other solid fats in that it already contains millions of fine bubbles to aid in leavening. Unprocessed solid shortening has a glassy, unappetizing appearance. For a more appealing snow white, processors bubble nitrogen, a nonreactive gas, into the shortening. Shortenings are, by volume, about 12 percent fine nitrogen bubbles. This means that shortening will make a lighter cake. Also, many shortenings have emulsifiers such as mono- and diglycerides added for better distribution of the fat in the batter. Better distribution of fat means better distribution of the air bubbles and a better-textured cake.

Shortening has an ideal texture to beat for volume and aeration, and even at warm room temperature there is no danger that it will melt and lose its air-holding ability. Butter from the refrigerator must warm slightly to be beaten, and it can melt if the room is too warm. And, of course, butter does not contain the 12 percent of volume in fine bubbles that shortening has. Nevertheless, there is nothing like the taste of real butter, so butter is normally the fat of choice of fine cake bakers. I must say, however, I have been amazed at how good cookies made with butter-flavored shortening are, so it could be a good choice for cakes, too—I haven't tried it.

Margarines have all the disadvantages of butter without its flavor advantage. Spreads, with their high water content, gums and starches, and major differences from brand to brand, behave in unpredictable ways and do not contribute much to taste.

Fats as tenderizers Oil coats flour proteins well and prevents them from absorbing liquid from the batter to make gluten. Cakes made with oil can be not only tender but very moist, too. When you want a cake or muffins really moist, think of oil. Excellent carrot cakes can be made with oil, such as the Carrot Oil Cake in the 1975 edition of *The Joy of Cooking*. Oil is also frequently used to make moist, tender muffins. Oil does not have air-holding ability to aid in leavening, so the eggs and any other thick ingredients like fruit purees in the batter must perform that task.

Sugar: the sweet moderator

Like fat, sugar tenderizes and moisturizes cake. Sugar prevents the flour proteins from joining to make tough gluten by combining with each of them. Glutenin and sugar combine to form a soluble protein-sugar compound, and gliadin and sugar combine in the same manner. We saw sugar as a great tenderizer in pie crust and cookies, and it plays the same role in cakes. Since sugar can stand in for fat as a tenderizer, it is an important ingredient in reduced-fat baked goods (page 180).

Different sugars have different characteristics in cooking. Honey, which is 42 percent fructose, absorbs water from the atmosphere and makes cakes and muffins good keepers. Under the same conditions, baked goods made with honey will stay moist longer than those made with sugar. Brown sugar contains some fructose too, and baked goods made with it retain moisture well.

Syrups offer an advantage in that the sugar is already dissolved. The dissolved-sugar method of mixing a cake is described on page 150. This method produces good crust color, a tender crust, excellent aeration, fine texture, and a tender crumb.

Syrups are very thick, so with the addition of egg whites they can help hold air bubbles and help with aeration. Corn syrup also gives additional flavor, which is a big help in reduced fat baking.

Corn syrup (glucose) browns at lower temperatures than other sugars. Depending on the specific situation, this enhanced browning may or may not be an advantage.

Flour: strength and structure

For a tender cake, pick a low-protein flour. Cake flour, pastry flour, and Southern bleached all-purpose flours are all low in protein and make tender cakes. National-brand bleached all-purpose dependably has an intermediate protein content and is preferable to an unknown local brand all-purpose, which may be quite high in protein. Most unbleached flours and bread flours have high protein contents and should be avoided for cakes.

Cake flour has numerous advantages for fine texture. Many bakers would not think of using anything other than cake flour. But there are differences of opinion, and some bakers strongly prefer a low- to moderate-protein all-purpose flour. Cake flour gives a definite taste difference. Some object to the fine texture that you get with cake flour and deliberately opt for the coarser texture of bleached all-purpose. Use of unbleached all-purpose or bread flour is a bad idea. All bread flours and most unbleached flours have a higher protein content, meaning more gluten and tough cakes unless you add excess fat and sugar. Here, again, it is a matter of personal preference. You should try different flours and make your own decision.

Cake flour   Cake flour will make a finer-textured cake for many reasons:

1. It is a low-protein flour (about 8 grams protein per cup), so it will help produce a very tender cake.

2. It is very finely ground, so it produces better texture.

3. It is chlorinated, which means that it is bleached with chlorine gas and deliberately left slightly acidic. This gives cake flour several advantages over nonchlorinated flour:

- The acidity causes cakes made with cake flour to set slightly sooner, producing cakes with a finer texture. Rose Levy Beranbaum points out that lower-pH (more acidic) batters produce a sweeter, more aromatic quality in cakes.
- Chlorination enhances the starch's ability to absorb water.
- Fat sticks to chlorinated starch but not to starch from the same wheat that has not been chlorinated. Since all the air bubbles are in the fat, this leads to a more even distribution of the bubbles, which produces a finer texture.

Cake flour in the United States is chlorinated; not so, however, in many countries. For example, cake flour in Japan is not chlorinated, and chlorinated flour in the United Kingdom is available only as self-rising, with baking powder (leaveners) already added. This means the flour sometimes has too much leavener for a successful cake. Rose Levy Beranbaum had this problem when adapting *The Cake Bible* for the United Kingdom. Nick Malgieri (author of *Nick Malgieri's Perfect Pastry*) offered Rose a solution—use the leavened chlorinated flour for fine texture and add a small amount of non-self-rising flour to correct for the excess of leaveners. Rose said that she had to make other changes, too, such as an increase in sugar.

Egg whites and yolks: separate roles

As always, eggs are not just eggs—whites are an incredible drying and leavening agent, and yolks are nature's great emulsifiers for creamy texture. Do not limit yourself to using whole eggs. If a cake or muffins are dry, cut an egg white. Go with two yolks instead of the whole egg and add a little more sugar and fat. Shortened cakes with a large amount of sugar and liquid may require extra emulsifiers. You will see many cake recipes that contain yolks alone. On the other hand, if you have a cake that's a soggy mess, you may need to add an egg white.

Whites are excellent leavening agents, too. Frequently some whites are beaten and folded into the batter during mixing as described in "Combination Conventional and Sponge (Whipping) Method" (page 150). However, as I relate in the story of the cake that is a soufflé (page 237), if you use all beaten whites in a cake, it will fall like a soufflé. When egg whites are beaten, they no longer contribute to structure as raw eggs do, so you also must have enough raw egg, milk, and flour proteins to cook and set the batter, or the cake will fall. It is true that meringues and angel food cakes are all beaten egg whites; however, they are foam cakes—essentially fat free, very high in sugar, and basically a cooked egg white foam.

Baking powder and baking soda: less is more

You might think if a cake falls or does not rise well that you have not used enough baking powder. However, just the opposite is usually true. When you have too much baking powder, the bubbles get too big, run into each other, float to the top, and—*pop!*—there goes your leavening. Many cake and muffin recipes have too much leavening. Also, remember that baking soda is four times as strong as baking powder. The general rule is 1 to 1¼ teaspoons *baking powder* per cup of flour in a recipe. This would be ¼ or a breath over ¼ teaspoon *baking soda* per cup of flour in the recipe.

Rules of thumb: 1 to 1¼ teaspoons baking powder to 1 cup flour, ¼ teaspoon baking soda to 1 cup flour.

These general rules are just that—general. Many variables determine the proper amount of baking powder to be used, and you need to adjust the general rules for your own cooking conditions. More leavening can be used if the recipe calls for a lot of heavy ingredients like chopped fruits. When you change pan size, the amount of baking powder should be altered. For a larger pan, in which the batter will not be as deep, you will need less baking powder than for a smaller pan with deeper batter. Rose Levy Beranbaum has charts in *The Cake Bible* that give exact amounts of leavener for different batter amounts and pan sizes.

Balancing a shortened cake recipe

The major ingredients in cakes must be in balance. A change in any one requires balancing changes in the others. Eggs, flour, and milk contain proteins that set (coagulate) with heat. They form the structure of the cake, but they are also the tougheners.

Sugar and fat, on the other hand, slow down or prevent these proteins from setting and are tenderizers—or "structure wreckers." Sugar combines with the two gluten-forming proteins, glutenin and gliadin, to prevent gluten formation. So sugar can limit how much tough gluten will form. Sugar also slows down the coagulation, or joining together, of egg proteins. This, too, limits the firm-set protein structure. Fats coat the gluten-forming proteins and prevent their joining together with each other and with water to form gluten.

The trick, then, is to use enough sugar and fat to produce a delicate, tender, melt-in-your-mouth structure, but not so much that the cake falls and is a soggy mush. You can think of this as a balance with flour, eggs, and milk on one side, sugar and fat on the other. The following two sets of formulas reflect this balance, which professionals follow *almost* religiously. They know that it will be a waste of time to test a recipe that isn't within about 20 percent of these conditions. You do not have to be exact, but you do need to be in the ballpark.

The two types of shortened cakes (regular and high-ratio) have slightly different formulas. Many modern cakes fall into the high-ratio category. They have a slightly higher ratio of sugar to flour than a regular cake (which has more flour than sugar or equal weights of flour and sugar). High-ratio cakes also contain extra emulsifiers in the form of egg yolks or in the shortening.

Formulas for regular shortened cakes

1. The weight of the sugar should be equal to or less than the weight of the flour.
2. The weight of the eggs should be equal to or greater than the weight of the fat.

3. The weight of the liquids (eggs and milk) should equal the weight of the flour.

Let's examine the recipe for one of the most successful regular cakes of all time—the classic pound cake—to see how it conforms to the formulas. In strict adherence to a classic recipe, which calls for a pound of flour, a pound of sugar, a pound of butter, and a pound of eggs, all three conditions are satisfied:

1. Sugar (1 pound) equal to or less than flour (1 pound)
2. Eggs (1 pound) equal to or greater than butter (1 pound)
3. Liquid—the eggs—(1 pound) equal to flour (1 pound)

Converting to volume rather than weight units, this gives the following recipe:

1 pound butter (8 ounces/cup)	= 2 cups
1 pound sugar (7 ounces/cup)	= 2¼ cups
1 pound eggs (1.7 ounces/large egg)	= 9 large
1 pound flour (4.6 ounces/cup)	= 3½ cups

Formulas for high-ratio cakes

Formulas for the popular high-ratio cakes are slightly different:

1. The weight of the sugar should be equal to or greater than the weight of the flour.
2. The weight of the eggs should be greater than the weight of the fat.
3. The weight of the liquid (eggs and milk) should be equal to or greater than the weight of the sugar.

These are the rules if you want a successful cake. This is not to say that master bakers do not make alterations. But you can bet that master cake bakers probably started their breathtaking creations with the math. They first work out a balanced recipe and then alter this recipe for specific differences in the particular cake they are preparing.

Adding leaveners to the mix

In addition to following the rules of balance for shortened cakes in developing recipes, cooks must pay close attention to the amount of leavening. The general rule is that 1 to 1½ ounces of baking powder are required for each pound of milk, water, or cream (liquid other than eggs) in the cake. Because eggs are natural leaveners, a cake that contains no liquid other than eggs (pound cake, for example) may manage with no additional leaveners.

For cakes that do contain liquid other than eggs, another way to estimate the correct amount of leavening is from the amount of flour. You need 1 to 1¼ teaspoons baking powder for each cup of flour or ¼ teaspoon baking soda per cup flour.

For cakes that contain large amounts of butter, remember that about 19 percent of the weight of butter consists of water and milk solids with about 18 percent water. A small amount of extra leavener can compensate for this additional liquid, which tends to make a cake more compact, by opening up the structure and tenderizing a pound cake.

Getting from the numbers to the sublime cake

With the formulas, you can work out recipes that meet your own requirements. As for me, I love very moist, sweet cakes, and I frequently make them with sour cream or cream cheese. Cream cheese makes a fairly dense, moist cake—almost like a pound cake—that slices beautifully. However, I want a basic cake with just a little more open texture, as you get from milk or buttermilk. I want an excellent-tasting moist, sweet, thick 9-inch cake that I can slice into three layers for simple cakes. Here is my reasoning.

Buttermilk produces basic cakes with a wonderful taste, so I'll go with that. Since I want the cake really sweet and moist, I want a high-ratio cake with more sugar than flour. I'll have to have some extra emulsifier for this much sugar, so I'll go with yolks in addition to whole eggs. I want it really moist. Oil makes the moistest cakes because it coats flour protein so well and does not allow water-absorbing gluten to be formed. So I want some of my fat from oil, but I want butter for its taste and aerating ability. How can I balance all of this?

Sugar weighs about 7 ounces per cup; with 1⅓ cups sugar, the weight of sugar equals 9.3 ounces.

APPROXIMATE WEIGHT-VOLUME EQUIVALENTS OF BASIC CAKE INGREDIENTS

Volume	Ingredient	ounces	grams
Sweeteners			
cup	Sugar—superfine and granulated	7	200
cup	Light brown sugar, packed	7.5	213
cup	Confectioners' sugar	4	113
cup	Corn syrup	11.5	326
tbsp	Corn syrup	0.7	20.3
cup	Honey	11.7	333
tbsp	Honey	0.75	20.8
Fats			
cup	Butter	8.0	227
tbsp	Butter	0.5	14
cup	Shortening	6.7	190
tbsp	Shortening	0.4	11.8
cup	Oil	7.5	212
tbsp	Oil	0.46	13.3
Eggs			
cup	Large whole, out of shell	1.75	50
cup	Large yolk	0.65	18
cup	Large white	1.06	30
Dairy			
cup	Milk, buttermilk	8.5	242
cup	Sour cream	8.5	242
cup	Whipping cream	8.1	232
Flour*			
cup	Cake, dipped	4.6	130
cup	Cake, spooned	4	112
cup	Cake, sifted	3.5	99
cup	All-purpose, dipped	5	145
cup	All-purpose, spooned	4.25	120
cup	All-purpose, sifted	4	114
cup	Bread, dipped	5.6	160
Other Dry Ingredients			
cup	Cocoa, dipped	3.3	95
cup	Cocoa, spooned	3	85
cup	Cocoa, sifted	2.6	75
cup	Cornstarch, spooned	4.4	125
tbsp	Cornstarch, spooned	0.27	7.8

*The only way to be accurate with flour is to weigh the flour over and over in the manner that you measure volume and take an average.

From personal measurements and "Average Weight of a Measured Cup of Various Foods," Home Economics Research Report No. 41, U.S. Department of Agriculture [1977]. There are many variables in going between weight and volume so these values are therefore approximate.

Cake flour measured by the dip-and-scoop method weighs about 4.6 ounces per cup; with 1½ cups flour, the weight of the flour equals 6.9 ounces. This satisfies the first requirement for a high-ratio cake:

1. The weight of the sugar is equal to or greater than the weight of the flour: 9.3 ounces is greater than 6.9 ounces.

    Now, how much eggs and fat am I going to need? I know that for Requirement 3 the weight of the buttermilk and eggs has to be at least 9.3 ounces (the weight of the sugar). If I use ½ cup buttermilk (4.3 ounces), I must have about 5 ounces of eggs. One whole large egg weighs about 1.7 ounces, and a yolk weighs about 0.65 ounce. With 2 eggs (3.4 ounces) and 3 yolks (1.95 ounces), the weight of the eggs is 5.35 ounces. This brings the total liquid weight to 9.6 ounces, which is greater than my 9.3 ounces of sugar. Perfect! Now I need about 5 ounces of fat to satisfy Requirement 2.

    Oil weighs 0.47 ounce per tablespoon, so the weight of ⅓ cup of oil is about 2.3 ounces. With 4 ounces (8 tablespoons) of butter at about 81 percent fat, I will have about 3.2 ounces of fat from the butter. This gives a total fat weight of 5.5 ounces, which is just right.

2. The weight of the eggs is about the weight of the fat: 5.35 ounces is close to 5.5 ounces.

    I have actually worked out my last requirement already. With ½ cup buttermilk (about 4.3 ounces) and 2 eggs and 3 yolks (5.35 ounces), the weight of the liquid is 9.65 ounces:

3. The weight of the liquid (eggs and milk) is equal to or greater than the weight of the sugar: 9.65 ounces is greater than 9.3 ounces.

The following amounts of flour, sugar, fat, eggs, and liquid meet all three requirements, and chances are excellent that I have a working recipe:

    1½ cups cake flour
    2 large eggs plus 3 large egg yolks
    ½ cup buttermilk
    1⅓ cups sugar
    ¼ pound (1 stick) butter plus ⅓ cup oil

In the section that follows, I use this recipe to give examples of the two more frequently used mixing techniques, the creaming method and the two-stage method. This is an old-time rich moist cake—in no way low-fat (see page 181 for Mellow Moist Low-Fat Chocolate Cake).

Cake-making techniques

Priorities—light and airy or velvety smooth, melt-in-your-mouth tender?

Mixing methods play a major role in cake texture. Frequently there are no real rights or wrongs in cooking. As the saying goes, "One man's meat is another man's poison." Many times "right" is a matter of personal preference. From the very beginning you may as well do some soul searching and decide what kind of cake person you really are. Do you love a feather-light, airy cake, or do you like fine, close, silky texture and melt-in-your-mouth tenderness enough to give up a little lightness for that smoothness?

If *lightness* is your first concern, you should choose a mixing method, like creaming, that gives prime importance to volume and aeration. Take care that the oven temperature is not so high that it sets the cake before it reaches the volume you want. On the other hand, if you are a *texture* person, you should choose the two-stage method, which prevents gluten development, and a higher oven temperature so that those bubbles won't expand too much and ruin the fine texture.

Mixing the cake

Different mixing methods produce different kinds of cake. The goals in mixing are to achieve a uniform blending of all ingredients, to incorporate a maximum number of air cells for volume and texture, and to develop a minimum amount of gluten for tenderness, texture, and volume.

Each of the currently used mixing methods has both advantages and disadvantages. Although most texts describe essentially the same methods, it can be confusing because a wide variety of names has been applied to the same method. I have included several typical names for each:

- Creaming, sugar-shortening, sugar batter, or conventional method
- Two-stage, blending, pastry blend, or flour batter method
- Single-stage, dump, one-bowl, or quick-mix method
- Muffin method
- Combination conventional and sponge (whipping) method
- Dissolved-sugar method

Creaming method In this popular method—also called *sugar-shortening, sugar batter,* or *conventional method*—the sugar and butter are creamed together to incorporate those vital air bubbles into the fat. The eggs are then beaten in (incorporating more air and adding emulsifiers from the egg yolk) to provide better incorporation of the air-bubble-filled fat into the batter. Finally, the dry ingredients (flour, baking powder, salt, etc.) are added alternately with the rest of the liquid ingredients (milk, vanilla, etc.). This method has the advantage of incorporating the maximum amount of air bubbles into the fat for the greatest volume.

An inexperienced cake baker can go wrong during several steps in this method. Possible pitfalls are the temperature of the butter, eggs, sugar, and mixing bowl; insufficient creaming; insufficient blending of baking powder and flour; and development of too much gluten.

Butter and bowl temperature The temperature of the butter is very important for maximum incorporation of air. Butter can go from rock-hard to melted within a fairly narrow temperature range. Experts disagree on the exact ideal temperature, but most recommend between 65° and 70°F (18° and 21°C). Bruce Healy, a classic French pastry expert, says that butter actually melts between 67° and 68°F (about 20°C), so his preferred starting temperature for the butter is 65°F (18°C). Bruce has done extensive experimentation with the time and batter volume at each stage and the final cake volume. Maximum aeration of the butter and the butter and sugar are the vital steps. He says the length of time of beating after the

eggs are added and the blending of flour and liquid are not as important as a thorough creaming of the butter and sugar. Cake expert Susan Purdy, author of *Have Your Cake and Eat It, Too,* agrees with Bruce 100 percent that the creaming of the butter and sugar is the vital step for volume.

Since this creaming step is so important, rinsing the bowl and beater in ice water is a good idea. Using a heavy metal bowl hot from the dishwasher can lead to disaster. To prevent the butter from starting to melt, stop creaming briefly and dip the bowl in an ice water bath to cool it down.

Bowl, kitchen, and ingredient temperatures present no problem at all in aerating shortening since it remains the same consistency over a wide temperature range and already contains 12 percent fine nitrogen bubbles.

Temperature of the eggs Most books recommend room-temperature eggs, but not all experts agree on the exact temperature here either. Bruce Healy says that his experimentation shows slightly, but definitely, reduced volume with cold eggs. Eggs from the refrigerator can be warmed fairly rapidly by placing them in their shells in a bowl of hot tap water.

Insufficient creaming With butter, most of the air bubbles in the cake are created in the creaming step. Baking powder only enlarges bubbles already in the dough. Shortening that is already aerated, as just mentioned, will produce a light cake even with a poor job of creaming, but not butter. In addition to using 65°F (18°C) butter, the length of beating time, the speed and type of mixer, and the bowl and room temperature are all important to get the fat and sugar very light and airy. Flo Braker, author of *The Simple Art of Perfect Baking,* recommends creaming for 4 to 5 minutes and Carol Walters 6 to 10 minutes. Hand beaters frequently take several minutes longer to cream well than a heavy-duty mixer and require you to move the mixer around the bowl and scrape down the sides several times for complete creaming.

Insufficient blending of leaveners and flour This is my personal Waterloo. I am always in a hurry, and it seems like such a minor step. But if you do not sift the leavens, flour, and salt together several times for even distribution, the cake can have a velvety texture in general but numerous unsightly large holes.

Development of gluten When adding flour and liquid alternately, you can develop gluten, which makes the cake tough or leads to tunnels. The first addition of flour will be well coated with fat and not form gluten, but once the liquid is added uncoated flour proteins can combine with milk to form tough gluten. I like to add a lot of the flour in that first addition. Once the liquid is added, you must limit the mix-ing or you can develop gluten. Simply overbeating the batter at this point can develop gluten.

When you're using the creaming method for a high-ratio cake, the batter can curdle if you add too much liquid at once. (Switch from the water-in-oil emulsion that you want to an oil-in-water emulsion—see "Two Kinds of Emulsions," page 296.) This is immediately remedied with the addition of more flour and causes no real problem other than fright when you see the curdled mess.

Creaming Method— Shirley's Basic Moist Sweet Cake, Version I

MAKES ONE 9-INCH LAYER

what this recipe shows

A balanced recipe (balanced weights of tenderizing and toughening ingredients) creates a successful cake.

The creaming method of mixing produces a light cake.

Dry ingredients must be well sifted for even texture.

Correct butter, sugar, egg, bowl, and room temperatures, as well as creaming time, contribute to lightness.

1¹⁄₃ cups sugar

Nonstick cooking spray with flour, such as Baker's Joy, or 1 tablespoon shortening and 1 tablespoon flour, to grease pan

Parchment or wax paper

1¹⁄₂ cups cake flour

1¹⁄₂ teaspoons baking powder

¹⁄₂ teaspoon salt

¹⁄₂ cup buttermilk at room temperature

1 teaspoon pure vanilla extract

¹⁄₄ pound (1 stick) unsalted butter at about 65°F (18°C)

2 large eggs at room temperature

3 large egg yolks at room temperature

¹⁄₃ cup vegetable or other mild-flavored oil (see Note)

1. Place the mixer bowl and whisk beater in the freezer. Measure the sugar into a zip-top plastic bag, seal, and place in the freezer to chill for about 20 minutes.

2. Place a shelf at the top of the lower third of the oven and preheat to 350°F (177°C).

continued

3. Grease a 9 × 2-inch round cake pan by spraying with nonstick cooking spray with flour or by rubbing the pan with shortening, shaking the pan with 1 tablespoon flour, and shaking out the excess. Insert a 9-inch parchment or wax paper circle in the pan and lightly spray or grease and flour.

4. Sift the flour, baking powder, and salt onto a sheet of wax paper. Resift at least once more onto another piece of wax paper. Measure the buttermilk in a glass measuring cup and add the vanilla.

5. Cream the butter on medium speed in a mixer with the whisk until light in color, about 3 minutes. Add the sugar in a steady stream with the mixer running. Continue beating the butter-sugar mixture for 3 to 4 minutes, scraping down the sides of the bowl once. Add the eggs and yolks one at a time, beating on medium speed for about 30 seconds after each addition. Continue to beat until the mixture is light and airy looking, another 1 to 2 minutes.

6. Remove the bowl from the mixer and stir in the oil. Fold in half of the flour mixture with a large rubber spatula. Scrape down the sides of the bowl, then fold in half of the buttermilk-vanilla mixture. Fold in the remaining flour and scrape down, then the remaining buttermilk-vanilla mixture.

7. Pour the batter into the prepared pan. Smooth the batter with the rubber spatula, leaving the edges a breath higher than the center. Bake until a toothpick or cake tester inserted an inch from the center comes out clean, about 35 minutes. The sides should just begin to pull away from the pan when you place the cake on the rack to cool.

8. Let the cake cool in the pan for 10 to 20 minutes. Tap the sides of the pan on the counter to loosen the cake or run a small spatula around the edge. Spray a cooling rack lightly with nonstick spray and invert the cake onto it. Peel off the parchment liner, then replace the liner on the cake with the sticky side up. Invert back onto the original cooling rack so that the cake is now right side up. Cool completely before storing or icing.

Note: *Walnut, hazelnut, or almond oil is good to use if you would like a subtle flavor note. Make sure the oil is not off tasting since nut oils turn rancid faster than some.*

Raspberries and Cream Cake

MAKES 8 SERVINGS

I have used mascarpone and whipped cream together ever since I discovered mascarpone, but when I tried cookbook author Michele Scicolone's recipe with honey I added that to this heavenly mixture. I love it and use it on fruit whenever I can. This is truly a gourmet raspberry shortcake.

what this recipe shows

Fat particles, which can melt if warm, line the bubbles in whipped cream. Very cold cream, bowl, and beaters and a cool room make whipping cream easier, and it is less likely to fall.

Care must be taken to whip cream until fairly stiff but not so stiff that it will turn to butter when folded into the other ingredients.

2 packages (10 ounces each) frozen
    raspberries, defrosted but not
    drained

3 tablespoons Chambord or other
    raspberry liqueur

1 recipe Shirley's Basic Moist Sweet
    Cake, either version (page 143
    or 146), or a plain cake of your choice

Double recipe Mascarpone Cream
    (page 245)

1 package (½ pint) fresh raspberries
    (optional)

1. Stir the defrosted raspberries and Chambord together in a medium mixing bowl and set aside.

2. When the cake has cooled, slice horizontally into 3 layers. Place the bottom layer, cut side up, on the serving dish or on a cardboard cake circle. Spoon half of the raspberry mixture, including the liquid, onto the layer. Spoon about a quarter of the Mascarpone Cream on top in large blobs and spread out carefully. Place the middle cake layer on top of the iced bottom layer. Spoon on the remaining raspberry mixture and ice as before. Place the top layer, cut side down, on the cake and ice generously with the remaining Mascarpone Cream. Refrigerate for at least 2 hours before serving. Best if refrigerated overnight. When ready to serve, arrange fresh raspberries on top. Serve cold.

Coconut Cake with Sour Cream Icing

MAKES ONE 9-INCH 3-LAYER CAKE

This cake is reminiscent of good old soggy Southern cakes at reunions and church dinners-on-the-grounds when I was a child in Georgia. It is from my sister, Joyce Hutcheson, a former food professional, who can make food taste really good in a hurry—a talent that has served her well as a busy working mom.

> **what this recipe shows**
>
> Turning the cut sides of the bottom and middle layers up and puncturing them encourage the soaking of icing into the layers.
>
> Using sour cream instead of whipped cream cuts the extreme sweetness of this cake.

1 recipe Shirley's Basic Moist Sweet
    Cake, either version (page 143 or
    146), or a plain cake of your choice

1½ cups sugar

1 large carton (16 ounces) sour cream

18 ounces (12-ounce and 6-ounce
    packages) flaked, sweetened frozen
    coconut

1 teaspoon pure vanilla extract

1. About 30 minutes before icing the cake, stir together the sugar, sour cream, coconut, and vanilla, reserving 3 tablespoons coconut to sprinkle on the top later. Let the mixture stand refrigerated for 30 minutes.

2. While the cake is still a little warm, slice horizontally into three even layers. Place the bottom layer, cut side up, on the serving dish or on a cake circle. Hold two sharp round toothpicks between your thumb and index finger, about $^1/_2$ inch apart, and carefully, without tearing the cake, poke holes all the way to the bottom across the entire layer. Ice the prepared bottom layer generously with between a quarter and a third of the icing.

3. Place the middle layer on top of the iced bottom layer. Prick full of holes as before and then ice. Place the top layer, cut side down, on the cake and ice generously with the remaining icing. (Do *not* make holes in the top layer.) Sprinkle the reserved coconut over the top. Refrigerate for at least 2 hours before serving. Best if refrigerated overnight. Serve cold.

Two-stage method In this method—also called *blending, pastry blend,* or *flour batter method*—the flour alone or with all of the dry ingredients (sugar, flour, baking powder, salt, etc.) is blended with all of the fat before the eggs and the other liquid ingredients are added.

Blending the fat and the flour coats the flour proteins with fat so that they cannot join with each other and with liquid to form gluten. A melt-in-your mouth, literally fall-apart-tender cake with a smooth, velvet texture is the result. This method is particularly well suited to high-ratio cakes. While it does not produce quite as good aeration as the creaming method, it produces cakes so tender that the mouthfeel is a sensation of lightness as the cake dissolves in your mouth. If you aerate by using appropriate mixing times, lightness is no problem with this method.

The drawbacks to this method are that the cake can be slightly heavier, and sometimes the cake can be too tender and fall apart.

Two-Stage Method— Shirley's Basic Moist Sweet Cake, Version II

MAKES ONE 9-INCH LAYER

what this recipe shows

Mixing all the fat with the dry ingredients greases the proteins, which prevents gluten formation and produces an extremely tender cake. This method does not aerate as well as the creaming method, but the cake has melt-in-your-mouth tenderness.

Nonstick cooking spray with flour, such as Baker's Joy, or 1 tablespoon shortening and 1 tablespoon flour, to grease pan

Parchment or wax paper

2 large eggs at room temperature

3 large egg yolks at room temperature

6 tablespoons *and* 2 tablespoons buttermilk ($^1/_2$ cup *total*)

1 teaspoon pure vanilla extract

$1^1/_2$ cups cake flour

$1^1/_3$ cups sugar

$1^1/_2$ teaspoons baking powder

$^1/_2$ teaspoon salt

$^1/_4$ pound (1 stick) unsalted butter, softened

$^1/_3$ cup vegetable or other mild-flavored oil (see Note)

1. Place a shelf at the top of the lower third of the oven and preheat to 350°F (177°C).

2. Grease a 9 × 2-inch round cake pan by spraying with nonstick cooking spray with flour or by rubbing the pan with shortening, shaking with 1 tablespoon flour, and shaking out excess. Line the pan with a 9-inch parchment or wax paper circle, then lightly spray or grease and flour the circle.

3. Stir the eggs, yolks, 6 tablespoons buttermilk, and the vanilla together in a medium bowl.

4. Mix the flour, sugar, baking powder, and salt in a mixer with the whisk on low speed for 30 seconds. Add the butter and oil and the remaining 2 tablespoons buttermilk. Mix on low speed to moisten the dry ingredients. Increase to medium speed and beat for 1½ minutes. Scrape down the sides. Add a third of the egg mixture and beat for 20 seconds. Repeat until all of the egg mixture is incorporated. Scrape down the sides with each addition. Pour the batter into the prepared pan and bake for about 35 minutes or until a tester inserted within an inch of the center comes out clean and the cake springs back when pressed lightly in the center. The cake won't shrink from the sides of the pan until after it's out of the oven.

5. Let the cake cool in the pan for about 10 minutes. Tap the sides of the pan on the counter to loosen the cake or run a small spatula around the edge. Invert the cake onto a cooling rack sprayed with nonstick cooking spray and peel off the parchment liner, then replace the liner on the cake with the sticky side up. Invert back onto the original cooling rack so that the cake is now right side up. Cool completely before storing or icing.

Note: *Walnut, hazelnut, or almond oil is good to use if you would like a subtle flavor note. Make sure the oil is not off tasting since nut oils turn rancid faster than some.*

Single-stage method In this method—also called *dump, one-bowl,* or *quick-mix method*—all the dry ingredients except the baking powder are sifted together, then combined with all the wet ingredients. The batter is blended with a paddle beater on low speed for 1 to 2 minutes, on a higher speed for several minutes, and finally on low speed for 2 minutes. The baking powder is added near the end of the beating time. Another version of the single-stage method is to combine all the ingredients except the eggs and then beat with the whisk on high speed for 1 minute. Finally, the eggs are stirred in on low speed for 30 seconds.

This method really works only if you use a shortening with emulsifiers—the shortening to provide aeration since there is no creaming step, the emulsifiers to help disperse the fat. Advantages of this method are its simplicity and speed, not quality.

Muffin method In this method, the eggs, milk, and melted fat are combined and then stirred together with the dry ingredients. This is a poor method of cake mixing since it gives uneven dispersion of ingredients, large cells, and a coarse crumb. When a coarse crumb is not objectionable—for instance in hearty muffins or cornbread—this method is fine.

Good-for-You Apple Bran Muffins
with Walnuts and Orange Zest

MAKES 14 MUFFINS

These muffins contain oat bran for soluble fiber, carrots and apples with vitamins and pectin for more soluble fiber, and nonfat dry milk for calcium. What a boon that something this good for you can also be so delicious!

Mixing the dry ingredients in one bowl and the wet in another and then stirring them together results in a slightly coarse crumb that is fine for hearty muffins.

It is not really coating nuts, grated carrots, etc., with flour that holds them up in the muffin; it is the fact that the batter is thick enough. In a thin batter, nuts and fruits settle to the bottom whether they have been coated with flour or not.

1 cup chopped walnuts

1 tablespoon butter

1/4 teaspoon salt

1 1/2 cups oat bran, not wheat bran

3/4 cup unbleached all-purpose flour

1/3 cup nonfat dry milk

3/4 cup packed dark brown sugar

1 tablespoon ground cinnamon

2 teaspoons salt

1 tablespoon baking powder

1 medium to large carrot, grated (about 1 cup)

1 Granny Smith apple, peeled, cored, and coarsely chopped

Finely grated zest of 2 oranges

1 large egg

2 large egg whites

1 cup buttermilk

1/4 cup canola, corn, walnut, or almond oil

3 tablespoons canned crushed pineapple, drained

Nonstick cooking spray

1. Arrange shelf just below the center of the oven and preheat the oven to 350°F (177°C).

2. Roast the walnuts on a baking sheet for 10 to 12 minutes. While the nuts are hot, stir in the butter and sprinkle with salt. Set aside.

3. Raise the oven temperature to 450°F (232°C).

4. Combine the oat bran, flour, dry milk, brown sugar, cinnamon, salt, and baking powder in a large mixing bowl. Work any lumps out of the sugar with your fingers if necessary. Add the grated carrot, apple, orange zest, and nuts. Stir to combine.

5. Beat together the egg, the whites, buttermilk, oil, pineapple, and vanilla in a medium mixing bowl. Pour the wet mixture into the dry mixture and stir to combine.

6. Spray a standard-size muffin tin with nonstick cooking spray, grease well, or use paper liners. Spoon 1/3 cup muffin batter into each cup; this should just fill each cup to the brim. You will have

enough to fill a twelve-muffin pan with enough batter left for a few more muffins. If filling two to three cups in a six-muffin pan, fill the empty spaces with water so that they will not burn. Bake until browned, about 20 to 25 minutes. Tap the sides of the pan on the counter to loosen the muffins and turn out onto a rack to cool. Serve warm or at room temperature.

All-Time Favorite Sour Cream Cornbread

MAKES 10 SERVINGS

Cornbread can be considered a savory cake, and certainly it is one in which you do not mind a coarse crumb. In this very easy recipe, the sour cream and creamed corn make very moist cornbread—a favorite Southern recipe, of which there are many variations. One is published as Mrs. Dean Rusk's cornbread. Nathalie Dupree has a version called "Snackin Cornbread" in her book, *New Southern Cooking*.

what this recipe shows

Oil, sour cream, and creamed corn make this cornbread extremely moist.

3 large eggs

1 1/2 cups canned creamed corn

1 1/2 cups (about 14 ounces) sour cream

3/4 cup corn, canola, or vegetable oil

1 1/2 cups cornbread mix or self-rising cornmeal, slightly packed

3/4 teaspoon salt

1/2 teaspoon baking powder

Nonstick cooking spray

3 tablespoons butter, melted

1. Preheat the oven to 425°F (218°C).

2. Beat the eggs slightly in a medium mixing bowl. Stir in the creamed corn, sour cream, and oil. Add the cornbread mix, salt, and baking powder. Stir to blend well. Spray a 9-inch skillet with an ovenproof handle with nonstick cooking spray (see Note). Pour in the batter.

3. Place the skillet on a burner over medium-high heat for 1 minute. Then place on a shelf in the upper third of the oven. Turn the oven down to 375°F (191°C), and bake for 35 to 40 minutes. Slide under the broiler, about 4 inches from the heat for 45 to 60 seconds to brown the top. Watch carefully. Brush the top with melted butter for a shiny finish.

Note: *Instead of a skillet, spray a 9-inch round cake pan with nonstick cooking spray. Pour the batter into the pan and place on a shelf in the upper third of the oven. Turn the oven down to 375°F (191°C) and bake for 40 minutes. Brown under the broiler as directed and brush with melted butter.*

Combination conventional and sponge (whipping) method This method combines the creaming method with the sponge method, in which you beat the eggs (or egg whites) with a portion of the sugar for additional aeration. This can produce lightness and good volume; however, when you beat egg whites, you denature and partially "cook" the whites. As mentioned earlier, they are no longer the strong contributor to structure that they were, so you need enough raw egg, milk, and flour proteins to cook and set the batter.

This method is outstanding when you need maximum volume. It is sometimes used in traditional pound cakes that are fine-grained and heavy since they are made without baking powder.

Dissolved-sugar method In this method the sugar is dissolved in half its weight of the liquid. Next, the dry ingredients (flour, salt, and baking powder) and fat (emulsified shortening unless extra yolks are used) are added and mixed for 5 minutes. Then the remaining liquid, flavoring, and eggs are blended in on low speed for 1 minute. Using dissolved sugar produces a better crust color and a more tender crust. This method provides excellent aeration by blending the dry ingredients with fat and fine texture and tenderness by coating the flour proteins with fat.

Baking the cake

The oven temperature and the type of cake pan both influence the way a cake bakes.

Oven temperature: precision matters Oven temperature affects the texture and shape of the cake. You should assume that your oven temperature control is *incorrect* (alas, even those that were recently calibrated). I find in traveling around that most ovens are off by 50° to 100°F (28° to 56°C) and that many also have a very uneven distribution of heat. For many types of baking and roasting, a watchful eye will produce good results even though the oven temperature is incorrect. But in cake baking the wrong temperature can lead to disaster.

Two accurate oven thermometers are a good investment. With both thermometers in the oven, check your oven at different locations at several temperature settings. Then, relocate your thermometers and check at those settings again. This will give you an idea of how your oven really works.

At lower temperatures (300°F/149°C), the cake must remain in the oven longer, and consequently the air cells have more time to expand and enlarge. They may even run together to produce very large cells before the egg, milk, and flour proteins coagulate and set the cake. The result is a coarse-textured cake. Most cake recipes recommend a higher temperature (350°F/177°C) for a finer texture.

At 375°F (191°C), the cake sets even sooner, with smaller air bubbles, and you get an even finer-textured cake. However, at too high an oven temperature, the outer edges of the cake will set while the liquid center is still rising, producing those great volcano cakes. While this is undesirable for cakes, it's perfect for muffins, which *should* look like volcanoes. If your muffins are not peaking as much as you would like, you should try a higher oven temperature.

In general, thin (layer) cakes can be cooked at higher temperatures than thick cakes such as Bundt, loaf, etc. The lower temperatures allow the center of a deep cake to cook before its outside becomes overcooked and overbrowned.

Bake-Even Strips—remedy for uneven cooking You may have seen a baker dampen a dish towel, fold it over and over until it is narrow like a flat rope, then tie it around a cake pan just before placing the pan in the oven. As the cake is baking, water evaporates from the towel and cools the outer edges of the pan, which would ordinarily get the hottest. This is a great technique, not just for cakes but for anything baked in a large pan, where the edges brown and dry out before the center is done. Try it for large pans of dressing at Thanksgiving and for the Old-fashioned Grated Sweet Potato Pudding (page 338).

Wilton makes easy-to-use, reusable strips called Bake-Even Strips that are ready to soak in water and pin around pans (see Sources—Bake-Even Strips, page 477). They are inexpensive and extremely helpful for more even baking.

Is it done? The first thing to suspect if a cake falls in the center is that it is not completely cooked. If a piece of broom straw, a toothpick, or a thin wire inserted into the center of the cake comes out dry, with no wet crumbs on it, the cake is done. When you cut into a fallen cake, you can readily tell whether it is dry or very moist—undercooked—in the center. If it is dry, the probable cause of the fall was too much leavening.

Also, some cakes are intentionally undercooked because the goal is a very moist cake; Alice Medrich's Queen of California Cake, which appeared in *House and Garden* in 1980, was just such a cake. Alice uses a short cooking time, and says that a cake tester inserted only 1 inch from the outer edge should come out wet. The cake rises and sets for a space of about 1 inch from the edge, the rest remaining somewhat sunken. Alice says "not to worry." Simply take your hand and press down the risen edge so that it is level with the center.

All pans are not equal Pan size, color, material, and texture are all important in the exacting art of cake baking. Choosing the correct pan size for the amount of batter is very important. If there is too little batter in a pan, and the cake does not rise to the top, the sides shield the top and it does not bake properly. Too much batter, on the other hand, overflows the pan and also yields inferior texture.

The batter should fill the pan from one-half to two-thirds full. If you have employed a mixing method that you know yields outstanding aeration and good baked volume, fill nearer half than two-thirds. To estimate the volume of an odd-shaped pan, fill the pan with water, measure how much water it took, and realize that you are going to need enough batter for about two-thirds that amount.

The size of the pan can create or solve some baking problems. Tunnels running between the outside and the center of a cake form more easily in deeper batters at higher oven temperatures. A switch from an 8-inch pan to a 9-inch pan may solve the problem. Other causes of tunnels are too high an oven temperature and too much gluten development.

Darker pans absorb heat in the oven better than light pans. Cakes in dark pans bake faster; however, dark pans create a dark crust. Dull pans absorb more heat than shiny, reflective pans. The material from which the pan is made determines its ability to conduct and distribute heat evenly. Glass pans bake even faster than dark metal pans because the batter receives radiant energy through the clear sides as well as heat from conduction. Most cake bakers prefer heavy, dull aluminum, straight-sided pans.

Shortened cakes require well-greased pans or greased *and* lined pans. Some recipes advise lining the bottom of the pan with parchment and greasing with shortening, then dusting with flour. There are excellent nonstick cooking sprays that contain lecithin and flour. An even spraying of these does a superb job.

at a *glance*
Cake problems

Problem	Probable Cause

Cake falls	Undercooked
	    Too much baking powder or soda
	    Most of the egg whites beaten to a foam

Volcano cake	Oven temperature too high
	    Pan too small
	    Recipe not balanced

Tunnels in cake	Oven temperature too high
	    Cake was overmixed, and too much gluten developed
	    Batter too deep in the pan

Cake tough	Too much gluten developed for one or more reasons:
	    Flour too high in protein
	    Too little sugar
	    Too little fat
	    Cake overmixed after flour and liquid are together

Cake crumbly	Not enough gluten developed for one or more reasons:
	    Too much sugar
	    Too much fat
	    Too little liquid
	    Cake undermixed
	    Improper mixing procedure

Cake grainy	Oven temperature too low Poor mixing procedure

Cake heavy	Oven temperature too high
	    Poor mixing procedure
	    Too much sugar and/or fat

Fruit or nuts sink	Batter not thick enough

High-altitude adjustments

Because there is less air pressure on top of cakes baked at high altitudes, liquids evaporate more easily, and gas bubbles in the cake rise to the top and pop more readily. There is then less liquid in the cake, and the other ingredients become concentrated—for example, you may end up with too much sugar. Too much sugar prevents the cake from setting. Losing the air bubbles means no bubbles for the leavening to inflate, causing the cake to fall—alas, a soggy, heavy cake.

The two problems are the same regardless of cake type, but different cake formulas require different solutions. Some experimentation on what works best for a specific recipe at a particular altitude will be required. Even the experts do not make identical recommendations.

For baking powder (shortened or butter) cakes or high-ratio cakes, Ernst J. Pyler's book *Baking Science and Technology* recommends a 15 percent reduction of leavening at 2,000 feet (just over $1/8$ teaspoon less per teaspoon called for in the recipe) up to a 60 percent reduction at 8,000 feet (over $1/2$ teaspoon cutback per teaspoon). However, the USDA's general high-altitude recommendations are to use $1/8$ teaspoon less leavening per teaspoon called for (a reduction of 12.5 percent) at 3,000 feet, up to a decrease of $1/4$ teaspoon per teaspoon (25 percent) at 7,000 feet.

For an altitude of 7,000 feet (Denver is only 5,280 feet in altitude), the USDA recommends a reduction of $1/4$ teaspoon per teaspoon called for, and *Baking Science and Technology* recommends a reduction of about $1/2$ teaspoon per teaspoon. Trial and error are necessary to arrive at the exact procedures best for a specific location.

The problem is slightly less for angel food cakes, sponge cakes, etc. (cakes without chemical leaveners such as baking powder or baking soda). A reduction in sugar compensates for the liquid loss and lets the eggs coagulate sooner to hold the precious bubbles or to prevent their enlarging to such an extent that the texture is extremely coarse. Some authorities recommend an increase in oven temperature of 25°F (14°C) to promote faster setting, finer texture, and less bubble loss.

The solution to the problem of liquid loss varies according to the cake formula, but an increase in whole eggs or egg whites helps in two ways. First, it provides an increase in liquid to make up for liquid loss, and, second, more eggs provide additional proteins to set the cake more quickly. *Baking Science and Technology* recommends a 2.5 percent increase in eggs at an altitude of 3,500 feet, up to a 15 percent increase at 7,500 feet.

With the liquid loss, sugar concentration becomes higher, making the cake set more slowly and therefore lose bubbles over a longer period. To solve this problem, you may either increase the amount of flour in the cake—recommendations run from a 2 percent increase in flour at 3,500 feet to a 10 percent increase at 8,000 feet—or decrease the amount of sugar. You will probably have to experiment with your specific recipe to arrive at the best solution.

Richly browned, crisp, nongreasy fried food is a taste joy but, unfortunately, an accomplishment with many challenges.

The incomparable taste of crisp, crusty fried food

Deep-fat frying

What really happens in frying? Let's examine French fries. When the potato sticks hit the hot fat, there is an instant *sssssszzzzzzzz*. The sudden high heat turns moisture near the surface of the potatoes to steam. The steam rushes out and sizzles as it hits the hot fat. Water in the center of the fry rushes outward to replace the water that has been lost. The steam rushing out cools the surface so it does not char. As the rapid rush of steam from the fry slows, the surface heats up and begins to brown. A little hot oil seeps into some of the steam channels and heats the starch in the center of the fry just enough for it to absorb water, swell, and cook through.

A perfect fry loses most of its free internal moisture so that it will not become limp. But it does not lose the water in the cooked puffed starch granules. The fry is a beautiful golden brown—crisp on the surface, firm, with a perfectly cooked center. Only a small amount of oil has been absorbed on and near the surface.

In an overcooked fry, or one cooked in deteriorated oil, when all the moisture has steamed off, hot oil rushes in and heats the center of the fry so hot that the puffed starch granules pop and lose their moisture too. There is no outward pressure of steam. Fat is literally sucked into the fry. It becomes greasy and sunken.

If the fry is cooked in oil that is too hot, the surface will crust and seal before much moisture is removed. As it cools, it becomes limp and soggy. The same principles apply to fried foods in general.

As you can see, frying is not a simple cooking technique. Water turns to steam, starches swell, proteins cook, sugars caramelize, and crust is formed. All of the following affect the product: the condition of the fat, the type and flavor of the fat, the moisture content of the food, the amount of surface for the volume of food, the porosity of the food (which influences how fast the water and/or fat can move through the food), the temperature and amount of food and fat, the cooking time, and whether the food has been precooked. There are so many variables in perfect frying that even scientists specializing in the field admit that frying is as much art as science.

The frying fat

Whether solid or liquid, saturated or unsaturated, fats deteriorate. In frying, the condition of the fat is as important as or even more important than the type of fat.

Top priority: purity

Products of deterioration are devastating in frying fats. Foods fried in deteriorated fats are extremely greasy, dark in color, and off tasting. Deterioration products lower fats' surface tension and cause them to soak into foods much faster than fresh fats. Proteins and sugars from previously fried foods contaminate used oil so that food fried later is dark in color. Foods can taste from mildly rancid to inedible. Some of these products of rancidity are toxic, and others pose a danger to health. Oxidized fats have been implicated in the initial lesions in cardiovascular disease. Deteriorated fats also become thick and can coat the fryer and reduce the amount of heat getting to the fat. In addition, such fats are a major safety risk.

Fats deteriorate by combining with oxygen, moisture, and food particles. High temperatures cause rapid breakdown of fats.

Fat combines with oxygen Even before frying begins, oxygen contaminates fat on the surface. Most commonly, the heat comes from the bottom, whether in a deep saucepan on the stovetop or in a fryer. As the fat next to the heat gets hot, it becomes less dense, rises to the top, comes into contact with oxygen in the air, and starts deteriorating. Fresh fat sinks to the bottom, gets hot, rises to the top, starts combining with oxygen, and so on. As soon as you start heating fat, the process of oxygen contamination and deterioration speeds up.

Oils for commercial use contain trace amounts of certain silicones, which form a film on the surface of the oil, preventing direct contact with oxygen in the air. This greatly slows deterioration and also reduces foaming. At one time oils sold in the grocery store for home use had a trace of silicones added. However, public concern over additives caused manufacturers to stop including this useful protective agent.

Moisture and food particles The liquid coming out of foods is not just water but may contain proteins, dissolved sugar, and other food substances that both contaminate the fat and enhance browning of the crust of food fried later in that fat. This is why food fried in fresh oil will be lighter in color than food fried for the same time and at the same temperature in used oil. Broken-off food particles not only contribute to discoloration but fall to the bottom and burn to produce terrible off flavors. To prevent fallen particles from burning, commercial fryers have heating rods just a little above the bottom of the fryer.

Deteriorated fats and rancidity

I do not reuse home-use fats and do not advise doing so. Quality and taste are compromised. Also, with home-use fats, which are not designed for reuse, there are major health concerns with rancidity and major safety risks.

Fats become rancid when they combine with other substances like oxygen or water. The short-chain fatty acids produced in these reactions taste very bad and can wreck the palatability of any food. Some of these rancid compounds are actually poisonous, and some have been implicated in hardening and blocking of the arteries. Research indicates that these oxides and peroxides of fat may play a role in the initial stages of cardiovascular disease (see page 178). In a paper presented at the Institute of Food Technologists' 1988 convention, Dr. Paul Addis, a cardiovascular researcher at the University of Minnesota, urged food technologists to make every effort to reduce conditions that can lead to rancidity in foods.

Enzymes, heat, light, and certain metals all promote rancidity, while antioxidants can slow it down.

Enzymes Enzymes in certain foods enhance the reactions of fats with water. Butter and cream contain such enzymes. The terrible taste of rancid butter is caused by butyric acid, a short-chain fatty acid formed when fat combines with water. Even butter that is just beginning to turn rancid can wreck the taste of a dish. Some chefs insist on salt-free butter so that they can easily smell if it is on the verge of rancidity. (Salt lengthens the shelf life of butter, but some feel that salt can hide the beginning smell and taste of rancidity, making it more difficult to identify butter that will give an off taste to a dish.)

Proper refrigeration of butter and cream slows down these enzymes' production of rancid compounds but does not totally prevent it.

Heat and light Energy in the form of heat, or just sunlight, can cause the breakdown of unsaturated fats. Even oil kept in a tightly sealed bottle in a cool place can become rancid from sunlight alone. For this reason, some processors pack oils and olive oils in cans. Crisco tried to market its oil in a brown bottle for better shelf life; unfortunately, consumers refused to accept the brown color. You should store fats in a cool, dark place. In my warm Southern climate, I refrigerate oil, shortening, and lard. Also remember that it is always best to buy fats in small quantities to have a fresh supply.

Metal catalysts Both iron and copper can catalyze (speed up) oxidative rancidity in foods. For this reason, stainless steel is the preferred material for fryers that maintain fats at high temperatures for long periods of time. However, there is no problem with using a heavy iron skillet for short-term frying, particularly if the fat is not reused.

Antioxidants Fortunately, nature has given us a helping hand in preventing oxidation and the resulting rancidity of fats. Many vegetable oils contain naturally occurring vitamin E or other antioxidants that protect them from rancidity. Safflower oil, on the other hand, is so unsaturated that it turns rancid very fast; antioxidants may be added to it in processing to give better shelf life. You should refrigerate safflower oil.

Deteriorated fats and safety

Even one single use of fat *at a high temperature* can decrease by 100°F (56°C) the temperature at which the fat bursts into flame (the flash point). This means that a perfectly safe fat, with a flash point somewhere around 500°F or so (260°C or so), might burst into flame at a dangerously low temperature after just one use at a high temperature.

This happened to me once while teaching a variation of Grace Chu's Millionaire Chicken, which calls for a huge cloud of rice sticks on top. When I teach this recipe, I fry the rice sticks (done in very hot fat—

425°F/218°C) early in the class and leave the assembly for the end, simulating the way home cooks would do it for company.

At the end of one class, I was doing a final stir-fry and turned to my assistant for oil. The oil in the wok in which we had earlier fried the rice sticks was sitting on a back counter. She ladled up a couple of spoons of this. I heated the *empty* wok very hot, and drizzled the oil in around the upper edges. The instant that the oil from my ladle hit the hot, empty wok, flames shot up! I placed the lid on the wok, which quickly extinguished the flames. Had there been a large amount of oil, and had the wok lid not been handy, this could have been a dangerous situation. It certainly impressed me with the potential dangers of used oil.

Never use frying fat twice. Even fat with a high flash point can burst into flame at a dangerously low temperature after just one high-temperature use!

Szechwan Chicken Salad

MAKES 6 TO 8 SERVINGS

This recipe started with Grace Chu's Millionaire Chicken from *The Pleasures of Chinese Cooking,* one of my favorite dishes—the crisp lettuce, the limp chicken, and the hot spiciness of the sauce, all in one bite. I added a crunchy rice-noodle cloud and hoisin sauce among other changes.

what this recipe shows

Heating the oil to the high temperature necessary to make the rice noodles cook makes it dangerous to reuse.

Oil is heated with ginger, scallions, and peppers to extract flavors, then cooled before adding the garlic and more delicate hoisin sauce.

6 cups water

6 chicken breast halves, with bone

1 medium yellow onion, quartered

2 ribs celery, cut into 1-inch pieces

1 head iceberg lettuce

1 slice fresh ginger, about 1 inch in diameter by 1/3 inch thick

1 tablespoon Szechwan peppercorns

5 scallions, green parts included, chopped

1/8 teaspoon red pepper flakes

1/4 cup vegetable oil

3 cloves garlic, minced

3 tablespoons hoisin sauce (see Notes)

3 tablespoon soy sauce

3 tablespoons dark corn syrup

3 to 4 cups fresh peanut or vegetable oil for deep frying

1 layer (about 4 x 5 inches) rice noodles (rice sticks)

1. Bring the water (enough to cover chicken pieces), chicken pieces, onion, and celery to a boil in a large, heavy saucepan over high heat. When the water begins to boil, turn it down to a low simmer. Simmer for 3 minutes. Cover and remove from the heat. Let stand 20 minutes. Remove the chicken from the water and let cool. Tear the meat with the grain into thin shreds.

2. Slice the lettuce into medium-thin shreds with a stainless-steel knife. Cover a large serving platter with the sliced lettuce. Spread the chicken evenly over the lettuce. Cover with plastic wrap and refrigerate until serving time.

3. Heat the ginger, peppercorns, scallions, and red pepper flakes in the oil in a small saucepan until bubbling hot. Let cool, then add the garlic, hoisin sauce, soy sauce, and corn syrup. When ready to serve, pour this spicy sauce over the lettuce and chicken and toss gently to coat.

4. Heat fresh oil in a wok until it almost begins to smoke (425°F/218°C), then drop the rice noodles into the hot oil. The noodles expand so much that they literally jump up out of the wok. With two large spatulas, remove the noodle cloud instantly from the oil and drain on an unrecycled brown paper bag or paper towels. This hot oil is dangerous; turn off the heat immediately and let it cool. Do *not* reuse this oil (see Notes). To serve, place the noodle cloud on top of the sauced chicken and lettuce.

Notes: *Hoisin sauces differ greatly in taste from brand to brand. The success of this dish depends on using a good hoisin. My favorite is Koon Chun.*

The noodle cloud may be prepared early in the day and left on the counter.

The whole dish can be prepared ahead and assembled at the last minute. Prepare the chicken and lettuce, cover, and keep refrigerated. Prepare the sauce, which may be left at room temperature. Cook the rice noodles and leave them at room temperature. To assemble, pour the sauce over the chicken and lettuce and toss. Top with the noodle cloud.

SAFETY IN DEEP-FAT FRYING

Deep frying is probably the most dangerous kitchen operation. A fryer should always be a minimum of 3 inches taller than the fat with the food in it. This will give you a few seconds of reaction time should problems arise. Restaurants have hoods with automatic fire extinguishers over the frying areas. Home cooks should keep a good working fire extinguisher in an easy-to-reach spot not too close to the stove. If the extinguisher is too close to the stove, the flames may prevent your reaching it.

When a pan of oil bursts into flame, you may have the urge to pick it up and walk out of the house with it, but it is very easy to fall or to drop such a dangerous burden. Many terrible fires have been started in just this way. Try to smother the fire by placing a lid or a large pot over it.

You can extinguish small fires by sprinkling with salt or baking soda to smother. Water, unless it is in a fine mist, is usually not effective on an oil fire. The oil simply floats on the water and continues to burn. Also, the water can explosively turn to steam and spread the hot oil. A very fine mist of water is much more effective at smothering flames than a thick stream.

Fire spreads faster than you can imagine. Move as fast as you can to extinguish flames, including crying out for help. A few precautions like a working fire extinguisher can give you peace of mind and may save your home. Keep a big wok lid or stockpot lid propped by the stove or keep a cylinder of salt on the stove—these precautions can help quickly extinguish a fire.

Which fat to use?

It should be clear by now that a suitable fat should have a smoke point considerably higher than the planned frying temperature and should be fresh and contain few products of deterioration. Frying fats can either be solid or liquids. A product fried in fat that is solid at room temperature has a slightly hazy, translucent appearance at cool room temperature, while foods fried in an oil will look glossy and transparent. In products like doughnuts, a solid fat would be preferred since icings and glazes will stick to it better than on a slippery oil.

As far as stability is concerned, one single double bond in a fatty acid (point of unsaturation) increases the oxidation (deterioration) rate tenfold. A second double bond—like those in polyunsaturated oils—makes the oil one hundred times more susceptible to oxidation. It is no wonder that in the past processors used saturated fats. Rancid products are dangerous to our health too. New crops like high oleic sunflower oil and low linolenic soybean oil maximize single double bonds and minimize double and triple double bonds, making for more stable and healthful oils. The more healthful unsaturated fats can be used if the oil is not going to be reused. Considerations like flavor and smoke point may be more important than saturation.

In highly seasoned food, the characteristic flavor of a fat may not be noticeable, but in delicate foods it is. In refined oils it may be difficult to identify the oil's source. With less refined oils, it may be easy to detect the nut, grain, seed, or fruit from whence it came. Even in refined oils, peanut and corn oils have slightly nutty overtones, while soy and canola have flavor notes closer to a bean.

A list of smoke points for some vegetable oils appears on page 159. These temperatures are for fresh oil, but oils differ within the same type, so the temperatures are only approximate. The amount of free fatty acids (see page 176) really determines when a fat will smoke or burst into flames. With a single use at high temperature, a fat can form so many free fatty acids that it can burst into flames at a temperature 100°F (56°C) lower than its normal smoke point.

Vegetable oils As you can see from the table, most refined vegetable oils have relatively high smoke points and are a good choice for frying. Cooks frequently use peanut, corn, canola, or soy oil.

Some vegetable oils (like extra-virgin olive oil or sesame oil) have intense flavors. They are also natural, very complex mixtures of many different fats, some of which break down at higher temperatures. They are not suitable for everything and are not used for deep-fat frying. A dash here and a dash there of intense-flavored oils like sesame oil and extra-virgin olive oil are excellent as flavorings and perfect in even larger amounts for some dishes but not for high-temperature frying.

Monounsaturated vegetable oils (olive oil and canola) Like most vegetable oils, olive oil and canola oil have relatively high smoke points, 410°F (210°C) for "pure" olive oil, and 437°F (225°C) for rapeseed (canola) oil. Extra-virgin and virgin olive oils have lower smoke points. In spite of its high smoke point, canola seems to break down with frying at higher temperatures and can leave a varnishlike coating on pans. Some processors make a canola-corn oil blend that fries better than pure canola.

Canola oil and some of the refined olive oils are mild in flavor and can be used in most foods without interfering with the taste. Virgin and extra-virgin olive oils have a stronger taste, which is an enhancement in some dishes but undesirable in others. From a health point of view, olive oil and canola oil are vegetable oils with a high percentage of natural monounsaturated fat.

Polyunsaturated vegetable oils (safflower oil) These oils have relatively high smoke points and are suitable for frying. Their handicap is that with their multiple double bonds they readily react (one hundred times faster than saturated fats) with oxygen and moisture and become rancid. Dishes prepared with these oils, as well as the oils themselves, can become rancid easily. They are suitable for dishes that are prepared and eaten right away or refrigerated promptly. In warm climates, you should refrigerate these oils for better shelf life.

APPROXIMATE SMOKE POINTS FOR FRESH HOME-USE OILS		
Type of Oil	°F	°C

(Note that these values are for fresh oil and can drop dramatically with a single use.)

Type of Oil	°F	°C
Sunflower	392	200
Corn (refined)	410	210
Olive ("pure")	410	210
Peanut	410	210
Sesame	410	210
Soybean	410	210
Rapeseed*	437	225
Grapeseed	446	230
Cottonseed	450	232
Safflower	450	232
Almond	495	257
Rice Bran	500	260
Avocado	520	271

*Canola oil comes from strains of rapeseed with low erucic acid content.

Mark Lake and Judy Ridgway. *The Simon & Schuster Pocket Guide to Oils, Vinegars and Seasonings.* New York: A Fireside Book by Simon & Schuster, 1989.

Daniel Swern (ed.). *Bailey's Industrial Oil and Fat Products,* 4th ed. New York: John Wiley and Sons, 1979, as reproduced in Michele Anna Jordan. *The Good Cook's Book of Oil & Vinegar.* Reading, Mass.: Addison-Wesley Publishing Company, 1992.

The good news with polyunsaturated oils is that there are new plant strains for high oleic sunflower oil and low linolenic soybean oil that maximize the single double bonds and minimize the double and triple double bonds, producing more stable and healthful oils.

Saturated vegetable oils (coconut oil) Saturated vegetable oils like coconut oil are very stable. They do not get an off taste easily, and foods fried with them do not turn rancid easily. Saturated oils are not absorbed by foods as easily as other oils, and they produce a crisp fried product with a good shelf life. It is no wonder that food producers have found these fats attractive. From a health standpoint, these oils are good in that they do not turn rancid easily, but bad in that some of the saturated fats raise serum cholesterol.

Vegetable shortenings Most solid shortenings available in grocery stores can safely be used only at lower frying temperatures. Some of the solid vegetable shortenings have mono- and diglycerides added. These emulsifiers make shortenings superior for use in cakes, but they also lower the smoke point (to about 370°F/188°C), thus making the shortenings less suitable for high-temperature frying.

Shortening blends—vegetable and animal fats

In the past, restaurants and food companies generally used specially formulated solid shortenings for commercial frying because they are relatively high in saturated fats that do not react or combine with other substances as easily as unsaturated fat. These hydrogenated fats are more durable and have much better shelf life than unsaturated fats. Because of this, neither the fat itself nor the food fried in it turns rancid as quickly or easily. However, because of consumer pressure, industry and restaurants are switching to unsaturated fats. The issue is complex because rancid food products that may result from unsaturated fat are dangerous, too (see page 155). Thank goodness for the new crops like high oleic sunflower oil and low linolenic soybean oil.

Margarines and spreads

Spreads are all over the place as far as the amount and type of fats they contain are concerned. To be labeled regular *margarine,* the product must have a certain amount of fat (80 percent), but *spreads* can have any amount of fat. In general, these products are poor for frying. Some of them do not melt until very hot, and others come totally apart with the slightest heat. Some of the spreads have very high water contents—over 50 percent.

From a health point of view, the firmer the more saturated. Solid margarines are firm and solid because they are hydrogenated. This means that they now contain a high percentage of saturated fat or the trans configuration of monounsaturated fat (see page 176), even though the initial product was unsaturated vegetable oil. Some of the fats produced by hydrogenation elevate serum cholesterol. Some of these products may be good for you, but some are not.

Butter

Nothing tastes like real butter. Unfortunately, butter is not good for frying because the milk solids in butter settle to the bottom and burn, even at *low* frying temperatures (about 250°F/121°C). To solve this problem you can clarify butter by removing the solids to get pure fat and then use it for frying.

Clarified butter and ghee

Clarified butter (pure butter fat with the solids removed) is not as flavorful as real butter with milk solids, but it certainly has a more buttery taste than oils, and you can fry with it without burning.

To clarify butter, I used to melt the butter until it foamed, skim all the foam from the top, and place the butter in the refrigerator. The pure fat solidified. I could then simply pick it up in a block, leaving the whey and soluble milk proteins in the liquid underneath.

I changed my procedure after Paul Prudhomme asked me to clarify butter for him for a cooking demonstration. He kept an eye on me as I melted the butter and carefully removed all the foam on top. When I started for the refrigerator to chill and solidify the fat for easy removal, he yelled at me, "Don't you dare put my hot butter in that refrigerator! It will pick up the smell of everything in there."

He was right, of course; fats are great carriers of flavor, and warm fat, in particular, would pick up the smell and taste of everything in the refrigerator. A better way to clarify butter is to melt, remove foam, and then carefully pour off the butter fat, leaving the liquid whey and milk solids behind. This clarified butter can then be used for frying.

Butter is clarified in India (and called *ghee*) because of lack of refrigeration. Since clarified butter has had the other dairy products removed from the fat, it will not spoil like unrefrigerated butter. In India, in the clarifying process, the butter is heated until *all* the water boils off and all the solids settle to the bottom and brown, giving the butter a nutty flavor.

Chicken, duck, and goose fat

With our health fears, we may have overreacted to all animal fat. Chicken fat contains about 45 percent monounsaturated fat, 21 percent polyunsaturated fat, and 30 percent saturated fat. It does contain 85 mg cholesterol per 100 grams. Duck fat has a similar profile, with 49 percent monounsaturated fat, 13 percent polyunsaturated fat, and 33 percent saturated fat. Goose fat is 57 percent monounsaturated fat, 11 percent polyunsaturated fat, and 28 percent saturated fat. These fats add good flavor to foods, and duck fat is prized in French cuisine. Some of the hydrogenated vegetable oils in margarines probably elevate serum cholesterol more than fats from fowls.

Lard Like chicken, duck, and goose fat, lard is a blend and is not all bad. It is 45 percent monounsaturated fat, 11 percent polyunsaturated fat, and 39 percent saturated fat. It does contain 95 mg cholesterol per 100 grams, which is less than butter at 219 mg per 100 grams.

The food to be fried

The type of food, its moisture content, its size and shape, and its porosity (which controls how fast moisture and fat flow through) all influence its frying.

Moisture content of the food

Different foods contain different amounts of moisture, and even different varieties of the same food can vary considerably in moisture content. Low-moisture Northwest-grown Russet Burbank potatoes (Idahos) contain between 20 and 23 percent solids, while most new potatoes and East Coast varieties have a much higher moisture content. The difference is important in determining proper cooking time and temperature. When a potato is cooked at a high temperature, a lot of moisture may be trapped inside the crust and, as it cools, it becomes limp; however, if the fry runs completely out of moisture during frying, fat is pulled in.

With small thin pieces of food like French fries, which have a lot of surface for the amount of food and can run out of moisture fast, an extra minute of cooking time or a few degrees higher in temperature can mean greasy food. This is not to say that a low-moisture potato itself is undesirable. The finest fries are made with high-starch, lower-moisture potatoes like Northwest-grown Russet Burbanks.

Ideally, as the starch absorbs water and cooks, the interiors of fries dry by the time the crust forms. There is no moisture trapped in the fry to make it soggy. There is little margin for error, but if you can get a combination of frying time and temperature to just dry but not to the point of grease absorption, the fries will be extraordinarily crisp and wonderful.

With thick pieces of food, like fried chicken, the fat needs to be at a temperature low enough for the food to cook through before it runs out of moisture and starts absorbing fat.

Surface area

Ideally, food to be fried should be uniform in size so that all pieces are done at the same time. Small, thin pieces have a lot of surface area for the volume of food and lose moisture rapidly. Because of this large surface area and rapid moisture loss, thin sticks will absorb fat much faster than the same weight of potatoes cut into thick pieces. Thin pieces cook through fast and should be fried a short time at higher temperatures.

Ounce for ounce, thin potato sticks will absorb fat much faster than thick ones when deep-fried.

Larger pieces of food that do not run out of moisture so fast can be cooked for longer times at lower temperatures to allow the center to cook through before the surface burns.

Porosity

Pieces of dense food must be fried more slowly than pieces of light, porous food, through which water and/or oil can move rapidly to conduct heat and cook faster.

Maintaining the heat while frying

The temperature of the fat, the temperature of the food, the amount of fat, and the amount of food are all intertwined in producing a well-fried food. At high temperatures, foods brown quickly and must be removed from the fat well before they run out of moisture, soak in fat, and get greasy. The shorter the cooking time, the less fat is absorbed. So, indirectly, a high fat temperature prevents greasy products. High temperatures are ideal for small, thin pieces of food.

However, larger pieces of food in high-temperature fat burn on the outside before the inside cooks.

The fryer needs to have enough hot fat to maintain the temperature when the food is added. Too much cold food dropped into a small amount of hot fat reduces the temperature drastically. Throughout the entire frying time, the fat needs to be hot enough to keep the moisture in the food at a boil and producing steam. If pressure from steam is not pushing out, grease will be pulled in.

Even though the fat is hot before the food is added, the food is cold and sometimes wet. Unless there is a large amount of hot fat, the temperature of the fat is lowered instantly (adding cold food is like adding lumps of ice), and any moisture on the food vaporizes and evaporates from the surface of the oil and cools it even more. The food surface should be as dry as possible. Dry food and a large amount of hot fat make it much easier to maintain the temperature.

The amount of food added to the fryer at one time is important for maintaining the temperature. If so much cold food is added that the fryer cannot recover fast to bring the temperature back up, grease can be sucked into the food. Fast-food restaurants may produce greasy food at rush time if their employees succumb to the temptation to overload the fryer.

Quick-recovery fryers are a blessing to the fast-food industry; they heat so fast that one can get away with a little deviation from good frying practices and still produce nongreasy food. Home fryers are available with thermostatic controls that approximate the commercial fryers. Since they contain a smaller volume of fat, they will be subject to wider temperature swings but, if not overloaded, can give good results. Use a fryer that is 3 inches deeper than the oil with the food in it and bring the fat back up to temperature before another batch is added.

For thin food like French fries, the Potato Board recommends frying at 390°F (199°C), while the Idaho Potato Commission recommends 375°F (191°C). For precooking, cook fries at 350°F (177°C) until the surface begins to take on a pale color, then remove and allow to drain on a cooling rack. The second frying should be between 375°F (191°C) and 390°F

(199°C) until the desired color is reached. Most books recommend frying larger pieces of food like chicken at 350°F (177°C) to 365°F (185°C). Among restaurateurs who fry good chicken, some like slightly lower temperatures for large, high-moisture pieces of dark meat, which take longer to cook, and 350° to 365°F (177° to 185°C) for the drier white meat, which should be cooked in a shorter time. Outstanding results require trial and error using your food, your fat, and your specific fryer to arrive at the optimum time and temperature.

Cooking time

Again, the shorter the frying time, the less fat is absorbed. Try to fry as fast (at as high a temperature) as possible to get the food cooked through. Cooking time depends on the fat and food temperature and amounts, the size, shape, porosity, and moisture content of the food, and whether it has been precooked. With doughs and batters, time is additionally dependent on the ingredients.

Precooking

Frequently food is partially precooked, which allows it to be fried at a high temperature for a very quick brown, crisp crust. Fast-food restaurants precook nearly everything to enable them to brown and serve food very fast.

The low temperatures required to cook large pieces of food all the way through may not provide the desired deeply browned, flavorful crust. Partially precooking at a low temperature (350°F/177°C), then frying at a high temperature (390°F/199°C) for a short time, reheats the food and produces a flavorful crust.

French fries can be precooked in boiling water (water blanching) or in oil (oil blanching). Precooking at lower temperatures in oil is preferred since water contributes to the frying fat's deterioration and the evaporating water lowers the temperature of the oil.

Breadings, batters, and doughs

A batter is a liquid mixture with flour (wheat, corn, rice, barley—any grain flour) or starch (corn, potato, rice, etc.) and seasoning that a food is dipped into before frying. Batters may be leavened or unleavened.

A breading is a dry mixture of a flour, starch, crumbs or flakes, and seasoning applied to moistened or battered food before frying. Any cereal flour can be used—wheat, corn, rice, or barley to name a few. Breadings can be a starch like wheat, corn, or rice or any of a variety of crumbs and flakes—bread crumbs, cracker crumbs, cereals like oatmeal, potato flakes—and even dry sauce mixes.

Fried doughs range from hush puppies to apple fritters.

Size, texture, color, browning rate, moisture and oil absorption potential, and the character of the food need to be considered when selecting the ingredients for a breading, batter, or dough.

Size and texture—A fine-crumb breading or surface will absorb less fat than a coarse-crumb breading, but a coarse-crumb batter or coating will get crisper.

Color and browning rate—Darker-colored ingredients will darken a dough or coating. A combination of sugar and protein in a dough or coating speeds browning. This can be good or bad. Sugar in a coating on onion rings makes the rings brown before the onion can cook and become sweet. For increased browning, add protein and/or sugar to the dough/coating. Corn syrup (glucose) browns at even lower temperatures than other sugars. To limit browning, make a dough or batter more acidic.

Moisture and oil absorption—Both fat content and texture of doughs/coatings influence fat absorption. The higher the fat content of a dough/batter, the more fat that it will absorb. For a light, dry, strong batter, egg white is ideal. If you need more tenderness, you may actually need a little oil absorption, which some yolk or fat in the batter will give. Lecithin in the egg yolk can combine with cooking fat and produce spattering and foaming. A fine-crumb breading will absorb less fat than a coarse-crumb breading. However, a high-fat batter tastes richer, and a coarse crumb batter or coating will get crisper.

Character of the food—The type and depth of the coating should relate to the character of the food being fried. A light coating should be used with thin, delicate food, like a thin fish fillet. The light flour coating on the Snapper Fingers with Smoked Pepper Tartar Sauce (page 165) is an example of a light coating. Large, heavier foods can accommodate thicker batters. The taste of the food, not of its coating, should always dominate.

Adhesion of batter and breading Since many foods, such as chicken skin, are very slick, it is often difficult to get batters and breading to stick to them. Adhesion is influenced by these factors:

Dusting—Food can be dusted with flour for a good dry surface before applying the batter. Batter will stick to a dry surface much better than to a wet, slippery surface.

Consistency of batter (viscosity)—The thickness of a batter is a major factor in determining how well it will adhere to food. In general, thicker batters adhere better.

Temperature of batter—The temperature of the batter affects how thickly it will coat and also is a factor in adhesion. A cold, thick batter will stick much better than a warm, thin one. When applied, the batter should be between 40° and 60°F (4.4° and 15.6°C) for best adhesion.

Leavening of the batter—Whether to use a leavened or an unleavened batter is another decision. Leavened batters are also called *tempura-type batters* and give an airy, puffed coating. Again, the choice of coating should be related to the nature of the food and the dish that you are preparing.

When to bread or batter Usually food should be breaded or battered just before frying and not let stand after coating. An exception would be fried ice cream. After coating the ice cream heavily with its coating, it is returned to the freezer to freeze very firm. Fried ice cream must go directly from the freezer into the hot fryer and come right back out.

at a *glance*
Deep frying

What to Do	Why
Use a fat with a smoke point much higher than the frying temperature.	For safety and prevention of off taste.
Use a fat whose flavor notes are comparable with the food being fried.	For good flavor.
Use frying fat that is in good condition.	Deteriorated fats have off tastes.
	Deteriorated fats smoke and can burst into flames at lower temperatures than clean fats.
	Deteriorated fats soak into foods much more easily than fresh fats and produce greasy products
	Deteriorated fats coat fryers and utensils with baked-on polymers.
Food to be fried together should be as uniform in size as possible.	So that all pieces will be done at the same time.
For excellent French fries, use russet or Russet Burbank (Idaho) potatoes.	These low-moisture, high-starch potatoes produce nongreasy, extremely crisp fries with proper frying time and temperature.
Be sure the surface of food to be fried is as dry as possible.	Water contaminates the frying fat. Water evaporating from the food lowers the fat temperature.
Temperature should be appropriate for the thickness of the food.	Small, thin pieces of food can be fried at high temperatures, while large thick pieces need lower frying temperatures so that the center is cooked before the outside is overdone.
Add a small amount of cold food to a large quantity of hot fat.	Maintaining frying temperature helps limit grease absorption.
For very crisp French fries, precook at 350°F (177°C) until fries just begin to color, drain, and, when ready to serve, brown at 390°F (199°C).	This cooks fries through ahead, then allows a fast reheating to get surface browning and crisping.

Snapper Fingers with Smoked Pepper Tartar Sauce

MAKES 20 HORS D'OEUVRE SERVINGS, 5 TO 6 MAIN-COURSE SERVINGS

I first had chef/restaurateur Charlie Hyneman's Smoked Pepper Tartar Sauce with soft-shelled crawfish at an elegant brunch featuring top Mississippi cooks and their dishes at the governor's mansion. I think that it's the best tartar sauce that I have ever had, the perfect foil for these crisp, browned bites with tender, juicy centers. Try them as finger-food hors d'oeuvres or as a main course with very crisp slaw, scalloped potatoes, and hush puppies or All-Time Favorite Sour Cream Cornbread (page 149).

what this recipe shows

A light coating is appropriate for delicate fish fillets.

Seasoning both marinade and coating adds depth of flavor.

1 pound fresh red snapper, tilapia, or orange roughy fillets, cut into finger-size strips (see Note)

2 cups buttermilk

$1/4$ teaspoon cayenne

1 cup bleached all-purpose flour

$1/2$ teaspoon salt

$1/4$ teaspoon ground white pepper

3 to 4 cups fresh vegetable or peanut oil for frying

1. Soak the snapper fingers in buttermilk and cayenne in a freezer-type zip-top plastic bag in the refrigerator for at least an hour.

2. Mix the flour, salt, and white pepper and spread out on a plate. Drain the fillets and dredge them in the flour mixture. Shake to remove any excess. Deep-fry or panfry in fresh oil. Drain on an unrecycled brown paper bag or paper towels. Serve immediately with Smoked Pepper Tartar Sauce (recipe follows).

Note: *If you are fortunate enough to get fresh soft-shelled crawfish (crayfish), this is excellent for them also.*

Smoked Pepper Tartar Sauce

MAKES 1¼ CUPS

Once you taste this sauce, you may never go back to ordinary tartar sauce.

what this recipe shows

Chipotles add a little heat and fascinating flavors.

Caper juice and capers enhance flavors.

continued

1 cup Homemade Mayonnaise
(page 289) or store-bought

2 teaspoons finely minced onion

1/2 teaspoon fresh lemon juice

1 tablespoon well-drained capers

2 teaspoons caper juice

2 dashes Tabasco sauce

1 tablespoon chopped chipotle
(see Note)

Dash paprika for color

In a food processor with the steel knife, blend together the mayonnaise, onion, lemon juice, capers, caper juice, and Tabasco. Add the chipotle and mix with one or two quick on/off pulses only. Mix in the paprika. Keep refrigerated until used.

Note: *If you're using chipotles (smoked jalapeño peppers) canned in adobo sauce, wipe off most of the sauce with a paper towel before chopping.*

Sautéing and stir-frying

Some books do not distinguish between sautéing and pan-frying and, except for Asian cookbooks, do not usually even mention stir-frying. I like to classify sautéing and stir-frying together, with pan-frying as a different technique. Sautéing and stir-frying both require rapid movement and turning of the food in a hot pan containing a small amount of fat. The cooking time is usually very short. On the other hand, pan-fried foods are typically larger pieces, are frequently battered or breaded, are cooked in more fat, and are turned only once during cooking.

The term sautéing comes from the French verb "to jump": You keep the food jumping. You use both hands with your thumbs up to hold the pan handle, shake the pan, and toss the food with a flip of the pan to turn it. A small amount of cooking fat is used, or if the food contains fat or has a fat layer, no fat at all may be used. The sauté pan has curved edges to facilitate the shaking and tossing, while the frying pan has straight, higher sides to better contain the deeper level of fat.

Stir-fry is frequently prepared in a wok with curved sides for easy rapid movement and tossing of food. Heat the empty wok first, then pour a small amount of oil around the sides and instantly add, toss, and stir the food. Woks are normally larger than sauté pans and frequently do not have a long handle. The food is tossed with a large spoon or spatula. If you have a wok with a long handle, you can toss food in a stir-fry just as in sautéing.

Both sautéing and stir-frying are done in very hot pans. A quick sauté is frequently used to enhance the flavor of herbs or aromatic vegetables or to crisp the outside of food before the cooking is completed by some other method, or sautéing can be used to completely cook thin slices of meat.

Stir-fried Chicken Steamed with Gin

MAKES 4 TO 6 SERVINGS

Green peppercorns, thyme, and gin give rich flavors to tender chicken slices that are spooned hot onto a cold bed of shredded lettuce. This low-fat dish has exciting contrasts—hot and cold, delicate and intense flavors. The platter is quite attractive when garnished with cilantro and thin dark green onion slices.

what this recipe shows

Marinating overnight in a mild acid like yogurt or buttermilk tenderizes the chicken.

Chicken will continue to cook for a few minutes after removing from the pan, so it has to be removed before it is totally cooked.

When chicken is not overcooked, it is very tender and juicy.

6 boneless chicken breast halves

2 cups plain yogurt or buttermilk

1/4 teaspoon cayenne

1/2 head iceberg lettuce, very thinly sliced into shreds

3 medium shallots, chopped

1/2 teaspoon green peppercorns packed in brine, crushed

3 tablespoons mild olive oil (such as Berio, Berillo, Plaginol)

1/3 cup gin

1 1/2 cups seedless red or black grapes pulled from the stem

1/2 cup Chicken Stock (page 267) or canned chicken broth

2 tablespoons dark corn syrup

2 teaspoons fresh thyme leaves or 1/2 to 3/4 teaspoon dried

1 teaspoon instant chicken bouillon (see Note)

1 tablespoon cornstarch stirred into 1/2 cup chicken stock or canned chicken broth

6 scallions, green parts included, sliced into thin rounds

6 sprigs cilantro, stems included, coarsely chopped

1. Slice each chicken breast across the grain into four slices and marinate in yogurt and cayenne, refrigerated overnight in a large freezer-type zip-top plastic bag.

2. Spread the shredded lettuce evenly over a serving platter and set aside.

3. Rinse the yogurt from the chicken, drain well, and pat dry.

4. Sauté the shallots and peppercorns in olive oil in a large skillet about 1 minute over high heat. Add the chicken and stir-fry briefly, until still pink in spots. Pour the gin over the chicken and cover for 1 minute to steam. Remove the cover. Immediately remove the chicken to the serving platter with a slotted spoon. Spread out on top of the shredded lettuce. Add the grapes to the skillet and stir-fry for about 5 seconds. Add 1/2 cup chicken stock and scrape to loosen any stuck particles. Add the corn syrup, thyme, and instant bouillon. Stir in the cornstarch mixture. Cook for several minutes, stirring constantly. When it thickens, immediately pour the sauce over the chicken. Garnish with scallion slices and cilantro and serve immediately.

Note: *The bouillon strengthens canned broth, but it would not be needed in strong homemade stock.*

at a glance
Stir-frying

What to Do	Why
Food should be cut into small enough pieces that it will cook through rapidly.	Large pieces will not get done in the brief cooking time.
Have sauce and all other ingredients ready before starting to cook.	Cooking goes very fast.
Foods that are to be cooked at the same time should be approximately the same size.	Pieces the same size will cook in about the same time.
Heat the empty wok first.	Having the metal hot before the food touches it reduces sticking.
Drizzle a small amount of oil around the upper edge of the wok.	Drizzling the oil in around the upper edge rapidly oils a large frying surface.
Toss food or use large spatulas to move food about rapidly while cooking.	Moving food fast cooks it evenly.

Pan-frying

In an illuminating demonstration for the International Association of Culinary Professionals in New Orleans, Paul Prudhomme showed how something simple like depth of oil can produce a significant difference between deep-fried and pan-fried foods. Starting with one fish fillet, he dredged the seasoned fish in seasoned flour, shook off the excess, dipped it in a milk wash (one egg beaten in a quart of milk), dipped it back in the flour, and shook it again. Then he cut the fillet in half and pan-fried one while his assistant deep-fried the other half—both to a nice golden brown.

When we compared the two carefully, we noticed tiny dark specks on the pan-fried fish, where the fillets had touched the bottom of the pan and caramelized. Then we tasted and found that the pan-fried fish actually tasted a little sweeter.

Just as in deep frying, many factors affect the quality of pan-fried food, from the pan you use to how often you turn the food.

A heavy pan is a must

Use a heavy pan that conducts heat well, such as a heavy aluminum pan or an iron skillet. I once handed Paul, one of the great masters of pan-frying, a thin pan with a brushstroke coating of copper on the bottom. He almost threw it at me but laughed and said, "If this is the kind of pan you use, no wonder you burn things!"

The level of fat: don't skimp

The fat should come about halfway up the side of the food so that the food has a little buoyancy. With any less fat, the food will rest heavily on the bottom of the pan during the entire cooking time, and it can burn more easily than if it just touches the bottom lightly as it partially floats.

Fat: again, purity

You cannot buy directly from the manufacturer as chefs like Paul Prudhomme do, but you can be sure of fresh oil by buying in small quantities and storing the unused oil tightly closed in a cool, dark place. For high-quality dishes, do not reuse oil.

The all-important seasoning

Season each element of the dish—the food, the flour, and the breading—so that every bite you take will be seasoned. As Chef Paul explained in his demonstration, he seasoned the fish and the flour and would have seasoned crumbs as well if he had been using them. Your seasoning can be salt, pepper, cayenne, herbs, spices—whatever flavors you have selected for the dish.

Breading or batters

Considerations in the selection of batters and breading for pan-fried foods are very similar to those for deep-fried foods (see page 162). The breading should be appropriate for the thickness of the food. A very thin, delicate fish fillet should have a light, delicate coating such as that just described. Thicker, heavier food can have a heavier coating—flour, egg or milk wash, and crumbs or a batter coating.

First side to fry

The first side that you fry will be the most attractive. Fry the presentation side first. For a fish fillet, this would be the rib side, not the skin side.

Turning during frying

Food should be turned only once during pan-frying. When you are frying the first side and you see little bubbles of moisture popping through the breading on the top side, you should have turned the food a second before. I know this is like saying "If you get to the corner, you've gone too far!" Still, the split second you see a bubble, turn the food.

Finishing in the oven

For a very crisp coating, finish frying in a hot oven. Turn a fish fillet, give the pan a few seconds on the heat, then place the pan in a hot oven. This keeps the top hot and crisp while the bottom side is still frying, as the next recipe shows.

Fillets with Pecans in Creole Meunière Sauce

MAKES 6 SERVINGS

This is an adaptation of a dish that Paul Prudhomme made famous at Commander's Palace as Trout with Pecans. Trout has a double rib cage, as does salmon, and the fillets will always have a row of pin bones, so I have suggested some other fillets that may be easier for home cooks. The final dish itself is very simple, but there are a number of parts to the sauce. These many components give you the incredible complexity and the memorable taste that you get from Chef Paul's dishes. Prepare all of the components ahead.

what this recipe shows

Each element of the dish is seasoned for flavor in every bite.

The Creole Meunière sauce is similar to a butter sauce and will break if you get it too hot or refrigerate it.

Clarified butter or fresh oil for
   pan-frying
6 fresh mild fish fillets (1 to 1½
   pounds), such as flounder, sole, or
   orange roughy
2 tablespoons *and* 1 tablespoon
   Seasoning Mix (page 171;
   see Note)—3 tablespoons *total*
1 cup all-purpose flour

2 large eggs
1 cup whole milk
Pecan Butter (page 171)
Creole Meunière Sauce (page 171)
¼ cup parsley sprigs, minced,
   to garnish
½ cup Roasted Pecans (page 171),
   chopped, to garnish

1. Preheat the oven to 350°F (177°C).

2. Heat ¼ inch of clarified butter in a heavy skillet with an ovenproof handle to about 350°F (177°C) over medium-low heat while flouring the fillets.

3. Rub the fillets well with 2 tablespoons Seasoning Mix. Stir together the flour and 1 tablespoon Seasoning Mix in a shallow pan. Dip the fillets in seasoned flour and shake off any excess.

4. Beat the eggs and stir into the milk a little at a time in a medium mixing bowl. Dip the floured fillets in this egg wash. Let any excess drip off. Then dip them back in the flour. Place the fillets in the hot oil, presentation side (rib side) down. Fillets should be browned well and then turned once only. Leave on the heat for about a minute after turning. Then place immediately in the oven to finish frying and to crisp the top, several minutes.

5. Remove the fillets to a serving platter, spread with Pecan Butter, top with Creole Meunière Sauce, and garnish with parsley and chopped roasted pecans. Serve immediately.

Note: *As a substitute, you can use Chef Paul Prudhomme's seafood seasonings, available in many grocery stores.*

Seasoning Mix

1½ teaspoons dried oregano

1½ teaspoons dried thyme

1½ tablespoons paprika, preferably recently opened (see Note)

3 tablespoons onion powder

1 tablespoon garlic powder

1½ teaspoons black pepper

1½ teaspoons cayenne (see Note)

1 tablespoon salt

Blend all ingredients. Keeps well tightly sealed in a jar.

Note: *Paprika and cayenne turn brown and lose flavor upon standing. For best taste, use freshly opened. Store tightly sealed jars in a freezer-type zip-top plastic bag in the freezer for best shelf life.*

Roasted Pecans and Pecan Butter

1 cup pecan pieces, ½ cup for Pecan Butter and ½ cup to garnish

2 tablespoons butter

¼ teaspoon salt

Juice of ½ lemon (about 1½ tablespoons)

½ teaspoon Worcestershire sauce, preferably Lea & Perrins

4 tablespoons (½ stick) butter, softened

1. Preheat the oven to 350°F (177°C) and roast the pecans on a baking sheet for 12 to 15 minutes. While the pecans are still hot, stir in 2 tablespoons butter and salt. Reserve ½ cup to garnish.

2. Process ½ cup pecans to a meal in the food processor with the steel knife. Add the lemon juice, Worcestershire, and butter. Process to blend well.

Creole Meunière Sauce

1 cup Fish Brown Sauce (page 172)

1 teaspoon Worcestershire sauce, preferably Lea & Perrins

Juice of ½ lemon (about 1½ tablespoons)

4 tablespoons (½ stick) room-temperature butter

2 tablespoons chopped parsley

Bring the brown sauce to a boil in a medium-size, heavy saucepan over high heat. Add the Worcestershire sauce and lemon juice. Remove the pan from the heat and whisk in the butter and then add the parsley.

Fish Brown Sauce

1 tablespoon fresh peanut or
   vegetable oil

1^1/$_2$ tablespoons all-purpose flour

2 cups Fish Stock (page 269) or clam
   juice

^{1}/$_2$ onion, chopped

^{1}/$_2$ tablespoon Worcestershire sauce,
   preferably Lea & Perrins

1 tablespoon tomato puree

1 bay leaf

^{1}/$_4$ teaspoon dried oregano

^{1}/$_4$ teaspoon garlic powder

^{1}/$_8$ teaspoon black or white pepper

^{1}/$_8$ teaspoon cayenne

2 tablespoons Burgundy or other full-
   flavored red wine

Salt to taste

1. Prepare a roux by heating the oil to medium-hot in a heavy skillet. Whisk in the flour and continue whisking on medium heat until the roux is a deep reddish brown. Remove from the heat and reserve to add later.

2. Bring the fish stock, onion, Worcestershire sauce, tomato puree, bay leaf, oregano, garlic powder, pepper, and cayenne to a boil in a medium-size, heavy saucepan over medium heat. Turn the heat down to very low and simmer for an hour, stirring occasionally. Whisk the roux into the stock and simmer for 30 minutes more. Remove from the heat and remove and discard the bay leaf. Stir in the Burgundy. Taste and add salt if needed.

Hunter's Chicken with Pasta

MAKES 8 SERVINGS

Pan-fried, thin, browned chicken breasts and a slightly spicy, herbed tomato sauce are superb with pasta. Serve with crusty bread (Crusty French-Type Bread, page 24) and a salad (Mixed Greens and Oranges with Brandied Dressing, page 186).

> **what this recipe shows**
>
> Each element of the dish is seasoned for good flavor in every bite.
>
> Chicken breasts are removed before they are completely done since they continue to "cook" outside the pan.
>
> Vodka dissolves and distributes flavor components of the ingredients in the sauce.

8 boneless chicken breast halves, flattened with a heavy pan or meat pounder

1 teaspoon *and* ¹/₄ teaspoon *and* ¹/₂ teaspoon *and* 1 tablespoon salt (about 1¹/₂ tablespoons *total*)

¹/₂ teaspoon *and* ¹/₄ teaspoon *and* ¹/₄ teaspoon white pepper (1 teaspoon *total*)

¹/₂ cup all-purpose flour

¹/₄ cup olive oil

3 bay leaves

1 teaspoon dried thyme

1¹/₂ large onions, chopped (about 1¹/₂ cups)

1 tablespoon fresh or 1 teaspoon dried marjoram

2 teaspoons dried or 2 tablespoons fresh thyme

¹/₄ teaspoon red pepper flakes

2 cups homemade Chicken Stock (page 267) or canned chicken broth

3 cans (14.5 ounces each) diced tomatoes

4 quarts water

10 to 12 ounces thin spaghetti or fettuccine

3 medium tomatoes, peeled, seeded, and chopped

2 tablespoons vodka

12 leaves fresh basil, sliced into narrow shreds

Finely grated zest of 1 lemon

¹/₄ cup flat-leaf parsley, chopped

6 scallions, green parts included, sliced into thin rounds

1. Preheat the oven to 200°F (93°C) and then turn off.

2. Sprinkle the chicken breasts lightly with 1 teaspoon salt and ¹/₂ teaspoon white pepper on both sides. Stir together ¹/₂ cup flour, ¹/₄ teaspoon salt, and ¹/₄ teaspoon white pepper in a shallow pan. Dredge the chicken pieces in the flour mixture, then shake off any excess.

3. Sauté the chicken in olive oil with the bay leaves and 1 teaspoon dried thyme in a large skillet, about 1 minute per side, until almost done. Reserving the drippings in the skillet, remove the chicken to a platter, cover with foil, and place in the oven to keep warm.

4. Add the chopped onions, marjoram, thyme, ¹/₂ teaspoon salt, ¹/₄ teaspoon white pepper, and red pepper flakes to the drippings and cook over medium-high heat a minute or two. Add the chicken stock and canned tomatoes, then turn heat down to low and simmer at a very low boil for 30 minutes to reduce to about half of the original volume.

5. During the last 15 minutes, while the sauce is reducing, bring the water with 1 tablespoon salt to a boil in a large pot and stir in the pasta. Stir continually for 1 full minute. Cook for the time directed on the package or until just tender. Drain but do not rinse.

6. Remove the bay leaves from the sauce and discard. Add the fresh tomatoes and vodka. Boil vigorously for 1 minute, then add the pasta, toss, and cover. Let stand for 1 to 2 minutes. Taste and add salt and pepper if necessary. Stir in the fresh basil and lemon zest and spoon onto a serving platter. Top with the chicken breasts and garnish with chopped parsley and sliced scallions. Serve immediately.

at a *glance*
Pan-frying

What to Do	Why
Use a fat with a high smoke point.	For safety and prevention of off tastes.
Use a fat whose flavor notes are compatible with the food being fried.	For good flavor.
Use fresh frying fat.	Deteriorated fats have off tastes.
	Deteriorated fats smoke and can burst into flames at lower temperatures than fresh fats.
	Deteriorated fats soak into foods much easier than fresh fats and produce greasy products.
	Deteriorated fats coat fryers and utensils with baked-on polymers.
Use a heavy pan that conducts heat well.	Thin, poorly conducting pan can have hot spots and burn foods.
Use enough fat to come halfway up the sides of the food.	If the fat is deep enough, the food will have some buoyancy and will not rest heavily on the pan the whole time.
Use a breading or batter that is appropriate for the thickness of the food.	A thick batter or breading on thin food can totally mask the taste of the food.
Season all elements of the food to be fried.	If all elements are seasoned, every bite will be flavored.
Fry the presentation side first.	The first side fried looks best.
Turn the food only once—just as moisture bubbles pop through the breading on the top side.	Bubbles of moisture pop through just as the side down is perfectly fried.
Finish frying the last side in a hot oven.	For a very crisp coating.

A little starch-and-water paste makes a meringue that does not shrink and cuts beautifully. The little yellow flecks in the meringue are pieces of lemon zest. Egg yolks make the filling silky smooth and a flaky crust was made using the Simple Flaky Crust recipe, page 110.

See Lemon on Lemon on Lemon Meringue Pie
(using 1 ½ times the meringue and filling), page 243.

The apple pie bottom and top crust are baked separately for a crisp texture. The filling is cooked to the desired texture on top of the stove, then poured into the pie shell and crowned with the top crust, ready for eating.

See Big-Chunk Fresh Apple Pie in Flaky Cheese Crust, page 121.

Wet doughs make light, moist breads.

See Touch-of-Grace Biscuits, page 77.

Notice the flakiness of the crust made by using the technique described on page 107. Baking soda softens cell walls, so dates soaked in boiling water with a little soda tear apart for excellent distribution of flavor. Note the dark pieces of date on the apple slices.

See Big-Chunk Fresh Apple Pie in Flaky Cheese Crust, page 121.

These two brioches, made with the same amounts of the same ingredients, are very different in volume and texture. The brioche on the left was developed by employing breadmaking techniques, making for a lighter, more open-textured loaf. In the dense, tender, cakelike loaf on the right, gluten formation was prevented by using pastry techniques.

See The Ultimate Brioche I: Light, Airy Breadlike Brioche, page 15, and The Ultimate Brioche II: Buttery Cakelike Brioche, page 17.

Cookies made by the same formula can be crisp and flat (on the left) or soft and puffed (on the right) by selecting different flours and fats.

See Chocolate Chip Cookies, page 132.

Egg yolks produce sensationally smooth chocolate dishes.

See The Secret Marquise, page 467.

The wet method of making caramel (actually Microwave Caramel) produces a medium-dark caramel-candied coating on thin strips of orange peel.

See Orange Slices with Cinnamon Candied Peel, page 428.

Soufflés that break all the rules—they are baked in a hot oven in well-greased muffin tins, then dumped into hot cream, sprinkled with cheese, and reheated. They repuff and are magnificently light and creamy.

See Incredibly Creamy Soufflés, page 238.

Anthocyanins, the red compounds in strawberries, grapes, cherries, and red cabbage, are water soluble and can fade and become washed-out looking when soaked in a water-based liquid, as the berries on the bottom were. The berries on the top were soaked in a balsamic vinegar–brown sugar sauce that stained them a deeper red.

See Sweet-Tart Fresh Strawberries, page 343.

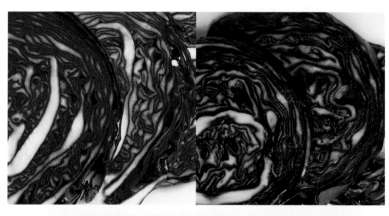

Anthocyanins must be acidic (as on the left) to be bright red. When they are alkaline (as on the right) they turn blue.

See Keeping the Reds Red, page 333.

Egg whites are substituted for two of the whole eggs in the pâte à choux, and bread flour is used to make extraordinary light, crisp puffs with a dry interior.

See Cocktail Puffs with Escargots, page 230.

Cauliflower cooked in water with a little white vinegar is snow white in color. The broccoli is bright green when cooked under seven minutes. The plate is warm, not hot, to prevent separation of the hollandaise.

See Broccoli and Cauliflower in Artichoke Cups with Hollandaise, page 303.

Plates sauced with beurre blanc are just warm not hot (under 136°F/58°C), or the heat will cause the sauce to break.

See Triple-Layered Fillets in Golden Beurre Blanc, page 298.

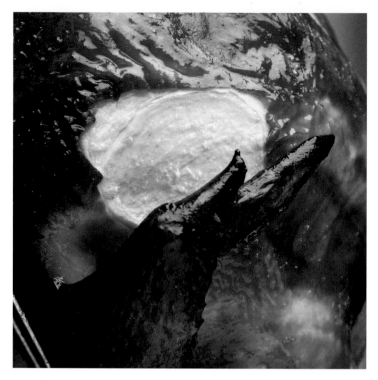

Brining produces a sensationally juicy roast chicken.

See Juicy Roast Chicken, page 389.

London broil cooked no more than medium rare and sliced thinly makes a flavorful but possibly tough cut tender and juicy.

See London Broil, page 378.

Fat structure and nutrition

A fat is a compound containing many carbon and hydrogen atoms. Three of the carbons are in a row, and each of these three has a fatty acid hooked onto it. These fatty acids are chains of carbons hooked to each other with hydrogen attached to many of them. Fatty acids can be short with only a few carbons in the chain or medium or long.

Triglycerides, diglycerides, and monoglycerides

Because there are three fatty acids, one hooked onto each carbon in a fat, the technical name for a fat is *triglyceride*. (The triglyceride level on a blood chemistry report is the level of fats in the blood.) When you heat fats, they may come apart. One of the fatty acids may break off and become a free fatty acid. The fat is then left with only two fatty acids and is called a diglyceride (*di* for "two"). If two fatty acids break off and become free fatty acids, the fat is left with only one fatty acid and is now called a monoglyceride (*mono* for "one"). You may have read these terms on a shortening container: "Mono- and diglycerides added for moister cakes. . . ." Mono- and diglycerides are not some horrible chemicals. You can think of them as simply fats that are broken apart.

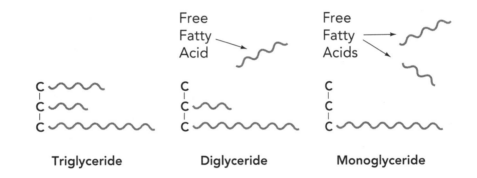

Triglyceride   Diglyceride   Monoglyceride

Fatty acids

A fat can contain three short-chain fatty acids, three medium-chain fatty acids, or three long-chain fatty acids, or any mixture of the three (two short and one long, one short and two medium, one medium and two long, etc.). The types of fatty acids that a fat contains are important to us for a number of reasons. Both the length of the chain and the way the carbons are linked to each other influence how a fat affects your body.

Saturated and unsaturated fatty acids and fats

Carbon atoms have four bonds or links that they can use to join to other atoms. In the chains of carbon atoms in a fatty acid, if the carbon is joined to its neighboring carbon on each side and has its other two bonds hooked onto hydrogen, all of its four bonds are tied up, and it is said to be *saturated*.

Carbon does not always hook to four separate atoms. Sometimes it forms a double bond with one of its neighboring carbons. When a fatty acid contains one or more carbons with double bonds like this, it is said to be *unsaturated*. If it contains only one double bond, it is *monounsaturated* (like olive oil and canola oil); if it contains several double bonds, it is *polyunsaturated* (like some vegetable oils).

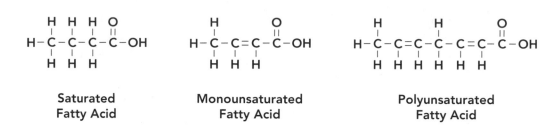

Saturated Fatty Acid

Monounsaturated Fatty Acid

Polyunsaturated Fatty Acid

Since saturated fats have all four bonds totally tied up, they are reasonably stable and have little tendency to combine with anything. But unsaturated fats, with loose double bonds, can very easily drop one of the bonds and latch on to anything and everything interesting that they encounter. For this reason, unsaturated fats easily combine with oxygen, or metals, or whatever and can easily become rancid and off tasting. Fats with one double bond combine ten times more readily than saturated fats, while fats with two double bonds combine one hundred times more readily.

Saturated fats, which do not have a tendency to combine with things, have much better shelf life. Their greater stability is a primary reason why saturated fats like coconut oil have been favorites of food processors. They taste good and yield products that do not go rancid easily. Food processors hydrogenate unsaturated fats by bubbling hydrogen through the fat to combine with the carbon at the double bonds. This converts unsaturated fats into saturated or partially saturated fats that will have better shelf life.

Free fatty acids

The fatty acids that break off when fats come apart are called *free fatty acids*. The preceding drawings of a mono- and diglyceride show free fatty acids breaking off. These free fatty acids contaminate fat and lower its smoke point and its flash point. A fat that is high in free fatty acids will burst into flames at a relatively low temperature.

Trans and cis monounsaturated fatty acids

The natural form of a fat is a shape called the *cis configuration*. When monounsaturated fats are made more stable by partial hydrogenation, the fat formed is in a different shape that is called the *trans configuration*. Presence of the trans monounsaturated fats used to appear on food labels as *monounsaturated*, but they act in the body like saturated fats and can raise blood serum cholesterol.

Some saturated fats and trans configuration (hydrogenated) monounsaturated fats elevate serum cholesterol. Natural (nonhydrogenated) monounsaturated fats lower LDLs, which are "the bad guys" and do not lower HDLs, which are "the good guys" (see page 179). Polyunsaturated fats lower both LDLs and HDLs.

Remember, natural fats are blends containing mono- and polyunsaturated fats and saturated fats. There are many kinds of vegetable oils. Some, like olive oil and canola oil, contain predominantly natural monounsaturated fats; some, like safflower oil, contain predominantly polyunsaturated fats; and some, like corn, soy, and peanut oils, are blends of predominantly mono- and polyunsaturated fats. New strains of oil crops now produce oils with more monounsaturated oils. Margarines, spreads, and vegetable shortenings use various plant oils. There also are animal fats: butter, clarified butter, and ghee; chicken, goose, or duck fat; and lard. And there are vegetable and animal fat blends.

You can check the table for the approximate percentage of monounsaturated, polyunsaturated, and saturated fats and of cholesterol in some typical fats. Different analyses of the same type of natural fat can differ, and definitions of saturation also vary, so these percentages should be considered only as general guides. As you can see, many fats that are thought to be bad have high amounts of mono- and polyunsaturated fats—good-for-you fats. Perhaps our attitudes toward fats have been oversimplified.

PERCENT MONOUNSATURATED, POLYUNSATURATED, AND SATURATED FATS AND CHOLESTEROL IN SELECTED NATURAL OILS AND SOLID FATS			
Type Oil/Fat	Monounsaturated %	Polyunsaturated %	Saturated %	Cholesterol mg/tbsp
Hazelnut	78	10	7.4	0
Olive	74	9	14	0
Canola	58	36	6	0
Peanut	46	32	17	0
Lard	46	12	40	12
Chicken fat	45	21	30	11
Beef tallow	42	4	50	14
Palm	37	10	50	0
Butter oil	29	4	62	33
Corn	25	59	13	0
Soybean	24	58	15	0
Sunflower	20	66	11	0
Cottonseed	18	52	26	0
Safflower	12	75	9	0
Coconut	6	2	87	0

USDA Agriculture Handbook No. 8-4, 1979.

Health concerns and low-fat cooking

Fats are essential

Any comments on fats and human health should be prefaced with "According to what we know now." A great deal of research is ongoing, and there are frequent announcements of new findings. Often these contradict earlier theories.

Fats are not villains. We have to have certain fatty acids (linoleic, linolenic, and the fatty acid derived from these two, arachidonic) in our diet to live. It is the high amounts of certain fats that Americans in general need to cut. In our zeal to reduce fat intake and remove fat, we have a tendency to forget that a little fat is absolutely necessary for life. It is through the fat that we eat that we get the essential fatty acids and the fat-soluble vitamins. There are also fats that have the potential for prevention and treatment of disease.

Good-for-you fats

Fats that contain saturated fatty acids of medium-chain length (eight to twelve carbons) are typically called *medium-chain triglycerides (or MCT)* and have unique and remarkable pharmaceutical properties. These fats do not digest like fats at all. They do not go through the normal lengthy metabolic pathways in the body that other fats do. They act like a carbohydrate and go straight to the liver for instant use.

Doctors have used these fats, which are concentrated sources of instant energy, for many years. They are ideal for at-risk patients and have been used in premature infant formulas and in hospitals for patients with major metabolic problems, Crohn's disease, cystic fibrosis, or patients who have just undergone major surgery.

Food scientists are now combining these easy-to-digest, healthful fats with other good-for-you fats. For hospital patients, in the past, nutritionists have simply stirred together a small amount of fat containing the essential fatty acids along with this instant-to-digest fat. Now food scientists are chemically combining these good fats to produce a single fat with many healthful properties. These fats provide a steady flow of energy. The medium-chain triglycerides, which go directly to the liver, provide instant energy, while the essential fatty acids are slow to digest and provide energy later.

Food scientists now include Omega 3, 6, and 9 fatty acids in structured lipids for their healthful qualities. Omega 3 enhances immune function, lowers serum triglycerides and LDL cholesterol, and reduces blood clotting and risk of coronary heart disease. Omega 6 and Omega 9 are essential fatty acids. Scientists now have structured fats with optimum blends of these healthful fats. In a clinical study, there were 70 percent fewer infections and a 22 percent reduction in the mean hospital stay in patients given structured lipids and fish oil.

These structured fats are now believed to lessen cancer risk in two ways. High-fat diets are linked to cancer, but medium-chain triglycerides are exempt from this link. A five-year study by the American Health Foundation showed that medium-chain triglycerides do not promote tumors. Other studies indicate that medium-chain triglycerides may beneficially damage tumors. Dietary regimes high in polyunsaturated fat to lower serum cholesterol can place patients at a greater risk for some types of cancer. This link may be associated with suppression of the immune function. With structured fats, you do not suppress the immune function. Structured lipids also promote a better nitrogen balance. This can lead to weight gains and improvements in critically ill patients.

Fats and cardiovascular disease

About half of all the deaths in the United States and in many other industrialized nations are caused by fatty-cholesterol deposits in arteries, which increase the risks of strokes and heart attacks.

The story of these fatty-cholesterol deposits is anything but simple. In his paper presented at the 1988 convention of the Institute of Food Technologists, Dr. Paul Addis outlined three distinct phases in cardiovascular disease: lesions, plaque buildup, and thrombosis.

Lesions Injury to the inside layer of the cells of the blood vessels is the beginning of it all. Research in the United States and in Japan indicates that these initial injuries may be caused by oxidation products of cholesterol and fatty acids. These are also products of rancidity. About eighty of these auto-oxidation products of cholesterol have been identified, and seven or eight cholesterol oxides are also found in rancid food products. This is reason enough to approach rancidity in foods from a new perspective.

Plaque buildup Once the lesions form in the lining of the blood vessels (injury stage), then plaque buildup begins. Thickening and fatty streaks occur in the lining of the vessels and deposits of waxy cholesterol adhere to the lining of the blood vessels, partially blocking the flow of blood. The major factor at this stage of cardiovascular disease is the blood level of cholesterol—more specifically, the blood levels of LDLs (low-density lipoproteins) and HDLs (high-density lipoproteins), as well as the ratio of these two to fat (triglycerides). What are these LDLs and HDLs?

Some saturated fatty acids and trans configuration monounsaturated fatty acids (page 176) stimulate the liver to produce cholesterol and raise the level of cholesterol in the blood. Cholesterol is not strictly a fat, but it is a fatlike substance. For fat to be carried in the bloodstream, the liver wraps fat in soluble proteins. These bundles of fat, cholesterol, and protein are called *lipoproteins* (*lipo* means "fat"). Fat is light weight and puffy.

If there is not much fat in the bundle, it is small and dense and is called a *high-density lipoprotein* (*HDL*). These are the "good guys." They are small and have room to pick up debris. They help carry cholesterol and fat from around the body back to the liver to be eliminated.

If there is a lot of fat in the lipoprotein bundle, it is light, large, and not dense and is called a *low-density lipoprotein* (*LDL*). These are the "bad guys." The LDLs with their big load of fat and cholesterol are the ones that deposit cholesterol and fat in blood vessels around the body.

For healthy blood vessels, we need to keep our HDL level up and our LDL level down. Endurance-type exercise on a regular basis increases our HDL level and can change our HDL/triglyceride (fat) ratio to healthier status. Alcohol in small amounts (no more than two drinks a day) increases HDL cholesterol, but it is not as beneficial as exercise, which both increases HDL and lowers fat (triglycerides). Sex hormones also affect the HDL and LDL levels in the blood; this is one reason that more men have heart attacks than women. Lower estrogen can cause a decrease in beneficial HDL, and this is why older women, whose estrogen levels decrease after menopause, are at increased risk of cardiovascular problems.

Thrombosis Clots and spasms can occur whether there is any buildup in the artery or not. Research indicates that this is related to the ratio of Omega 3 fatty acids (in fish oils) to Omega 6 fatty acids (in vegetable oils). Too high an intake of Omega 6 and too low an intake of Omega 3 can lead to thrombosis.

We used to think that we should switch to polyunsaturated fat in our diets because it lowered serum cholesterol. Then we learned about good cholesterol—high-density lipoproteins (HDLs)—and bad cholesterol—low-density lipoproteins (LDLs). We found out that monounsaturated fats were better for us because they lower LDLs (bad guys) but do not lower HDLs (good guys).

Now, just when we thought we had a good grasp on the situation, we have to revise our thinking. We knew that monounsaturated fat was good for us and saturated fat was bad for us. We became conscious label readers and tried to eat right.

Alas, it is not that simple. Research now shows that some saturated fats metabolize or work in the body like monounsaturated fats and lower serum cholesterol or, at least, do not raise it. Some of the fats that we thought were bad guys are actually good guys. To make matters even more complicated, there are some monounsaturated fats (the trans configuration monounsaturated fatty acids) that act as saturated fat in the body and raise serum cholesterol.

At the 1991 Institute of Food Technologists' convention, Dr. Scott Grundy, a medical doctor and noted cardiovascular researcher, made a plea to label products not with amounts of saturated, monounsaturated, and polyunsaturated fats, but to label them with the amounts of fats that raise serum cholesterol, amounts of fats that lower serum cholesterol, and amounts of fats that do not affect serum cholesterol. This type of labeling would certainly make our life simpler. Dr. Grundy has made these recommendations to the FDA.

It is confusing and upsetting to learn that we can't count on all monounsaturated fat to be good and all saturated fat to be bad, but there is good news in this, too. One of the saturated fatty acids that research indicates does not raise serum cholesterol is stearic acid, one of the main saturated fatty acids in butter. Butter still has cholesterol, but it may not be as bad for us as we thought. Julia was right all along!

Fats and cholesterol Doctors and nutritionists agree that a high intake of certain fats raises the blood level of cholesterol in most people. Our own bodies produce most of the cholesterol in our bloodstream (75 to 80 percent). Eating certain saturated fats and monounsaturated manmade trans configuration fats encourages our body to produce more cholesterol. There is still disagreement on how much influence eating foods containing cholesterol has on the blood cholesterol level in most people.

Some research indicates that dietary cholesterol may have more influence on serum cholesterol than earlier thought, but the studies were with a limited number of participants. At present there are no definitive answers as to how much dietary cholesterol alone, the fats that elevate serum cholesterol (certain saturated fats, trans configuration monounsaturated fats) in combination with dietary cholesterol, or the fats that elevate serum cholesterol alone elevate the blood level of cholesterol. In studies such as the Edington study, where fat levels were kept low, increases in cholesterol in the diet did not increase levels of cholesterol in the blood.

Out of all we presently know about fats, one thing is clear. Americans need to cut their high fat intake. To quote Dr. Addis, "Fat is where it's at." Reducing fat intake, increasing fiber in the diet, and getting more exercise may be the secrets of good cardiovascular health.

Challenges in cutting fats in recipes

To cut fat from our diets we need to understand the roles of fats in different cooking situations and then look at other ingredients and techniques that can do fat's jobs as well as at some low-fat cooking techniques.

The overwhelming problem in removing fat, whether it is in baked goods, sauces, frying, or cooking in general, is flavor. I never realized the contribution of fat to flavor until I started working with reduced-fat recipes. I'll address this ongoing problem after discussing the principles of fat reduction.

Cutting fat in baking

In baking, fats tenderize, aerate, act as a spacer between layers of dough to create flakiness, carry flavor and nutrients, and determine cookie shape by the way that they melt. What substitutes are available for each of these jobs?

Tenderizing In baking, fat coats flour proteins to prevent their forming gluten. In this way fat is an excellent tenderizer. For this tenderizing role, sugar is an excellent substitute. When sugar is present in amounts above 2 tablespoons per cup of flour, the two proteins in flour that normally join together with each other and water to form gluten join with sugar instead. Good low-fat recipes using high-sugar-content ingredients like dried fruit (the "infamous" chocolate prune cakes) or corn syrup have been published in newspapers. In the Mellow Moist Low-Fat Chocolate Cake that follows, I use fat-free sour cream and pureed sweet potato. Master pastry chef Chris Northmore at the Cherokee Town and Country Club in Atlanta taught me the pleasures of using sweet potatoes in low-fat dishes. They have much better flavor than dried prunes.

Aerating In cakes, muffins, pancakes, and many baked goods—products leavened with baking powder and baking soda—fat is of major importance as an aerator (see page 232). Fortunately, there are other thick, syrupy textured foods that will hold air bubbles when you beat them—egg whites for one. You can beat air into many thick mixtures like syrup and pureed fruit, but egg whites are outstanding aerators. The Good-for-You Apple Bran Muffins with Walnuts and Orange Zest (page 148) use egg whites to leaven.

Mellow Moist Low-Fat Chocolate Cake

MAKES ONE 9-INCH LAYER

I set out to design a low-fat chocolate cake with these principles in mind: master pastry chef Chris Northmore's use of sweet potato puree in a luscious chocolate cake, Alice Medrich's emphasis on how important a little real chocolate is, and what a nice texture fat-free sour cream can contribute. The result is a very moist, rich, full-flavored cake. It's hard to believe it's low in fat.

what this recipe shows

Brown sugar gives the chocolate a fudgy taste.

Sweet potato puree contributes good taste as well as supplying sugar to tenderize.

A single egg yolk contributes emulsifiers for good texture.

Keeping egg whites low keeps the cake moist and avoids the rubbery texture of low-fat products.

Nonstick spray with flour, such as Baker's Joy, or 1 tablespoon shortening and 1 tablespoon flour to grease pan

5 tablespoons Dutch process cocoa powder

1 cup fat-free sour cream

1 large egg

1 large egg white

2/3 cup (about 5 ounces) baked sweet potato, pureed

1 tablespoon pure vanilla extract

1 ounce unsweetened chocolate, melted

3/4 cup cake flour

1/2 cup granulated sugar

1/2 cup packed light brown sugar

3/4 teaspoon baking powder

1/4 teaspoon baking soda

1/2 teaspoon salt

Nonstick cooking spray

1. Preheat the oven to 350°F (177°C).

2. Grease a 9 × 1 1/2-inch round cake pan by spraying with nonstick cooking spray with flour or by rubbing with shortening, shaking with 1 tablespoon flour, and shaking out any excess. Line the bottom of the pan with a parchment or wax paper circle, then grease and flour the circle.

3. Sift the cocoa through a strainer onto the sour cream and stir together in a medium mixing bowl. Add the egg and egg white. Beat at least 1 to 2 minutes on high speed. Add the sweet potato puree, vanilla, and melted chocolate. Beat in well.

4. Combine and sift together the cake flour, sugars, baking powder, baking soda, and salt in a medium mixing bowl. Add half of the dry ingredients to the cocoa mixture and stir in. Add the remaining dry ingredients and stir in well. Scrape down the sides with each addition. Pour the batter into the prepared pan and smooth the surface with a rubber spatula. Bake for about 25 minutes, until a cake tester inserted near the center comes out clean and the cake springs back when pressed lightly in the center. Do not overcook.

5. Let the cake cool in the pan on a cooling rack for 10 minutes. Invert the cake onto a cooling rack sprayed with nonstick cooking spray. After cooling, reinvert onto a serving platter.

Acting as a spacer In pastry, fat plays a role as tenderizer, as explained, and it also plays a role in creating flakiness. Fat acts as a spacer to hold thin layers of dough apart in the hot oven just long enough to begin to set. Steam from the dough then puffs the layers apart. Joy Dawson, an excellent cooking teacher in New Philadelphia, Ohio, has a way to create low-fat flakiness. She sprays thin sheets of phyllo with butter-flavored nonstick cooking spray and sprinkles fine crumbs between the layers. Sugar or chopped toasted oatmeal is excellent sprinkled between the layers.

Determining cookie shape The way that a fat melts helps to determine the shape of cookies. A fat like butter that is a solid at one temperature, then melts when just a little hotter, causes cookies to spread. On the other hand, some fats (shortenings and many of the fat substitutes) do not change texture over a wide range of temperatures. These fats will hold their shape long enough for the dough to set, and cookies made with them do not spread.

Looking for low-fat cookies? Then forget about crisp sugar cookies.

You can make soft, cake-like cookies easily with little or no fat, using sugar or syrup, pureed fruit, grains like oatmeal to replace some of the flour, and egg white. Crisp sugar cookies are probably beyond the reach of low-fat cooking.

Cutting fat in sauces

Fat thickens sauces by forming emulsions, prevents dairy proteins from curdling, and carries flavor and nutrients. Other ingredients can do these jobs, too.

Thickening Fats thicken sauces like hollandaise, béarnaise, beurre blanc, beurre rouge, or even simple reduction sauces, by forming emulsions. Purees do an excellent job of low-fat thickening.

Purees in fact thickened the earliest known sauces. Ground bread crumbs or ground nuts thickened stocks for sauces of antiquity. Early chefs even had their own variety of anchovy paste (ground-up small, dried fish) to thicken and flavor sauces. Through the years, Weight Watchers has used purees (either crumbs or vegetable purees) for thickening sauces in many of its recipes. Currently, pureed vegetable sauces are in vogue in upscale restaurants. (For more about purees as sauces, see page 252.) My Fire and Ice—Spicy Grilled Chicken Fingers (page 258), with a sauce of pureed peppers, is a low-fat dish, and Strictly American "Eat Your Veggies" Spaghetti (page 253) can be made low-fat by using low-fat ground beef or low-fat smoked sausage. The Shrimp Bisque that follows uses pureed rice to make a thick, rich-tasting, low-fat soup. Pureed rice is also excellent for thickening sauces.

Some chefs use a puree thinned with a little strong stock and add just a small amount of fat for flavor. A sauce made this way is delicious and relatively low in fat.

Shrimp Bisque

MAKES 8 SERVINGS

It is hard to believe that this thick, creamy, rich-tasting soup is low in fat.

what this recipe shows

Shrimp shells make an excellent-flavored stock.

Rice cooked in the stock from the shrimp is permeated with good shrimp flavor.

Onions, a small amount of jalapeño, and numerous herbs add complexity and depth of flavor.

Pureed rice and shrimp give a rich thickness to the bisque.

Dry sherry contributes its own flavor and also acts as a flavor enhancer by dissolving and distributing alcohol-soluble flavors from other food components.

1 small to medium onion, chopped
(about $^1/_2$ cup)

1-inch piece of jalapeño pepper, seeded
and chopped

1 bay leaf

1 tablespoon mild olive or vegetable oil

$^1/_2$ teaspoon dried thyme leaves

2 teaspoons seafood-boil seasoning
(such as Old Bay)

1 teaspoon salt

5 cups water

$1^1/_3$ pounds shrimp (any size) in shells

$^1/_3$ cup rice

About 2 cups low-fat or skim milk

2 tablespoons dry sherry (optional)

1. Sauté the onion, jalapeño, and bay leaf in oil in a large saucepan over medium heat until the onion is soft. Stir in the thyme, seafood seasoning, and salt and continue to cook for about 1 minute. Add the water, turn up the heat, and bring to a boil.

2. When the water is at a full rolling boil, add the shrimp. For large shrimp, time 5 minutes from the time that you put the shrimp in the water (3 to 4 minutes for medium or small shrimp). Remove the shrimp with a slotted spoon to a strainer and rinse with cold water to stop the cooking. Set aside and reserve the cooking water.

3. Add the rice to the cooking water. Bring to a boil, stir well, cover, and reduce the heat to low. Cook covered for 30 to 45 minutes, until the rice is very soft. Remove from the heat and set aside. Remove and discard the bay leaf.

4. Peel and, if desired, devein the shrimp. Place in a food processor with the steel knife. With a slotted spoon, transfer three large spoonfuls of rice to the food processor and puree the mixture. Add the remaining rice and cooking water in two batches, pureeing after each addition. Pour the puree into a heavy soup pot.

5. Add the milk until the bisque is of desired consistency. Reheat briefly, stir in the sherry, if used, and serve hot.

Preventing dairy products from curdling If you try to replace heavy cream in a sauce with low-fat milk or yogurt, you can get a mass of curds. Cream has a high fat content that coats the few proteins in the cream and prevents coagulation. In low-fat milk or low-fat yogurt, there is very little fat to coat proteins, and there are lots of proteins. So, when you heat low-fat dairy products, these proteins join together and produce a mass of curds.

Starch is an ideal fat replacer here. Page 213 explains starch's role in preventing proteins from joining. You can make excellent starch-bound sauces with good flavored fat-free stock, low-fat dairy products, and a little cornstarch. (See page 273 for complete information on starch-bound sauces.)

Cutting fat in pan-frying
Fat prevents sticking in frying. A hot pan, nonstick pans, nonstick cooking sprays, and covering techniques all help with this problem.

The importance of a hot pan One of the secrets of nonstick cooking is to briefly heat the empty pan before you add anything. (Page 205 gives more details about sticking.) Paul Prudhomme points out that if you have a hot pan (350°F/177°C), even a boneless chicken breast without a drop of oil will stick only momentarily, then release.

Nonstick pans Outstanding pans with nonstick coatings are now available. Early pans with nonstick coatings were thin, and the metal rippled with heat, but now many companies produce excellent heavy pans with effective nonstick coatings. These pans make nonstick frying possible quite easily with only a light spray of the nonstick lecithin cooking sprays such as Pam or Mazola.

Covering food in the pan Chopped onions, celery, or other vegetables that need to be softened before being combined with other ingredients can be cooked beautifully in a nonstick pan that has been sprayed with nonstick cooking spray. Cover the food with a piece of wax paper or parchment that has been slightly crumpled so that you can press it to fit in the pan directly on top of the food. This technique prevents moisture loss from the food, and the onions or celery cook in their own juices. If you place a lid on a pan, the lid is a couple of inches away from the food and there is considerable moisture loss.

Cutting fat in deep-fat frying

Until the FDA approves some of the many products for low-fat frying now under development, the best that you can do is to observe the guidelines for nongreasy deep-fat frying, page 164.

Low-fat cooking methods

Many cooking methods do not add fat and actually remove fat from foods. Steaming, grilling, broiling, roasting, baking, and braising are all low-fat cooking techniques. Steamed chicken breasts can be lovely if they are not overcooked. A medium boneless chicken breast will steam nicely in about 7 minutes, and then you can top it with a low-fat topping like Gremolata (page 410) or a sauce such as in Fire and Ice—Spicy Grilled Chicken Fingers (page 258). Grilling is fast and easy with modern gas and electric grills. You can marinate foods in low-fat marinades or coat with low-fat dressings and grill for quick cooking.

Commercial products for fat reduction

Home cooks have only indirect access to products used by large companies to give their low-fat foods the smooth, creamy mouthfeel of fat. For example, the home cook may not be able to buy maltodextrins, but we can buy nonfat sour cream that contains maltodextrins. With this nonfat sour cream we can then make cold low-fat products with rich, creamy mouthfeel just like the big processors. This Low-Fat Lemon-Ginger Ice Cream is so fabulously creamy, you will think you are eating the most expensive super-high-fat brand.

Low-Fat Lemon-Ginger Ice Cream

MAKES 8 SERVINGS

Chris Northmore, master pastry chef at the Cherokee Town and Country Club in Atlanta and an expert in reduced-fat fine desserts, introduced me to the use of nonfat sour cream in ice cream. I bless him for it.

what this recipe shows

Maltodextrins in the nonfat sour cream give ice cream a rich, creamy mouthfeel.

Intense ingredients like lemon zest and ginger add to the flavors provided by the lemon yogurt and vanilla.

³/₄ cup lightly packed light brown sugar

2 large cartons (16 ounces each) nonfat
sour cream

1 carton (8 ounces) nonfat or low-fat
lemon yogurt

3 tablespoons light corn syrup

¹/₈ teaspoon salt

2 teaspoons pure vanilla extract

Finely grated zest of 1¹/₂ lemons

3 tablespoons minced candied ginger

Push the brown sugar through a large-mesh strainer to remove lumps. Place the brown sugar, sour cream, yogurt, corn syrup, salt, vanilla, zest, and ginger in a blender or food processor and process for several seconds to blend well. Chill the mixture for 10 minutes in the freezer compartment of the refrigerator. Freeze in an ice cream maker according to the manufacturer's directions.

Fats and flavor

All foods contain fats. In fact, on a weight basis watermelon has twice as much fat as cantaloupe, although both values are very low. Fats dissolve and release some of the fat flavors in foods into the sauce or surroundings. Without a little fat, these flavors remain locked in the food.

Fats not only release and carry flavors; they also hold flavors in the mouth. You can have an intense water-soluble flavor, but it lasts only an instant. Fats coat the mouth and hold flavors for complete and rounded tastes. When you think about how the oil of a hot chili seems to coat your entire mouth, you realize how fat coats and holds flavor. This flavor releasing, carrying, and holding ability is difficult if not impossible to replace.

In baking, you can rely on other intensely flavored ingredients like cocoa powder (as in the Mellow Moist Low-Fat Chocolate Cake, page 181) or orange or lemon zest (as in the Good-for-You Apple Bran Muffins with Walnuts and Orange Zest, page 148). It is easier to make a good-tasting low-fat chocolate cake than a plain white cake, which can be flat and uninteresting unless intense flavors like citrus zest are added.

For sauces, intensely flavored ingredients or a small amount of fat that is infused with a strong flavor, as demonstrated in the Chili Oil (page 328) with Steamed Broccoli (page 327) or the Roasted Asparagus with Lemon-Chili Oil (page 330), allow you to reduce the fat and yet still have good flavor. Alcohol, as discussed later, is also an excellent flavor releaser.

For food in general, the problem is the same—it is difficult, if not impossible, to replace fat's flavor releasing, carrying, and holding abilities. Food that can be flavorful will taste flat without a little fat to release and prolong some of its taste components in the mouth.

Low-fat shouldn't mean no fat—most successful fat replacement recipes retain a small amount of fat to carry nutrients and flavor.

Your best bet for successful low-fat dishes is to replace fat in its technical roles but still use a small amount of fat for flavor.

Alcohol and flavor A sometimes overlooked flavor enhancer is ethyl alcohol (ethanol). In addition to water-soluble and oil-soluble flavors, some flavors are alcohol-soluble. (A good example is pure vanilla extract.) In fact, alcohol dissolves most water- and oil-soluble flavors and some that neither oil nor water dissolves. By adding a small amount of alcohol, you are possibly releasing additional flavors. You will notice a little vodka in many of my tomato sauces. Vodka is usually 40 or 50 percent alcohol, depending on whether it is 80 or 100 proof. Even when most of the alcohol is cooked off, it still gives tomatoes a burst of flavor. In addition to containing alcohol for dissolving and dispersing flavors, whiskeys and liqueurs have intense flavors of their own. When used without heating, alcohol evaporates in the mouth for an intense flavor burst.

A tiny bit of brandy or cognac in the Brandied Dressing is not detectable but gives the salad a flavor boost in the mouth. With a full-flavored dressing like this, you can use less dressing for an excellent salad.

Mixed Greens and Oranges with Brandied Dressing

MAKES 8 SERVINGS

Orange sections, roasted almonds, and a full-flavored dressing with brandy give this salad distinct flavors. Orange sections make the salad even more cooling and refreshing—an ideal accompaniment to heavier meat dishes or perfect for luncheons.

what this recipe shows

Cold, crisp, dry greens minimize the amount of dressing needed.

Tossing the salad with the dressing just before serving prevents wilting.

Salting the salad adequately gives good flavor.

Soaking canned mandarin orange sections briefly in sugar-ginger water minimizes any canned taste.

1 small head romaine lettuce

1 head Boston or Bibb lettuce

1/3 cup almonds, slivers or slices

1 tablespoon butter

1/4 teaspoon salt

1 cup hot tap water

1 walnut-size piece of ginger, cut into very thin slices

2 tablespoons sugar

4 ice cubes

2 small cans (11 ounces each) mandarin orange sections, drained and rinsed with water (see Note)

Brandied Dressing (page 187)

1. Rinse the lettuce in cold water, then let it stand in a large bowl or in the sink in cold water with ice cubes for 30 minutes. Spin dry and pat dry. Store in zip-top plastic bags with two damp paper towels. Squeeze the air from the bags and seal. Keep refrigerated until ready to prepare the salad.

2. Preheat the oven to 350°F (177°C). Spread the almonds on a baking sheet and roast in the oven until lightly browned, 5 to 8 minutes. While the almonds are hot, stir in the butter and salt.

3. Stir the hot water, ginger slices, and sugar together in a medium mixing bowl, then place the bowl in the freezer for 5 minutes to cool. Stir in the ice cubes for further cooling, then add the orange sections. Soak for at least 5 minutes.

4. When ready to serve, tear the lettuce into pieces. In a large mixing bowl, toss the lettuce well with enough Brandied Dressing to coat all the leaves. Lift the orange sections from the ginger water with a slotted spoon, leaving the ginger behind, and place them on a paper towel for a minute to drain. Add the orange sections to the salad. Toss very gently to combine and add a little more dressing if needed. Taste and add salt if needed. Remove the salad to a serving platter, sprinkle with roasted almonds, and serve immediately.

Note: *Fresh orange sections soaked in the ginger water are excellent if you have time for the preparation.*

Brandied Dressing

MAKES ABOUT ²/₃ CUP

2 medium shallots

1¹/₂ teaspoons salt

¹/₄ teaspoon white pepper

1 teaspoon light brown sugar

1 teaspoon Dijon mustard

1 teaspoon brandy or cognac

1 tablespoon balsamic vinegar

¹/₂ cup mild olive oil or vegetable oil

With the steel knife in the workbowl and the food processor running, drop the shallots down the feedtube onto the spinning blade to mince. Add the salt, pepper, brown sugar, mustard, brandy, and vinegar. Process to blend. With the food processor running, drizzle in the oil, very slowly at first, until emulsified.

Here are two fish recipes that use intense flavors from ingredients like mustard and horseradish.

Salmon Fillet with Sweet, Grainy Mustard Crust

MAKES 4 SERVINGS

Reduced apple juice gives a distinct sweetness to grainy mustard and fresh dill to make an intriguing spicy coating on a salmon fillet.

what this recipe shows

Reduced apple juice adds complex taste to grainy mustard.

2 cups apple juice

¹/₂ cup coarse-grain mustard

4 sprigs fresh dill, finely chopped

1 tablespoon oil to grease pan

1³/₄-pound salmon fillet (see Note)

¹/₄ teaspoon salt

¹/₈ teaspoon white pepper

Several sprigs fresh dill to garnish

continued

1.   Preheat the oven to 375°F (191°C).

2.   Bring the apple juice to a boil in a medium skillet over medium-high heat and boil vigorously to reduce to less than ¼ cup. (You really want only about 1 to 2 tablespoons of apple juice.) Stir the mustard into the reduced apple juice. Heat and stir until the mixture is the consistency of the mustard before you added the apple juice. Remove from the heat. Stir in the fresh dill.

3.   Lightly grease a medium baking sheet with oil. Rub the salmon with salt and white pepper and place on the baking sheet. Coat the fillet with the apple juice–mustard mixture. Bake for 10 to 15 minutes, depending on the thickness of the salmon. Place on a serving platter, garnish with sprigs of fresh dill, and serve immediately.

Note: *Salmon is a high-fat fish, but it does contain healthful fats—unsaturated fats and Omega 3 fatty acids. You can switch to haddock or cod for a much lower-fat dish.*

Horseradish Meringue–Crusted Fillets

MAKES 4 SERVINGS

The Mayo Clinic has used meringues as toppings in many of its low-fat recipes. This particular dish was inspired by a recipe by Molly O'Neill in *Cooking Light* magazine. With a puffed, lightly browned, cloudlike topping, this dish is a beauty and is so good that you won't miss the fat.

what this recipe shows

Intense flavors—horseradish, lemon zest, and dill—make a major contribution to the dish when you have little fat to carry and hold flavors.

Nonstick cooking spray

½ cup part-skim-milk ricotta

½ cup grated fresh horseradish or 1 jar (5 ounces), well drained

½ teaspoon salt

4 large egg whites

1 teaspoon sugar

Finely grated zest of ½ lemon

8 sprigs fresh dill, chopped (about 1 tablespoon)

4 (3 to 4 ounces each) mild fish fillets (sole, flounder, tilapia, scrod, orange roughy, etc.)

4 sprigs fresh dill to garnish

1.   Preheat the oven to 425°F (218°C).

2.   Spray a baking dish large enough to hold the fillets with nonstick cooking spray.

3.   Process the ricotta until creamy in the food processor with the steel knife, add the horseradish and salt, and process with a few quick on/off pulses to blend.

4. Beat the egg whites in a medium bowl until they form soft peaks. Whisk in the sugar and continue to whisk until they form stiff peaks. Stir several spoonfuls of beaten egg whites into the ricotta mixture to lighten. Then gently fold the lightened ricotta mixture into the rest of the beaten whites. Fold in the lemon zest and chopped dill.

5. Place the fillets in the prepared baking dish and spread with meringue. Bake for 8 to 10 minutes, until the topping is puffed and lightly browned. Remove to a serving platter, garnish with fresh dill sprigs, and serve immediately.

at a glance
Roles of fats and possible replacements

Most successful fat replacement replaces fat in its technical roles but retains a small amount of fat to carry nutrients and flavor.

Baking
- Tenderizing: Replacement—sugar (pureed fruit, honey, molasses, etc.).
- Aerating: Replacement—egg whites.
- Acting as a spacer in pastry to provide flakiness: Replacement—not really—but, can spray phyllo with butter-flavored nonstick spray and sprinkle with crumbs and/or sugar, top with several more layers of phyllo done in same way for flaky pastry.
- Helping to determine cookie shape: Choose the fat for the shape you want.
- Carrying nutrients and flavor: Replacement—use small amount of fat to carry nutrients and flavor and use intense-flavored ingredients like lemon zest, dried fruits, cocoa, brandies, etc.

Sauces
- Thickening: Replacement—purees.
- Preventing curdling: Replacement—starch.
- Carrying nutrients and flavor: Replacement—use small amount of fat to carry nutrients and flavor and use intense-flavored ingredients like lemon zest, dried fruits, cocoa, brandies, etc. Alcohol is helpful in dissolving and carrying flavor, as are liqueurs and brandies which have intense flavors of their own.

Frying
- Preventing sticking in pan-frying: Replacement—nonstick sprays.
- Conducting heat in deep frying. Currently no replacement, but research is under way.
- Providing flavorful, crusty coating: Replacement—flavorful crumb coatings that you bake on.

Cooking in general
- Carrying and holding flavors in the mouth: Replacement—do everything you can to add flavor to low-fat dishes.

the recipes

Safe Meringue

Deviled Eggs with Caviar

Eggs Sardou

Baked Eggs—A Little Luxury

Company Scrambled Eggs

Jack, Avocado, Sprouts, and Bacon Omelet

Omelet Soufflé with Dark Cherries and Mascarpone Honey Sauce

Crème Anglaise

White Chocolate Floating Islands

Pastry Cream (Crème Pâtissière)

Lemon Curd

Mesmerizingly Smooth Flan

Rose Levy Beranbaum's Cordon Rose Cream Cheesecake

White Chocolate Icing for Warmer Conditions

Wedding Cheesecake

Savory Herbed Cheesecake

Pâte à Choux

Cocktail Puffs with Escargots

Chocolate Soufflé with White Chocolate Chunks

Incredibly Creamy Soufflés

Lemon on Lemon on Lemon Meringue Pie

Raspberries with Mascarpone Cream in Walnut-Oat Meringue

Great Flower Cake

eggs unscrambled

eggs create the airy lightness of a soufflé. Their yolks are the secret behind decadently smooth flans and irresistibly moist chocolate cakes. Extra egg whites make extraordinarily crisp, dry cream puffs.

Yet for all their wonder eggs present many challenges to the cook. Why is the filling of a banana cream pie that was firm when you put it into the refrigerator soupy the next day? Why do hardcooked eggs sometimes have an ugly green coating on the yolks? Why do scrambled eggs stick to the pan? Why do the whites of poached eggs float about in thin wisps? With a little egg knowledge—how important freshness is, how egg proteins cook, what causes sticking, how egg white foams work—you can make eggs work for you to create exactly the textures that you want, whether smooth and creamy, dry and crisp, or light and airy.

Eggs perform many cooking tasks, from holding foods together to leavening. But precisely because they are such a basic ingredient, it's easy to overlook how important egg know-how is to fine cooking. Knowing how to use eggs—not just whole eggs but also whites and yolks separately—can mean the difference between an ordinary and an extraordinary dish.

Eggs in some of their many roles appear in other chapters. Eggs as emulsifiers are in Chapter 4, eggs used to prevent sugar crystallization in Chapter 7, eggs in shortened cakes in Chapter 2, and eggs as washes and glazes in Chapter 1.

at a glance
What eggs do

Add nutrients	Eggs are high in protein (both whites and yolks) and provide valuable nutrients such as vitamin A, calcium, magnesium, iron, and riboflavin.
Color and flavor	Golden egg yolks improve the appearance as well as the flavor and texture of many dishes. Eggs also improve color by providing proteins for the browning reaction.
Dry	In baking, egg whites can make dishes like cream puffs dry and crisp. This can be a drawback in baked goods that you want moist, like muffins.
Emulsify	Natural emulsifiers in egg yolks hold fat and water together as emulsions in many foods, including sauces like hollandaise, béarnaise, and mayonnaise. Emulsifiers are also the secret to the smooth, creamy texture of many dishes.
Glaze	Eggs, egg whites, egg yolks, and egg washes brushed on breads and pastries and baked add a rich, shiny glaze.
Filter or clarify	Egg whites clarify stock to produce clear consommé.
Hold	Egg proteins hold sauces, muffins, cakes, pies, custards, quiches, casseroles, and meat loaf together. For low-fat cooking and baking, egg whites alone can be used.
Leaven	Eggs aerate meringues, cream puffs, popovers, and certain cakes. They lighten muffins and pancakes.
Prevent crystallization	Egg whites control crystal size or prevent crystal formation in candy-making.

The anatomy of an egg

Chicken eggs are essential to the propagation of their species and are important to our species as a food and as a food ingredient that performs culinary miracles.

Inside the shell

Within the eggshell are two membranes that separate at the large end of the egg to form the air cell. Inside these membranes lies the albumen (the white), made up of an outer thin white, a thick white, and another layer of thin white. Immediately surrounding the colorless yolk membrane is a chalaziferous layer. This layer turns into opaque fibrous cords (chalazae—the white strings in eggs) on each end of the yolk that stretch through the white to hold the yolk in place. Each chalaza is like a shock cord holding the yolk and preventing it from bumping against the shell. Prominent chalazae are a sign of freshness. There may be a small blood spot in a tiny number of eggs. This blood does not affect the cooking quality or the nutrition of the egg. If you wish, you may remove the spot with a spoon as you would a bit of eggshell.

The eggshell

An eggshell in itself is a miracle. It is coarse and granular with as many as 17,000 pores. To prevent bacteria from entering, the pores contain protein fibers and the shell is covered with a natural protein protective coating called the *cuticle*.

As a child, I loved to gather eggs in my aunt Ruth's huge hen yard. It was like an Easter egg hunt. There were eggs in patches of weeds, behind the pillars that held up the henhouse, and right out in the open. Some of the eggs were dirty, and I thought I was being helpful by washing them. The eggs I washed spoiled within two days because I had, unknowingly, washed off nature's protective coating. Commercial egg processors wash eggs after gathering, but they immediately spray them with colorless, tasteless mineral oil to replace the natural coating.

A lack of minerals in a hen's diet will cause the hen to lay eggs with very thin shells that break easily. My grandmother kept crushed oyster shells (a good source of calcium) in the hen yard for the hens to peck. Modern processed food for chickens contains adequate mineral content to produce good shells.

An egg is an egg is an egg

You may have heard that brown eggs are better for you than white eggs, but the truth is that color does not affect nutritional value, cooking characteristics, or quality. The breed of chicken determines the color of the egg. Rhode Island Reds and several other breeds lay brown eggs.

If you want to know what color eggs a hen will lay, check her ears. Carefully push the feathers aside on the sides of the hen's head, and there are the ears. White ears mean white eggs, and reddish brown ears mean brown eggs.

I was astounded when Marie Murphy came in from the henhouse at Marie and Joe's Whileaway Farm near Atlanta with a basket of pale green and pale blue eggs. I made a beeline for the henhouse. Their Araucana hens were beautiful colors ranging from brown to pinkish pale brown to smoky gray. Sure enough, some of the chickens had pale green ears and some had pale blue ears.

Eggs come in different sizes as well as colors. Standard sizes are small, medium, large, extra-large, and jumbo—plus peewee, which you don't normally find in stores. All the recipes in this book use large eggs, as do most recipes. Here is a table with various size, weight, and volume relationships.

STANDARD EGG SIZES				
Size	Ounces per dozen	Number in 1 cup		
	Whole	Yolks	Whites	Whole
Jumbo	30	11	5	4
Extra-large	27	12	6	4
Large	24	14	7	5
Medium	21	16	8	5

How to tell if an egg is fresh

As an egg ages, many physical changes take place that make a major difference in how the egg cooks. Freshness affects poaching, hardcooking, frying, and even the beating of egg white foams. And certainly freshness affects flavor. Just how freshness affects some dishes is explained with that particular method. We have good ways to determine an egg's freshness both in and out of the shell.

In the shell

As eggs age, carbon dioxide that was dissolved in the egg white seeps out through the pores in the shell, and oxygen and gases in the air seep in. (This loss of carbon dioxide makes the egg more alkaline and therefore more susceptible to attack by bacteria.) The older the egg, the more air has seeped in and the larger the air cell.

An easy way to tell if an egg is fresh is to place the egg in its shell in a bowl of tap water. If the egg lies flat on the bottom, it has a small air cell and is quite fresh. If the egg stands up and bobs on the bottom, it has a larger air cell and is not as fresh. If the egg actually floats to the surface, look out! It may be rotten. You may never need to check an egg. We have good-quality, reasonably fresh eggs in our markets. With any luck, you may never in your lifetime run into a rotten egg.

Out of the shell

Experts tell us that the best way to crack an egg is on a hard flat surface. There is less shell shatter if you do not crack an egg on an edge, so you will be less likely to have bits of shell in your eggs. Hit the egg firmly but not forcefully on the side and pull the shell apart with your thumbs. If you have ever watched children learning to cook, you know how difficult it is to crack an egg. I sometimes feel I've spent a good part of my life looking for an edge to crack eggs on. Now, alas, I find I was wrong all along.

The best way to crack an egg? On a hard flat surface—the shell shatters less, so you get fewer bits of shell in your eggs.

Crack an egg onto a plate. If the white is thick and does not spread widely, the egg is fresh. A fresh egg's white is nearly all thick. As the egg ages, this thick part of the white deteriorates into runny white. Notice the two chalazae (the cords at each end of the yolk). Again, prominent chalazae are another sign of a fresh egg.

Membranes in the egg also change with age. The yolk membrane becomes weaker and eventually breaks. Look at the yolk. If it stands up, the yolk membrane is strong and the egg is very fresh. If the yolk is flat and spread out, you have an older egg. If you've cracked an egg carefully but the yolk broke anyway, the egg could have been so old that the yolk sac was broken even before you cracked the shell.

Judging freshness in the store

Egg graders classify eggs with quality standards that reflect both their age and their care. (*Agriculture Handbook No. 75,* available from the USDA, explains egg grading standards.) To do this, graders use a technique known as *candling,* in which they examine each egg over a strong light. This way graders can see into the egg and judge the size of the air cell as well as the condition of both the white and the yolk.

Eggs with small air cells, thick whites, strong yolk membranes, and good shells are very fresh, of high quality, and in excellent condition; they are graded AA. Grade AA eggs, available in upscale markets, usually cost a little more than other eggs, but their high quality produces better food, and a longer shelf life saves on spoilage and waste. For restaurateurs and caterers, who can easily purchase AA eggs, they may be a better buy, for both quality and shelf life. The next grade, Grade A, is what is available in most retail supermarkets.

Storing eggs to preserve freshness

Commercial processors store eggs in refrigerated chambers. At one time some processors piped in carbon dioxide to slow down the natural carbon dioxide loss. With developments in refrigeration and rapid distribution systems, however, they no longer need or use this method.

The American Egg Board recommends storing eggs in the original carton on a refrigerator shelf rather than in a compartment on the refrigerator door. This keeps the eggs stable and prevents their being jarred every time the door is opened and closed. The carton also keeps the eggs upright with the broad end (air cell end) up and protects them somewhat from refrigerator odors.

When refrigeration is not available, people have found ingenious solutions to keep eggs. One of my students grew up in Italy during World War II. Her family stored eggs in a cave where it was cool and packed them in a chalky powder to preserve them. The powder was probably calcium carbonate, which gave off carbon dioxide that seeped into the egg shells and slowed deterioration.

Another student, who worked as a chef on a racing yacht, would go to a farm on the day of departure and buy just-gathered eggs. To prevent carbon dioxide loss, she immediately coated the eggs with petroleum jelly. This slowed deterioration and kept the eggs in good condition for long periods at sea. As the eggs got older, their poaching quality went first, and then it became difficult to fry them without breaking the yolks. The older eggs were fine for scrambling or other dishes.

Eggs and salmonella

Salmonellosis is a foodborne infection of the gastrointestinal tract caused by *salmonella* bacteria. A few outbreaks of salmonella food poisoning from bacteria in raw or undercooked eggs have been reported in the United States, primarily confined to the Northeast and Mid-Atlantic states. The problem is, a hen that appears perfectly healthy can have colonies of salmonella bacteria in her ovaries, and yolks of eggs that this hen lays may or may not contain the bacteria. The contamination is not from the shell or cracks in the shell but from the yolk itself. You can have a perfectly clean, crack-free egg that contains salmonella.

There have been rare cases of whites containing salmonella. The salmonella content on the yolk must have been so high that a significant amount was left on the white in the area where it touched the yolk. In one documented incident, the eggs had not been refrigerated and were left quite warm, so the bacteria multiplied. Meringues made with these eggs were not cooked but slipped under the broiler for a minute to brown the tops. Then the pies were left to stand at room temperature.

Food dose irradiation could completely eliminate the salmonella problem, but this is not a solution consumers seem willing to accept at this time. Common sense, proper refrigeration, and proper cooking techniques are powerful tools for combating salmonella. The American Egg Board and various writers have devised procedures for killing salmonella in egg yolks. For Swiss and Italian meringues, which call for heating the whites, but not quite high enough to ensure safety, a safe procedure similar to that for yolks can be used. I now use Alice Medrich's procedure (page 196) for soft meringues to ensure safety of whites, and I have converted all my recipes that contain uncooked or partially cooked yolks to my own favorite procedure for eliminating salmonella in egg yolks.

Safe Egg Yolks

Heat 2 egg yolks and $1/4$ cup liquid from the recipe—for example, lemon juice and water in mayonnaise or cream in truffles—and $1/2$ teaspoon sugar in a small skillet over very low heat, stirring and scraping the bottom of the pan constantly with a spatula. At the first sign of thickening, remove the pan from the heat, but continue stirring and dip the pan bottom in a larger pan of cold water to stop the cooking. Use in the recipe instead of raw yolks.

Destroying bacteria is a matter of both time and temperature. You do not even have to get the yolks to the salmonella instant-kill temperature of 160°F (71°C). Holding them at slightly lower temperatures for several minutes is just as effective. Yolks are pasteurized by holding them for 3.5 minutes at 140°F (60°C), which is not very hot—hot tap water is in this range. Eggs don't scramble (cook) until about 180°F (82°C). So there is some margin here.

Safe Meringue

MAKES TOPPING FOR ONE 8- TO 10-INCH PIE

I love vanilla-flavored meringue on pies, so this is a generous six-egg recipe. I have incorporated Alice Medrich's technique from *Chocolate and the Art of Low-Fat Desserts* into my meringue recipe. I also include a little cornstarch for tenderness and prevention of shrinking.

what this recipe shows

Diluting egg whites with water and adding sugar will enable careful heating of the whites to 160°F (71°C), the instant-kill temperature for salmonella.

Swollen starch added to the beaten whites produces a tender, smooth-cutting meringue that does not shrink or bead easily.

Meringue made with egg whites heated in this manner is an excellent stable meringue.

6 large egg whites	1 tablespoon cornstarch
2 tablespoons water	1/3 cup cool water
3/4 teaspoon cream of tartar	1/8 teaspoon salt
3/4 cup sugar	1 teaspoon pure vanilla extract

1. Stir the whites, water, cream of tartar, and sugar together well to break up whites (try not to create foam since it cooks at a lower temperature) in a medium-size stainless-steel bowl. Heat 1 inch of water to a simmer in a medium skillet and turn the heat off. Run a cup of hot tap water, place an instant-read thermometer in it, and place it near the skillet. Place the metal bowl of egg white mixture in the skillet of hot water and scrape the bottom and the sides of the bowl constantly with a rubber spatula to prevent the whites from overheating. After 1 minute of constant scraping and stirring, remove the bowl of egg whites from the hot water and place the thermometer in the whites, tilting the bowl so that you have about 2 inches of white covering the thermometer stem. If the temperature is up to 160°F (71°C), beat until peaks form when the beater is lifted. If necessary, place the bowl of whites back in the hot water and scrape constantly in 15-second increments until the temperature reaches 160°F (71°C). Rinse the thermometer in the hot water in the skillet (to kill salmonella) and replace in the cup of hot water after each use.

2. Sprinkle the cornstarch into a small saucepan, add the cool water, and let stand 1 minute. Then stir well. Bring the water and cornstarch to a boil, stirring constantly. The mixture will be thick and slightly cloudy. Let it cool for a couple of minutes, then whisk 1 to 2 tablespoons of the cornstarch mixture into the meringue and continue adding and beating in until all is incorporated. Whisk in the salt, vanilla, and any other desired flavoring such as lemon zest. Set aside while preparing the pie filling.

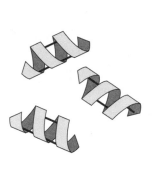

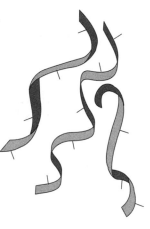

**Natural
Proteins**

**Denatured
Proteins**

**Coagulated
Proteins**

How eggs cook and how to cook eggs

Egg yolks and egg whites are both high in protein. These proteins change from liquid to solid as they cook.

Natural proteins are separate individual units, shaped like coils or springs or complex wads of coils. They have links from one section of the coil or wad to another that hold the wad together in a tightly bound individual unit. They can be long, rather straight springs, or curled-around wads. Each one is separate, so there is room for light to pass between them. You can see this yourself when you crack open an egg. You can see through the egg white because the proteins are separate units. Raw chicken breast, fish, and raw shrimp also have a translucent appearance because light can travel between the proteins.

When proteins are heated or exposed to air or acids, the bonds across their springs break and the proteins unwind. The proteins straighten out, with their bonds sticking out to the sides. This protein has changed from its natural form and is now termed a *denatured protein*.

Almost immediately, this denatured (unwound) protein with its bonds sticking out runs into another denatured protein with its bonds sticking out, and they link together to form a delicate three-dimensional mesh. This is what happens when your egg in the pan goes from translucent to opaque white. With the proteins linked together, there is no longer space between individual proteins for light to pass through, and the egg white becomes a solid white gel.

The same unwinding and linking of proteins takes place when you cook any meat. This is easy to see in light meats like chicken or fish. Heat makes the individual proteins in the raw translucent food unwind and hook together to become a snowy solid white. At this point the proteins are still hooked together loosely so that water attached to and surrounding the proteins is held in the mesh of the linked proteins and keeps the egg white, chicken, or fish moist and tender.

The longer you heat proteins and the higher the temperature, the tighter the bonds of the protein mesh become, squeezing out the water. Tender, moist, juicy proteins become dry and leathery. Gentle heating is the real secret to cooking proteins, whether eggs or meat. This principle applies no matter how you cook eggs—softcooked or hardcooked (egg experts prefer to use this term as a reminder that the eggs should not be cooked at a hard boil), poached, fried, or scrambled.

Gentle heating is the real secret to cooking proteins, from eggs to meat.

Cooking eggs in the shell

One would think that eggs cooked in their shells should be the simplest of all, but there can still be problems. Hardcooked eggs can be difficult if not impossible to peel, and the yolk can turn an unappetizing gray-green.

Softcooked eggs

Getting softcooked—also called *softboiled* or *coddled*—eggs to the desired doneness every time is not easy. For hardcooked eggs I like to use the cold-water start for convenience. But for softcooked eggs I find I have better luck cooking them exactly as I want by using the boiling-water start.

Softcooked Eggs

Warm one or two eggs per person to room temperature by putting them in a bowl of hot tap water for 4 to 5 minutes. Fill a saucepan with enough water to cover the eggs by 1 inch and bring to a boil. Stir 1 teaspoon salt into the water. Carefully lower the eggs into the water, bring back to a boil, then turn heat down to a low simmer. At $2^1/_2$ to $3^1/_2$ minutes, the eggs are cooked, but some of the white is still soft to runny. At 4 minutes, the white is firm and the yolk runny. (For *oeufs mollets,* medium-soft, cook for $5^1/_2$ to 6 minutes.) Hold the eggs briefly under cold running water to stop the cooking. Peel and use as you wish (over crumbled saltine crackers is a favorite) or place each egg in an egg cup, broad end up. Cut off the top $^1/_2$ inch with a sharp knife or egg scissors. Eat the egg from the shell with a small spoon. If you are serving the eggs in egg cups, you can use very fresh eggs for excellent taste, but you will not be able to peel them.

Hardcooked eggs

Hardcooked eggs that you normally need to peel are better made with older eggs (7 to 10 days old). Here are some tips for getting perfect hardcooked eggs.

To make hardcooked eggs easier to peel: Egg researchers have found that ease of peeling is related to pH, a measure of acid/alkaline levels. Older eggs, which have lost some of their carbon dioxide, are more alkaline, with a pH of 8.7 to 8.9 or higher. Hardcooked older eggs are easier to peel than fresh eggs. Because you want to keep this high level of alkalinity, you should not add vinegar to the cooking water, as some people recommend. You can add a bit of salt, though. It makes the egg white coagulate faster to seal any cracks in the egg. (Quick cooling also helps a little with peeling.)

To prevent eggs from cracking: First inspect each egg for small but visible cracks. Warm refrigerated eggs for 4 to 5 minutes in hot tap water. Intact eggs at room temperature are less likely to crack in the hot water than cold eggs. Also, cooking the eggs at a low simmer reduces the likelihood of cracking.

As to piercing the shell, the jury is still out. Some scientists, notably the chemist Arthur Grosser, believe that it allows the expanding gases to escape from the air cell as it is heated. This sounds very reasonable, but with fresh eggs researchers reported that in batches of about five dozen eggs each those batches that were not pierced had an average of only 5 percent eggs cracked. Batches that were pierced had an average of 55 percent cracked! In fresh eggs, a hole in the shell may encourage a crack instead of preventing it. The American Egg Board notes cautiously, "Piercing also often produces hairline cracks in the shell."

With older eggs the story was different. For 28-day-old eggs, piercing prevented cracking. Fresh eggs are porous and have small air cells, so it probably is not necessary to pierce them. As an egg gets older, with the shell less permeable and the air cell larger, piercing may help to prevent cracking. I look at the date on the carton and bother to pierce only really old eggs, close to the expiration date.

To prevent a green layer from forming: It is accepted that the ugly green layer on the yolk is the result of iron in the yolk combining with sulfur (hydrogen sulfide) in the white to form green iron sulfide. Author Harold McGee recently did some experiments in which he got the yolks green all the way through, indicating that the yolk alone contains enough iron and sulfur to produce the green iron sulfide. While the egg is cooking, heat speeds up this chemical reaction. The longer the egg cooks, the greater the chance for discoloration, so watch the time carefully. Quick cooling also helps prevent the layer from forming.

To center the yolk: In some recipes that use hardcooked eggs—deviled eggs, for instance—centered yolks are preferred. What you're really looking for, of course, is perfectly formed, even shells of white to use as a receptacle for a filling.

Researchers have found that eggs stored on their sides have the most consistently centered yolks (eggs stored large end down came in second). Some recommend cooking eggs in an upright position—you may have seen little wire holders or racks designed to hold an egg or several eggs upright in the pot.

At the Cordon Bleu Cooking School in London, students are instructed to twirl the eggs for several minutes at the beginning of cooking. Researchers have confirmed the validity of this technique, and rollers are used to hardcook eggs commercially. Few home cooks have the patience for twirling eggs. You can do only one at a time—which makes it out of the question for a large batch—and it's easy to lose track of which have and which haven't been twirled. I prefer another technique. The night before hardcooking eggs, I turn the carton—tightly closed—on its side. If you have an egg rack, cooking them standing up is effective, too.

For hardcooking, don't choose the freshest eggs—they're harder to peel than the older ones.

To store hardcooked eggs: You can leave hardcooked eggs in their shells for refrigerator storage. It is slightly more difficult later to remove the shell than it is immediately after cooking. Roll the egg gently back and forth on the counter, pressing down just enough to crack the shell all over. Then hold it under running water and start peeling at the large end.

If you are making many hardcooked eggs to use soon after cooking, here is a trick an old deli chef who used to hardcook 50 to 100 eggs a day taught me. Rinse the eggs in cold water several times and pour off the water. Leave the eggs in the pan and shake it vigorously to bump and bang the eggs against one another. Peel the shells under running water. It's amazing how they practically peel themselves this way.

There are two schools of thought about how to cook hardcooked eggs—the cold-water start and the boiling-water start. I prefer a cold-water start because I don't always have time to warm the eggs before cooking.

Hardcooked Eggs

For a cold-water start, place the eggs in a heavy saucepan large enough to hold them in one layer and cover with $1\frac{1}{2}$ inches cold tap water. Partially cover the pot and bring to a full rolling boil. Turn heat down to low and leave on the heat, covered, for 30 seconds, then remove from the heat and let eggs stand in the hot water for 15 minutes. Rinse under cold running water for about 5 minutes.

For a boiling-water start, warm the eggs by placing them in a bowl and covering them with very hot tap water. Let stand about 4 minutes, drain, and cover again with very hot tap water. Let stand again for about 4 minutes. Bring a heavy saucepan full of water to a boil. Gently lower the eggs into the water. Reduce the heat to a very low simmer (about 180°F/82°C) and simmer, uncovered, for 15 to 18 minutes. Rinse under cold running water for about 5 minutes.

Deviled Eggs with Caviar

MAKES 10 TO 12 APPETIZER SERVINGS

Deviled eggs are a must at Southern church dinners-on-the-grounds and camp meetings. This is a little more sophisticated version, with lemon zest and a touch of vivid red caviar.

what this recipe shows

Placing the securely closed carton of eggs on its side the night before cooking centers the yolks.

Salt in the water makes leaking white cook fast and seals cracks on any egg that may break.

Letting the eggs stand in hot water for 15 minutes cooks them slowly for greater tenderness.

Running cold water over the eggs at the end of cooking cools them quickly to help prevent greening of the yolk surface and makes peeling easier.

Banging the eggs against each other in the pan gently cracks the shells and makes them easier to peel.

Keeping caviar cold improves its taste.

---

12 large eggs, at least a week old

1 tablespoon salt *and* 1¼ teaspoons
   salt (about 1½ tablespoons *total*)

3 medium shallots, minced

2 tablespoons butter

½ cup sour cream

⅓ cup Homemade Mayonnaise
   (page 289) or store-bought

⅛ teaspoon cayenne

¼ cup finely chopped fresh chives

Finely grated zest of 3 lemons

3 tablespoons red caviar, well chilled

1. The night before the eggs are to be cooked, seal the carton with a piece of tape and turn on its side to center the yolks.

2. When ready to cook, place the eggs in a medium saucepan. Add water to cover by 1½ inches. Add 1 tablespoon salt. Partially cover the pot and bring to a full rolling boil. Turn heat down to low and leave on the heat, covered, for 30 seconds, then remove from the heat and let eggs stand, covered, in the hot water for 15 minutes. Pour off hot water and rinse eggs under cold running water for 5 minutes. Pour off water and shake the pan to bump the eggs against each other until all eggs are well cracked. Cover with cold water.

3. Peel the eggs under running water, rinse them, and cut in half lengthwise. Transfer the yolks to a bowl. Cover and chill yolks and whites separately for 1 hour.

4. Sauté the shallots in butter in a medium skillet over medium-high heat until soft, about 2 minutes. Set aside. Mash the yolks with a fork. Mash in the sour cream. Add the mayonnaise, 1¼ teaspoons salt, and cayenne. Whisk until well blended. Whisk in the shallots.

5. Fill the egg white halves with yolk mixture, extending the mixture over part of the white. Generously sprinkle each stuffed egg with chives, leaving a little yellow showing around the edges. Sprinkle each half with lemon zest. Cover and refrigerate. When ready to serve, spoon a bit of cold caviar onto each half. Serve well chilled.

at a *glance*
Hardcooked eggs

What to Do	Why
Use older eggs.	Older eggs have a higher pH (lower acidity) and are easier to peel.
Add salt to the water.	Salt speeds up denaturing and thus causes faster cooking of white at a crack so less white feathers into the water. Salt also slightly raises the boiling point of the water and speeds cooking.
Time cooking carefully and rinse immediately in cold water for about 5 minutes.	High heat and overcooking produce a green layer on the egg yolk.

Cooking eggs out of the shell

Cracking a raw egg opens a world of culinary wonders. Poaching or frying an egg is just the beginning. Eggs can be simply scrambled or made into omelets with an infinite variety of fillings; they can be separated and recombined in soufflés, both sweet and savory, or used to make creamy flans.

Poached eggs

Poached eggs are easy to prepare and versatile—you can serve them on toast with hollandaise or a lusty wine sauce, on a bed of warm spinach, or atop a glorious Wiener Schnitzel—yet many people shy away from making them. The yolks can break, the whites spread out uncontrollably in the water, and it's difficult to tell when the eggs are done. Perfectly poached eggs are possible using the following techniques:

- Use fresh eggs. The fresher the egg, the stronger the yolk membrane and the thicker the white. The yolk is less likely to break, and the thick white clings better to the yolk.
- Add vinegar and salt to the cooking water. Acids such as vinegar make proteins in egg whites unwind (denature) faster. Anything that makes the egg white cook faster will cut down on feath-

ering. Salt also promotes denaturing and faster cooking of the egg.

- Adjust water temperature and watch cooking time. Water temperature is the major factor controlling the texture of cooked eggs, and cooking time has an effect as well. If you heat proteins gently, they unwind (denature) and join together loosely with their neighboring denatured proteins. Water between the proteins or attached to the proteins is held in a moist, tender network. If the water is too hot or the cooking time too long, the protein mesh tightens, squeezing out the water, and the proteins become tough and leathery. For successful poaching, slip the eggs one by one into gently boiling water. As soon as the outsides of the whites start to set, lower the heat just below a simmer and cook until the whites set. Or cover the pot and remove it from the heat once the whites start to set.

Eggs can be poached in liquids other than water—white wine, stock, or even cream. You would, of course, not use salt and vinegar as you do in water.

The most delicately tender poached eggs I ever ate were at a small Greek restaurant just outside the gates of West Point in upstate New York. I asked the owner how they were prepared. He brought his wife out of the kitchen. She did not speak English, but with a lot of sign language and her husband as translator, she managed to explain that she had a big heavy pot and a lot of boiling water. She lowered the eggs into the boiling water, covered the pot and continued to heat for less than a minute, then removed it from the heat. The eggs cooked slowly standing in the large heavy pot of hot water.

Poached Eggs for One or Two People

For up to four eggs, fill a nonstick 8-inch skillet a little over half full of water (about 2 cups), add 2 teaspoons vinegar and ³/₄ teaspoon salt, and bring to a slow boil. Break an egg into a saucer and slip it into the water. (If you are poaching more than four eggs in a larger skillet, arrange them in a clockwise pattern so you will know which you put in first.) Just as the water comes back to a boil, reduce to a low simmer. When the eggs begin to set, take a spatula and gently run it under each one to release it from the bottom. If a yolk breaks, do not disturb it; it will seal by itself. Cook until the whites are firm. When the eggs are done, lift them out with a slotted spoon. Rinse them briefly in hot water to remove the vinegar. Drain well and serve.

Poached Eggs for a Crowd

You may have noticed the pretty teardrop shape of some restaurant poached eggs. This is the shape formed as the egg gently sinks to the bottom of a tall pot of simmering water. If you are preparing poached eggs for a crowd, you'll find this restaurant procedure very useful.

Fill the tallest stockpot you have with water and bring it to a boil. It must be at least 12 inches tall for this to work. Add 3 tablespoons vinegar and 1 tablespoon salt for every 4 quarts of water. Bring to a full boil. Then crack eggs, one at a time, and slip into the water. When all the eggs are added, reduce the heat to a medium simmer. Eggs will partially cook as they sink to the bottom and then float back up. When they reach the top, they are nearly done. Remove an egg with a slotted spoon. Touch the white gently near the yolk to make sure it is firm. Put it back in the pot if it is not; put it in a bowl of ice water if it is. When all the eggs are done and in the ice water, trim any thin streamers from each. Then place them in a fresh bowl of ice water and refrigerate until ready to use.

To reheat, place the eggs, two or three at a time, in a strainer and lower them into simmering water for about 45 seconds. Drain and serve.

at a *glance*
Poached eggs

What to Do	Why
Use very fresh eggs (Grade AA if possible).	The white of a fresh egg is thick and holds together. A fresh egg's yolk membrane is strong and will not break easily.
Start in boiling water, then reduce to a low simmer or cover and remove from heat.	Boiling water will set the outside of the egg and prevent the white from spreading. Low heat will produce a tender egg.
Use vinegar and salt in the water.	Both acid (vinegar) and salt make the white of the egg cook (coagulate) faster so it does not spread.
When preparing poached eggs ahead, plunge the eggs into ice water after poaching, store in ice water in the refrigerator, and reheat by lowering eggs into simmering water.	Ice water stops the cooking so that the white is cooked but the yolk is runny. Ice water and water for reheating also remove vinegar.
Use a heavy nonstick pan.	Eggs are less likely to stick.

Eggs Sardou

MAKES 8 APPETIZER OR 4 MAIN-COURSE SERVINGS

Poached eggs are the basis for many classic dishes, like Eggs Benedict. Eggs Sardou, a New Orleans specialty, with spinach, artichokes, and hollandaise, is one of my favorites. This is the rich, classic version. You can prepare an excellent lower-fat version by using $1/2$ cup high-quality reduced-fat mayonnaise in place of the cream and rationing the amount of hollandaise on the serving. This can be made in individual dishes or in a large serving dish for a brunch buffet. Allow one egg per person as an appetizer, two as a main dish.

what this recipe shows

Bright red means cayenne is fresh; browned cayenne is old, has a dried straw taste, and can spoil a dish.

The cream cuts the heat of the cayenne and mellows any harshness of the spinach.

continued

1 tablespoon butter

1 small onion, chopped

1 can (14 ounces) artichoke bottoms, drained, rinsed twice, halved, and each sliced $1/4$ inch thick

$1/8$ to $1/4$ teaspoon cayenne, preferably recently opened

$1/8$ teaspoon freshly ground white or black pepper

$1/8$ teaspoon freshly grated nutmeg

$1/2$ teaspoon salt

1 cup heavy or whipping cream

2 packages (10 ounces each) frozen chopped spinach, thawed and drained

$1/3$ cup freshly grated Parmesan

8 poached eggs (page 202), reheated if necessary

1 to 2 cups Quick Blender Hollandaise (page 302)

1. Heat the butter in a medium skillet over medium heat and sauté the onion until soft. Add the artichokes, cayenne, pepper, nutmeg, and salt. Stir well. Add the cream and boil for 2 to 3 minutes to reduce slightly. Place the spinach in a clean dish towel and squeeze to remove moisture. Stir the spinach into the cream mixture. Add the Parmesan. Taste and add salt if needed.

2. Spoon a generous portion of spinach into individual serving dishes or spread all of it in the bottom of a large serving dish. Place the hot egg or eggs on top. Cover with hollandaise and serve immediately.

Baked Eggs—A Little Luxury

MAKES 1 OR 2 SERVINGS

Baked eggs are special—a white ramekin with a pale gold cheese crust that gives way to a creamy mixture of crumbs and egg white moistened with yolk, spiked with nutmeg and a little nip from hot sauce. I make them on leisurely Sundays as a luxurious little treat.

what this recipe shows

The slightest change—like using a different ramekin or adding a bit more cream—alters the time needed to get the white totally set but the yolk runny.

Cream softens the heat of the hot sauce.

2 tablespoons butter

$1/4$ cup dried bread crumbs

2 large eggs

Freshly grated nutmeg

Hot pepper sauce

Salt and pepper

$1/4$ cup heavy or whipping cream

2 tablespoons grated Gruyère

1. Preheat the oven to 350°F (160°C).

2. Grease two small ramekins or custard cups with butter. Distribute the bread crumbs evenly around the sides and bottom of the cups. Crack an egg and place it in the center of each cup. Sprinkle each egg with a pinch of freshly grated nutmeg, a dash or two of pepper sauce, and a light sprinkling of salt and pepper. Spoon 2 tablespoons of cream on top to distribute the seasonings. Sprinkle each with Gruyère. Bake for 8 to 10 minutes—the time depends on your oven and how you like your eggs.

Over easy: cooking fried eggs, scrambled eggs, and omelets with ease

Fried eggs, scrambled eggs, and omelets have similar problems—a tendency to stick to the pan and toughness. With frying there's also the problem of excessive fat absorption. How do you make tender, moist eggs that do not stick to the pan or absorb too much fat?

Sticking is a real problem, not just in frying or scrambling but in all types of egg dishes. In general, proteins stick much more than vegetables, but the problem is exacerbated because egg is a liquid when it goes into the pan and can get down into any nicks or rough spots. Even very smooth pans have minute holes or irregularities. Then, when the pan is heated, the egg literally cooks into the pan. Meats leak juices full of proteins that do the same thing.

To prevent eggs from sticking:

- Use a heavy pan made of a metal that is a good conductor of heat like copper or aluminum or at least has a thick layer of such metal. It will not have hot spots and will cook evenly.
- Use a pan with a very smooth surface, a well-seasoned skillet, or a pan with a nonstick coating.
- Use nonstick cooking spray. Such sprays contain lecithin, nature's incredible greasing agent, as their active ingredient. They are now available in butter or olive oil flavors as well as plain vegetable oil.
- Warm the empty pan before you add the fat. When you warm the pan, the metal expands and seals some of the minute holes. Besides, if the pan is already warm when the liquid protein hits it, the protein cooks on the surface before it gets down into the pan itself. The pan should be very warm but not so hot that the eggs brown as soon as they hit.

Again, gentle heating combined with gentle treatment is the secret of tenderness in all protein cooking. When eggs are overheated, their protein bonds tighten and squeeze out the moisture. The result—dry, leathery, tough eggs.

You can control moisture in two ways. Using low heat and avoiding overcooking prevent the egg proteins from squeezing out moisture. A hot pan, particularly a heavy one, will continue to cook and dry the eggs even off the heat. For moist eggs, remove the pan from the heat before the eggs are as dry as you want.

Fried eggs

Fried eggs are a world unto themselves. Before the days of cholesterol terror, fried eggs—with or without bacon, ham, or sausage, not to mention grits or home fries—were a favorite American breakfast. They still are, though we indulge less often. In other parts of the world, fried eggs are more likely to be eaten for lunch or as a light supper with spinach or pan-fried potatoes. They are excellent with asparagus in season. Fried eggs can be cooked sunny side up (not turned) or over easy (turned once). They can even be semipoached by adding a spoonful of water to the pan and covering it for the last few minutes of cooking.

Frying is one of the simplest, and possibly the most abused, ways to cook an egg. The temperature at which you fry an egg determines both the texture and the amount of fat that the egg absorbs. An egg fried at a very high temperature becomes brown, crisp, and tough. The white encapsulates little bubbles of fat. (My son loves them like this!) Fried in this way, an egg absorbs so much cooking fat that its total fat content can be up to 20 percent. An egg fried gently over low heat (275°F/135°C) absorbs so little fat that its total content is only 12 percent—just 2 percent above that of a poached or hardcooked egg.

Temperature also makes a tremendous difference in the texture—from soft and moist to leathery and dry. A low temperature will give you less fat absorption *and* a tender, moist egg.

Fried eggs

What to Do	Why
Use a heavy pan and/or one made of a good heat conductor or a heavy pan with a nonstick coating.	Even distribution of heat prevents hot spots.
Warm the empty pan, remove from heat, and spray with nonstick spray. (Do not spray the pan while on the burner.) Then add oil, butter, or clarified butter.	Warming the pan first expands metal, which, along with nonstick cooking spray, helps prevent sticking.
Use low heat.	Slow cooking produces a tender, moist egg that does not absorb much fat.

Scrambled eggs

Everyone knows what scrambled eggs are—well, no, not really. People know what *their* scrambled eggs are. Crack eggs into a bowl, lightly beat them with a fork, and then pour them into a greased pan. After that, scrambled eggs become a very personal matter. Some people stir constantly over very low heat for a very moist small curd (French *oeufs brouillés* are done this way in the top of a double boiler); others stir only after the eggs have begun to set.

Adding water to beaten eggs (up to about 2 tablespoons per egg) will make them puff when scrambled as the water turns to steam. There is some disagreement about adding milk. Some cooks believe that adding milk to the beaten eggs (1 to 1½ tablespoons per egg) makes scrambled eggs creamier. Others argue that the milk toughens the eggs because the milk proteins coagulate with heat just like the eggs. Heavy cream, on the other hand, has more fat and less protein than milk. French chefs often stir in a little cream at the end, when the eggs are almost done, for a very rich, creamy dish.

Years ago, my former husband and I opened a boys' boarding school. Thank goodness, we started with just a few boys. I knew very little about cooking, and scrambled eggs were my nemesis. I would frantically stir, trying to keep the eggs from sticking. They just seemed to stick all the more. I would scramble a dozen eggs, but with my frantic stirring and sticking, they came out just a small pile of knotty mess.

Fortunately, my German mother-in-law, Anna Hecht, came to visit and showed me how to scramble eggs.

First she warmed the pan. When the pan was warm, she added a little oil (to raise the temperature at which butter burns) and butter. She tilted the hot pan to coat the bottom and up the sides a little. She quickly poured in well-beaten eggs. If they were cooking too fast, she lifted the pan from the heat, but she did not stir. When the eggs began to set, she took a big spatula and pushed the curds from the bottom to one side. The remaining uncooked eggs spread over the bottom. Again she let the eggs stand and puff before gently piling the curds to one side. When the eggs were almost set—still slightly wet—she removed the pan from the heat and turned the eggs out onto a platter. By the time she got them on the platter they were perfectly cooked.

Company Scrambled Eggs

MAKES 4 TO 6 SERVINGS

Cream cheese and a little mayonnaise keep these scrambled eggs creamy and moist. They are flavored with mild goat cheese, herbs, and sweet onion. For a more detectable cheese taste, add up to an ounce more Montrachet. Broiled tomato slices with lemon zest are a perfect accompaniment and make a handsome presentation.

what this recipe shows

Preheating the pan helps prevent sticking.

Nonstick cooking spray prevents the sticking that results when unclarified butter burns.

Sautéing shallots very briefly gives them a slightly mellower taste.

Allowing the eggs to stand and puff before gently pushing the curds to one side produces large, soft curds.

Removing the eggs from the heat while moist prevents overcooking because they will continue to cook and dry in the hot pan.

1 small package (3 ounces) cream cheese

1 ounce mild goat cheese, such as Montrachet

8 large eggs

2 tablespoons Homemade Mayonnaise (page 289) or store-bought

1/8 teaspoon cayenne

Freshly grated nutmeg

1/4 cup coarsely chopped basil leaves

3 tablespoons minced chives

Nonstick cooking spray

2 tablespoons butter

2 medium shallots, minced

1/2 teaspoon salt

Broiled Tomato Slices (page 208)

Several sprigs basil for garnish

1. Combine the cream cheese, Montrachet, and one of the eggs in a food processor fitted with the steel knife or in a blender. Blend well. Add one more egg and blend. Transfer to a medium mixing bowl and whisk together the cream cheese mixture and the remaining six eggs. Whisk in the mayonnaise, cayenne, nutmeg, basil, and chives.

2. Warm a 10-inch nonstick skillet over medium heat. Hold the pan away from the stove and spray with nonstick cooking spray. Add the butter to the pan and return to the heat. Tilt to spread melted butter over the bottom of the pan. Add the shallots and salt and sauté briefly. Pour in the egg mixture and let stand on medium heat about 30 seconds. Remove the pan from the heat, let stand 1 minute, then gently scrape the curds from sides and bottom.

3. Return the pan to the heat. After about 1 minute, repeat the procedure. Continue until the eggs are cooked almost to your liking. They will continue to cook after they are removed from the heat. Spoon the eggs into the middle of a warm serving platter, surround with broiled tomatoes, and garnish with basil sprigs. Serve immediately.

Broiled Tomato Slices

4 firm ripe tomatoes, sliced $1/2$ inch thick	Finely grated zest of 2 lemons
1 teaspoon salt	3 tablespoons butter
$1/2$ teaspoon ground white pepper	$1/2$ cup dry bread crumbs

Line a large baking sheet with foil. Spread the tomato slices on the baking sheet and sprinkle with salt, pepper, and lemon zest. Melt the butter in a medium skillet and stir in the bread crumbs. Sprinkle over the tomato slices. Place the tomatoes under the broiler, 4 to 5 inches from the heat, and broil several minutes to brown crumbs lightly, watching carefully. Serve hot.

at a glance
Scrambled eggs

What to Do	Why
Use a heavy pan and/or one made of a good heat conductor. Heavy pans with a nonstick coating are excellent for scrambling.	Even distribution of heat reduces burning and sticking.
Warm the empty pan, remove from heat, and spray with nonstick cooking spray. (Do not spray the pan while on the burner.) Then add oil, butter, or clarified butter.	Warming the pan first expands metal, which, along with nonstick cooking spray, helps prevent sticking.
If using butter, keep the temperature low. Or use a mixture of oil and butter or clarified butter.	Butter burns and sticks at relatively low temperatures. Oil raises the smoke point and burning temperature somewhat. Clarified butter (butter oil) burns at a higher temperature than butter.
Do not scramble too many eggs for your pan and burner size.	You need to be able to stir without splashing egg out of the pan. You will not get even heat if the pan is much larger than the burner.
For light, large curds, let eggs stand and puff, then gently push curds to one side.	Frequent stirring creates fine curds.
For creamy fine curds, stir eggs constantly in the top of a double boiler or on very low heat.	Extremely low heat and constant stirring create small curds.
Remove the pan from the heat before eggs are completely done.	Retained heat from the pan will continue to cook eggs.

Omelets

Basically, omelets are glorified scrambled eggs. They start out the same way, and then, when the eggs are partially set but still juicy, they are folded over. The inside finishes cooking on the plate. Herbs are sometimes added for extra flavor.

Omelets are often filled before folding, with everything from ham and cheese to salsa. The so-called Western omelet, with ham, green peppers, onions, and sometimes cheese folded into the eggs, may still be America's most popular omelet. Sweet dessert omelets like the omelet soufflé on page 211 are sometimes filled with fruit and accented with liqueur.

Omelets are ideal for a light lunch or quick supper, and they make elegant fast desserts. The big fear in making omelets has been sticking, but fear no more, we have solutions. Toughness can also be a problem, but the biggest challenge, perhaps, is technique. A classic French omelet takes 30 to 50 seconds. You have to be well rehearsed to know exactly what to do.

Omelet techniques Use a well-seasoned heavy pan or one with a nonstick coating—an omelet pan or a 7- to 8-inch skillet with sloping sides is good. Heat the pan just until warm, remove from the heat, and spray with nonstick cooking spray. (If you're making several omelets in a row, spray only the first time.) Add butter, clarified butter, or butter and oil, tilting the pan to coat the bottom and up the sides, and immediately add the beaten eggs. When the eggs begin to set, push them up to the side and let the uncooked eggs flow to the bottom. When the eggs are nearly cooked but still wet, shake the pan to free the omelet from the bottom and side of the pan. If it's sticking, loosen it with a spatula. If you are right-handed, scoot the omelet over to the left edge of the pan. The edge of the omelet should be at the edge of the pan. With the spatula, fold the right third of the omelet over the center third. Tilt the pan and place the left edge of the pan with the omelet against it on the plate. Turn the pan upside down to fold the omelet out of the pan and over on itself.

To prevent toughness, use eggs that are not refrigerator cold. Place the eggs in a bowl with hot tap water and let stand for 4 to 5 minutes. Cooking an omelet is so fast that if the eggs are cold the eggs on the bottom of the pan will overcook and become tough before the eggs above them can get warm and start to cook.

To master technique, practice and have everything ready. Practice until you get the knack of folding the omelet out onto the plate. It won't take long, and you and your family can enjoy all those imperfect but delicious omelets made while learning.

Omelet soufflés Another way to prepare omelets is to partially cook them on the stove, then place them in a hot oven to finish cooking. This makes a beautifully puffed omelet, or *omelet soufflé*. The Jack, Avocado, Sprouts, and Bacon Omelet that follows can be converted to a magnificent puffed omelet. Save the sprouts out of the filling mixture, then, in Step 4, while the omelet is still fairly wet, sprinkle the filling mixture without the sprouts over half of the omelet. (It will puff so that you can fold it only in half, not thirds.) Place it on the middle shelf of a preheated 400°F (204°C) oven until well puffed and the cheese has melted. Remove from the oven, add the sprouts, and fold in half. Finish with sour cream and tomato slices as directed.

For a dramatic, even greater puff, you can beat the yolks and whites separately, then combine them. My dessert Omelet Soufflé with Dark Cherries and Mascarpone Honey Sauce (page 211) is just such a recipe.

Jack, Avocado, Sprouts, and Bacon Omelet

MAKES 1 TO 2 SERVINGS

I know that they must serve this omelet in heaven. The melted mellow Monterey Jack, crisp crumbled bacon, avocado, and sprouts with a little sour cream on top and ripe tomato slices on the side are truly marvelous. Add a slice of hearty, grainy toast for a completely delicious meal.

what this recipe shows

Warming the eggs in hot tap water helps prevent toughness.

Heating the empty pan and spraying it lightly with nonstick cooking spray prevents sticking.

3 large eggs

1/4 cup grated Monterey Jack cheese

1/4 avocado, cut into 1/2- to 3/4-inch chunks

1 strip crisp cooked bacon, crumbled

1/4 cup alfalfa sprouts

Sprinkle of white pepper

Nonstick cooking spray

1 to 2 tablespoons butter

2 to 3 tablespoons sour cream

3 slices ripe tomato, 1/2 inch thick

1. Place the eggs in a bowl with hot tap water and let stand for 4 to 5 minutes.

2. In a small mixing bowl, toss the cheese, avocado, bacon, and sprouts together with a fork.

3. Beat the eggs with white pepper in a medium mixing bowl. Place the filling mixture, a serving plate, and a spatula near the burner.

4. Heat a well-seasoned or nonstick omelet pan or small skillet until warm. Remove from the heat and spray with nonstick cooking spray. Add the butter, tilting to coat the bottom and up the sides of the pan as it melts. Return the pan to the heat. Beat the eggs up again and add to the pan. When the eggs begin to set, push them to the side and let the uncooked eggs flow to the bottom. While the eggs are still slightly runny, shake the pan, and loosen the bottom of the eggs with a spatula if necessary, until the eggs are free. Shake again until the omelet is up the left side (if you are right-handed) of the pan with the omelet edge touching the left edge. Spread the cheese-avocado filling evenly across the middle of the omelet. (See page 209 to convert to an omelet soufflé.)

5. With a spatula, fold the right third of the omelet over the filled center. Tilt the pan and place the left edge of the pan with the omelet against it on the plate. With a quick flip, turn the pan upside down to fold the omelet out of the pan and over on itself. Spoon the sour cream on top of the omelet. Arrange the tomato slices overlapping beside the omelet and serve immediately.

Omelet Soufflé with Dark Cherries and Mascarpone Honey Sauce

MAKES 4 SERVINGS

This lightly sweetened, beautifully puffed dessert omelet has dark cherries contrasted against pale creamy mascarpone. A delicate hint of orange from the Grand Marnier and zest adds depth and complexity.

what this recipe shows

Beating egg whites separately and then folding them with the yolks creates a puffed soufflélike dish.

Heating the empty pan and spraying it lightly with nonstick cooking spray prevents sticking.

1 cup pitted dark sweet fresh, frozen, or
  canned cherries, drained

1 tablespoon honey

3 tablespoons Grand Marnier

3 large egg yolks

1 tablespoon water

1 tablespoon *and* 3 tablespoons *and*
  2 tablespoons sugar
  (6 tablespoons *total*)

1/8 teaspoon salt

4 large egg whites

Finely grated zest of 1 orange

Nonstick cooking spray

2 tablespoons butter

Mascarpone Honey Sauce
  (page 212)

1. Preheat the oven to 400°F (204°C).

2. Place the cherries, honey, and Grand Marnier in a medium mixing bowl and stir to coat cherries.

3. Beat the egg yolks with the water, 1 tablespoon sugar, and salt in a medium bowl to mix well. In a separate bowl, beat the egg whites until they form soft peaks. Beat in 3 tablespoons sugar and the grated zest. Stir a spoonful of whites into the yolk mixture, then gently fold in the remaining whites.

4. Warm a 9- or 10-inch well-seasoned or nonstick ovenproof skillet (or cover the skillet handle with foil) over medium-high heat. Remove from the heat and spray lightly with nonstick cooking spray. Add butter and return to the heat, tilting to coat the bottom and slightly up the sides of the pan. Add the egg mixture and smooth the top, leaving it slightly higher in the center. Reduce the heat and cook for 2 to 3 minutes, shaking the pan gently.

5. When the eggs turn golden around the edges, scatter the cherries and any juice on top. Sprinkle with the remaining 2 tablespoons sugar. Place in the oven and bake until puffed and golden, about 5 minutes. Serve immediately with Mascarpone Honey Sauce.

Mascarpone Honey Sauce

MAKES ABOUT ½ CUP

½ cup mascarpone (see Note on
    page 245) or sour cream

2 tablespoons honey

Stir the mascarpone and honey together in a small bowl. Refrigerate until ready to serve.

at a *glance*
Omelets

What to Do	Why
Warm the eggs in hot tap water for 4 to 5 minutes.	Cold eggs make a tough omelet.
Have all ingredients ready by the stove before starting to cook.	Omelet making goes fast; there is no time to stop and prepare ingredients.
Use a well-seasoned heavy pan and/or one made of a good heat conductor or with a nonstick coating.	To prevent sticking and burning.
Warm the empty pan, remove from heat, and spray with nonstick cooking spray. (Do not spray the pan while on the burner.) Then add butter, clarified butter, or butter and oil.	Warming the pan first expands metal and helps prevent sticking.
Practice!	Flipping out a perfectly folded omelet is a skill that improves rapidly with practice.

Using eggs to thicken and set food

Eggs are used to thicken and set an almost infinite number of dishes. Custard is a good example. Many of the principles underlying custards apply to literally hundreds of other dishes containing eggs.

A custard can be any liquid that is thickened or set into a soft gel by eggs; even zabaione, Marsala wine thickened with egg yolks, falls into the custard category. The custards that we are most familiar with, though, contain flavored milk and/or cream and sometimes sugar.

Custards can contain either cream or milk or both. They can be made with whole eggs, whole eggs plus yolks, or just yolks. For remarkable creamy texture, use yolks, which contain natural emulsifiers like lecithin and create a satiny smooth texture.

Custards fall into two major categories—stirred custards and baked custards. Stirred custards—like crème anglaise (boiled custard), pastry cream, lemon curd, and zabaione—are stirred during the setting period. The constant movement breaks bonds as the proteins try to set, creating a saucelike texture. Baked custards, including cheesecake, flan, and quiche, are left undisturbed during the setting period. The eggs set into a solid but still soft gel.

My earliest custard memories are of my mother making boiled custard to take to Grandpa. My step-grandmother was a teetotaler who wouldn't let *Life* magazine in the house because it was dripping with whiskey ads. The only booze my poor Grandpa ever got was the big slug of bourbon that my mother put in his boiled custard.

The great custard mystery

The two classic stirred custards are crème anglaise (also called *English custard, custard sauce, soft custard,* or *boiled custard*) and pastry cream (*crème pâtissière*). These two custards start with primarily the same ingredients—flavored milk and/or cream, sugar, and eggs—and yet they are quite different. Crème anglaise is heated carefully in a double boiler. If it is overcooked, you end up with scrambled eggs in a watery juice. Pastry cream is simply heated and stirred until it comes to a boil and gets thick.

The only difference is a little starch. Crème anglaise is made without the starch, and it thickens at about 160°F (71°C) and curdles at 180°F (82°C)—not much room for error. Pastry cream, on the other hand, is made with some form of starch, usually cornstarch or flour; it doesn't even thicken in the 180°F (82°C) range. Instead, at a temperature somewhere close to boiling (212°F/100°C), the custard simply gets thick.

Stirred custards *without* starch, like crème anglaise, lemon curd, and zabaione, require very low heat—or use of a double boiler—and constant stirring. Baked custards, which are also without starch, need a water bath. On the other hand, stirred custards *with* starch can be made directly on the heat and baked custards with starch can be baked at moderate temperatures without a water bath.

How can just a little starch make such a big dif-

ference? Natural proteins are individual units held tightly together by bonds across their coils (page 197). When these proteins are heated, they unwind with their bonds sticking out, run into other unwound proteins, and link together to form a delicate three-dimensional mesh that transforms liquids like milk or cream into a gel. If you continue heating this gel, the bonds tighten and squeeze out the liquid, resulting in scrambled eggs and juice.

When starch is present, however, the denatured (unwound) proteins do not join together. The exact mechanism that causes this behavior is not known. It may simply be physically that the starch expands and acts as "boulders" standing in the way, simply keeping the proteins from getting to each other. In any case, the result is that when starch is present, eggs do not scramble at the temperature at which they normally do. When the mixture gets hot enough (somewhere close to boiling), everything simply joins together—starch to protein, protein to protein, etc. This starch-egg connection makes a big difference in the correct technique to use in preparing custards.

The starches available to the home cook prevent eggs from coagulating. Some other starches do not. I used egg yolks and cornstarch to make a sweet-hot honey mustard. During development of this as a commercial product, we tried to find the perfect starch and even tried one "modified" starch that did not prevent coagulation at all. We ended up with 150 gallons of scrambled eggs! Why?

Dr. Peter Barham of the University of Bristol suggested that perhaps starches that do not expand until they reach higher temperatures (such as some modified starches) are not getting in the way enough to block the proteins from coagulating. I also found out that some starches are modified with acid and are fairly acidic. This would make the eggs denature faster and may contribute to the eggs' scrambling. At this time, modified starches are not available to the home cook, and available starches like cornstarch, flour, arrowroot, tapioca, and potato starch do prevent coagulation, so this is not a problem unless you have a commercial product.

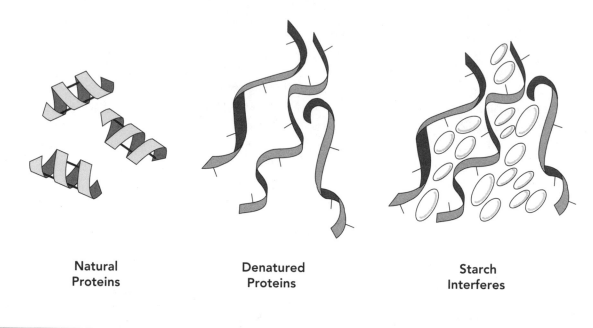

Natural Proteins   **Denatured Proteins**   **Starch Interferes**

DOUBLE BOILERS

Placing a dish over boiling water or in hot water is an excellent way to control the temperature. Under normal atmospheric pressure, water never gets any hotter than its boiling point—212°F (100°C) at sea level, lower temperatures at higher altitudes. It boils off before it can reach a higher temperature.

To cook over boiling water, you do not have to use a traditional double boiler—a saucepan that nests into another but sits several inches above its bottom. Any bowl or saucepan that rests securely on top of another will do. The bottom of the top bowl or pan must not touch the boiling water; it should sit well above it. The custard should be heated by steam only. For dishes that are not strongly acidic, my favorite double boiler is a medium-small copper bowl that rests securely on top of a medium saucepan. The copper is an outstanding heat distributor. Stirring constantly, I can keep the temperature of custard very even. I find this perfect for crème anglaise.

Stirred custards

In addition to watching for starch and deciding whether you need a double boiler, ingredients and techniques influence the custard, too.

Getting the eggs to set   The trick to any custard is getting the eggs to thicken and set properly. You will need to face the following challenges:

Time and temperature matter. Many different things can wreck a simple crème anglaise: heating it too fast, heating it too long, or not stirring it enough to distribute the heat. Eggs can be frustratingly complex, and you can't even count on their thickening at the same temperature every time. This thickening—linking together of proteins—is caused by a combination of time and heat. If you slowly heat an egg mixture without starch over low heat, the eggs begin to bond and thicken the mixture when the temperature reaches the low 160s°F (70s°C). Eggs do not go to hard curds—scrambled eggs—until about 180°F (82°C). You still have time to take the custard off the heat and let it cool down out of the danger zone if it starts turning into scrambled eggs. However, if you heat the eggs very fast over high heat, thickening does not begin in the 160s°F (70s°C). It may not even begin until about one degree before the custard turns to scrambled eggs. In this case, there is simply not enough time to save it.

The ratio of eggs and yolks to milk also counts. The more eggs, or eggs and extra yolks, the more unwound proteins to grab each other and the faster the custard sets. The eggs-to-milk ratio depends on the kind of custard you need. One whole egg or two egg yolks per cup of milk or cream will just barely thicken. One egg per cup is very thin. More eggs or yolks are normally used to ensure reasonable thickening. As you can see in my Crème Anglaise (page 216), I use 5 yolks for $1\frac{1}{2}$ cups liquid.

Sugar, on the other hand, has just the opposite effect. Too much sugar slows down or can prevent the setting of custards. Sugar molecules are fairly large and block the way of proteins trying to find each other.

Acidic ingredients, such as acidic fruits and vegetables, speed up setting. Remember how adding vinegar to the poaching water makes the egg whites set faster (page 201)? In much the same way, even mild acids present in nearly all fruits and vegetables will cause custards to set faster.

Alpha amylase and pastry cream: party time for a starch-loving enzyme. I still remember the first (disastrous) coconut cream pie I made many years ago. I heated and stirred the starch, sugar, and milk/cream until it was thick, adding a little extra starch to make sure it would thicken. The recipe said "Remove the pot from the heat, let cool a few minutes, stir in the egg yolks, then return to the heat again and cook until thick."

I had already had my share of egg-curdling disasters. Not wanting to take any chances with my pie filling, I allowed the custard to cool almost completely before I stirred in the yolks. The custard was so thick that even when I added the egg yolks a spoon stood up in the pot. I assumed that the only purpose for reheating the filling was to make it thicker, but because I was afraid of ending up with a curdled filling, I decided it was thick enough.

I spooned the filling into my crust, allowed it to cool, and piled on a 2-inch whipped cream topping. The pie went into the refrigerator to chill. The next evening when I cut into the pie, it was a soupy mess under the cream topping. The filling ran all over the place. What happened?

Raw egg yolks contain a starch-loving enzyme called *alpha amylase,* which must be killed by heating the yolks, or it can destroy a starch-based custard. In the unheated filling, the alpha amylase in the raw egg yolks went wild and had an all-night feeding frenzy, wiping out my nice firm starch gel. Egg yolks in starch custards must be heated almost to boiling to kill this enzyme. This is a common problem with cream pies, and, unfortunately, many cooks and chefs are unaware of it.

How to prevent a skin from forming You have probably had a skin form on hot chocolate. That skin is actually casein, a nutritious protein found in milk, which has dried out because of evaporation from the hot surface of the milk. Removing the skin does little good since more casein will dry out and a new skin will form, to say nothing of the fact that you are also removing nutrients. With hot chocolate, the simplest solution is to add marshmallows, which melt on the surface. They contain an egg white foam and gelatin, which coat the surface and prevent the casein from drying.

A casein skin will form on the surface of any hot dairy product, including hot crème anglaise or pastry cream. To prevent this, you can dot the hot surface with butter: Hold a stick of butter in its wrapper at one end and touch the other end to the hot surface in a number of places, using less than a teaspoon for a large pan. The resulting thin film of fat on the surface prevents the casein from drying out. If you don't want to add any fat, place a piece of plastic wrap directly on the surface to prevent the cream from being exposed to the air.

Here are examples of both types of stirred custards—one without starch and one with starch. The Crème Anglaise (without starch) is then used in the White Chocolate Floating Islands (both recipes follow). The Pastry Cream, or Crème Pâtissière, is a stirred custard with starch. It makes an excellent base for a fruit tart and is used on page 116 as the base for the Renaissance Classic Fruit Tart. It can also be used as the filling for cream puffs or éclairs, and with flavorings—coconut, chocolate, banana—it is the basis for all cream pies.

Crème Anglaise

MAKES ABOUT 2 CUPS

1 vanilla bean (see Note)

1 cup whole milk

$^1/_2$ cup heavy or whipping cream

5 large egg yolks

$^1/_3$ cup sugar

Salt

1. Pry a vanilla bean open with the tip of a knife and scrape some of its grainy black interior into a medium saucepan. Add the vanilla bean itself, milk, and cream. Heat over medium heat until the mixture just begins to steam. Stir the egg yolks, sugar, and a generous pinch of salt together in a bowl. Drizzle the hot milk mixture into the yolk mixture, whisking constantly. Strain the yolk mixture into the top of a double boiler.

2. Bring about $1^1/_2$ inches of water to a boil in the bottom of the double boiler. Make sure the water does not touch the top pan. Stir the custard mixture constantly until it just begins to thicken. Remove from the double boiler and continue stirring. Dip a large spoon into the custard, then remove it. The custard should be thick enough to leave a line when you drag your finger across the middle of the spoon. This is what recipes mean by "thick enough to coat a spoon."

3. Transfer the custard to a bowl. Dot the surface with butter or cover with a piece of plastic wrap touching the entire surface of the custard. Refrigerate until needed.

Note: *Vanilla beans can be rinsed, dried, and reused. They will have less flavor after you have scraped out the insides but can still contribute a rich vanilla taste to a dish.*

White Chocolate Floating Islands

MAKES 6 SERVINGS

Floating islands is a classic elegant dessert. Back when I was rich, I used to order it whenever we were at the St. Regis in New York. Once my former husband was surprised when I failed to notice that Rex Harrison with his distinctive voice was in the booth behind us and Salvador Dali with a number of people asking for his autograph was in the booth in front of us. I was so thrilled with the wonderful floating islands that I was unaware of the goings-on.

what this recipe shows

Superfine sugar dissolves fast and makes excellent meringues.

Meringues are flavored delicately with vanilla and salt.

4 large egg whites

4 teaspoons water

$^1/_2$ teaspoon cream of tartar

$^1/_2$ cup superfine sugar (see Notes)

1 teaspoon pure vanilla extract

Pinch of salt

2 recipes Crème Anglaise (above)

4 ounces white chocolate, finely chopped

1 recipe microwave caramel (page 425)

1. Stir together the egg whites, water, cream of tartar, and sugar in a medium stainless bowl (try not to create foam since foam cooks at a lower temperature). Heat 1 inch of water to a simmer in a medium skillet and turn the heat off. Run a cup of hot tap water, place an instant-read thermometer in it, and place it near the skillet. Place the metal bowl of egg white mixture in the skillet of hot water and scrape the bottom and the sides of the bowl constantly with a rubber spatula to prevent the whites from overheating. After 1 minute of constant scraping and stirring, remove the bowl of egg whites from the hot water and place the thermometer in the whites, tilting the bowl so that you have about 2 inches of white covering the thermometer stem. If the temperature has reached 160°F (71°C), using a mixer or by hand, beat until peaks form when the beater is lifted. If necessary, place the bowl of whites back in the hot water and scrape constantly in 15-second increments until the temperature reaches 160°F (71°C). Rinse the thermometer in the hot water in the skillet (to kill salmonella) and replace in the cup of hot water after each use. Beat in the vanilla and salt (see Notes).

2. Use two large serving spoons to shape six to eight oval meringues of beaten egg whites and drop them into simmering water in a large skillet. Poach until firm, turning each once after 2 to 3 minutes. Remove from the water with a slotted spoon and drain on paper towels.

3. Prepare the crème anglaise and, while it is hot, stir in the white chocolate. Pour the mixture into a warm shallow serving bowl, float the meringue ovals on the custard, and drizzle with hot caramel. Serve at room temperature or chilled.

Notes: *Because these meringues are poached, they are probably safe even if you just beat the whites with cream of tartar and add sugar 1 tablespoon at a time after peaks form. However, since the brief poaching may not kill salmonella, I have used Alice Medrich's alternative procedure in the recipe.*

If you do not have superfine sugar, process granulated sugar in a food processor with the steel knife for 1 or 2 minutes.

Pastry Cream (Crème Pâtissière)

MAKES ABOUT 2 CUPS

1 vanilla bean	Salt
1 cup whole milk	3 tablespoons cornstarch
1/2 cup heavy or whipping cream	5 large egg yolks
1/3 cup sugar	

1. Pry a vanilla bean open with the tip of a knife and scrape some of its grainy black interior into a medium saucepan. Add the vanilla bean itself, milk, and cream. Heat over medium heat until the mixture just begins to steam. Stir the sugar, a generous pinch of salt, and the cornstarch together in another medium saucepan. Remove the vanilla bean from the hot milk (see Note). Drizzle the hot milk into the sugar mixture, whisking constantly. Return the saucepan to the heat and cook over medium heat, stirring constantly, until the mixture thickens.

2. Stir the egg yolks together in a bowl. Stir about ¼ cup of the hot mixture into the yolks, then scrape the yolk mixture into the saucepan. Return to the heat and bring to a gentle boil, stirring constantly until the custard becomes thick and smooth.

3. Transfer the custard to a bowl. Dot the surface with butter or cover with a piece of plastic wrap touching the entire surface of the custard. Refrigerate until needed.

Note: *Vanilla beans can be rinsed, dried, and reused. They will have less flavor after you have scraped out the insides but can still contribute a rich vanilla taste to a dish.*

at a *glance*
Crème anglaise and pastry cream

What to Do	Why
For crème anglaise, use a double boiler.	Water that does not get hotter than 212°F (100°C) provides a controlled temperature surrounding the delicate starch-free custard.
For pastry cream, use a regular saucepan.	The custard may be brought to a boil over direct heat because it contains starch.
Bring pastry cream back almost to a boil after stirring in egg yolks.	Heat will kill enzymes in the raw egg yolks that break down starch and thin the custard.
Add all flavorings while pastry cream is still warm.	Once the custard has set to a firm gel, more stirring will cause thinning.
Cover the warm crème anglaise or pastry cream surface with plastic wrap or dot it with butter.	This will prevent a dried casein skin from forming on the custard surface.

Custards made without milk Any liquid thickened with eggs, not just milk, is, strictly speaking, a custard. Zabaione and lemon curd are good examples. My experience with different lemon curd recipes shows how important each ingredient can be in selecting the right cooking method.

One day Roland Mesnier, the White House pastry chef, asked me to prepare his lemon curd recipe by heating it on high, bringing it to a boil, and straining it. Knowing that the recipe had no starch and fearing it would turn to scrambled eggs easily, I questioned, "High?" Chef Mesnier stood there with his arms folded across his chest and repeated, "On high!"

I obeyed. The split second I saw bubbles, I

strained the lemon curd into a cold bowl. Sure enough, I had a tablespoon of scrambled eggs left in the strainer, but in the bowl I had a thick lemon curd made in minutes. Chef Mesnier knew that his recipe had enough sugar to prevent the eggs from scrambling instantly and could be cooked over high heat. A busy chef won't spend 10 minutes doing something that can be done in one.

In addition to the eggs, other ingredients in a recipe—sugar, starch, acid—have a major influence on which method you can use safely. In experimenting with different recipes for stirred custard, you will find that a recipe with no starch and a small amount of sugar, like crème anglaise, is almost impossible to prepare without a double boiler. Even if you can cook it over direct heat, the custard has a much better texture when prepared in a double boiler. To save time, you can get away with having the water boil rapidly— but only if the cooking vessel is well above it.

I need to use a double boiler with my Lemon Curd recipe, because it has different amounts of sugar, eggs, and lemon juice from Chef Mesnier's recipe.

Lemon Curd

MAKES ABOUT 2½ CUPS

If you've never eaten lemon curd on a soft, light biscuit, stop everything and make this recipe right now. While it's cooling, you can make a batch of Touch-of-Grace Biscuits (page 77) to go with it. Lemon curd can also be used to fill tarts by itself or folded into an equal portion of whipped cream. Individual tart shells filled with lemon curd and cream and topped with fresh blueberries, or thin lemon curd tarts iced with a layer of melted chocolate, are superb.

what this recipe shows

Sugar slows the setting of the egg yolks.

A double boiler controls the heat and prevents the eggs from scrambling. Heating slightly more slowly in a double boiler also produces a better texture.

The lemon curd is removed from the heat when it reaches the thickness of a light to medium cream sauce since it becomes thicker as it cools.

5 large egg yolks	2 cups sugar
Finely grated zest and juice of 4 lemons	¼ pound (1 stick) butter

1. Combine all ingredients in the top of a heavy double boiler or in a heavy mixing bowl (not copper because of the acidity) and place over a pot of boiling water. Make sure the top vessel is not touching the water. Whisk ingredients constantly or scrape the bottom with a large flexible spatula until thickening begins to occur. With the water at a low boil, this will take about 5 minutes. Remove from the bottom of the double boiler and continue to whisk. The heat of the pan should be enough to thicken the custard until it coats the spoon (see Step 2 on page 216).

2. Strain the curd into a cool bowl. Cover with plastic wrap touching the surface or dot with butter and chill. Keeps well for 2 weeks, refrigerated.

Baked custards

Most of the caveats for stirred custards apply to baked custards as well. Both time and temperature influence setting, and ingredients have the same effect on both types of custards. For a delicate, barely set custard, use 1 large whole egg or 2 egg yolks per cup of milk or cream. For a quiche, which normally uses 2 cups of milk or cream, use 3 large whole eggs or 2 eggs and 2 yolks. The extra egg or yolks will ensure a good set since 1 large egg per cup is very soft. As with stirred custards, sugar slows the setting, while acidic fruits and vegetables speed up the setting. A timbale (a small unmolded custard) can turn into an overcoagulated mess, full of watery holes, when made with fruits and/or vegetables. For soft, creamy timbales, either cut back on the amount of vegetables (the source of acid) or add a small amount of cornstarch.

You can easily see for yourself the influence of different ingredients on the setting time of baked custards. Prepare a double batch of custard without starch (Crème Anglaise, page 216) to the point where the mixture is strained. After straining, divide the custard into four equal portions. To one portion add an extra egg, to another add extra sugar, leave one plain, and to the last add some acidic fruit such as blueberries. Pour each portion into marked custard cups and bake in a water bath. The custard with the extra egg will set first, then the one with the fruit, then the original recipe, and finally the custard with the extra sugar.

Just as there are many kinds of stirred custards, baked custards come in many sizes and shapes. A baked custard can be a flan like the Mesmerizingly Smooth Flan that follows, a quiche, or even a cheesecake, as well as dessert custard cups and savory timbales.

Mesmerizingly Smooth Flan

MAKES 8 SERVINGS

This flan is unbelievably silky smooth. I have taught the recipe many times in Texas to people who had been making flans for years, and I consider it a great compliment that these experts switched to my recipe.

what this recipe shows

Using corn syrup and a little lemon juice prevents the caramel from crystallizing.

Extra egg yolks are the secret of the velvety smooth texture of this flan.

A towel in the bottom of the water bath prevents overcooking on the bottom of the flan.

CARAMEL

1 cup sugar

$^1/_3$ cup water

2 tablespoons light corn syrup

4 drops fresh lemon juice or vinegar

CUSTARD

2 vanilla beans

3 cups milk

1 cup heavy or whipping cream

$1^1/_4$ cups sugar

$^1/_4$ teaspoon salt

4 large eggs

7 large egg yolks

1. Combine the ingredients for the caramel in a heavy unlined pot or an iron skillet over medium-high heat. Stir initially to dissolve sugar, but stop stirring or shaking as water evaporates. Heat until the sugar turns light to medium brown. Carefully pour into a 2-quart soufflé dish, heavy ceramic casserole, or eight individual custard cups. Immediately tilt the dish or dishes to spread caramel over the bottom and part way up the sides. Set aside while preparing the custard.

2. Preheat the oven to 350°F (177°C).

3. Pry each vanilla bean open with the tip of a knife and scrape some of its grainy black interior into a large heavy saucepan. Add the vanilla beans themselves, milk, cream, sugar, and salt. Simmer over low heat about 5 minutes. Remove from the heat.

4. Whisk together the eggs and yolks in a large mixing bowl. When the milk has cooled to warm, remove the vanilla beans (see Note) and slowly whisk the milk into the eggs. Strain the custard into each caramel-coated dish.

5. Prepare a water bath. Fold a terry-cloth towel and place it in the bottom of a roasting pan at least 1 inch larger on all sides than the custard dish. Place the custard dish or dishes on the towel in the pan.

6. Pull the middle oven shelf out slightly. Place pan on the shelf. Carefully pour enough nearly boiling water into the pan so that the water is at least 1½ inches deep. Bake a large custard until just set, about 1¼ hours. Individual custards will take about 20 minutes. A knife inserted into the center should come out clean. Remove from the water bath immediately and refrigerate at once to chill.

7. If serving unmolded, loosen flan from the dish by running a small spatula just around the upper edge. Place a serving platter over the flan. Invert very quickly to avoid spilling the caramel syrup.

Note: *Vanilla beans can be rinsed, dried, and reused. They will have less flavor after you have scraped out the insides but can still contribute a rich vanilla taste to a dish.*

WATER BATHS

Custards with starch do not need a water bath, but those without some form of starch do. Water baths, or bain-maries, for baked custards and other delicate dishes cooked in the oven need to be large enough and deep enough to provide temperature protection. If you have a flan in a 9¾-inch pan and place it in a 10-inch water bath, the thin rim of water around the flan will evaporate fast and may not provide full temperature control. A 12-inch round pan or a 12-inch-wide roasting pan would be appropriate for a water bath. Placing a thick towel in the bottom of the water bath pan provides protection from heat on the bottom.

Most of the time you need to fill a water bath with very hot—nearly boiling—water. The safe way to do this is to pull out the oven shelf and place the water bath pan containing the custards on it. Pour nearly boiling water from a kettle to fill the water bath, then carefully slide the pan and shelf back into the oven.

There are two methods for baking custards in water baths. One is to cook the custard until it sets, then remove it from the oven and the hot water bath. The other is to remove the custard from the oven before it is completely set and leave it in the hot water to finish cooking. You may use either method successfully. Be warned, however: If you completely cook the custards and then also leave them in the hot water, you may have the little tunnels with watery juice that are caused by overcooking. This is called *syneresis.*

Cheesecake: a custard by any other name

Cheesecakes and other egg-thickened dishes develop a different texture when flour or another starch is added to the custard mixture. The amount of starch added determines the extent of the change—for a small addition, the change in texture will hardly be noticeable. You can clearly see the textural difference caused by starch in two of the cheesecake recipes that follow—Rose Levy Beranbaum's Cordon Rose Cream Cheesecake (below) and the Savory Herbed Cheesecake (page 225).

Rose's dessert cheesecake contains no starch and should be baked in a water bath or at a very low temperature; it has a marvelous creamy, smooth texture. The herbed cheesecake, on the other hand, contains 3 tablespoons of flour—not much, but enough to make the texture more like that of set pastry cream than the silky smoothness of Rose's. It does not have to be baked in a water bath.

Why cheesecakes crack Cheesecakes crack because they are overcooked. I don't think there's anything in cooking as deceptive as the doneness of a cheesecake. When a cheesecake is completely cooked, a 3-inch circle in the center is still wobbly and shaky, and it looks undercooked. I must have made Rose's cheesecake over fifty times and still, when I pull it from the oven, I think, "Something must have gone wrong this time. With such a juicy-looking center, it can't be done." I have to force myself to put it in the refrigerator overnight. After chilling, it is the perfect consistency every time.

Rose Levy Beranbaum's Cordon Rose Cream Cheesecake

MAKES 12 SERVINGS

This extraordinary cheesecake comes from my friend Rose Levy Beranbaum, author of *The Cake Bible*. I add amaretto, an almond-flavored liqueur I'm very fond of, with the vanilla. If you prefer a lemony flavor, omit the amaretto and use 3 tablespoons of lemon juice as Rose does.

Rose often decorates her cheesecake with a white chocolate icing (although it is delicious without any embellishment). Rose uses 4 tablespoons of butter in her icing, but I find that makes it dangerously soft where I live in the warm South. My White Chocolate Icing for Warmer Conditions follows the cheesecake recipe.

what this recipe shows

Custard that contains no starch must be baked in a water bath or a very-low-temperature oven (just below 200°F/93°C).

Lack of starch makes the texture satiny smooth and creamy.

Cheesecakes are very deceptive and look undercooked when they are done.

CRUST
14 chocolate wafers
3 tablespoons sugar

3 tablespoons butter, melted
Nonstick cooking spray

FILLING

2 packages (8 ounces each) cream
   cheese

1 cup sugar

3 large eggs

2 teaspoons pure vanilla extract

$1/4$ teaspoon salt

$1/4$ cup amaretto

3 cups sour cream

White Chocolate Icing for Warmer
   Conditions (page 224)

1. Crush the chocolate wafers in a plastic bag with a rolling pin or in the food processor with the steel knife. Stir together the wafers, sugar, and melted butter in a medium mixing bowl. Line the bottom of an 8 × 3-inch round cake pan with a parchment or wax paper circle. Spray the lined pan with nonstick cooking spray and press in the crumb crust.

2. Preheat the oven to 350°F (177°C).

3. Blend the cream cheese and sugar well in a food processor with the steel knife or in a large mixing bowl with a beater. Be sure to get out all the lumps at this stage. Add the eggs, one at a time, and blend well after each addition. Add the vanilla, salt, and amaretto. Blend well. Blend in the sour cream. Pour into the prepared pan.

4. Prepare a water bath. Fold a terry-cloth towel and place it in the bottom of a roasting pan at least 1 inch larger on all sides than the cake pan. Place the cake pan on the towel in the larger pan.

5. Pull the middle oven shelf out slightly. Place the pan on the shelf. Carefully pour enough nearly boiling water into the pan to come at least 1 inch up the side of the cake pan. Bake the cheesecake for 45 minutes. Do not open the oven. Turn off the oven and leave the cheesecake in for 1 hour more. The cheesecake will not look done, and a 3-inch circle in the center will still wiggle. Have faith. Remove it from the oven and refrigerate overnight.

6. Since the cheesecake is in a cake pan, you have to invert it twice. Cover a baking sheet with plastic wrap and set aside. Turn a burner on low and place the cheesecake pan on the burner to heat the bottom of the pan just seconds for easy removal. Run a knife around the edges and jar the pan on one side. Place the baking sheet on top of the cheesecake and invert. Peel off the parchment. Place a pedestal cake dish on the bottom and invert again. Refrigerate.

7. Spread with White Chocolate Icing or serve cold as is.

Variations: *For this amaretto cheesecake, I sometimes make an almond macaroon cookie crust, using 10 almond macaroons, ground, and 3 tablespoons melted butter but no sugar.*

Instead of icing the cake, you can brush the sides with melted apple jelly and press in roasted almond slices.

White Chocolate Icing for Warmer Conditions

MAKES ENOUGH TO ICE AN 8-INCH CHEESECAKE

> **what this recipe shows**
>
> The addition of cream cheese makes temperamental white chocolate into an easily spread-able icing that is smooth yet firm.

8 ounces white chocolate, chopped or grated

1 package (8 ounces) cream cheese

1. Bring 1 inch of water to a boil in a large skillet, remove from the heat, and let stand for 3 minutes. Place the chocolate in a well-dried stainless or metal bowl and cover tightly with plastic wrap that does not come down the sides of the bowl far enough to touch the water. Place the bowl of chocolate in the hot water and let stand 5 minutes. Remove from the water bath, dry the bottom well, and stir the melted chocolate with a dry spatula. If not completely melted, re-cover the chocolate and return it to the hot water bath for a few minutes longer.

2. Beat the cream cheese until smooth in a food processor or mixing bowl. Beat in the melted chocolate. Spread evenly on cold cheesecake. Fill a pastry bag fitted with your chosen tip and pipe a border if desired.

Wedding Cheesecake

If you are ambitious, Rose's cheesecake makes a unique wedding cake. I have made this cheesecake as a three-tiered wedding cake for three of my children and for a close friend. I prepare seven times the cheesecake recipe and make three round layers, 12, 10, and 8 inches, each 3 inches deep. (Rose makes 12-, 9-, and 6-inch layers, which are easier to handle because the upper layers weigh less, but I couldn't find a 6-inch pan.) I have been able to prepare the batter all at once in a friend's large Hobart mixer, but normally it has to be done in several batches. You can do two batches at once in a large mixer. Practice by making the cheesecake and inverting and icing it. This does take time and care, but every time I made it the guests said it was the best wedding cake they had ever had.

2 recipes crust for Rose Levy Beranbaum's Cordon Rose Cream Cheesecake (page 222)

7 recipes filling for Rose Levy Beranbaum's Cordon Rose Cream Cheesecake (page 223)

5 recipes of White Chocolate Icing for Warmer Conditions (above)

1. Preheat the oven to 350°F (177°C).

2. Line the bottom of a 12 × 3-inch round cake pan with a parchment or wax paper circle. Spray the lined pan with nonstick cooking spray and press in the crumb crust.

3. Prepare a water bath. Fold a terry-cloth towel and place it in the bottom of a roasting pan at least 1 inch larger on all sides than the cake pan. Place the cake pan on the towel in the larger pan.

4. Pull the middle oven shelf out slightly. Place the roasting pan with the filled cake pan on the shelf. Carefully pour enough nearly boiling water into the larger pan to come at least 1 inch up the side of the cake pan. Bake the cheesecake for 55 minutes. Do not open oven. Turn off the oven and leave the cheesecake in for 1 hour more. Cheesecake will not look done; a 3-inch circle in the center will still wiggle. Have faith. Remove it from the oven and refrigerate overnight.

5. Bake the two smaller layers in a similar manner, 50 minutes for the 10-inch layer and the 8-inch layer. If you're using 9-inch and 6-inch pans, the 6-inch cake will get a little overdone if baked 50 minutes with the 9-inch, but if you put it in the front or in a section of the oven that you know is slightly cooler, it usually works. To bake separately, bake the 9-inch layer for 50 minutes and the 6-inch for 45 minutes. Let each layer stay in the oven for 1 hour, then refrigerate.

6. Since the cheesecake layers are in cake pans, you have to invert them twice. Cover a baking sheet with plastic wrap and set aside. Turn a burner on low and place the 12-inch layer with the crumb bottom on the burner to heat the bottom of the pan just seconds for easy removal. You will not need to heat the two smaller layers. Jar the pan on one side and then the other. Cover the layer with the plastic-covered baking sheet and invert. Peel off the parchment. Cut cardboard circles just a breath smaller than each layer and cover the circles with foil. Place the 12-inch circle on the bottom and invert the layer onto it. Refrigerate until ready to ice. Jar the pan on several sides to release. Invert each onto its correct-size prepared foil-covered circle and refrigerate until ready to ice.

7. Ice each layer, garnish with a fluted edge if desired, and refrigerate to set, about 20 minutes.

8. Take a large plastic straw and measure it to be a breath taller than the thickness of the 12-inch bottom layer. Cut 6 straws to this length. Place one straw in the middle of the layer and five others in an 8-inch-diameter circle around it. Repeat the measurement for the 10-inch layer and cut six straws to this length. Place a straw in the middle of the layer and five straws in a 6-inch-diameter circle around it. With two large flat spatulas, carefully lower the 10-inch layer so that its cardboard base rests on the straws of the 12-inch layer and is centered over it. Finally, again with spatulas, carefully set the 8-inch layer on top.

9. Glue a lace ribbon over a solid ribbon. Measure around each cake pan and cut a piece of this double ribbon to fit around each layer. Carefully arrange each layer's ribbon around the base of the layer to hide any gaps. Put baby's breath around the bottom and a sentimental bouquet of the tiniest rosebuds you can find on the top. Refrigerate until serving time.

Savory Herbed Cheesecake

MAKES 15 TO 20 SERVINGS

This unusual hors d'oeuvre cheesecake with fresh herbs in the filling and lemon zest in the crust is deep brown on top with golden sides. On a pedestal cake dish encircled with a garland of fresh herbs, or garnished with a cluster of tomato roses and fresh basil leaves, it attracts attention instantly. It is excellent for large parties and a favorite of caterers because it is dramatic and unusual and neither expensive nor difficult to prepare, and it can be made several days ahead.

continued

what this recipe shows

Starch in the form of flour in this custard produces a firmer texture and enables the cheese-cake to be baked at 325°F (163°C) without a water bath.

The large ratio of butter to flour in this crust makes it so tender that it is almost crumbly.

CRUST

1 cup bleached all-purpose flour

1/2 teaspoon salt

Finely grated zest of 1 lemon

1/4 pound (1 stick) butter, cut into 8 pieces

3 tablespoons fresh lemon juice

1 large egg yolk

Nonstick cooking spray

FILLING

3 packages (8 ounces each) cream cheese

3 tablespoons bleached all-purpose flour

4 large eggs

1 teaspoon salt

1/2 teaspoon hot pepper sauce

2 tablespoons fresh lemon juice

1 1/2 teaspoons minced fresh oregano or 1/2 teaspoon dried

1 1/2 teaspoons minced fresh tarragon or 1/2 teaspoon dried

1 tablespoon chopped fresh basil or 1 teaspoon dried

1 teaspoon minced fresh rosemary or 1/3 teaspoon dried

1/2 cup chopped parsley

1 cup freshly grated Parmesan (about 3 ounces)

4 shallots, finely chopped

3 large tomatoes

3 sprigs fresh basil

1. For the crust, process the flour, salt, lemon zest, and butter in a food processor with the steel knife for several seconds. Some butter should remain in lumps. Add the lemon juice and egg yolk. Process with a few on/off pulses just to mix. Spray an 8- or 9-inch springform pan with nonstick cooking spray. Crumble a third of the dough into the bottom of the pan and press out evenly. Press the remaining dough around the sides of the pan. Place in the freezer while you mix the filling.

2. Preheat the oven to 400°F (204°C). Place an oven shelf slightly above center.

3. Process the cream cheese until smooth in a food processor with the steel knife. Add the flour and 1 egg, process, and scrape down sides of bowl. Add the remaining eggs, one at a time, processing after each egg. Add the salt, hot sauce, lemon juice, herbs, Parmesan, and shallots. Process just to blend well.

4. Pour into the prepared pan and bake for 15 minutes. Turn the oven down to 325°F (163°C) and bake until the top is medium brown, about 50 minutes more. Cool for 10 minutes on a cooling rack. Loosen the crust from the springform by running a thin knife between the crust and pan. Open the springform ring and remove. Let come to room temperature, at least 1 hour. Refrigerate if not serving immediately (see Note).

5. Carefully peel the tomatoes deeply with a sharp paring knife, in one long piece of peel if possible. Wind the peel into a tomato rose. Arrange three roses on top of the cheesecake with basil leaves to garnish. To make it easy for guests to serve themselves, precut portions by cutting a partial inner circle about 1½ inches from the edge, then cut ¾-inch slices from the edge to this circle. Serve at room temperature.

Note: *This cheesecake can be prepared 1 or 2 days ahead and refrigerated.*
Variation: *Decorate with fresh herbs and edible flowers instead of tomato roses. Wrap the cheesecake in garlands of fresh thyme sprigs, for example, interspersed with flowers from flowering herbs or other edible flowers.*

at a glance
Baked custard

What to Do	Why
Use at least one large whole egg or two egg yolks per cup of milk or cream.	This is the minimum amount for a very delicate set.
Use a water bath that is lined with a terry-cloth towel and is about 1 inch larger all around than the custard pan.	Water baths control temperature and prevent overcooking. A thick towel provides protection from heat on the bottom.
Remove custard from the water bath as soon as it is done and refrigerate.	Custard sitting in hot water will continue to cook and will overcook.

Eggs in baking

Eggs perform many different functions in baked products (page 89). In a savarin, the high egg content provides a porous but strong structure so that the yeast dough can be soaked with a rum and sugar syrup and not disintegrate. The same is true of génoise, which is normally soaked with a liqueur syrup. Pâte à choux (used for cheese puffs, cream puffs, and éclairs, among other things), popovers and Yorkshire pudding, some fritters, and even hush puppies all depend on eggs as both leavener and setting agent.

Many cakes, like génoise, sponge cake, and angel food cake, depend solely on eggs for leavening and are covered in this chapter. Other cakes, muffins, and pancakes depend on eggs along with baking powder for their leavening. These are covered in more detail in other chapters.

In baking, whole eggs are not always the best choice. Sometimes you need yolks for creaminess, and sometimes you need whites to dry and crisp. I solve many of the problems that test kitchens and restaurants call me about by simply replacing some or all of the whole eggs in a recipe with whites or yolks. Hush puppies at a certain Baton Rouge restaurant were too greasy and too heavy. Egg whites to the rescue! The recipe called for a dozen whole eggs. I got the chef to measure their volume in a convenient container and mark it. Then I told him to use six whole eggs and add whites to fill the container to the marked volume. The hush puppies were light and dry and crisp on the surface—no longer heavy or greasy.

Indeed substituting either yolks or whites for some or all of the whole eggs can make the difference between a dish that is just OK and one that is really extraordinary. We saw how egg yolks can make custards silken and creamy. Now we'll see how egg whites dry and lighten baked goods.

Pâte à choux

Pâte à choux is an extremely versatile dough. Baked, it makes such elegant savory dishes as bite-size cheese puffs or a ring of large puffs (*gougère*) or my Cocktail Puffs with Escargots (page 230) as well as sweet pastries like éclairs, cream puffs, and profiteroles. It can also be fried and even boiled.

Its ingredients are simple—only butter, salt, water, flour, and eggs. However, the right amount of each ingredient is vital to a successful dough. The trick is to add as many eggs as you can for lightness but not so many that the dough is too thin to shape and rise. With so few ingredients, each one counts, as does technique.

Fat You can make pâte à choux with butter, lard, shortening, margarine, or even olive oil. The amount does matter, however, and the more fat, the tenderer the puffs. Remember that some fats like butter and margarine are only about 80 percent fat. For the normal $1/4$ pound of butter, use only $6^{1}/_{2}$ tablespoons of lard, chicken fat, shortening, or oil.

Flour High-protein flour, such as bread flour, forms good elastic sheets of gluten and produces lighter puffs. High-protein flour also absorbs more water so that the initial water-butter-flour dough is firm enough to add at least three large eggs or two large eggs plus two to three whites. Low-protein flour does not absorb water as well; a dough made from this flour will be thinner, and fewer eggs can be added. The puffs will be heavier with low-protein flour from both fewer eggs and less gluten. Dough puffs much better when there are strong elastic sheets of gluten to hold in the steam.

The flour should be added all at once so the starch can swell instantly in the hot water. If you add only a little at a time, the water will be hot enough to make some of the starch swell correctly but not that added later.

Eggs Both the number of eggs and whether they are whole or just whites are important. Ordinary pâte à choux recipes call for whole eggs, and the results are ordinary. Large puffs are frequently gooey inside in spite of techniques such as puncturing to dry out the interior. A much lighter and crisper puff is made by replacing one or two whole eggs with whites. Egg whites are great drying agents. The dough must be firm enough to pipe or spoon into firm balls, or the puffs will not rise. Using less water initially ($3/4$ cup instead of 1 cup) will allow you to add more eggs for lighter puffs.

Sticking Proteins in the liquid eggs in the dough can get into minute crevices and literally cook into the pan. Supersmooth nonstick baking sheets and pans or those wonderful Teflon nonstick sheets (see Sources—Nonstick sheets, page 477) are a big help, as are nonstick cooking sprays. If you are caught in a sticky situation without your trusty nonstick cooking spray, use the old double-grease technique (page 90). You can also use well-greased foil (or parchment paper, except in a convection oven). If you are using an egg wash on baked puffs or pastries, be careful not to get any between the puff and the pan and to wipe up any spills.

Heat from the bottom For pâte à choux to puff up properly, the most heat must come from the bottom of the oven. This allows the eggs in the dough to expand and the dough to rise before the tops become so cooked and firm that they stop the puffing action. I love the way that Anne Willan, noted cookbook writer and owner of La Varenne cooking school, puts it: "They should be cooked in a rising oven." One way to achieve a "rising oven" is to preheat the oven to a lower temperature, place the puffs in, then turn the oven up to the correct baking temperature for maximum puffing. After the puffs are fully risen, the oven needs to be turned down so that they will not burn as they dry thoroughly. (Popovers, which have the same flour-to-liquid ratio as pâte à choux and are solely dependent on eggs as leaveners, have the identical baking problems and should be baked the same way or placed in a cold oven and then have the heat turned on to the correct temperature.)

To bake and dry a large puff thoroughly seems to take forever, but if you do not get the puff dry inside, it will fall before your eyes. Generally speaking, small puffs cook through reasonably fast, 20 to 40 minutes, depending on size. A large ring will take over 1 hour, again depending on size. These approximate times include the lower-temperature drying times.

Cutting a small slit in the side of each puff when it comes out of the oven allows steam to escape and helps to dry the interior.

Piping and shaping pâte à choux You can pipe pâte à choux directly onto a baking sheet, making round puffs or cylinders for éclairs with a pastry bag fitted with a ½-inch plain round tip. You can also spoon the dough on for puffs, using a teaspoon or tablespoon, depending on the size you want. Professional pastry chefs have their own techniques for piping and shaping pâte à choux. It was a joy to watch Chef Albert Jorand, who taught baking at La Varenne, pipe perfectly spaced puffs or éclairs. He insisted they be spaced evenly for even cooking. When he finished piping puffs, he would oil the back of a fork and gently press each on the top from one angle and then another. They would be slightly flattened and have a light crisscross pattern from the fork. This, he said, makes them rise more evenly.

Uses Pâte à choux has wide uses both as a savory, as in the Cocktail Puffs with Escargots (page 230) and cheese-filled or cheese- and meat-filled puffs or rings, and as sweet small or large puffs filled with pastry cream or rings, like the classic Paris-Brest or gâteau Saint-Honoré. Peggy O'Donnell, an excellent cooking teacher in Land O' Lakes, Florida, makes unusual dessert loaves by piping a layer of sweetened pâte à choux on top of rectangles of flaky pastry, topping with almond paste and almonds, and baking to golden perfection.

Pâte à Choux

MAKES ABOUT 30 HORS D'OUEVRE PUFFS

what this recipe shows

Having the water-butter mixture boiling when incorporating the flour causes the starch to swell instantly, absorbing all the liquid.

Using some egg whites instead of whole eggs makes the puffs drier and crisper.

Beating eggs and whites in a little at a time avoids lumps of dough floating in eggs and keeps the dough at the proper consistency.

Piping the same-size puffs and spacing evenly on the baking sheet produces more even baking.

Preheating the oven only to a medium temperature and then raising it produces heat from the bottom.

Turning the oven down after the puffs are well risen lets them dry without burning.

Cutting slits in the cooked puffs allows steam to escape and the puffs to dry further.

¾ cup water	3 large egg whites
¼ pound (1 stick) unsalted butter	1 large whole egg, stirred
½ teaspoon salt	Butter to grease pan
1 cup bread flour	1 large whole egg, beaten (optional)
1 large whole egg	

continued

1. Preheat the oven to 300°F (149°C).

2. Bring the water, butter, and salt to a boil in a heavy medium saucepan. Add the bread flour *all at once* and stir in well. This will make a dough ball that pulls away from the pan. Let cool for several minutes.

3. Beat the dough by hand or in a food processor with the steel knife for 1 minute to cool it further. Remove the cover after beating if using the processor. You will see steam coming out. Continue to cool until the dough is no longer hot enough to cook the eggs. You can touch the dough with your finger to be sure it's not piping hot. Beat in the egg. As soon as it is incorporated, add the egg whites and process or beat in. Add about half of the stirred egg and process or beat in well. This may be all the egg you can add. The dough must remain thick enough to spoon or pipe it into balls that will hold their shape. If the dough is still firm, add the rest of the stirred egg, beating well.

4. Spoon or pipe the dough onto a well-greased heavy baking sheet. Make small (about 1 inch), evenly spaced balls for individual puffs. With a greased fork, gently press the top of each puff to level it slightly. Glaze the puffs with a beaten egg if a glossy surface is desired, taking care not to get any glaze between the puff and the baking sheet.

5. Place the baking sheet in the lower half of the oven. Turn the oven up to 450°F (232°C) and bake until well puffed and browned, about 10 to 15 minutes. Turn the oven down to 300°F (149°C) and bake 15 to 20 minutes longer to dry well.

6. Cut a small slit in the side of each puff shortly after you remove them from the oven to let steam escape and dry the inside. Cool on a rack. Puffs can be used right away or frozen. They are moist and contain eggs; if they are kept in a warm, moist spot, they will mold within a day.

Cocktail Puffs with Escargots

MAKES 24 COCKTAIL PUFFS

Once when I was teaching in Santa Barbara, California, Julia and Paul Child came over to join us in eating the goodies prepared in class. After Julia had three little puffs, she asked, "Shirley, what are these? They are delicious." "Escargot," I replied. She said, "No wonder they're so good."

what this recipe shows

High-quality salted butter and a little freshly grated nutmeg are the secret of fine snail butter.

2 cans (7 ounces each) medium snails (see Note)

2 shallots

4 cloves garlic

1 cup parsley sprigs

1/2 pound (2 sticks) lightly salted butter, softened

Dash hot pepper sauce or cayenne

Pinch of freshly grated nutmeg

Salt

Freshly ground white pepper

1 recipe Pâte à Choux (page 229), formed into 1-inch balls, baked, and cooled

1. Preheat the oven to 450°F (232°C).

2. Drain the snails and place on a paper towel. Chop the shallots, garlic, and parsley by hand or in a food processor with the steel knife. Add softened butter, hot sauce, nutmeg, and ¼ teaspoon salt and blend in. Taste and add salt and pepper as needed. (Snail butter can be prepared 1 or 2 days ahead and kept tightly sealed in the refrigerator.)

3. Cut tops from the puffs, place a snail or snails in each, add a generous teaspoon of snail butter, and replace the top. Warm puffs in the oven just long enough to melt butter, about 4 minutes. Serve immediately.

Note: *Canned jumbo snails, twelve to a can, are the most commonly available. If you have jumbo snails, you may have to cut them in half. With small snails, add two or three to a puff.*

at a glance
Pâte à choux

What to Do	Why
Use bread flour for higher, lighter puffs.	Bread flour lightens puffs in two ways. It has more gluten-forming proteins, and it absorbs more water, allowing the use of more eggs.
Add flour all at once to boiling liquid.	Flour stirred in all at once will quickly swell the starch and dry the dough.
Cool dough a little before beating in the eggs.	If the dough is too hot, it may cook the eggs.
Replace at least one whole egg with two whites for crisp, dry puffs.	Egg whites are excellent drying and crisping agents.
When adding eggs, stop while the consistency is firm enough to hold its shape.	Puffs will not rise if the dough is too thin.
Preheat the oven to a lower temperature, then turn up to the cooking temperature.	Heat from the bottom makes the puffs rise before the top crust sets and holds them down.
Cut a small slit in the side of each puff shortly after you remove it from the oven.	This allows steam to escape and the puff to dry inside.

Using eggs as aerators

Eggs are the secret to cloudlike soufflés, feather-light meringues, and delicate cakes. Whole eggs, egg whites, and egg yolks all capture and hold air bubbles in many foods, making them light and airy. Whole eggs hold air and steam and contribute to the leavening in génoises and shortened cakes. Both egg yolks and whites lighten sponge cakes. Egg white foams are the sole leavening in angel food cakes.

Egg white foams

How do egg white foams work? When you beat egg whites, air makes some of the proteins in the egg whites unwind (page 197). The bonds holding together the natural protein coils pop apart. The proteins unwind and straighten with their bonds sticking out. The unwound (denatured) proteins bump into neighboring unwound proteins, and they join together, forming a lining or reinforcement around each air bubble.

In a perfect egg white foam, the proteins link together loosely and stay soft, moist, and elastic. When you place the dish containing a perfect foam in the oven, the air bubbles get hot and expand beautifully. The soufflé, sponge cake, or meringue rises until the heat cooks and sets (coagulates) the proteins around the bubbles.

If you overbeat an egg white foam or leave it exposed to air, the bonds between the proteins tighten and squeeze out the moisture. As that soft, moist spiderweb of loosely bonded proteins begins to dry, it loses its elasticity and becomes lumpy and inflexible. In such a foam, the protein lining of the air bubbles is already set; it is stiff and dry. When you fold this foam into a batter, the air bubbles cannot expand with heat. They are already firm and set before they go into the oven. The soufflé, cake, or meringue will rise little, if at all.

Older egg whites whip easier and to greater volume, but fresh whites produce a more stable foam and better meringues.

You want to beat as much air into the whites as possible, but you must keep the delicate protein network that lines each bubble soft, moist, and elastic. Nearly everything matters when you prepare an egg white foam. How do you get a perfect egg white foam, and how do you know when you have it?

The right white Old, thin egg whites beat up faster and easier than very fresh, thick ones and also give slightly more volume than fresh whites. Fresh egg whites, however, make a more stable foam, one that will hold up better in a soufflé or cake. Frozen egg whites, if they have not been pasteurized (heat treatment of eggs prolongs beating time), whip more eas-

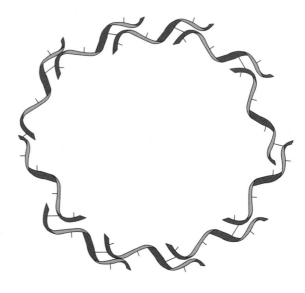

**Air Bubble Lined
with Denatured Proteins**

ily than fresh and make good foams and cakes of high quality. Dried egg whites that have been pasteurized require a longer beating time. Poor-quality eggs will result in a cake of reduced volume.

Some chefs say that you can whip egg whites out of the refrigerator just fine. That's true. But the colder the egg whites, the longer it takes to beat them to a good foam. Beating time is directly related to temperature.

No yolk in the whites Since even a trace of egg yolk will deflate egg white foam, use the three-bowl technique to separate eggs. Eggs are easier to separate when warm, but the yolk breaks more easily. Crack an egg, let the white drop into a small bowl, and place the yolk in another bowl. If the yolk did not break and the white is absolutely pure, pour the white from the small bowl into the larger bowl in which it will be beaten. Continue this until all the eggs are separated. If you break a yolk, even if you do not see any yolk in the white, throw out that white and wash the small bowl rather than contaminate the big bowl of whites with a minute trace of yolk.

Spotlessly clean utensils Beat egg whites in a scrupulously clean copper, glass, ceramic, or metal bowl with a clean whisk. The tiniest trace of fat can wreck an egg white foam. Fats in egg yolk and olive oil are worse than others. Fat slows down foaming and dramatically reduces volume. An oily whisk or a seemingly clean plastic bowl (plastic can hold hidden traces of oil) can be devastating to an egg white foam.

The right whisk or beater Beaters with many tines will incorporate air into egg whites much faster than those with a few. Select a large, flexible balloon whisk with lots of tines for easier hand beating. The type of mixer also makes a difference. A mixer with beaters that do not stay in the same place in the bowl incorporates air into a foam faster and better than one with stationary beaters. Some mixers are designed so that the beater or beaters continuously move around the edge of the bowl. If you are using a handheld mixer, move it around the bowl for best results.

Salt for flavor only Salt decreases the stability of an egg white foam and causes it to lose moisture and dry out. Salt is a disadvantage technically, but for flavor it is an advantage. If it is needed for flavor, add a minimum amount.

Water for softness Water increases the volume of egg white foam almost as much as if you added an equivalent amount of egg white. It makes a cake softer. You can replace up to almost a quarter of the total volume of egg whites with water. The drawback is that it produces a less stable, softer foam. Bakers use this information to their advantage. When Rose Beranbaum wanted to soften snow-white layers for a wedding cake (they were essentially hard meringues), I suggested water. It worked like a charm.

Water makes puffier, softer egg white foam— try it in cakes.

Cream of tartar or vinegar for stability Adding acid makes egg whites denature faster and speeds up the unwinding of proteins around the air bubbles in an egg white foam. This reinforces the air bubbles faster and creates a good foam that will hold up until heat can coagulate the proteins and set the soufflé, cake, or meringue. Many recipes call for either cream of tartar—a mildly acidic salt produced from grapes—or white vinegar to be added to the egg whites at the beginning of beating. The usual amount is 1/8 teaspoon cream of tartar or 1/8 teaspoon of distilled white vinegar per egg white.

Sugar for stability When you beat sugar into egg white foam, it is like coating the unwound proteins in syrup so they cannot dry out. If you add the sugar early in beating, it takes longer to get a foam of good volume. You can get a good-volume foam faster by waiting until soft peaks begin to form before adding the sugar. Take care not to add the sugar too late. The foam can dry and lose its elasticity. It is better to add it too soon and have to beat longer than to add it too late, after the foam has started to dry.

Finely ground sugar, sometimes called *superfine, baker's, bar,* or *berry sugar,* dissolves much faster in the foam than granulated sugar. Some pastry chefs feel that it gives a superior-textured foam. If superfine sugar is not available at your grocery store, make your own by processing regular granulated sugar in a food processor for a minute or two. Most pastry chefs do not like confectioners' sugar except for hard meringues (it makes very light hard meringues) because it contains cornstarch.

When to stop beating The biggest single mistake people make when beating egg whites is to overbeat. Rarely are egg whites overbeaten by hand, but it is very easy to do in a mixer. When the egg whites are overbeaten, the delicate protein linings of the air bubbles lose their elasticity and dry out. Egg white foams are more stable and more effective before they reach maximum volume. I was trained to beat whites until "they no longer slipped in the bowl." But now I feel that it is better to stop while they still slip just a little. When a soft peak is left that does not drop over when you lift a portion of the beaten egg whites, tilt the bowl and see how much the foam slides. If there is a lot of unbeaten white in the bottom, it will slide about very fast. The time to stop is when there's a small amount of liquid left and the foam slips just a little when the bowl is tilted.

The copper bowl advantage

For years food scientists ignored the copper bowl, insisting that chefs just imagined that it improved egg white foams. It was a great blow to me when I discovered that, in fact, there's no sizable difference in the volume of raw egg whites beaten in a copper bowl compared to those beaten in a noncopper one. There is, however, a major difference in the volume of the cooked dish! How can this be?

Several years ago, I asked three of my associates at Nathalie Dupree's Rich's Cooking School in Atlanta to help me prove or disprove the advantages of using a copper bowl. We measured sixteen equal portions of egg whites. Each of us beat the whites in four different ways: in a stainless-steel bowl with a heavy duty mixer, in a glass bowl with a hand mixer, in a stainless-steel bowl with a whisk, and in a copper bowl with a whisk.

I had read that you got a third more volume with the copper bowl, but to our great disappointment the sixteen batches of beaten egg whites were about the same. We carefully folded each portion of beaten egg whites into equal portions of soufflé base and baked the sixteen soufflés. To our astonishment, the four soufflés containing whites beaten in the copper bowl were twice as high as all the others. I bought my first copper bowl that day.

When writing an article on eggs for *The Cook's Magazine,* I wanted to find out why beating egg whites in the copper bowl made such a difference. I asked Dr. Ing C. Peng, a top egg researcher then at Purdue University, what could be happening. As he explained, the particular protein in egg whites that links together around the air bubbles is conalbumin, which loves to combine with copper. Dr. Peng said that if the egg whites were in contact with the copper for more than a minute or two, this combining might take place. Then you would no longer have conalbumin around each air bubble in the foam; you would have copper-conalbumin. Copperconalbumin is much more stable, does not dry out as easily, and even has a higher temperature of coagulation than plain conalbumin.

Reasoning that there had to be a scientific reason for cooks' two-hundred-year-long preference for copper bowls, food science writer Harold McGee decided to test whether the egg white foam lost less liquid when beaten in a copper bowl. If the whites were not holding up well, the proteins would tighten up and squeeze out the water that they held. If egg white foam beaten in a glass or stainless bowl lost more water faster than egg white foam beaten in a copper bowl, the copper-bowl foam had to be staying moist, soft, and elastic longer and therefore would perform much better in cooking.

That is exactly what happened. The foams beaten in glass and noncopper metal bowls all dripped over 11 milliliters of fluid in 10 minutes. The foam beaten in the copper bowl dripped less than 2 milliliters in 20 minutes. Over five times more liquid was lost in the foams beaten in glass or metal bowls in half the time.

Egg whites beaten in copper bowls stay soft and elastic. When they go into the oven, the air bubbles become hot and expand beautifully to produce higher soufflés. Also, copperconalbumin has a higher temperature of coagulation than plain conalbumin, so the egg whites have to reach a higher temperature before they become firm and set and maybe the bubbles can get a little bigger before they set. (The other egg proteins such as the globulins still set at the same lower temperature.)

Egg whites beaten in a copper bowl are more

forgiving than egg whites beaten in glass, ceramic, or other metal containers. Even if you overbeat or interrupt the beating and let your egg whites stand before adding sugar, actions that ordinarily dry out and make the proteins firm around the air bubbles, egg whites beaten in a copper bowl remain elastic and perform well.

Many pastry chefs swear by cream of tartar and feel that you get a more stable foam adding $1/8$ teaspoon cream of tartar per egg white at the beginning of beating than with the copper bowl or any other way. I would love to see a comparison of stability between using a copper bowl without cream of tartar and using whites with cream of tartar beaten in a glass or stainless bowl. A little cream of tartar and egg whites in a copper bowl may be the best route; however, this is not recommended since use of acidic ingredients in a copper bowl might dissolve more copper than is safe for consumption.

Soufflés

A soufflé is an intensely flavored base lightened with beaten egg whites and baked to puff to glorious heights. Soufflés can be very simple—vegetable puree lightened with beaten egg whites—or very complicated, with many ingredients. The ideal soufflé is light and airy yet still moist and creamy. This airy creaminess is created by perfectly cooked proteins just loosely joined, but it can be lost with overcooking. (Cold or frozen soufflés are not really soufflés at all. They are mousses, which are often held with gelatin.) Both ingredients and techniques play important roles in a successful soufflé:

The base Frequently a mixture of a thick cream sauce, a puree, and egg yolks, the soufflé base usually carries the soufflé's flavor. Since the base is diluted by the egg white foam, which is several times its volume, and further diluted by the rise, it should be intensely flavored. Smoked fish, country ham, prosciutto, strongly flavored greens—all these make good flavorings for savory soufflés. Grated cheese is another common savory flavoring, but it should be folded in as the soufflé is folded together. If you add the cheese to the hot base, it melts and can make the base so thick that it is impossible to fold it together with the egg whites without deflating them. Excellent flavors for dessert soufflés include chocolate, lemon, and orange zest plus Grand Marnier.

A base does not have to contain cream sauce or even egg yolks. For one of my favorite soufflés I substitute $1/2$ cup pureed creamed corn with 1 tablespoon honey and $1/4$ cup mascarpone or cream for $3/4$ cup milk in the soufflés in Incredibly Creamy Soufflés (page 238). As I fold the base and beaten egg whites together, I add $1/2$ cup diced country ham and a small finely chopped serrano chili pepper. Bake in individual Parmesan-coated soufflé dishes.

For a light, fragile soufflé, a thick puree is all that's needed. But for a soufflé with even brief holding ability, you need a thick base (the consistency of a very thick cream sauce or a pastry cream) that contains uncooked proteins. In a typical soufflé, these proteins are in the egg yolks and in the flour in the cream sauce.

The base is not only the source of the soufflé's flavor, it also determines the structure. The beaten egg whites in the foam are already unwound and loosely joined to each other. They are not free proteins that can contribute to holding the soufflé up. The proteins in the base must do this.

The egg white foam If you can beat a moist, elastic egg white foam (page 232), you can make a superb soufflé. Soufflés puff so magnificently because air trapped in the egg white foam expands in the hot oven. If the egg whites are elastic, they can expand without breaking. This allows the soufflé to rise until it gets hot enough for the eggs to cook and set. Overbeaten egg whites are dry and have lost their elasticity. They result in heavy soufflés.

A soufflé will rise straight up without a collar—as long as you have a thick base and properly beaten egg whites.

The whites should be just firm enough to form a peak that does not fall over when the whisk is lifted. They should still slip a little when you tilt the bowl.

The number of egg whites in a soufflé is somewhat arbitrary. Many soufflés call for an equal number of whites and yolks. Others add up to three extra whites.

Folding Once you have gone to a great deal of trouble to beat air into the egg whites, you don't want to knock it all out of them when you fold the soufflé together. The base must be thick to do its job of providing structure, but if you stir at least 1 cup of the beaten whites into the base you will lighten it enough to have two mixtures that can be folded together satisfactorily. Either pile the beaten whites on top of the base or push the whites to one side and pour in the base. Just don't put the heavy base on top of the whites—you'll press the air out of them.

Many different folding procedures work well. A wide spatula is an enormous help. Some cooks like to insert the spatula in the center of the bowl, drag it across the bottom to the edge, and bring it up and back to the center, spreading some of the contents from the bottom across the top. Others like to insert the spatula at the side of the bowl. Others like to move the spatula in a figure eight. Remember that you're folding, not stirring. It is not necessary to get every bit of white thoroughly incorporated. It is better to stop while there is still a patch or two of whites than to deflate by overworking.

Baking Like pâte à choux, soufflés need heat from the bottom to rise properly. You want as much rise as possible before the top cooks. Just placing the soufflé in the lower third of the oven usually does the trick. You can also preheat the oven a little lower than you want (about 25°F/14°C lower), then turn the oven up when you place the soufflé in. This will ensure more heat from the bottom.

Larger soufflés (1½- and 2-quart soufflés) need to be cooked at lower temperatures than small soufflés to avoid having the outside overcooked and dry before the inside is done. Some cooks chill the soufflé dish in the freezer for about 5 minutes before filling. The cold dish slows down the cooking of the outside while the center warms up.

If you have a thick base and properly beaten egg whites, there is no need for a parchment paper collar. The soufflé will rise straight up without it. For a dramatic soufflé that is puffed well above the dish, fill it to within ½ inch of the top. Place the soufflé dish or dishes in the center of a shelf in the lower third of the oven and position so there is good air flow around them.

SOUFFLÉ SIZES

I much prefer individual soufflés. They are quick, there are few problems of uneven cooking, and they are easy to serve. Because of the problem of overcooking edges, it is best to make soufflés no larger than 2 quarts. The common larger soufflé sizes are:

- 1 quart—two entree servings or three to four side-dish or dessert servings
- 1½ quarts—three to four entree servings or about six side-dish or dessert servings
- 2 quarts—four to five entree servings or about eight side-dish or dessert servings.

Soufflé recipes vary greatly, from those with equal numbers of egg yolks and whites to those that use one, two, or three more whites than yolks. Some even use more yolks than whites. Then there are my concoctions that have whole eggs thrown in. It is impossible to give anything but a rough estimate of how many eggs require a particular size soufflé dish. You sometimes find recipes for four egg yolks and seven whites that make four to six servings, but you'll probably use four or five large eggs for a 1-quart soufflé, about six large eggs for a 1½-quart soufflé, and eight or nine large eggs for a 2-quart soufflé.

Why soufflés fall Soufflés fall. That, unfortunately, is their nature. Are there ways to make a soufflé hold up just a little longer? I think so. When cookbook author Susan Purdy was working on low-fat baking recipes for her book *Have Your Cake and Eat It, Too*, she was having a problem with one of the cake recipes. Made with one egg yolk and several beaten whites, the cake puffed beautifully in the oven but fell as it cooled. The cake was behaving just like a soufflé because . . . well, it was essentially a soufflé.

I reasoned that except for one egg yolk all the eggs in Susan's cake—the beaten egg whites—were unwound (denatured) proteins lightly hooked together to form the egg white foam. There was only one egg yolk to set the cake, and it was not enough. A cake that size needed at least two eggs to set it. In place of the yolk, Susan added a whole egg plus a couple of unbeaten whites to the batter and reduced the milk a little to compensate for the added liquid. The cake held.

If this worked for Susan's cake, I figured it would work for a soufflé, which is set by egg yolks alone. It was certainly worth a try.

Zona Spray is a fine teacher who has been training chefs in French and Italian cuisine for many years. I decided to experiment with her Soufflé au Chocolat, an excellent classic soufflé recipe. I added two whole eggs to the base instead of one yolk. Then I added a little extra flour to the base and reduced the milk slightly so that the base would be the same consistency. I made one recipe the regular way and one recipe the altered way. Sure enough, the soufflés made with more eggs stayed up a significant several minutes longer, though they still fall eventually. I have used this same principle with the following Chocolate Soufflé with White Chocolate Chunks.

Chocolate Soufflé with White Chocolate Chunks

MAKES 8 SERVINGS

Frozen chunks of white chocolate melt and form little pockets in the dark creamy soufflé.

what this recipe shows

Adding extra eggs to the base increases stability.

2 tablespoons butter to grease dishes

2 tablespoons *and* 1/4 cup *and* 1/4 cup sugar (about 3/4 cup *total*)

1 tablespoon water

3 tablespoons Kahlúa

1 cup (6 ounces) chocolate chips, semisweet or sweet

2 teaspoons instant coffee

1/3 cup bleached all-purpose flour

Salt

3/4 cup milk

1 tablespoon pure vanilla extract

2 large eggs

5 large egg yolks

6 large egg whites at room temperature

3/4 teaspoon cream of tartar (optional)

6 ounces white chocolate, cut into 1/2-inch chunks and frozen

TOPPING

1 1/2 cups heavy or whipping cream

1/4 cup sugar

continued

1. Butter eight individual soufflé dishes. Divide 2 tablespoons sugar among the dishes and rotate to coat. Pour off any remaining sugar.

2. Heat the water, Kahlúa, chocolate chips, and coffee in a small saucepan over very low heat, stirring constantly, until the chocolate just melts. Set aside.

3. Blend the flour, ¼ cup sugar, and a pinch of salt together in a medium saucepan. Slowly whisk in the milk until smooth. Stir over medium heat for about 2 minutes to cook the flour, stirring constantly to avoid lumps. Remove from the heat. Add the vanilla and melted chocolate mixture.

4. Preheat the oven to 375°F (191°C).

5. Stir together the whole eggs and the yolks in a medium bowl. Spoon several tablespoons of hot chocolate mixture into the eggs and yolks to warm them. Then fold all of the egg mixture back into the warm chocolate mixture.

6. Beat the egg whites in a large bowl until foamy. Add the cream of tartar and continue beating until soft peaks form. (Omit the cream of tartar if using a copper bowl.) Add ¼ cup sugar, 1 tablespoon at a time, beating about 15 seconds after each addition. Beat until soft peaks form that do not fall over when you lift the beater. Stir a quarter of the whites into the warm chocolate mixture to lighten it. Gently fold in the remaining egg whites.

7. Turn the mixture into the prepared dishes. To add the white chocolate, fill the soufflé dishes halfway, divide the chunks of frozen white chocolate among the dishes, and add the rest of the soufflé mix to fill each dish. (Soufflés can be prepared in the baking dishes 1 hour in advance and covered with plastic wrap.) Bake the soufflés on a rack in the middle of the oven until well puffed, 25 to 35 minutes.

8. For the topping, whip the cream in a cold bowl with cold beaters until fairly firm. Stir in the sugar. Cream will thin when you add sugar. To serve, pass the whipped cream and let your guests punch a hole in the center of their soufflés and spoon whipped cream into the hole.

Incredibly Creamy Soufflés

MAKES 6 SERVINGS

These freestanding soufflés break the rules. They are baked in very well greased muffin pans, not ceramic dishes, in a 450°F (232°C) oven, an outrageously high temperature for a soufflé. And, then they are dumped out! They are turned out of the pan into a shallow dish of hot cream, sprinkled lightly with cheese, and slipped back into the oven until the cheese melts and the cream is bubbly. I think that once you have dumped out a soufflé you will be over any fear of soufflés forever.

what this recipe shows

Soufflés do not have to have textured sides—grated Parmesan or sugar—to rise.

Soufflés can be baked in metal as well as ceramic.

Soufflés can even be prepared a day ahead, dumped into the cream, then reheated when ready to serve.

Nonstick cooking spray

2 to 3 tablespoons softened butter to grease pans

3 cups *and* ¹/₂ cup heavy or whipping cream (3¹/₂ cups *total*)

3 tablespoons butter

¹/₄ cup bleached all-purpose flour

1 cup milk

¹/₄ teaspoon *and* ¹/₂ teaspoon salt (³/₄ teaspoon *total*)

¹/₄ teaspoon *and* ¹/₈ teaspoon white pepper (about ¹/₂ teaspoon *total*)

4 large eggs

¹/₂ teaspoon cream of tartar (optional)

¹/₂ cup grated Gruyère

1. Spray a twelve-muffin nonstick 3-inch muffin pan or two six-muffin pans with nonstick cooking spray, place in the freezer for at least 10 minutes, and grease with softened butter. Place the pan back in the freezer until ready to fill.

2. Boil 3 cups cream in a large saucepan or a 9 × 14-inch metal baking casserole over low heat to reduce by a third. Transfer the cream from the saucepan to a gratin or oval baking dish if necessary. Set aside.

3. Preheat the oven to 425°F (218°C). Place an oven shelf in the lower third of the oven.

4. Melt 3 tablespoons butter in a heavy saucepan, whisk in the flour, and cook over low heat for several minutes, stirring constantly. Remove from the heat. In another saucepan, heat the milk and ¹/₂ cup cream until hot. Slowly whisk into the flour mixture. Return the pan to the heat and cook, stirring slowly, until quite thick. Remove from the heat. Season with ¹/₄ teaspoon salt and ¹/₄ teaspoon white pepper. This is the soufflé base.

5. Separate the eggs carefully. Be certain that the whites are absolutely free of even a trace of yolk. Place the yolks in a medium mixing bowl. Place the whites in a copper bowl, if available, or a glass or other metal bowl. (Add cream of tartar if beating in a noncopper bowl.)

6. Stir a few tablespoons of the soufflé base into the egg yolks, then stir the yolks into the rest of the base.

7. Whisk the egg whites until soft peaks form that do not fall over when you lift the whisk. Egg whites should still slip just a little in the bowl. Stir a quarter of the beaten whites into the soufflé base. Then carefully fold the rest of the whites into the base. Fill cold prepared muffin cups with soufflé mix. Place the soufflés in the oven and turn it up to 450°F (232°C). Bake until the soufflés are well puffed but not brown, about 5 minutes.

8. Season the reduced cream with ¹/₂ teaspoon salt and ¹/₈ teaspoon white pepper. Move the gratin dish of hot cream to a trivet on a clean counter top. When the soufflés are risen and just set, turn the muffin tin upside down and dump the soufflés into the hot cream. If a soufflé falls onto the counter, scoop it up with a spatula and place it in the cream.

9. The soufflés will fall at first, but the cream will puff them up again. Sprinkle Gruyère over the top and return to the oven. Bake until the cream is bubbly and the cheese is melted and beginning to brown, about 4 minutes. Be sure to serve a generous portion of cream with each soufflé (see Note).

continued

Note: *These soufflés can be dumped into slightly reduced cream, then refrigerated and held, even overnight, and reheated when ready to serve. The hot cream repuffs them. They may not be quite as high as they were when served immediately, but the difference is hardly discernible.*

Variation: *Add any desired ingredients that do not require additional cooking to the cream right after you have dumped the soufflés in. Fresh crabmeat is wonderful.*

at a glance
Soufflés

What to Do	Why
Use a scrupulously clean bowl and beaters for egg whites.	Any oil or grease will wreck an egg white foam.
Use fresh egg whites.	Fresh egg whites take slightly longer to beat and yield slightly less volume than older whites, but they are more stable and make better soufflés.
Use room-temperature eggs.	The length of time it takes to beat egg whites is directly related to how cold the whites are. The colder the whites, the longer the beating time. Warmer eggs are also easier to separate.
Separate the eggs using the three-bowl method described on page 233.	Even a trace of egg yolk will wreck an egg white foam.
Use a copper bowl or add ⅛ teaspoon cream of tartar per egg white.	To get a more stable foam that will hold up better in cooking.
If the recipe calls for sugar, add when soft peaks form when the beaters are lifted.	Adding sugar too early reduces the volume and requires longer beating time.
Do not overbeat egg whites.	Overbeaten egg white foams become dry and rigid and will not expand satisfactorily in a hot oven.
Mix a cup of egg white foam into the base.	The base should be lightened so that it can be folded properly with the beaten whites.
Have an intensely flavored base.	All the flavoring is in the base of many soufflés.
Fold grated cheese in when folding egg whites and base together.	Melted cheese will make the base so thick that it can't be folded together with the whites.

Foam cakes

While all cakes are actually foams, only cakes with a high ratio of eggs to flour are called *foam cakes*. Some people use the term *sponge cake* for all foam cakes, but this causes confusion. Within the rather broad category of true sponge cakes there are three basic types of cakes: cakes with no fat at all, such as angel food cakes, meringues, dacquoises, and vacherins, which are made with egg whites; cakes whose only fat comes from egg yolks, like sponge cakes and ladyfingers; and cakes with eggs and some additional fat, like chiffon cakes and génoises. In all of these cakes, eggs act as aerators.

Angel food cake   Basically, angel food cake is baked meringue with flour. Egg whites are beaten into a foam with an acid, preferably cream of tartar, a flavoring, usually almond extract or vanilla, and some of the sugar in the recipe. Flour, double-sifted with the remaining sugar, is folded in. The cake is baked in an ungreased pan, most often a tube pan, though a loaf pan is sometimes used. Here are some tips:

- Make a good egg white foam as directed on pages 232–234, adding a bit of water for a softer, moister cake.
- Add cream of tartar at the very beginning. Researchers have found that it increases the volume of angel food cakes and whitens the crumb. Cakes made with cream of tartar also have finer grain and shrink less than cakes made with other acids, like acetic acid (vinegar) or citric acid.
- Pour the batter into an ungreased pan that has been rinsed with hot water. The fragile foam should not come into contact with any fat.
- Bake in a moderate oven until the cake springs back when pressed. Angel food cake must be cooked thoroughly before it is removed from the oven.
- Cool the cake thoroughly in the pan, inverted so that air can circulate freely around it. This gives the cake a chance to stretch and set. It will collapse if the pan stays hot or if the cake is removed while still warm.

Meringue, dacquoise, and vacherin

There are four common types of meringue: soft meringue, hard meringue, Swiss meringue, and Italian meringue. Each either uses a different ratio of sugar to egg whites or combines the egg whites and sugar in a different way.

Soft meringues, which are used to top pies and other desserts like baked Alaska, to fold into batters to lighten them, or to be part of a dessert like floating islands, have a basic sugar–to–egg white ratio of 2 tablespoons sugar per egg white. The whites are beaten in a copper bowl if possible or with cream of tartar added at the start of beating until soft peaks form. Sugar is beaten in a few tablespoons at a time until the peak does not fall when the beater is lifted.

To make a fail-safe soft meringue, I like to add a tablespoon of cornstarch that has been stirred into $^1/_3$ cup water and heated to form a thick gel. I beat this gel into the beaten egg whites a tablespoon at a time. This prevents the meringue from shrinking, lowers the chances of beading, and makes the meringue tender and easier to cut smoothly.

Hard meringues, such as dacquoises (meringue cake layers with nuts) and vacherins (meringue shells), contain twice as much sugar as soft meringues, $^1/_4$ cup per egg white. The egg whites and cream of tartar are beaten, and the sugar is beaten in. If nuts are used, they are folded in. The meringues are then spooned or piped onto baking sheets and dried in a low-temperature oven for several hours or overnight.

Swiss meringue sets very stiff and is used for icing and decorations. It is prepared by beating egg whites and confectioners' sugar over, but not touching, boiling water. When the egg whites and sugar reach 120°F (49°C), they are removed from the heat and beaten at high speed for 5 minutes. The speed is lowered, and the mixture is beaten for another 5 minutes, until very stiff.

Italian meringue is used on pie filling and puddings. It is prepared by heating sugar syrup to the hard-ball stage (248°F/120°C). This very hot syrup is drizzled into well-beaten egg whites until the meringue is cool and ready to spread.

Soft meringues   Sugar is a vital part of the meringue's structure. Sugar grabs water and removes moisture from the foam itself and causes it to set. Sugar initially dissolves in the moisture stolen from the foam and makes a syrup coating on the delicate protein network. Water evaporates from the syrup in the oven, and the network is left covered with fine, dry sugar. Meringues simply will not work without sugar.

What causes meringues on pies to bead? Not humidity or fast cooking but overcooking.

Choosing the right amount and kind of sugar is important. About 1½ tablespoons of sugar per egg white is as low as you can go, and this meringue is not as stable as that made with the usual ratio of 2 tablespoons per egg white. (Adding a little cream of tartar or lemon juice or vinegar helps give meringues stability.) For soft meringues on pies, superfine sugar is preferred.

Undercooking and overcooking are the most common problems with soft meringues on pies. Believe it or not, it is possible to overcook and undercook a meringue at the same time. Overcooking, not humidity or fast cooling, causes beading, the little sugary drops of moisture on baked meringue topping. Weeping, the watery layer between the meringue and the filling, is caused by undercooking. If you pile meringue onto a cold filling and cook it in an oven that is too hot, the top of the meringue can overcook and bead while the bottom remains undercooked and weeps. The filling has to be hot for the meringue topping to cook through.

When a meringue is undercooked, the uncoagulated (uncooked) egg white foam loses the moisture that it held, causing weeping as the meringue stands.

Sprinkling fine cake crumbs over a hot filling just before adding meringue prevents leakage from ruining your pie.

With a meringue pie, you can avoid weeping by beating the topping before making the filling. Once you have added the sugar to the egg white foam, you can let it stand without fear of its drying out. After you have spooned the piping-hot filling into the pie shell, cover it with the meringue.

Roland Mesnier, the White House pastry chef, sprinkles some fine cake crumbs over the hot filling just before he spoons on the meringue. If there is any leakage, the crumbs absorb it. I tried this. When I peeked under the meringue on my slice of pie, I could not see a single crumb and there was a fine dry interface between the meringue and the pie.

Overcooking, on the other hand, tightens the bonds between proteins and squeezes out water, which forms drops or beads. It also makes the egg whites tough. You might think that a low cooking temperature (325° to 350°F/163° to 177°C) would be the solution, but the interior of the meringue gets hotter when it is cooked at a low temperature for a longer time rather than at a high temperature for a shorter time. Actually, a high oven temperature (425° to 450°F/218° to 232°C) combined with a short cooking time of 4 to 5 minutes prevents beading because it does not get the interior too hot.

With the present possibility of salmonella in eggs (page 195), you would be wise to verge on overcooking meringues. The combination of lower temperature and longer time (though not my favorite) is considered the best approach, or you can use Alice Medrich's Safe Meringue (page 196).

Humidity is also a factor to consider when making meringues. There is a lot of truth to the old adage "Never make meringues on a damp day." Meringues have high sugar content—2 tablespoons of sugar per egg white for soft meringues and 4 tablespoons of sugar per egg white for hard meringues. Since sugar absorbs moisture from the atmosphere, meringues can absorb moisture and become soft if exposed to a humid environment.

I include some cornstarch in all my soft meringues. Starch performs the same magic on meringues as it does on custards. It prevents egg whites from overcoagulating just as it prevents whole eggs from curdling. Meringues with their high sugar content keep all the water tied up, so it is necessary to dissolve the cornstarch in water and heat it before beating it into the meringue. This lets the starch absorb water and swell before it goes into the meringue where no water is available to it. Meringues with starch are tender, cut beautifully, and do not shrink as much or overcook as easily as meringues without it. They are picture perfect.

Lemon on Lemon on Lemon Meringue Pie

MAKES 6 TO 8 SERVINGS

This is a refreshing conclusion to any meal. There is lemon zest in the crust, the filling, and the meringue for a fresh lemon taste in every bite. You can prepare the crust ahead, but the pie is best made the same day it is to be served.

what this recipe shows

Thickening the starch mixture before the lemon juice is added prevents acids from interfering with starch's swelling and thickening.

Reheating the filling after the eggs are added kills alpha amylase, the enzyme in yolks that can thin starch custards (page 215). Eggs will not curdle during this reheating because of the presence of the starch.

Egg yolks provide emulsifiers for a sensuously smooth filling.

1 prebaked Simple Flaky Crust (page 110), with 1 teaspoon finely grated lemon zest added

1 recipe Safe Meringue (page 196) with finely grated zest of 1 lemon folded in

LEMON FILLING

$1/2$ cup cornstarch

$1^1/2$ cups sugar

$2^3/4$ cups water

2 large eggs

6 large egg yolks

$1/4$ teaspoon salt

4 tablespoons butter

$1/3$ cup fresh lemon juice

Finely grated zest of 2 lemons

2 teaspoons pure vanilla extract

3 to 4 tablespoons very fine cake crumbs (see Note)

1. Prepare prebaked flaky crust and the Safe Meringue.

2. Preheat the oven to 325°F (163°C). Place an oven shelf in the lower third of the oven.

3. Prepare the filling by stirring together the cornstarch and sugar in a medium saucepan. Stir in the water and heat, over medium heat, stirring constantly, until thick. Whisk the eggs and yolks together in a medium bowl and stir in several tablespoons of hot filling to warm the mixture. Pour the egg mixture into the hot filling and continue to heat. Bring back to a boil and cook for 3 or 4 minutes, stirring constantly, to kill enzymes in the yolks that can thin the pie. Remove from the heat and stir in salt, butter, lemon juice, lemon zest, and vanilla.

4. Pour the filling into the prebaked crust. While the filling is piping hot, sprinkle with fine cake crumbs and cover with some of the meringue. Take care to spread the meringue so that it touches the crust all the way around. After you have covered the hot filling well, pile on the rest of the meringue and make decorative swirls with the back of the spoon. Bake until the meringue begins to brown lightly, about 30 minutes. Refrigerate, uncovered, for several hours before serving.

Note: *For fine cake crumbs, even a ground-up Twinkie or a cake doughnut will do.*

Hard meringues Hard meringues, like dacquoises and vacherins, must have a high sugar content, ¼ cup per egg white. The sugar robs the egg white foam of water so that in a slow oven the meringues become very dry and crisp. Hard meringues are prepared in the same way as soft meringues except for using more sugar and omitting the cornstarch. For very light hard meringues, use confectioners' sugar (which does contain cornstarch). Beat the foam until the peaks are glossy and the peak formed when you lift the beater does not fold over. Bake at a low temperature so the meringues can dry thoroughly without burning. They are best done by leaving them in the oven overnight at the lowest temperature setting.

Hard meringues make excellent low-fat pie crusts or dessert shells. These crusts can be kept several days if sealed tightly against moisture in the air. You can put them in a 200°F (93°C) oven for 30 minutes to redry them thoroughly.

Raspberries with Mascarpone Cream in Walnut-Oat Meringue

MAKES 6 TO 8 SERVINGS

This heavenly raspberry pie in a walnut-oat crust is derived from one made by Judy Falk, an outstanding Atlanta cook. Judy doesn't use Chambord and tops hers with whipped cream while I go for a mascarpone-honey topping.

what this recipe shows

Roasting enhances the flavor of the walnuts.

Cream of tartar helps produce a more stable meringue.

Confectioners' sugar makes the meringue crust lighter.

Since the meringue is thin and baked at 325°F (163°C) for 25 to 30 minutes, there is no danger of salmonella.

¹/₂ cup chopped walnuts

3 large egg whites

¹/₄ teaspoon cream of tartar (optional)

1 cup confectioners' sugar

¹/₄ teaspoon baking powder

¹/₄ teaspoon salt

³/₄ cup quick oats

¹/₂ teaspoon pure vanilla extract

Nonstick cooking spray containing flour, such as Baker's Joy

2 packages (10 ounces each) frozen raspberries, thawed

2 tablespoons cornstarch

1 carton (¹/₂ pint) fresh raspberries (if available)

2 tablespoons Chambord or other raspberry liqueur

1 recipe Mascarpone Cream (below)

1.   Preheat the oven to 325°F (163°C). Roast the walnuts on a baking sheet for 10 to 15 minutes. Set aside.

2.   Beat the egg whites and cream of tartar until soft peaks form when the beater is lifted. (Omit the cream of tartar if a copper bowl is used.) Beat in the sugar and baking powder. Fold in the walnuts, salt, oats, and vanilla. Spray a 10-inch pie pan with nonstick cooking spray containing flour and spoon the meringue onto the plate. Spread thinly over the bottom and pile higher around the edges to form a shell for filling. Bake until set and lightly browned, 25 to 30 minutes. Cool completely on a cooling rack.

3.   Drain the frozen raspberries in a medium strainer, reserving the juice. First whisk the cornstarch into a small portion of raspberry juice in a medium saucepan, then pour in all the raspberry juice and heat over medium heat, stirring constantly, until thickened. Let cool briefly, fold in the drained frozen raspberries and fresh raspberries, if used, and the Chambord. Spoon the filling into the meringue crust. Refrigerate to chill thoroughly, then spoon on Mascarpone Cream and serve cold.

Mascarpone Cream

¹/₂ cup mascarpone (see Note)

2 tablespoons honey

1 cup heavy or whipping cream

In a small bowl, whisk together the mascarpone and honey. Place a medium bowl and beaters in the freezer for 5 minutes. Whip the cream to soft peaks in a cold bowl with cold beaters. Fold the mascarpone mixture into the whipped cream. Chill until ready to serve. Serve cold.

Note: *Mascarpone can be ordered from The Mozzarella Company (see Sources—Cheese, page 477) if it is not available locally. A substitute can be made by heating 2 cups heavy cream to 180°F (82°C) in a stainless saucepan, then stirring in a heavy pinch (¹/₈ teaspoon) tartaric acid (available from pharmacists in drugstores). Stir for a full minute, pour into a clean container, and refrigerate overnight.*

Sponge cake, génoise, and chiffon cake Some people use the term *sponge cake* to refer to the whole category of foam cakes. Strictly speaking, though, a sponge cake is a foam cake that contains no leavening and no fat other than that provided by the eggs. Génoise, sometimes called *French butter sponge cake,* contains butter (usually clarified butter). Chiffon cakes are foam cakes containing oil.

Usually in a sponge cake, egg yolks are beaten with part of the sugar to incorporate as much air as possible. The whites are beaten separately with the rest of the sugar. The two are folded together carefully, and then the flour is folded in. Sometimes the flour is added to the yolk mixture first and the yolk mixture and the whites are folded together.

Making a distinction between génoise and sponge cake may be making a distinction where there is no difference. I have noticed often that one chef's sponge cake is another chef's génoise. In principle, though, with génoise the whole eggs or whole eggs and yolks are beaten together; rarely are the whites and yolks beaten separately. The flour is folded in carefully. One cup of the batter is mixed with the butter, which is melted or very soft but not hot. This mixture is then folded into the rest of the batter.

In a chiffon cake the oil and egg yolks are beaten into the flour and dry ingredients. This coats the flour proteins with fat to reduce gluten formation and makes the cake very tender. The egg whites are beaten separately and folded in.

Here are some tips for success with these three cakes—sponge, génoise, and chiffon:

Make the ribbon. Many of these cakes start by making the ribbon—that is, beating the egg yolks and sugar until the mixture is light in color and a drizzle of the mixture falls in a ribbon pattern that remains on top of the batter for seconds before sinking. Egg yolks are 31 percent fat, and in recipes that call for large amounts of sugar there may not be enough liquid in the yolks alone to dissolve it all. Roland Mesnier taught me an old chef's trick: Add a little warm water whenever you are ribboning egg yolks and sugar. A tablespoon or two will help the sugar dissolve.

For a génoise, most French-trained chefs beat eggs and sugar in a bowl in a water bath or double boiler until lukewarm (95° to 100°F/35° to 38°C). Others use an easier way that I think is better. Richard Grausman, author of *At Home with French Classics,* lets the eggs stand for a few minutes in hot tap water (as I do for softcooked eggs and omelets) to warm them. If you also warm another bowl to beat the eggs in, there is no need for the water bath. One problem beginners have is getting the eggs too warm and beating in too much air. Using hot tap water to warm both eggs and beating bowl solves this problem.

Use warm eggs. Heat speeds up the unwinding (denaturing) of egg proteins. Making the ribbon goes faster when the yolks are warm. Whites whip to a foam much more quickly when warm. Whole eggs beat to a well-aerated foam faster and better when warm, too. Using a copper bowl to beat the egg whites or adding cream of tartar also makes a better, more stable egg white foam, which, in turn, produces a better cake.

Choose the right sugar. Superfine sugar dissolves faster in egg yolks, whites, and whole eggs than regular granulated sugar does. The finer sugar produces finer-textured cakes and smoother meringues. Superfine sugar, also called *bar* or *berry sugar,* is widely available. If you do not have it on hand, process regular granulated sugar in the food processor for a minute or two.

Not all the sugar in a recipe is necessarily combined with the yolks. Sometimes part of the sugar is set aside to beat into the whites to stabilize them and a little is reserved to sift with the flour to prevent lumping.

A unique touch with sugar appears in Bert Greene's Special Sponge Cake recipe included in Rose Levy Beranbaum's *The Cake Bible.* Just before the sponge cake is placed in the oven, the top of the batter is sprinkled with sugar. This makes a sweet, crunchy crust.

As for the fat, in old génoise recipes melted butter was usually folded carefully into the batter, frequently with loss of air from the foam. The great French pastry chef Gaston Lenôtre introduced the technique of mixing the butter with 1 cup of batter and then folding this back into the rest of the batter.

This procedure is much more successful and widely used. Richard Grausman has another excellent idea: He suggests using softened butter and blending a small amount of batter with the butter. This makes the butter very easy to fold into the batter.

In a chiffon cake, on the other hand, the oil and egg yolks are beaten together and then beaten into the dry ingredients.

Pick the right pan. A large tube pan is ideal for sponge cake and chiffon cake. In a tube pan, a cake can cook from the center as well as from the outside. The tube pan allows more even heat distribution— especially important with sponge and chiffon cakes that have a large amount of batter. You can then get a thoroughly done foam cake without overcooking the outside.

Sponge cake and génoise can be prepared in layers too. It is important to have the correct amount of batter for the pan size, so follow recipe directions. Use a springform pan for easier removal.

To grease or not to grease? As with angel food cake, fat must not come in contact with the batter of sponge cake since fat can deflate foams. The pan should not be greased. Chiffon cakes, which contain oil, are somewhat different, but they too should be baked in ungreased pans. Some professionals recommend rinsing the pan with hot water just before fill-ing for angel food cake, sponge cake, and chiffon cake. The point is to leave some moisture behind on the pan to add steam to the cake.

A génoise is different. The pan should be lined with a parchment paper circle, greased well with shortening, floured well, and shaken to remove the excess flour.

Bake until done. A foam cake must be thoroughly done to hold its fragile foam network. It is ready to remove from the oven when the edges shrink from the sides. Another way to tell doneness is that the cake will spring back when touched.

Cool until set. Foam cakes, with their delicate air network, would settle and shrink badly if left to cool in a hot pan. They must be turned upside down while the foam is hot so that the cake stretches as it cools and firms. You can invert a tube pan on a bottle through the hole in the center. The bottle holds the pan upside down, so the cake has plenty of air to stretch, cool, and set. You can invert a cake or springform pan by placing it upside down on four inverted drinking glasses. When a foam cake has cooled thoroughly, remove it from the pan by running a small knife around the edges to free it. Then invert it.

A roulade is a sponge cake rolled like a jelly roll. Roulades, like the following recipe, make wonderful desserts and savories.

Great Flower Cake

MAKES ABOUT 10 SERVINGS

This is one of the most spectacular cakes imaginable. A spiral of dark nut roll and snow-white whipped cream is topped with thin slices of strawberries arranged like the petals of a big flower.

> ### what this recipe shows
>
> Roasting enhances the flavor of the pecans.
>
> Adding 1 tablespoon water when beating the yolks with sugar ensures dissolving of the sugar.
>
> Adding cream of tartar to the egg whites helps make a stable foam.

PECAN SPONGE

1¾ cups pecan pieces

Salt

Nonstick baking spray with flour, such as Baker's Joy, or shortening and flour to grease pan

6 large egg yolks

1 tablespoon water

1 cup superfine sugar (see Notes)

1 teaspoon baking powder

¼ teaspoon salt

1 teaspoon pure vanilla extract

6 large egg whites

¾ teaspoon cream of tartar (optional)

Confectioners' sugar

WHIPPED CREAM AND FRUIT

2 cups heavy or whipping cream

½ cup confectioners' sugar

1 tablespoon *and* 1 tablespoon Grand Marnier (2 tablespoons *total*)

1 quart fresh strawberries, each cut lengthwise into several slices

½ cup red currant jelly

1. Preheat the oven to 350°F (177°C) and roast the pecan pieces on a baking sheet for 10 to 12 minutes. Sprinkle lightly with salt. Let cool. Chop the pecan pieces in a food processor with the steel knife, using quick on/off pulses until finely chopped.

2. Preheat the oven to 400°F (204°C). Prepare an 11 × 17 × ½-inch jelly-roll pan by lining with foil, parchment paper, or a nonstick baking sheet liner (see Notes). Spray with Baker's Joy or grease with shortening, dust well with flour, and shake to remove any excess. Set aside.

3. Place the egg yolks in the bowl of a mixer with the whisk attachment. Add water and all but 2 tablespoons of the superfine sugar. Whip on medium speed until well aerated and light in color. Add the chopped pecans, baking powder, salt, and vanilla and stir in.

4. Beat the egg whites and cream of tartar in a clean mixer bowl until soft peaks form when the beater is lifted. (Omit the cream of tartar if a copper bowl is used.) Add the remaining 2 tablespoons sugar and beat in well. Spoon about a fifth of the beaten egg whites into the yolk mixture with a large spatula. Stir in well to lighten the yolk mixture. Add the rest of the whites to yolk mixture and fold in gently.

5. Spread the batter evenly in the jelly-roll pan and bake until springy to the touch, 6 to 10 minutes. Dust a clean dish towel heavily with confectioners' sugar, invert cake onto the towel, and carefully peel off the foil or liner. Let stand to cool thoroughly.

6. For the whipped cream, place a bowl and beaters in the freezer for 10 minutes to chill well. Place the cream in the freezer for 5 minutes. Whip cream to firm peaks. Stir in confectioners' sugar and 1 tablespoon Grand Marnier.

7. Spread the flavored whipped cream evenly in a thick layer on top of the cooled nut cake. With a serrated knife, slice the cake lengthwise into eight equal strips, about 1½ × 17 inches. The easiest way to do this is to slice the cake in half lengthwise, slice each half in half lengthwise, and then slice each piece in half again.

8. To assemble, take one strip and roll it up toward the whipped cream like a jelly roll. Now turn the spiral over on one side like a cinnamon roll and place it flat in the center of a serving platter. Take another strip and carefully curl it around the first one to make a larger spiral. Continue adding one strip at a time until all eight strips are used. The cake is now a spiral of dark cake and white whipped cream, about 1½ inches high and a little more than 10 inches in diameter.

9. Starting at the outside edge, arrange strawberry slices with the tips of the berries pointing outward, overlapping all the way around the outside edge. Arrange another circle just inside the first so that the inner circle overlaps the first circle about halfway. Continue with another overlapping circle inside the first two, and so on, until the entire cake is covered and looks like a great red flower blossom.

10. Warm the jelly just enough to melt and stir in 1 tablespoon Grand Marnier to thin. Glaze the entire top with jelly glaze. Refrigerate and serve cold.

Notes: *If you do not have superfine sugar, process granulated sugar in the food processor with the steel knife for 1 or 2 minutes.*

Nonstick baking sheets are available by mail order (see Sources—Nonstick sheets, page 477).
Variation: *Peaches, ripe mangoes, or plums may be substituted for the strawberries. Use an appropriate liqueur such as amaretto with peaches or mangoes or Chambord with plums.*

Great Flower Cake

1. Spread whipped cream

2. Cut into 8 strips

3. Roll a strip up like a jelly roll

4. Then place it on its side

5. Add one strip at a time to enlarge cake
   Wrap all 7 strips around the first one

the recipes

sauce sense

there are wonderful tales in French cooking of the accidental ingredient that elevated a sauce to a gastronomic delight. One that I love is of the old chef who hopes and prays for a visit from the famous restaurant reviewer. He knows that his food is the finest. If only the world could hear about him, diners would come to his tiny inn and save him from financial ruin.

It is late on a dreadful cold, rainy night. Alas, there will be no more diners. With sadness, the chef looks at the magnificent dish that he spent so much time and care preparing in hope that the reviewer would come. His beloved cat purrs at his ankles. He caresses the cat. "Tonight, my faithful friend, this creation of my heart will go to you."

As a good chef would do, he personalizes the dish for his cat by adding a little finely minced fresh catnip to the sauce. Just at that moment, a neighbor calls him away to help with an emergency.

A wet and weary traveler arrives at the inn. He tells the chef's wife he knows that it is late. He is so tired and hungry. Is there even a bowl of soup? The wife explains that a neighbor has just called her husband away but she will see if there is anything. She is pleased to find the just prepared hot, lovely dish and serves it to the traveler.

As fate would have it, the traveler is the famous reviewer. He raves over the incredible dish from this tiny inn. He has had this dish prepared by many famous chefs, but there is something—maybe an exotic herb?—in this dish that makes it the most remarkable he has ever had. Gourmets from all over come for this fantastic dish, and the inn is saved.

Unfortunately, in real life there is rarely a magic ingredient that creates a remarkable sauce. In sauce making, as in other cooking, the magic comes from classic techniques and good ingredients.

Taking a new look at some of these time-honored procedures should demystify them so we can make them as fail-safe as possible. We'll start with the simple sauces made of or thickened with purees that are usually low in calories and popular today in fine restaurants. Then it's on to French reduction sauces with their complex, intense flavors and the fine stocks that help make them. Next are the slightly more complicated but quite manageable starch-bound sauces, thickened with flour, cornstarch, or other starches. Our final challenge is mastery of the emulsion sauces—mayonnaise, hollandaise, and béarnaise— and the butter sauces, beurre blanc and beurre rouge.

Purees

A puree is any cooked or raw food that has been finely pulverized or mashed to a thick consistency. Purees may be used to thicken sauces or may stand on their own as a sauce. They may be savory or sweet. The simplest and oldest known sauces were thickened with fine bread crumbs. Even anchovy paste has its roots deep in the past. Small dried fish were ground up and used for both thickening and flavoring.

Just as they are very old, sauces made of or thickened with a puree are also very new. Some of the most in-vogue sauces in fine restaurants today are bright vegetable purees, seasoned well and thinned with a good stock. Chef Jean-Georges Vongerichten of JoJo restaurant in New York is noted for sauces that he prepares with his juicer. He began creating them, he explained, because his guests were asking for no sauce or sauce on the side. He wanted to make sauces that even fat-conscious diners could enjoy. Sauces of vegetable purees are brilliant in color (bright red bell pepper, yellow pepper, or orange carrot), and low in calories. They do require other intense-flavored ingredients, as do most low-fat dishes (see page 185). Nora Pouillon, of Nora's restaurant in Washington, D.C., thickens her vegetable sauces, for example, with pureed roasted shallots and garlic.

Mexican cuisine has many moles thickened with ground chilies, pumpkin seeds, sesame seeds, or such. India's cookery also uses pureed vegetables, as in pureed lentil and chicken dishes, as well as pureed spices, aromatics, chilies, and coconut in curry pastes and sauces.

Typical modern purees used as thickeners are beans mashed and returned to bean soup, pureed vegetables returned to pot roast gravy, pureed cooked rice in soup as in the Shrimp Bisque (page 182), or ground nuts as in Fresh Tomato-Basil Sauce with Roasted Almonds and Garlic (page 255).

Tomato sauces, pestos (Southeast Asian Herb Sauce, page 256), pureed fruit dessert sauces, and pureed vegetable sauces (Seafood with Rockefeller Sauce, page 259) are all purees used as a sauce.

Puree basics

Use a minimum of liquid. If there is too much liquid when pureeing, the food can get out of the way of the blade in the blender or food processor and avoid being pulverized. Add just enough liquid to keep the food pureeing well.

Do not puree seeds with a bitter taste. The chili seeds in Southeast Asian Herb Sauce (page 256) do not give an off taste. However, tomato seeds and some others are bitter when pureed. You can chew a seed to see how it will taste pureed.

Fight for flavor. As explained in Chapter 2, fats dissolve and carry flavors. Since vegetable purees are low in fat, they lack this ability. In pesto the problem is solved by adding the very thing that vegetables lack:

large amounts of fat—olive oil, Parmesan, and nuts. In the Rockefeller Sauce (page 259), cream plays the flavor-carrying fat role. Fire and Ice—Spicy Grilled Chicken Fingers (page 258) is low fat but flavorful because ingredients with intense flavors are used—jalapeños, lemon zest, and ginger—along with a small amount of fat.

Purees as thickeners

Purees are often used to thicken soups. A portion of the soup can be pureed and returned to the soup, or the whole soup can be pureed. Pureed cooked rice or potato is an excellent thickener. Vichyssoise, a puree of potatoes and leeks, is a good example. The low-fat Shrimp Bisque (page 182) shows off rice as a creamy thickener. I think pureed rice is too often overlooked. It makes excellent rich-looking and rich-tasting low-fat creamed vegetable soups.

Pureeing some of the ingredients in a sauce is a flavor-enhancing and easy way to thicken. The following two recipes illustrate how purees thicken two very different pasta sauces, the first with a carrot, the second with almonds, both with a little of the sauce pureed.

Strictly American "Eat Your Veggies" Spaghetti

MAKES 6 TO 8 SERVINGS

As a research chemist, I used to spend many late nights in the lab, where fellow worker Dick Wohl and I were often sustained by a spaghetti dinner we cooked up right on the premises. We boiled the spaghetti in a huge beaker over Bunsen burners while Dick's wife made the sauce in an electric frying pan. She felt that we should have our vegetables, so she started with an abundance of onions, celery, and peppers. The sauce was a little thin in those days. Now I thicken it with a pureed carrot, which cuts the acidity of the tomatoes. (Children love the mild, slightly sweet taste.) It can be prepared with a little meat or a lot of meat according to your taste. Serve with a bold salad or an antipasto platter, crusty bread, and a hearty wine.

what this recipe shows

Adding sweet pureed carrot lends complexity and subtle flavors to the sauce.

Pureeing some of the sauce ingredients makes a flavorful thickener for the sauce.

Cooking the pasta in salted water seasons the pasta internally. Oil, which would coat the pasta and prevent its absorbing the sauce, is omitted.

Stirring the pasta constantly for the first minute or two of cooking prevents the strands from sticking together.

Not rinsing pasta allows any starch left on the surface to help thicken the sauce.

Placing the pasta in the sauce, covering it, and allowing it to stand for a minute or two lets it absorb some of the sauce.

continued

3 tablespoons *and* 3 tablespoons olive oil (6 tablespoons *total*)

2 medium to large onions, chopped (about 3 cups)

2 tablespoons fresh oregano leaves or 1 tablespoon dried

1 teaspoon fennel seeds, crushed

4 ribs celery, chopped (about 2 cups)

1/2 small to medium green bell pepper, chopped

1 small to medium red bell pepper, chopped

2 small hot chilies with seeds, chopped

4 cloves garlic, minced

1 tablespoon light brown sugar

1 teaspoon *and* 1 tablespoon salt (4 teaspoons *total*)

1/2 teaspoon pepper

1 1/2 pounds Italian sausage, cut into 1/2-inch-thick rounds, or ground beef

1 medium to large carrot, scrubbed and quartered lengthwise

1 can (10 1/2 to 14 1/2 ounces) chicken broth

4 cans (14 1/2 ounces each) diced tomatoes

1 1/2 pounds spaghetti

1. Heat 3 tablespoons olive oil in a large skillet, add the onions, oregano, and fennel, and sauté until soft. With a slotted spoon, transfer most of the onions to a large heavy soup pot. Add the celery, peppers, chilies, and 3 tablespoons olive oil to the skillet and cook over medium-high heat for 2 minutes, stirring constantly. Add the garlic, brown sugar, 1 teaspoon salt, and pepper and cook less than 1 minute, stirring constantly.

2. With a slotted spoon, remove the pepper-celery mixture to the soup pot. Add the sausage to the skillet and cook over medium heat, stirring constantly, until the sausage is browned, about 3 minutes. Drain any excess fat. Transfer the sausage mixture to the soup pot. Add the carrot, broth, and tomatoes. Simmer, uncovered, over very low heat for 20 minutes, stirring frequently.

3. Transfer the carrot to a food processor with the steel knife and process about 20 seconds. Add 1/2 cup of the sauce and process for 30 seconds more. Add 1/2 cup more of the sauce and process to a thick puree. Stir puree back into the sauce. Taste and add salt, pepper, oregano, or fennel as needed. Leave the sauce in the pot. You can make the sauce ahead and refrigerate or freeze it. Sauce will keep for 2 to 3 days in the refrigerator or 1 month in the freezer.

4. Meanwhile, bring a large pot of water with 1 tablespoon salt to a boil. Add the spaghetti and stir constantly the first 1 to 2 minutes of cooking. Cook until tender, about 8 minutes. Reheat the sauce so that it is at a boil when pasta is done. Drain the pasta but do not rinse, and place in the pot with the sauce. Toss well and remove from heat. Cover and let stand for 1 minute. Serve immediately.

Fresh Tomato-Basil Sauce with Roasted Almonds and Garlic

MAKES 4 SERVINGS

In the tiny mountaintop village overlooking the Mediterranean, I enjoyed twelve glorious days of sun and pasta when I spoke at a conference in Erice, Sicily. Back home, before I even unpacked my suitcase, I was in the kitchen trying to duplicate a pasta sauce that I had loved—fresh tomatoes and basil slightly thickened with ground almonds. Since I didn't have those incredible Sicilian tomatoes, I added vodka to dissolve and distribute flavors. Serve this with a salad dressed with good olive oil, crusty bread, and a light red or even white wine.

what this recipe shows

Roasting nuts enhances flavor.

Briefly cooking the garlic produces a milder flavor.

Vodka dissolves and spreads alcohol-soluble flavors through the sauce.

Pureeing some of the sauce with the ground almonds provides flavorful thickening.

3/4 cup sliced or slivered almonds

1 tablespoon butter

1/2 teaspoon *and* 1 1/2 teaspoons *and*
   1 tablespoon salt (5 teaspoons *total*)

1 large onion, finely chopped

1 tablespoon fresh thyme or
   2 teaspoons dried

1 large bay leaf

1/2 teaspoon fennel seeds, crushed

1 teaspoon dried red pepper flakes

1/2 teaspoon pepper

3 tablespoons olive oil

3 large cloves garlic, minced

2 cans (14 1/2 ounces each) diced
   tomatoes

3/4 pound dried fettuccine

3 tablespoons vodka

Finely grated zest of 1 lemon

4 large ripe tomatoes, peeled, seeded,
   and coarsely chopped
   (about 2 pounds)

12 basil leaves, cut into slivers

1. Preheat the oven to 350°F (177°C).

2. Spread the almonds on a baking sheet and roast until lightly browned, about 10 minutes. While almonds are still hot, stir in butter and sprinkle with 1/2 teaspoon salt. When cool, place almonds in a food processor with the steel knife or in a blender and process several seconds to grind almost to a paste. Leave in food processor and set aside. Start heating a large pot of water for the pasta.

3. Sauté the onion, thyme, bay leaf, fennel seeds, red pepper, 1 1/2 teaspoons salt, and pepper in olive oil in a very large frying pan or casserole until onion is soft. Add garlic, cook briefly, then add canned tomatoes and simmer on low to medium heat for 10 minutes. Remove and discard the bay leaf.

continued

4. Add ¹/₂ cup of the tomato-onion mixture to the food processor with the almonds and process for several seconds. Add another ¹/₂ cup and process until pureed. Pour almond-tomato mixture back into the frying pan.

5. When the water boils, add 1 tablespoon salt and the fettuccine and stir constantly during the first 2 minutes of cooking. Cook until tender, about 8 minutes.

6. While the pasta is cooking, stir the vodka and lemon zest into the sauce and bring to a boil. When the pasta is ready, add fresh tomatoes and basil slivers to the hot sauce.

7. Drain the pasta, but do not rinse, and place in the pan with the sauce. Toss well. Cover and let stand for 1 minute. Serve immediately.

Purees as sauces

One of the most highly touted, much used (and overused) puree sauces to hit the United States in recent years is pesto. This paste of basil, garlic, and pine nuts was made originally by pulverizing the ingredients in a mortar with some olive oil. Nowadays it is more likely to be made in a food processor; the nuts may be walnuts or almonds or peanuts instead of pine nuts. The basil is often replaced with a combination of basil and parsley or other herbs, even cilantro or spinach, which help the sauce maintain a brighter green color. Tomatoes and hot chilies have found their way into pesto as well.

One problem with pesto is that when basil leaves are cut or mashed they lose their bright green color and turn blackish. Basil also reacts with a compound in flour to form a muddy brown substance, making pesto on pasta look really unattractive. Blanching the basil for over 30 seconds kills the enzyme that enhances this reaction, but some of the fresh basil taste is destroyed in blanching. The water in which the basil is blanched is noticeably green. Fortunately, adding a little lemon juice to the pasta-cooking water helps prevent the reaction between the basil and pasta and limits this discoloration.

Bruce Cost's unusual pesto takes a different route to a brilliant green, and the two recipes that follow illustrate just how vibrant a puree sauce can be.

Southeast Asian Herb Sauce with Soft Noodles

MAKES 4 SERVINGS

Bruce Cost's herb sauce has the brilliant green of new-growth leaves and is a far cry from the muddy color of most pestos. He combines the basil with mint and cilantro, which he prefers to call *coriander*, to give his pesto sauce an almost chartreuse color. I discovered that tossing the basil with lemon juice before chopping seems to aid in retaining the bright color. I altered Bruce's recipe to include this step.

This sauce is spectacular on pure white soft noodles made from soft winter wheat. They are available in Asian markets or in grocery stores as ramen noodles. I frequently use four small packages of the low-fat soft ramen noodles and discard the flavoring packets. If the sauce is too hot for you with two chilies, cut back to one or even a half. I like to serve this with a fruit salad or other fruit dish, like the Fresh Fruit with Ginger (page 319).

what this recipe shows

Adding cilantro and mint keeps the pureed sauce a brighter green than basil alone.

Tossing the basil with lemon juice slows the discoloration reaction of basil and between the basil and pasta.

Including the seeds and veins, which are the hotter parts of the chili, gives heat to this sauce.

1 cup peanut oil

1/2 cup raw peanuts

2 small hot green chilies, with seeds

1 tablespoon chopped fresh ginger

4 cloves garlic

1 1/2 cups basil leaves

3 tablespoons fresh lemon juice

1/4 cup mint

1/4 cup cilantro leaves and stems

1 1/2 teaspoons *and* 1 tablespoon salt
   (1 1/2 tablespoons *total*)

1 teaspoon sugar

3/4 pound soft wheat noodles or 4 small
   packages (3 ounces each) ramen
   noodles

1. Heat the oil in a medium saucepan over medium-high heat until it is almost smoking. Remove the pan from the heat and carefully stir in the peanuts. Let the peanuts cook in the hot oil until golden, then remove with a slotted spoon. Save the oil.

2. Process the peanuts to a rough paste in a food processor with the steel knife. Add the chilies, ginger, and garlic and process to blend. Toss basil and lemon juice together well in a small mixing bowl, then add to the processor. Process with several quick on/off pulses. Add mint and cilantro and several tablespoons reserved oil. Process to puree to a coarse paste. Add 1 1/2 teaspoons salt and sugar and process to blend. Scrape into a bowl. Stir in more reserved oil until you have a thick sauce. You may not need all the oil.

3. Meanwhile bring a large pot of water with 1 tablespoon salt to a boil. Add the noodles, stirring constantly the first minute, and cook until just tender, about 5 minutes. If using packaged noodles with flavoring, discard the flavoring packet. Drain. Serve sauce alongside or pour into the center of the noodles. Do not stir until immediately before serving. Serve warm or at room temperature.

Fire and Ice—Spicy Grilled Chicken Fingers

MAKES 6 SERVINGS

Mild, cooling, snow-white chicken is in vivid contrast to the fiery red sauce. Strips of chicken breast are marinated in half of the sauce, then grilled and served on a bed of greens. The other half of the puree serves as a dipping sauce.

what this recipe shows

Simmering in water before the sugar is added allows the bell peppers to soften since sugar slows or prevents softening of fruits or vegetables.

Simmering the jalapeños in sugar-vinegar water tames their heat.

Adding intense flavor components—peppers, ginger, and lemon zest—enhances low-fat sauce.

Adding a small amount of oil and alcohol (dry sherry) to dissolve and release fat-soluble flavor components in the ingredients enhances taste.

4 red bell peppers, seeded and chopped	Finely grated zest of 1 lemon
4 medium shallots	1/2 teaspoon ground cumin
1 to 1 1/2 cups water	1 teaspoon ground coriander
6 medium-size (2 to 2 1/2 inches) red jalapeños, split lengthwise and seeded (see Notes)	2 teaspoons salt
	2 tablespoon dark sesame oil
1/2 cup sugar	1 tablespoon dry sherry
1/4 cup white vinegar	6 skinless and boneless chicken breast halves
4 slices candied ginger, coarsely chopped	3/4 pound watercress or blanched kale for serving

1. Simmer bell peppers and shallots in a heavy saucepan in just enough water to cover for about 10 minutes, until peppers are beginning to soften. Stir in the jalapeños, sugar, and vinegar and continue simmering over low heat for 5 to 10 minutes. With a slotted spoon or strainer, transfer the pepper-shallot mixture to a blender or a food processor with the steel knife. Set aside the sugar-vinegar liquid.

2. Add the candied ginger, lemon zest, and 1 or 2 tablespoons of the sugar-vinegar liquid to the blender. Puree until smooth. Add the cumin, coriander, salt, sesame oil, and sherry and blend. Thin with the sugar-vinegar water to the consistency desired for a dipping sauce.

3. Cut the chicken breasts at an angle across the grain into 1/2-inch strips and marinate in half of the pepper sauce for at least 1 hour. Reserve the rest of the sauce as dipping sauce.

4. Grill the chicken fingers briefly, until lightly browned on all sides. Or spray a cooling rack with nonstick cooking spray, cover a baking sheet with foil, place rack on sheet and the chicken fingers on the rack, and broil 4 inches from the heat for about 2 minutes. Turn once and baste with more marinade. Broil 2 minutes more. The chicken strips should be moist and tender. Do not overcook. Boil the leftover marinade to use as a sauce if desired (see Notes).

5. Line a platter with watercress or blanched kale. Arrange the chicken around the platter with a small dish of the reserved sauce for dipping in the center. Spoon some of the reserved sauce over the chicken. Serve hot, cold, or at room temperature.

Notes: *If red jalapeños are not available, use green, leave the seeds in to give more flavor to the sugar-vinegar water, then remove the green jalapeños and discard. Just use the flavored water and the red bell peppers to keep the red color.*

If more heat is desired, leave some of the white veins in the peppers.

If you wish to use leftover marinade as a dipping sauce, add 1/3 cup of water and boil it for at least 4 minutes to kill any bacteria from the raw meat.

Seafood with Rockefeller Sauce

MAKES 4 SERVINGS OF OYSTERS/CLAMS, 6 SERVINGS OF SCALLOPS, OR 8 SERVINGS OF FISH, WITH ABOUT 3½ CUPS SAUCE

As red and hot as Fire and Ice—Spicy Grilled Chicken Fingers is, Rockefeller Sauce is green and mellow, ideal for oysters or clams on the half shell, scallops, or fish fillets.

what this recipe shows

Cream smooths any bitterness of the greens.

Parsley and watercress give a more complex, fresh taste than spinach alone.

Frozen spinach has been blanched, which removed some acids for a milder taste.

Using alcohol dissolves and releases flavor components in the ingredients to enhance taste.

2 ribs celery, cut into 1-inch pieces

6 scallions, green parts included, cut into 1-inch pieces

1/2 cup chopped parsley

1 cup watercress leaves

2 packages (10 ounces each) frozen chopped spinach, thawed, well drained, and squeezed against strainer to remove liquid

1 tablespoon butter

1/2 teaspoon cayenne

1 teaspoon salt

1/2 teaspoon black pepper

1 cup heavy or whipping cream

1 teaspoon Worcestershire sauce, preferably Lea & Perrins

1/4 cup anisette or Herbsaint

2 dozen oysters or clams on the half shell, 1½ pounds sea or bay scallops, or 8 fish fillets

continued

1. Process celery, scallions, parsley, and watercress in a food processor with the steel knife just to mince. Add chopped spinach and process to puree.

2. Melt butter in a large skillet over medium heat. Add cayenne, salt, and pepper. Add cream, raise heat to medium-high, and bring to a simmer. Remove from the heat and mix in well the Worcestershire, anisette, and pureed greens.

3. Put the clams or oysters in their bottom shells on a bed of rock salt in pie pans or arrange scallops or fish fillets in a lightly buttered baking dish. Cover seafood with sauce and bake in a preheated 375°F (191°C) oven for 10 minutes, until bubbly.

at a glance
Purees

What to Do	Why
Keep liquid to a minimum while pureeing.	For a fine texture.
Remove bitter-tasting seeds, like tomato seeds, before pureeing.	To avoid bitter, off tastes in sauce.
Pay careful attention to flavor in low-fat purees, adding intense seasonings and/or a small amount of fat or alcohol to dissolve and release fat-soluble flavors in the ingredients.	To avoid a bland sauce.

Reduction sauces

"Reduce! Reduce! Reduce! That is the secret of the great French sauces." So an old saying goes, and how true it is. Many French sauces start by deglazing the pan with wine, adding stock and reducing, then adding cream and reducing again.

Reductions can be anything from a complex sauce with intensely flavored stocks and tremendous richness and depth of flavors to a quick, delicious sauce that any home cook can prepare easily.

Reduction sauces are ideal for a quick sauté, a restaurant technique that's excellent for home cooks too. Restaurant cooking is pretty much confined to dishes that can be prepped ahead or done with great speed, like sautéing, poaching, and steaming. In a quick sauté, meat is sautéed and removed to a dish. Then the flavorful drippings from the sauté pan are deglazed (rinsed loose) with a little wine. Next some stock is added and reduced. At the end, cream is added and the sauce is reduced again. Voilà! A fantastic sauce in minutes. Veal scaloppine and pan-fried steak with red wine sauce are good examples of this method.

At a class for home cooks in Atlanta, Chef Joseph Lageder, formerly of The Ritz-Carlton hotels, demonstrated some excellent reduction sauces. Home cooks may not have the rich homemade stocks that restaurants do, but simple stocks made along with the dish or a poaching liquid can make fine sauces. The chef's reductions were made with simple stocks from bony pieces of the squabs we were preparing. Many cooks automatically do this when they make a little stock from the chicken neck and giblets while a chicken is roasting.

Poaching liquid is another example of a simple stock made during a dish's preparation and then used in the sauce. When fish is poached, some or all of the liquid, which is basically a fish stock, is boiled down to half its volume, then cream is added and reduced until it thickens slightly. The classic French recipe for flounder with spinach is a perfect example.

Flounder Florentine

MAKES 4 SERVINGS

I fix this for my husband when I need to bribe him. White flounder on a bed of bright green, barely wilted spinach is sauced with a classic reduction sauce at its best. This rich sauce with great depth of flavor is very easy to prepare and truly impressive. A rice pilaf makes a good companion. I also usually serve a small grapefruit salad or sometimes grapefruit and avocado wedges with lemon juice or Poppy Seed Dressing (page 286).

what this recipe shows

Intensifying the poaching liquid, containing wine and permeated with flavors from the flounder, and cream by reduction produces marvelous complex flavors.

Frozen spinach has been blanched, which removed some acids. If fresh spinach is heated any longer than to barely wilt, it exudes acids and can have bitter flavor notes.

1 teaspoon butter for pan and foil	2 tablespoons butter
3 medium shallots, chopped	¼ teaspoon *and* ½ teaspoon salt (¾ teaspoon *total*)
4 flounder or sole fillets, skinned (about 1 pound)	⅛ teaspoon freshly grated nutmeg
2 cups dry white wine	⅛ teaspoon *and* ⅛ teaspoon white pepper (¼ teaspoon *total*)
1 pound fresh spinach, large stems removed, washed well and dried, or 2 packages (10 ounces each) frozen leaf spinach, thawed and drained well	2 tablespoons coarse-grain mustard
	2 cups heavy or whipping cream

1. Preheat the oven to 300°F (149°C).

2. Butter a large skillet with an ovenproof handle. Spread shallots evenly on bottom. Lay the flounder on top of the shallots and barely cover with wine. Place a piece of buttered aluminum foil on top of the fish. Heat on medium-high heat for 1 minute, then place in the oven to finish poaching, about 10 minutes. Remove when fish turns white and is just beginning to flake.

3. In another large skillet, sauté the spinach very briefly just to wilt in 2 tablespoons butter. Season with 1/4 teaspoon salt, nutmeg, and 1/8 teaspoon white pepper. Arrange drained spinach around the outside edge of a platter.

4. Carefully remove and drain two flounder fillets with a wide slotted spatula. Place the fillets side by side on the platter inside the spinach ring. Spread fillets with mustard. Carefully lift and drain the other two fillets and place them on top of the mustard-coated fillets. Cover fish with foil again and place platter in warm spot.

5. Place skillet of poaching liquid over medium-high heat and boil to reduce to about 1 cup. Add the cream and continue to reduce until sauce thickens. Add 1/2 teaspoon salt and 1/8 teaspoon white pepper. Pour over fillets and serve immediately.

Reduction is not simply boiling off water—far from it. An intensification of flavors results from the evaporation of water during boiling, of course, but other major changes take place as well. Many of the acids and other substances in food have relatively low boiling points. These compounds evaporate during reduction just as water does. As these low-boiling-point substances are removed, the taste of the reduction changes.

Chemical changes also take place as the sauce gets hot. Heat makes molecules move faster and bounce into each other more often. Heat speeds up chemical change in general. As this happens, some components of the sauce break down, forming different compounds with different tastes. Some of these compounds then combine with others to create totally new compounds with totally new tastes. Big molecules are breaking apart, little molecules are hooking together—all kinds of things are going on.

Caramelization is just such a process and one of the many changes taking place during reduction. In a strict sense, *caramelization* refers to the melting of simple table sugar and its resulting breakdown into 120 or so different compounds (page 423). In a broader sense, however, *caramelization* can also refer to the burning or breaking down of any sugars in foods. The searing of onions until they are deep brown as in Roasted Sweet Onions with Balsamic Vinegar (page 344) can be called "caramelizing onions."

As a sauce becomes more concentrated during reduction, some of it sticks to the pan just above the liquid line. Stuck on the hot pan without liquid to keep the temperature down (water in an open pan cannot get hotter than 212°F [100°C], its normal boiling point), sugars and proteins in the sauce start turning very brown and producing some of those marvelously flavorful compounds that you get in caramelization. When the sauce is stirred or the pan is swirled, liquid is splashed up onto this flavorful caramelized part, and it is redissolved back into the sauce, adding more flavor. Lemon Veal with Cream is an example of a typical quick sauté-reduction sauce.

Lemon Veal with Cream

MAKES 6 TO 8 SERVINGS

This sauce has complex flavors from the pan drippings and stock, enhanced by sweet compounds formed in browning, and all dissolved, carried, and intertwined by the alcohol in the wine and the fat in the cream and frying fats. This is a classic rich reduction sauce, but you can easily convert it to a low-fat *velouté*-type sauce (see Variations). This dish can be prepared with flattened boneless chicken breast or flattened thin slices of pork tenderloin as well as with veal. I like to serve this with a light green vegetable like Roasted Asparagus with Lemon-Chili Oil (page 330) and a salad such as Mixed Greens with Walnuts (page 314).

what this recipe shows

Seasoning both veal and flour ensures full flavor.

Using unbleached flour, with its slightly higher protein content, for coating enhances browning.

Deglazing the pan with wine dissolves some flavor components that water cannot.

Boiling off the wine mellows any sharp alcoholic taste and reduces the possibility that the alcohol will cause the cream to curdle.

Using reduced wine, pan drippings, stock, and cream produces a sauce with rich, complex flavors.

8 thin slices veal (about 1 1/2 pounds) cut for scaloppine

Salt and white pepper for seasoning veal

1/3 cup unbleached flour

1/4 teaspoon salt

1/8 teaspoon white pepper

3 tablespoons vegetable oil

2 tablespoons butter

1 tablespoon fresh thyme or 1 teaspoon dried

1 bay leaf

4 medium shallots, chopped

1/2 cup medium-dry white wine, such as Chenin Blanc, Chablis, or Rhine wine

1 cup strong reduced Chicken Stock (page 267), or 1 cup canned broth with 1 teaspoon instant chicken bouillon stirred in

2 thin slices lemon

1 cup heavy or whipping cream

1 lemon, thinly sliced, for garnish

3 sprigs parsley, chopped, for garnish

3 scallions, chopped, for garnish

1. Season veal lightly with salt and white pepper. Season flour with 1/4 teaspoon salt and 1/8 teaspoon white pepper. Lightly coat veal with seasoned flour, shaking to remove excess. Sauté the veal in oil and butter in a heavy skillet over medium-high heat until barely browned, adding the thyme and bay leaf to pan. Do not overcook; 1 or 2 minutes on each side should be long enough. Remove veal slices to a warm platter.

continued

2. Add the shallots to the skillet and sauté briefly over medium heat. Add the wine and scrape up any stuck particles. Boil wine until reduced to about a third. Add the stock and 2 slices lemon. Continue to reduce to about half original volume. Add cream and continue to reduce to about 1 cup. Remove and discard the bay leaf.

3. Pour the hot sauce over veal slices. Garnish with thin slices of lemon and chopped parsley and scallions.

Variations: *For a lighter sauce, omit the cream and whisk in 1 tablespoon cornstarch stirred into 3 tablespoons stock. Heat, stirring constantly, until sauce thickens. This is now a starch-bound sauce, not a true reduction.*

To serve the meat over pasta or rice, you can expand the sauce by adding ¹/₂ cup more of both stock and cream and thickening with a slurry of 1 tablespoon cornstarch in ¹/₄ cup cold stock—another starch-bound sauce.

There's a story of a French chef who, in the traditional manner, had done a sauté, added stock and wine, reduced and added cream, and was now reducing again. He planned to add a little beurre manié to thicken his sauce, but before he could, he was called away. When he returned, his sauce had thickened without a drop of flour—and nouvelle cuisine flourless sauces had been born.

I doubt that nouvelle cuisine was really spawned in such a manner, but it's certainly true that flourless nouvelle cuisine sauces are simply reduction sauces. When you boil some of the water out of the cream, it thickens nicely, especially if the sauce is mildly acidic from the wine.

Good stocks and drinking-quality wines make the best sauces. Avoid cooking wines, which contain salt and are frequently of poor quality with an acidic off taste.

Basic reduction sauce problems

The sauce separates. Reduction sauces with cream or butter are really emulsions (water and fat blended together) and can have all the problems of typical emulsion sauces (page 284). In reductions, if the sauce is reduced

Remove reduction sauces from the heat and then reheat; water loss from holding them over low heat can cause separation.

and reduced, not enough liquid may remain to go between the fat droplets and maintain the emulsion. Free fat will form around the edges of the sauce and will float on top. You can save the sauce simply by whisking in a little liquid to restore an adequate liquid-to-fat ratio to hold the emulsion.

The cream curdles. If you do not boil the wine or alcoholic beverages in the sauce before adding cream, the cream can curdle. Fortunately, when you reduce the wine, the alcohol evaporates and the other compounds that can cause curdling undergo some chemical changes—then you can add cream and reduce without curdling.

The milk or yogurt curdles. Low-fat dairy products are not suitable for pure reduction sauces. Unlike cream, these products do not have enough fat to coat proteins and interfere with their coagulation. Acid and heat make individual proteins unwind (denature) and then join together (coagulate). Milk and yogurt have high protein contents and will curdle. If you want to use milk or yogurt instead of cream, stir a little starch into the low-fat dairy product; it will give the proteins some protection from curdling (see page 183 for details).

Stirring starch into low-fat dairy products like yogurt helps prevent curdling when heated.

at a glance

Reduction sauces

What to Do	Why
Use stocks and wines that complement or enhance the entree.	The sauce should not overpower the entree.
Use wine that is good enough to drink.	The wine affects the taste of the dish. Cooking wines contain salt and are frequently of poor quality and taste terrible.
Remove excess grease before adding wine or stock.	To avoid a greasy sauce.
Pour wine into the hot pan and boil.	To prevent curdling of cream.
Use cream or butter.	Low-fat dairy products will curdle without starch.
If sauce starts to separate, remove from heat and whisk in a tablespoon of water.	To get a proper liquid/fat balance to hold the emulsion.
Taste and correct seasonings.	Flavors will change as the sauce is reduced.
To hold, remove the sauce from heat and reheat. Do not hold over low heat.	Water loss from evaporation can cause the sauce to separate.

Stocks

The basic stocks used in reduction sauces are:

- Brown stock, made with beef and veal bones and aromatic vegetables that are well browned, plus herbs
- White stock, made with veal bones or veal and chicken bones, aromatic vegetables, and herbs
- Fish stock, made with fish bones and heads, shrimp shells and heads, or other seafood trimmings, along with lemons or wine, and herbs
- Chicken stock, made with chicken bones, aromatic vegetables, and herbs
- Game stocks, made with venison bones, aromatic vegetables, and herbs
- Vegetable stocks, used in vegetarian cooking

Stay away from using bones that have very strong, distinct flavors like lamb or duck unless the stock is intended solely for lamb or duck dishes.

The aromatic vegetables normally used in stocks are onions, leeks, celery, carrots, and occasionally garlic. Parsley, bay, and thyme are the usual herbs; sometimes peppercorns are included. Ripe or over-ripe vegetables are sweeter and have a softer cell structure and therefore contribute the most flavor.

Rigid recipes are not necessary for stocks. A little more of this or that is OK. I want to thank L'Academie de Cuisine for allowing me to use their recipes as the basis for those included here.

The vital factors in making good stocks are:

Use a tall, narrow pot to slow evaporation. Stockpots are tall and narrow to hold a lot of bones and other ingredients with a limited surface area so that you do not have to constantly replace evaporated water.

Start with cold water to extract the most flavor. If you plunge vegetables into hot water, the starches on their surfaces expand and gel. This slows extraction of flavor from the insides of the vegetables. Some stock recipes start with all the ingredients in cold water, while other recipes deliberately start with only the meat and onions in cold water. The other vegetables are added after the stock is boiling to contribute minor flavors.

Simmer only. It is very important to simmer stocks and not permit them to boil until after all the fat has been removed. If you boil a stock vigorously, the fat will emulsify or combine with the liquid and form a cloudy, fatty stock. Instead, you want the fat to remain separate and float to the top so that you can remove every bit of it. There are many times in cooking when you do want to make stable emulsions and get fat and liquid to stay together. This, however, is not one of those times.

Cool properly for safety. The major cause of food poisoning in the United States, the United Kingdom, and many other countries is improper cooling. Bacteria grow most rapidly between 40°F (4.4°C) and 140°F (60°C). The center of a hot stockpot, even when placed in a cold restaurant walk-in cooler, is going to stay in this temperature range for hours. In twelve hours, a single bacterial cell that divides every 15 minutes can produce 281 trillion bacteria!

To prevent bacterial growth, cool stock as quickly as possible. Improvise a cold water bath. Place two bricks or boards several inches apart in a deep sink.

Place the stockpot on the bricks so that water can flow under the pot and fill the sink with cold water. Run a fine stream of cold water constantly through the sink to cool the stock. Stir the stock occasionally. If the sink is large enough and the stockpot is small, you may be able to tilt the pot and run cold water down the outside of the pot to cool the stock more rapidly. Whatever you do, find some way to cool down the stock. Then refrigerate it immediately.

If you must refrigerate the stock without rapidly cooling it, pour the stock into small containers that will cool through more quickly. Dr. Frank Bryan, formerly at the Centers for Disease Control and one of the world's experts on foodborne disease, has what he calls the *4-inch rule*. He says that hot food should not be refrigerated in containers any deeper than 4 inches. Recently he amended that to 3½ inches.

This is not to say you should not refrigerate hot food. Whenever possible, hot food should first be cooled down rapidly through the temperature zone of maximum bacterial growth.

Boil any stock that has been stored before using it. If bubbles appear in a refrigerated stock, discard it immediately.

FREEZING STOCK

There are so many uses for good chicken stock and brown veal stock that I've given recipes to make a large quantity. It's so handy to have these stocks in the freezer. You can freeze in plastic freezer containers, or you can reduce the stock and freeze in ice cube trays. When the cubes are frozen, dump them into a freezer-type zip-top bag for keeping. It's great to be able to grab a cube or two for a sauce. However, if you don't want to store the stocks, you can easily halve the recipes.

Chicken Stock

MAKES ABOUT 7 QUARTS (SEE NOTE)

what this recipe shows

Using a tall pot limits evaporation.

Starting the meat and vegetables in cold water aids in extracting flavor.

Using ripe or overripe vegetables adds flavor since some of their insoluble pectic substances have already changed to soluble pectins.

Keeping stock at a low simmer and skimming it frequently prevents emulsification of the fat.

5 chicken carcasses, including necks, or about 4 pounds chicken necks and backs, no livers

4 onions, quartered

3 carrots, cut into 1-inch slices

2 leeks, white parts only, cut into $\frac{1}{2}$-inch slices

3 ribs celery, leaves included, cut into 1-inch pieces

6 mushroom stems (optional)

$\frac{1}{4}$ white turnip (optional)

8 quarts cold water

10 sprigs fresh thyme

2 bay leaves

10 parsley stems, leaves removed

6 white peppercorns

2 cloves garlic (optional)

2 sprigs fresh rosemary (optional)

1. If there is any blood on the chicken bones, soak them in a large bowl of very cold water for 20 minutes. Place the bones, onions, and carrots in a full 12-quart or larger stockpot. Place the leek slices in a large bowl of cold water and break into rings. Stir to let any dirt settle to the bottom of the bowl, then lift leek slices out and add to stockpot. Add the celery and mushroom stems and turnip if using. Cover with cold water. Bring just to a boil and skim off fat and foam. Reduce heat to simmer. Add thyme, bay leaves, parsley, peppercorns, and garlic and rosemary if using. Simmer, uncovered, for 3 to 5 hours, adding hot water as necessary to keep vegetables covered.

2. Degrease well, first by skimming off fat with a spoon, then a skimmer mop, or by dragging a piece of paper towel over the surface, then strain stock through a fine-mesh strainer into a large saucepan or soup pot. Cool immediately by placing the saucepan in a sink filled with ice water. Store in containers that have lids. Refrigerate immediately, leaving uncovered until completely cold, then seal. Stock will keep for 5 to 7 days in the refrigerator or up to 3 months in the freezer. Boil before using any stored stock.

Note: *Quantities may be halved if desired.*

Brown Veal Stock

MAKES ABOUT 7 QUARTS (SEE NOTE)

what this recipe shows

Roasting bones and onions adds color and intense caramelized flavors to the stock.

Using a tall pot limits evaporation.

Starting the meat and vegetables in cold water aids in extracting flavor.

Using ripe or overripe vegetables adds flavor since some of their insoluble pectic substances have already changed to soluble pectins.

8 pounds veal bones

5 pounds veal shoulder, cut into
   2- to 3-inch pieces

Leftover cooked veal scraps (optional)

4 onions, quartered

3 cloves garlic, halved

1 can (6 ounces) tomato paste

1/2 cup Madeira or red wine

1/2 cup *and* 8 quarts cold water
   (8 quarts plus 1/2 cup *total*)

2 leeks, white parts only, cut into 1-inch
   slices

3 carrots, cut into 1 1/2-inch pieces

2 ribs celery, leaves included, cut into
   1 1/2-inch pieces

1 turnip, cut into 3/4-inch slices

6 sprigs fresh thyme or 1 teaspoon dried

10 parsley stems

3 sprigs fresh rosemary

2 bay leaves

6 black peppercorns

6 cloves (optional)

6 dried juniper berries (optional)

1. Preheat the oven to 400°F (204°C).

2. Roast the bones and meat in a roasting pan with low sides or a jelly-roll pan until brown. Pour off the grease. Add the onions and garlic and continue roasting until onions, meat, and bones are well browned but not burned. Spread tomato paste over the meat and roast a little longer. Place meat, bones, onions, and garlic in a 12-quart or larger stockpot. Pour Madeira into the roasting pan and scrape up any bits stuck to the bottom. Pour into the stockpot. Rinse the pan with 1/2 cup cold water and add to stockpot.

3. Add 8 quarts cold water to stockpot. Bring to a boil, reduce heat, and skim off fat and foam. Place the leek slices in a large bowl of cold water and break into rings, then lift out with a slotted spoon. Add leeks, carrots, celery, turnip, thyme, parsley, rosemary, bay leaves, peppercorns, and cloves and juniper if using. Keep at a very low simmer, uncovered, skimming frequently, for 8 to 12 hours.

4. Strain stock through a fine-mesh strainer into a large saucepan or soup pot. Press meat and vegetables against strainer to squeeze out juices. Degrease, first with a spoon, then with a skimmer mop or by dragging a paper towel across the surface. Return to low heat and slowly reduce to desired consistency. Stock should not be reduced by more than half in 8 hours. Cool immediately by placing the saucepan in a sink filled with ice water. Store in containers that have lids.

Refrigerate immediately, leaving uncovered until completely cold, then seal. Stock will keep for 5 to 7 days in the refrigerator or up to 3 months in the freezer. Boil before using any stored stock.

Note: *Quantities may be halved if desired.*
Variation: *White Veal Stock—Do not roast the bones, meat, onions, and garlic; omit the tomato paste. Simmer for 8 to 12 hours as directed. White veal stock can be clarified for an excellent consommé (page 270).*

Fish Stock

MAKES ABOUT 2½ QUARTS

what this recipe shows

Sautéing the vegetables and bones enhances flavor.

Using both wine and water dissolves more flavors.

Ripe and overripe vegetables are sweeter and have more soluble components for better flavor.

2 onions, cut into ½-inch slices

1 rib celery, leaves included, cut into 1-inch pieces

4 mushroom stems (optional)

3 tablespoons butter

5 pounds fresh fish bones, preferably from flatfish like flounder or sole but not oily fish like salmon or bluefish; or use shrimp shells and heads

2 cups dry white wine

2½ quarts cold water

6 sprigs fresh thyme

10 parsley stems

1 bay leaf

6 white peppercorns

1. Sauté the onions, celery, and mushroom stems in the butter in a large stockpot. Add the bones or shells and sauté over low heat about a minute. Take care not to burn. Add the wine, water, thyme, parsley, bay leaf, and peppercorns. Bring to a boil, turn down to simmer, and simmer for 30 minutes.

2. Strain stock through a fine-mesh strainer into a large saucepan or soup pot. Remove all grease, first with a spoon, then with a skimmer mop or by dragging a paper towel across the surface. Cool immediately by placing the pan in a sink filled with ice water. Store in containers that have lids. Refrigerate immediately, leaving uncovered until completely cold, then seal. Stock will keep for 1 to 2 days in the refrigerator or up to 1 week in the freezer. Boil before using any stored stock.

Some years ago, Francois Dionot, the head of L'Academie de Cuisine in Bethesda, Maryland, prepared a memorable dinner for Julia Child and members of the Board of Directors of the International Association of Culinary Professionals. For the first course, he spooned boiling crystal-clear consommé over thin slices of raw sea scallops in hot soup bowls. The scallops cooked to perfect tenderness in the hot liquid, and the soup was garnished with slivers of carrots and chives. It was a spectacular dish.

Francois later told me that his day had been a nightmare and he had arrived to prepare dinner only minutes before the guests. Having no time to clarify the stock he had made especially for the soup, he grabbed whatever clarified stocks were in the freezer. There was not enough of any one, so he had to combine veal, chicken, and a little beef. Francois said that he, too, remembered it was a nice soup, and from that day on he never hesitated to combine stocks for depth of flavor in some dishes. Whether or not you decide to do the same, you'll have perfectly clarified stocks every time if you use Francois's procedure.

Clarified Stock (Consommé)

MAKES ABOUT 2 QUARTS

This is the procedure for clarifying veal or beef stock. For chicken stock, use ground chicken instead of ground beef.

what this recipe shows

Using ground meat and leeks enhances the flavor of the stock.

As the proteins in the egg white coagulate (cook) they trap fine particles in the stock.

Bubbling the stock through the coagulated egg-meat crust filters out any remaining particles for a very clear consommé.

2½ quarts veal or beef stock	2 leeks, white parts only, sliced
1½ pounds very lean ground beef	4 or 5 large egg whites

1. Pour the stock into a large saucepan and heat over low to medium heat until lukewarm.

2. Stir together the ground beef, leeks, and egg whites in a mixing bowl. Stir the mixture into stock, stirring constantly while heating to a full boil. The mixture will be very cloudy. Turn down to a simmer and stop stirring. On very low heat, simmer for 1 to 1½ hours.

3. Remove the crust from the surface along with any fat. Strain stock through a very-fine-mesh strainer or a strainer lined with a double layer of cheesecloth into a large saucepan. Cool immediately by placing the saucepan in a sink filled with ice water. Store in containers that have lids. Refrigerate immediately, leaving uncovered until completely cold, then seal. Stock will keep for 5 to 7 days in the refrigerator or up to 3 months in the freezer. Boil before using any stored stock.

at a glance
Stock making

What to Do	Why
Use a stockpot.	Stockpots are tall and narrow to slow water loss from evaporation.
Start with cold water.	For maximum extraction of flavor components from the bones, meat, and vegetables.
If there is blood on bones, soak them for 20 minutes.	To remove blood and prevent stock from clouding.
Use ripe or overripe vegetables.	They are sweeter, and many of the insoluble pectic substances that hold cells together have changed to soluble pectins, which dissolve and contribute more flavor.
Simmer only.	Boiling the stock causes the fat and stock to emulsify and make a greasy, cloudy stock.
Skim frequently.	Removing the foam as it forms helps prevent a cloudy stock.
Do not cover stock while cooking.	For slow evaporation and intensification of flavor.
Do not stir stock.	Stirring causes emulsification of fat and stock and makes stock cloudy and greasy.
Cool stock as fast as possible.	Bacteria grow most rapidly between 40°F (44°C) and 140°F (60°C).
Never cover hot stock after removing from the heat.	The space between the surface and the lid will remain hot and permit bacterial growth.

Complexity of taste

Paul Prudhomme says that in the first bite of a dish you should taste the main ingredient—in the first bite of a chicken dish, for instance, you should taste chicken—but, in each additional bite, you should detect something that you did not experience in the bite before. A dish should have such complexity that there are new, exciting taste qualities in every bite.

Some of the ways to achieve this complexity have already been covered—reduction, caramelization, and use of great stocks. Another important technique is layering.

In Seven-Bone Steak Gumbo, adapted here from *Chef Paul Prudhomme's Louisiana Kitchen,* Paul uses several techniques to achieve complexity. Early in the cooking, he sautés some of the okra until it is deeply browned and adds some of the onions. He adds stock a little at a time for semidry cooking, and finally he adds the other vegetables, the meat, the rest of the stock, and the rest of the okra and onions. This layering of ingredients—that is, adding the same ingredient at several different points while cooking—is the secret to excellent soups and gumbos. Some of the vegetables are cooked to a mush to add flavor, and some of the same vegetables can be crisp and bright in color.

Seven-Bone Steak Gumbo

MAKES 8 SERVINGS

This hearty gumbo is a meal in itself. Serve it with good bread and fruit or a salad. At Paul's New Orleans restaurant, K-Paul, along with your gumbo you might have Cajun popcorn (lightly battered, deep-fried crawfish tails) as an appetizer and Cajun martinis made with jalapeño-flavored vodka and served in pint jars. Prepare this recipe a day ahead for excellent flavors, but do not add the shrimp until you reheat.

what this recipe shows

Adding stock a little at a time keeps the mixture from burning but still fairly dry so that the vegetables are cooking and browning on the bottom of the pan at the same time.

Adding some of the okra and onions at the beginning of cooking and some later produces different flavors and different textures for depth of flavor.

2 1/2 pounds 7-bone steak (see Note)

2 tablespoons *and* 1 1/2 teaspoons *and* 2 teaspoons *and* 4 teaspoons Seasoning Mix (page 273)— 4 1/2 tablespoons *total*

1/2 cup bacon drippings or vegetable oil

1/2 cup all-purpose flour

4 cups *and* 4 cups okra, cut into 1/4-inch slices (8 cups *total*, about 2 pounds)

1 cup *and* 2 cups chopped onion (3 cups *total*)

3 tablespoons butter

4 bay leaves

1/2 cup *and* 1/2 cup *and* 1/2 cup *and* 6 cups beef stock or Brown Veal Stock (page 268)—7 1/2 cups *total*

2 cups chopped celery

2 cups chopped green bell pepper

2 cups peeled and chopped tomato

2 tablespoons seeded and chopped jalapeño (1 large or 2 medium)

1 tablespoon minced garlic

3/4 pound medium shrimp, peeled

3 cups hot cooked rice for serving

1. Cut meat into eight equal pieces. Rub 2 tablespoons Seasoning Mix well into all sides of the meat. Heat the bacon drippings in a large heavy skillet. Stir together the flour and 1 1/2 teaspoons of the Seasoning Mix in a plate or pie pan and dredge meat well on all sides. Brown meat and bone in hot fat, transfer to a large soup pot, and set aside.

2. Add 4 cups of the okra to the skillet. Fry, stirring frequently, over high heat until most of okra is dark brown, about 8 minutes. Add 1 cup of the onion, the butter, and 2 teaspoons of the Seasoning Mix. Cover and cook over high heat for 4 minutes, stirring often. Add the bay leaves and 1/2 cup of the stock. Cook for 4 minutes, stirring often. Add another 1/2 cup stock. Cook for 5 minutes, stirring frequently and scraping the bottom well if vegetables start to stick. Add another 1/2 cup stock and cook for 3 minutes, stirring occasionally. Add remaining 2 cups onion, all the celery and bell pepper, and the rest of the Seasoning Mix. Stir well. Cook for 5 minutes, stirring occasionally. Stir in tomato, jalapeño, and garlic.

3. Add to the large soup pot with the browned meat and bone. Add the remaining 6 cups stock. Cover and cook over high heat for 10 minutes. Add the remaining 4 cups okra and lower heat to a simmer. Cook, covered, until meat is tender, about 20 minutes. Stir frequently, scraping the bottom to make sure the gumbo is not sticking. Stir in the shrimp, cover, and remove from heat. Let stand 10 minutes.

4. Serve each person $1/3$ cup rice and a piece of meat, then cover with gumbo. As you serve, watch carefully and remove bay leaves.

Note: *The 7-bone steak is a cut of chuck near the neck. It is tough but has great flavor. It may be labeled chuck roast or some other name, but it can be identified by the large bone that is shaped like a 7 with an extension (or a backward 7, depending on which side you are looking at).*

Seasoning Mix

MAKES ABOUT $1/2$ CUP

2 tablespoons salt

2 teaspoons white pepper

$3/4$ teaspoon black pepper

1 tablespoon sweet paprika

1 teaspoon cayenne

2 teaspoons onion powder

$1 1/4$ teaspoons garlic powder

1 teaspoon dry mustard

$1 1/2$ teaspoons dried thyme leaves

1 teaspoon dried basil

Stir all seasonings together well in a small bowl. Store in a tightly sealed jar.

Starch-bound sauces

Adding starch is another way to thicken sauces. White sauce (cream sauce, béchamel), cheese sauce (Mornay), and meat gravies are all starch-bound sauces. Pastry cream and the filling for Lemon on Lemon on Lemon Meringue Pie in Chapter 3 are also made with starch.

Exactly what are starches? They are simply sugars hooked together. During the photosynthesis process, plants take in carbon dioxide and water and, with the aid of sunlight, make the simple sugar glucose. Starches are hundreds or even thousands of molecules of glucose linked together. This is nature's way of storing sugar for times when plants need it. Most green-leafed plants contain starch in one place or another. The seeds of grains contain starch to supply food for the new plant until it can get its roots down and its leaves up and start making its own food. The roots and tubers of some plants—tapioca and potatoes, for example—contain starch. In others the starch is in the fruit, as in bananas, or in the stem, as in sago.

Starches are the main source of carbohydrate in our diets. But starches also play a more subtle role in cooking. They thicken sauces, soups, and pie fillings, prevent curdling in sauces and custards, stabilize some salad dressings, act as binders in some foods, and form coatings or glazes, which help retain moisture.

How starches work

To the naked eye, starches look like fine powdered solids. They are not cells, and they are not crystals. Scientists call them *granules* and describe them as *semicrystalline* in structure. A starch contains layer on top of layer, like an onion, of tightly packed starch molecules.

Starches normally do not dissolve in cold water. As the water is heated, however, the molecules of starch begin to move more rapidly and their bonds weaken, allowing water to work its way into the softer areas of the starch granules between the harder, crystallike portions. This makes the granules expand or puff. The hotter the water gets, the more the granules absorb and the more they puff. At 90°F (32°C), a starch granule can soak in many times its weight of liquid. This soaking and puffing goes slowly at first, but when the granules are somewhat puffed it becomes easier for the liquid to seep in. Near the end, the granules swell very fast.

Finally, somewhere near the boiling point (a slightly different temperature for each kind of starch), the granules have absorbed enormous amounts of water. With some starches the granules have absorbed over one hundred times their weight in liquid. Finally, the granules pop, starch rushes out into the sauce, and suddenly it thickens. This thickening is caused by both the empty swollen granules and the free starch.

The way this looks when you are cooking is that you are heating and stirring, heating and stirring, and nothing happens. In fact, the granules are puffing, but you can't see this. Since it looks like nothing is happening, you may mix up a little more cornstarch into cold water and add it. Then, all at once, when you get to the temperature when the starch granules pop, you have absolute glue because you have added too much starch.

The secret of working with starch-thickened sauces is to bring them to a gentle boil first. Then decide whether they are thick enough and add more starch in cold water if necessary.

Two kinds of starch

Plants store glucose as starch in two different forms: in long straight chains (amylose) or in small branched shapes (amylopectin).

A plant's starch can contain from 1 to 28 percent of amylose, the long straight chained molecule. How much amylose a starch contains has a major effect on how it behaves in cooking. When the granules pop and starch rushes out into the sauce or filling, these long amylose molecules are more effective thickeners than the small amylopectin molecules. An even greater change takes place as the starch cools. Amylose bonds together to form a firm solid gel, while amylopectin does not.

High-amylose starches set up into an opaque gel that is firm enough to cut with a knife. Starches that are mostly amylopectin form a clear, thick, glossy coating.

You can see that sometimes you will need one kind of starch and sometimes the other. If you are making a coconut cream pie, you definitely need it firm enough to cut, but it doesn't matter if it is opaque. On the other hand, it would be a shame to have a cloudy covering on bright red cherries in a cherry pie.

There are a number of other differences between the two types of starch. Sauces made with a high-amylose starch fare badly when frozen and thawed. They become a dry, spongy mass surrounded by liquid. High-amylopectin starches freeze and thaw beautifully. Amylopectin granules swell and pop at lower temperatures than those of amylose, so their sauces thicken sooner. They thin slightly when cooled or reheated. High-amylopectin starches are also thinned by overheating or overstirring. This deflates the empty swollen granules, which are the main source of thickening for amylopectin. The long chains are also major contributors to thickening in high-amylose starches, so they behave differently.

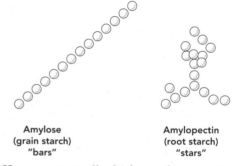

Amylose
(grain starch)
"bars"

Amylopectin
(root starch)
"stars"

How can you tell which starches contain more amylose? Fortunately, Mother Nature has simplified this for us. The ordinary grain starches like wheat and corn are high in amylose, about 26 percent, while the root starches like arrowroot and tapioca contain less, about 17 to 21 percent. Potatoes are a tuber, not a true root, and potato starch falls somewhere in between with about 23 percent amylose. Some types of cereal starches called *waxy* (for example, waxy cornstarch), unlike regular cornstarch, are very high in amylopectin (99 percent). These high-amylopectin starches and modified starches are the mainstay of the frozen food industry.

RETROGRADATION

A sauce made with a high-amylose starch sets into a gel because the amylose molecules rejoin together tightly after they have been heated to their gelation temperature and cooled. This rejoining of amylose is even more dramatic when not much water is present. There is very little free water in long-grain rice, for example. When the amylose that has been heated to gelation temperature cools and dries, it actually forms hard crystals. Think about how hard leftover Chinese take-out rice is the next day. It becomes perfectly soft when you reheat it—you're melting the amylose crystals—but when cold it's as hard as a rock.

This can pose a problem in rice pudding. If you have not cooked the rice for an hour in the old-fashioned way, and if you use long-grain rice, the rice in the rice pudding can become very hard when chilled because of the crystallized amylose. The easy solution is to switch to medium- or short-grain rice, both of which have less amylose than long-grain rice.

This recrystallizing of amylose is also one of the reasons that bread gets hard when it stales (page 93). The technical term for the recrystallization of amylose is *retrogradation*.

at a *glance*

Starches

Grain Starches	Root and Waxy Starches
(High amylose—up to 28%)	*(High amylopectin—up to 99%)*
Transparent at gelation temperature; slightly opaque when cool.	Transparent during gelation; transparent and glossy when cool.
Thickens at gelation temperature and becomes even thicker as it cools and sets into a firm gel that can be molded or sliced.	Becomes thickest at gelation temperature; thins some as it cools.
Thickens at just below the boiling point of water; can be held at this temperature without damage.	Thickens at relatively low temperature (167°F/75°C); may thin if overheated.
Withstands moderate mechanical action (stirring) without thinning while hot; once set into a solid gel, stirring thins it.	Sauce may thin dramatically if stirred vigorously after it reaches gelation temperature.
Reheats without thinning.	Thins when reheated.
Becomes spongy and leaks watery fluid when frozen and thawed.	Freezes and thaws well without change.

Occasionally you'll find yourself in a bind when trying to choose the right starch. In a cherry pie, for instance, you would rather use amylopectin, which remains crystal clear and glossy when cool, to show off the cherries. But you may want to eat this cherry pie hot with ice cream on top, and amylopectin thins when reheated. In an instance like this, you can do what some commercial pie makers do—use part amylopectin (tapioca or arrowroot) for transparency and part moderately-high-amylose starch (cornstarch) to prevent thinning during reheating.

The starch industry does this type of thing in a more systematic manner, offering tailor-made starches for food processing problems. These so-called modified starches, not available for home use, are designed to do particular things in a commercial food product.

Many cooks have limited themselves to flour and cornstarch as thickeners because they're readily available. I encourage you to try some of the many starches available in Asian markets. You may find tapioca in powdered form, pure wheat starch, and arrowroot at a fraction of the cost you'd pay in a supermarket. Also look for potato starch—it has less amylose than corn or wheat starch but more amylose than root starches. It makes a firm, clear gel that has many uses, such as thickening pie filling and making tender, gluten-free cakes.

Three methods for making starch-bound sauces

Each of the three classic methods for thickening sauces with starch—slurry, beurre manié (kneaded butter), and roux—does essentially the same thing: keep sauce granules separated so that they don't form lumps as they disperse in the liquid.

Suppose you have a saucepan full of hot liquid that you want to thicken. You scoop up a spoonful of flour and stir it in. Wham! Instant lumps. The starch granules on the outside of every lump of flour expanded and formed a waterproof gel around each lump as it hit the hot liquid. This tight swollen starch gel is a perfect waterproof wrapper that will keep each lump dry and firm and intact—a lumpy mess.

Since each starch granule absorbs an enormous amount of water and enlarges tremendously, the starch needs to be stirred constantly as it soaks in water. The granules need to be surrounded by a water supply, not packed together in the bottom of the pan, where they will expand and glue themselves together into big lumps. Stirring is important to keep the granules apart while they are puffing. Once the fragile starch granules are completely swollen, though—and this is especially true with amylopectin—overstirring will deflate them and thin the sauce.

Let's see how each of the three classic starch-bound sauce methods keeps the granules separated:

With a slurry, you stir cornstarch into cold water or stock to separate the starch grain by grain. When you are ready to thicken the sauce, stir the slurry again and then stir it into the pan or wok with hot food and hot drippings.

With beurre manié (kneaded butter), you knead together equal parts of flour and butter and then, when you are ready to thicken the sauce, you add a spoonful of this paste. As the butter melts in the hot sauce, it releases the starch grain by grain. The granules separate and are free to soak up the sauce and expand. You can make up a quantity of beurre manié and keep it on hand in the refrigerator.

With a roux, stirring flour into fat coats the starch granules with fat to separate them grain by grain, leaving each granule separate and free to soak up the liquid for the sauce. My grandmother used this method in making milk gravy. She fried chicken in a large iron skillet, poured off all the fat except for a few tablespoons, then stirred in a few tablespoons of flour. She cooked the flour in the fat over low heat, stirring constantly, for a few minutes. She added salt and pepper, then removed the pan from the heat and stirred in milk a little at a time. She placed the pan back on the heat and brought the gravy to a simmer to thicken it. With those good drippings and little browned bits of chicken it was the world's most wonderful milk gravy.

Roux If you've ever tasted flour straight from the package, you've experienced raw cereal taste. When making a roux, you remove this taste by cooking the flour after adding it to the fat or drippings in the pan as my grandmother did. Some chefs insist that you cook the flour for a full 10 minutes; others settle for cooking it for several minutes. Some feel that cooking the sauce after it has thickened does the job. Master teacher Pascal Dionot at L'Academie de Cuisine feels that subsequent cooking of the thickened sauce removes the raw taste and improves the flavor. In *Food Science,* Helen Charley says, "Flavor of gelatinized starch is improved by cooking the paste an additional five minutes on direct heat." Cooking either the roux or the thickened sauce or both is effective.

One of the old rules of cooking is to add cold liquid to a hot roux and hot liquid to a cold roux. If one or the other is cold, there is enough time to whisk well and distribute the starch before the sauce gets back up to the gelation temperature. If you let the starch get to its maximum swelling stage while the granules are still packed together, you can have a colossal lumpy mess.

Of course, rules are made to be broken—many cooks add a hot liquid to a hot roux. For instance, Giuliano Bugialli, the Italian cooking teacher and author, makes a balsamella (béchamel) by pouring hot liquid into hot roux and whisking vigorously. He gets a lovely thick cream sauce instantly. This makes a lot of sense. If the liquid you add to a roux is hot, you do not have to spend a long time heating the sauce to the temperature at which the starch granules pop and the sauce thickens. With both the roux and the liquid hot, the thickening takes place right away. The trick is to pour in the hot liquid and whisk like crazy for a few seconds. I have done this myself many times. It works fine when the amounts are not too large to whisk together rapidly (usually 4 or 5 cups of liquid maximum).

In a roux, the flour may be cooked over low heat so that it does not change color at all, or it may be cooked over higher heat so that it darkens. I was always taught that there were three basic roux—light, medium, and dark—but Paul Prudhomme uses seventeen! The youngest of twelve children, Paul was brought up in the bayou, where much of the time the family's food was what they could grow or catch. Often the only way that his mother could change the flavor of a dish was to change how she cooked the flour. The old French cooking methods of the Cajuns (Acadians) indeed make good use of roux.

Cajun cooks are experts in the subtleties of roux. For example, they know that with a dark roux you sacrifice thickening for flavor. The darker the roux, the less thickening capacity it will have. Paul Prudhomme once asked me why. To make his dark roux, Paul heats a heavy iron skillet until very hot, then adds oil and flour and stirs like mad. When the starch granules in the flour hit that hot oil, they are seared and scorched almost instantly by the high heat. The flour turns very dark brown, almost black. Much of the starch is broken down into dextrins (sugarlike compounds) and burned. This does not leave many intact starch granules to soak up moisture and swell as they normally do.

In recognition of these differences, some recipes

TAMING CAJUN NAPALM

A hot skillet, hot oil, and hot flour—that's an explosive mixture that literally erupts out of the skillet with the addition of a small amount of liquid. No wonder dark roux is sometimes called *Cajun napalm.* Obviously you must take great care in adding ingredients to a hot roux. The secret is to add a lot in a hurry. Stand back, position the dish with ingredients to be added as close to the hot pan as you can (pouring ingredients from a distance will spatter more), and dump in the ingredients all at once. A good quantity of cold or room-temperature food added all at once will lower the temperature of the pan and prevent splattering.

from New Orleans use both a dark roux for flavor and a light roux for thickening. If you get the flavor in the roux, says Paul, you'll have it throughout the dish. So for a good gumbo you may start with a dark roux for flavor and, after you add other ingredients and cook, add a light roux to thicken.

The darker the roux, the less its thickening power.

Slurry, roux, beurre manié—which method to choose? The way a dish is prepared frequently dictates how to thicken the sauce. With fried chicken or sautéed veal, you want to take advantage of the pan drippings, which contain the juices that came out of the meat while it was cooking as well as the little caramelized bits that broke off during cooking. Stirring in flour to make a roux is not only one of the fastest ways to create a sauce but may be the most flavorful—especially if you brown the roux a bit.

With poached fish, on the other hand, you end up with a pan of reduced liquid. To thicken this with a roux, you would have to use another pan. For a lot of sauce, a quicker solution would be to use a slurry. For a small amount of intensely flavorful sauce, as for Flounder Florentine (page 261), you could continue to reduce the liquid to make a reduction sauce.

Scientists have determined the exact temperature at which each of the different starches thickens sauces. Unfortunately, they haven't fully covered all the things that make starch sauces break (separate) or fail altogether. Which method of sauce making—slurry, kneaded butter, or roux—gives the sauce that's least likely to break?

Fillets en Papillote is one of my favorite dishes to prepare for company. I prepare a cream sauce with shrimp and crabmeat, then cool the sauce completely. A couple of hours before the party, I spoon the cold sauce onto fish fillets in individual parchment packages. I put the packets on a baking sheet back in the refrigerator. When I'm ready to serve, I place the baking sheet in the oven and heat the packets through thoroughly to cook the fish.

At first I thickened the sauce with a slurry and made it with heavy cream. This worked fine for me because I was careful about my reheating time. One of my students, though, told me that she had forgotten and left the fish packets in the oven a little too long and the sauce broke. The dish still tasted good, but it was not as pretty.

Because I have also had rich cream sauces break from being heated too long, I set about to make this recipe more fail-safe. Knowing that a high fat content has a major influence on how easily a cream sauce will break, I changed the heavy cream to part milk and part cream. You can boil a white sauce made with plain milk until it gets so thick that you can hardly stir it and it will not break; given the same treatment, a sauce made with heavy cream will break right away.

I also wondered about the method. Would the sauce hold up better if made with a slurry, with beurre manié, or with a roux? I prepared the sauce all three ways, made several marked packets containing each sauce, and baked them. The sauces with the slurry and the beurre manié broke first. The sauce made with roux held up much better.

Cooking the flour a bit in the roux causes some changes in the flour that help the sauce stand up better. The enzyme alpha amylase, which breaks down starch gels, is in flour as well as egg yolks. Helen Charley, writing in *Food Science,* says, "When wheat starch (flour) is used as a thickening agent, the sauce will be thicker if the flour is put into hot fat rather than dispersed in cold liquid. Heat inactivates alpha-amylase in flour, which would otherwise bring about some hydrolysis [breakdown] of the starch."

My experience reinforces Helen Charley's statement—you do have a more stable sauce when it is made with a roux.

Fillets en Papillote

MAKES 4 SERVINGS

This impressive company dish is a complete do-ahead that allows me to enjoy the party too. Browned parchment packets contain fish fillets topped with crab and shrimp in an herbed cream sauce. You can cut around the top edges of the parchment but leave it covering the packet. Show your guests how to roll back the top with their fork and let the wonderful aromas waft up to them. I usually serve this with Sizzling Broiled Tomatoes with Herbs (page 359), which can be done ahead and reheated with the fish, and a green vegetable like broccoli with Chili Oil or Roasted Asparagus with Lemon-Chili Oil (page 328 or 330).

what this recipe shows

Using a roux for the cream sauce makes a slightly more stable sauce.

Using some milk instead of all cream results in a more stable sauce that can withstand reheating.

Buttering both sides of the parchment makes the packets brown more for a handsome effect.

2 tablespoons melted butter for
    parchment

4 scallions, green parts included,
    chopped

2 medium shallots, chopped

3/4 teaspoon fresh thyme or 1/4 teaspoon
    dried

1 bay leaf

3 tablespoons *and* 2 tablespoons butter
    (5 tablespoons *total*)

3 tablespoons all-purpose flour

1/2 cup dry white wine

3/4 cup heavy or whipping cream
    (see Notes)

1/4 cup milk

6 ounces crabmeat, picked free of shell

1 cup diced peeled shrimp

4 (about 1 pound) thin mild fish fillets
    (such as flounder, sole, orange roughy,
    or tilapia)

1. Preheat the oven to 400°F (204°C). Cut four 8 × 11-inch pieces of parchment and brush on both sides with melted butter.

2. Briefly sauté chopped scallions, shallots, thyme, and bay leaf in 3 tablespoons butter in a large skillet. Add flour and cook over low heat, stirring constantly, for several minutes. Remove from heat. Remove and discard bay leaf. Whisk in wine a little at a time and place back on the heat for a minute, whisking constantly. Whisk in the cream and milk. Simmer, stirring constantly, until thick.

3. Sauté the crab and shrimp in 2 tablespoons butter in a medium skillet for about 1 minute (they won't be fully cooked). Stir the crab and shrimp into the cream sauce.

continued

4. Place a fish fillet on each piece of parchment and top with sauce. Fold parchment over and tightly roll up edges to seal. Bake for about 8 minutes to cook through the fish and lightly brown parchment (see Notes).

Notes: *This is a typical French cream sauce. A lower-fat version can be made with all milk.*

If you prepare the dish several hours ahead, cool the sauce completely before spooning it over the fillets and refrigerating. Increase cooking time to 10 minutes.

How much starch to use

In general, the amounts of flour per cup of liquid to use for a sauce are as follows:

Thin sauce: 1 tablespoon flour per cup

Medium sauce: 2 tablespoons per cup

Thick sauce: 3 tablespoons per cup

Other starches are completely different. For example, you need only a little over half as much cornstarch to achieve about the same thickness.

Other comparisons, which I found in the technical literature, were all related by weight to cornstarch. I purchased five types of starch, plus flour, and tried the amounts suggested. My amounts varied greatly from the literature. Here is a table of approximate amounts needed to thicken 1 cup of water to medium thickness that I developed. In James Beard's *Theory and Practice of Good Cooking,* he suggests essentially the same amounts of all the starches except for arrowroot, where his amount is half mine. In *Sauces,* James Peterson gives almost identical amounts for arrowroot and cornstarch, which agrees with my findings.

There was a tremendous difference in clarity among these gels. Flour made an opaque cream-colored gel, rice starch gave a very white gel, and cornstarch was cloudy but not as opaque as the other two. The gels from arrowroot, potato starch, and tapioca starch were clear.

APPROXIMATE AMOUNT OF STARCH NEEDED TO THICKEN 1 CUP OF LIQUID	
Starch	Amount for 1 Cup Medium Sauce
Grain Starches (Opaque Gel)	
Cornstarch	1 tablespoon + 1 teaspoon
Flour	2 tablespoons
Rice starch	1 tablespoon + $\frac{1}{2}$ teaspoon
Root and Tuber Starches (Clear Gel)	
Arrowroot	1 tablespoon + 1 teaspoon
Potato starch	$2\frac{1}{4}$ teaspoons
Tapioca starch (Asian market)	1 tablespoon + 1 teaspoon
Quick tapioca pudding (supermarket)	1 tablespoon + $\frac{1}{4}$ teaspoon

One day Rose Beranbaum, a perfectionist who keeps exact records and measurements, called and said, "Shirley, this dessert sauce recipe I'm working on now requires twice as much arrowroot as it did when I developed it a year ago. Is this possible?"

The answer was yes. We don't ordinarily think of things like cornstarch, arrowroot, and flour as losing their strength, and, in fact, these starches do have reasonably long shelf life. But Dr. Carl Hoseney, an expert in starch, confirmed that starches exposed to air over long periods of time combine with oxygen and lose some of their thickening ability. When storing starches, remember to keep them very dry in an airtight container. If you are using cornstarch that has been on the shelf for a couple of years, you may need to use more of it to thicken than the recipe indicates.

Differences among manufacturers can cause variations in products and therefore affect the thickness of a sauce, too. So can the specific type of starch. The flour in my comparison test was a low-protein all-purpose flour with a high starch content, but you would need more of a high-protein flour like bread flour. Also, note that the tapioca starch in the table is the starch powder sold in Asian markets. The quick tapioca pudding is the fine-textured type available in supermarkets, not the one with coarse granules.

Other ingredients in starch sauces—salt, sugar, and acid Other ingredients in starch sauces affect the thickness, too. Salt raises the temperature at which sauces thicken. In large amounts, acids (like lemon juice or vinegar) and sugar can prevent starches from swelling and gelling.

When dissolved in water, many natural compounds break into positive and negative particles called *ions*. All salts—table salt, sea salt (which is a blend of many salts), and baking soda, to name a few—do this. Any salt will slightly raise the temperature at which sauces thicken. Even the small amount of salt in hard tap water can have a slight effect on starch-bound sauces.

Sugar has a greater influence than salt. Too much sugar can prevent a lemon pie from thickening enough to set. Sugar ties up water and deprives the starch of the water it needs to expand and burst. Some commercial processors add only part of the sugar and all of the starch along with most of the water and heat until the starch thickens. Then they add the rest of the sugar dissolved in the rest of the water to ensure that the starch has gotten enough water to puff and thicken well. You can do the same thing at home if you have a pie with a high sugar content.

Some sugars have a greater effect on thickening than others. Table sugar (sucrose) and the sugar in milk (lactose) are two that have the greatest effect on starch sauces and gels. This means you may have trouble thickening a starch sauce or custard that uses evaporated milk, which has a high lactose content.

> Large amounts of lactose can inhibit thickening—avoid evaporated milk in custards.

Sugar also raises the temperature at which sauces thicken and makes the fragile swollen starch granules more likely to deflate when stirred. When that happens, the sauce thins. In a gel like a pie filling, sugar produces a more tender gel and improves clarity and gloss.

Acids like lemon juice and other fruit juices have an even greater influence on sauces than sugar. They reduce the thickness of starch sauces and gels by causing the starch granules to disintegrate too soon, preventing the starch mixture from forming the tangled network of molecules that makes the mixture thick. Adding acids at the end of the cooking period minimizes the changes they cause to the starch and thus minimizes thinning. This is why the lemon juice is usually stirred in after a lemon pie filling has thickened but while it is still hot. If you let it cool and set completely, you will permanently wreck its structure and thickness by stirring.

When making a filling for a lemon meringue pie, then, wait until after you have heated the starch and thickened the filling before adding the lemon juice. If you add too much of it before thickening, the filling may not gel.

Other problems with starch Here are three other problems to face when thickening with starch:

1. *Starches will not thicken a strongly acidic sauce.* I once overheated a beurre blanc, a highly acidic sauce made with wine vinegar and butter (page 297). The sauce broke, and I thought I would thicken it with a little starch. No way. No matter how much starch I added and heated to a boil, the sauce simply would not thicken.

    When a starch slurry or beurre manié is added to an acidic sauce, the unswollen starch is immediately exposed to acid and loses most of its thickening ability. It may be possible to achieve some thickening by adding a very thick sauce—made either from a roux or a slurry—in which the starch is already cooked and swollen.

2. *Once a starch gel has cooled completely and totally set, stirring breaks the starch linkages.* Again, that's why you stir the lemon juice into pie filling while the filling is still hot. With pastry cream (page 217), you can wait until the custard cools a little before adding the flavoring—vanilla, Grand Marnier, or whatever. To prevent thinning, though, the flavoring must be stirred in while the custard is still hot.

3. *A skin will form on top of custards or sauces containing milk or dairy products when they cool.* This skin, the same as you get on hot chocolate, is casein, a protein found in milk, which has dried out on the surface. To prevent surface drying, dot butter on the hot surface; it will melt and form a thin glaze of fat. Or cover the sauce with plastic wrap with the wrap touching the surface.

Classic starch-bound sauces

The two classic starch-bound sauces of French cookery are béchamel, made from a white roux and flavored milk, and velouté, made from a white roux and a white stock. More precisely, when cream is added to a béchamel it becomes cream sauce, though in modern usage béchamel is known as *white sauce* or *cream sauce* whether it has cream or not. When cheese is added to a béchamel, it becomes a cheese sauce or sauce Mornay. There are dozens of other variations on both béchamel and velouté.

Add a little cheese Stirring cheese into hot liquids is tricky—the cheese can curdle or become stringy—unless starch comes to the rescue. As long as there is starch in the sauce—as with béchamel or velouté—and the cheese is stirred in after removing the sauce from the heat, the cheese should not curdle. But if the sauce is stirred vigorously after the cheese is added, particularly Swiss, it still may become stringy. In *The Way to Cook,* Julia Child says that stringiness often can be overcome by using James Beard's recommendation to stir in a little lemon juice or dry white wine while heating over low heat with frequent stirring until smooth.

Dr. Anthony Blake, director of food science and technology for the international flavor and fragrance company Firmenich SA, lives in Switzerland and is a lover of fondues. He explained to me that wine is usually acidic enough to prevent stringiness but that the citric acid in lemon juice is even more effective. It actually combines with the calcium in the cheese. In the mozzarella sauce for Fettuccine with Mozzarella, Mushrooms, and Tomatoes (page 283) I had a real problem and used both lemon juice and starch for a smooth sauce.

White Sauce (Béchamel)

For 2 cups of sauce, heat just to a simmer in a medium saucepan 2 cups milk, 1 small bay leaf, the leafy end of 1 celery rib, 1 small onion, quartered, with 1 clove stuck into one of the quarters, and 2 crushed peppercorns. Remove from the heat and let stand for 5 minutes. To make the roux, melt 4 tablespoons ($1/2$ stick) butter in a medium saucepan. Add $1/4$ cup flour and cook over low heat, stirring constantly, for 2 to 3 minutes. Do not let brown. Remove from the heat. Strain the warm milk into the roux and whisk vigorously. Place back over medium heat and cook, stirring constantly. When sauce comes to a low simmer, turn the heat very low and cook at least 5 minutes, preferably 10, stirring frequently. This makes a medium-thick sauce. For a thin sauce, use only 2 tablespoons each flour and butter; for a thick sauce, use 6 tablespoons each. Season with salt and white pepper.

For cream sauce, replace 1 cup of milk with cream or add $^3/_4$ cup heavy or whipping cream and simmer for 10 minutes longer, stirring frequently.

For cheese sauce, after the sauce is removed from the heat, add $^1/_2$ cup grated cheddar, Monterey Jack, Gruyère or other Swiss, 2 tablespoons lemon juice or dry white wine, and $^1/_4$ teaspoon each cayenne, dry mustard, and freshly grated nutmeg. Stir just to blend.

For sauce Mornay, after the sauce is removed from the heat, add $^1/_4$ cup grated Gruyère, $^1/_4$ cup grated Parmesan, and $^1/_4$ teaspoon each cayenne and freshly grated nutmeg. Stir just to blend.

Fettuccine with Mozzarella, Mushrooms, and Tomatoes

MAKES 6 SERVINGS

In Sicily, restaurants serve excellent fettuccine with a nonstringy mozzarella sauce. They have the advantage of very fresh mozzarella. To get a nonstringy sauce with store-bought mozzarella I had to pull every science trick I could, including starch and lemon juice. This is a delicious pasta and very fresh tasting with just-warmed tomatoes.

what this recipe shows

Using both dried and fresh mushrooms imparts a distinct mushroom taste.

Flour in the sauce prevents the cheese from curdling.

Lemon juice, containing citric acid, cuts the typical stringiness of mozzarella for a smooth sauce.

The tomatoes are not cooked but simply warmed by the hot sauce for a fresh tomato taste.

6 tablespoons dried mushrooms
   (cèpes, porcini)

$^1/_3$ cup hot water

1 tablespoon *and* $^1/_2$ teaspoon salt
   ($3^1/_2$ teaspoons *total*)

2 gallons water

2 tablespoons butter

3 tablespoons flour

$^1/_8$ teaspoon white pepper

1 cup heavy or whipping cream

1 cup milk

2 tablespoons lemon juice

$^3/_4$ cup grated mozzarella

$^1/_2$ teaspoon red pepper flakes

$^1/_4$ cup olive oil

$^1/_2$ pound fresh mushrooms, sliced

$^1/_2$ cup country ham, finely chopped

4 medium tomatoes (about 2 pounds),
   peeled, seeded, and coarsely chopped

18 ounces fettuccine

continued

1. Soak the dried mushrooms in the hot water in a small bowl.

2. Add 1 tablespoon salt to the 2 gallons water in a large pot and put over high heat to boil.

3. While the water is heating, melt the butter in a large skillet over low heat. Stir in the flour, $^1/_2$ teaspoon salt, and pepper and simmer over low heat for 2 minutes. Remove from the heat and whisk in the cream a little at a time, then the milk. Heat over medium heat, stirring constantly, until smooth. Sprinkle the lemon juice over the mozzarella and whisk in.

4. Stir the pepper flakes into the olive oil in another large skillet over medium heat. Stir in the sliced mushrooms. Sauté briefly, add the tomatoes, cook less than a minute, and remove skillet from the heat. Lift the dried mushrooms out of the bowl with a slotted spoon and stir into the skillet. Pour all but the last tablespoon or so of the soaking liquid into the skillet. Take care not to pour in any sand. Stir the mushroom mixture, ham, and tomatoes into the mozzarella sauce.

5. Stir the fettuccine into the boiling water and continue stirring for the first 2 minutes of cooking. Cook for the time recommended on the package. Reheat the sauce. Drain the pasta and stir into the skillet with the sauce. Cover and let stand for 1 minute. Serve immediately.

Velouté sauces are enjoying something of a revival because they contain fat-free stock and are lower in calories and fat than cream reduction sauces. Since few households have homemade stock, true veloutés are confined mostly to restaurant cooking. Smothered Chicken with Ham and Almonds (page 417) is an example of a modern velouté using canned broth.

Sauce Velouté

For 2 cups of sauce, bring 2 cups flavorful reduced stock to a boil in a medium saucepan over medium heat. Remove from the heat and set aside. Melt 4 tablespoons (½ stick) butter in a medium saucepan over low heat. Whisk in ¼ cup flour and cook 2 to 3 minutes, whisking constantly. Remove from the heat. Add hot stock and whisk vigorously. Return to medium heat and cook, whisking constantly, until sauce thickens. Turn heat to low and cook 30 minutes, stirring frequently.

Emulsion sauces

Years ago, a friend called me in a state of great excitement: "Shirley, I just had the best sauce I ever tasted. I wrote down the name and hope that you can teach me how to make it!" He spelled out H-O-L-L-A-N-D-A-I-S-E. The first time you have a good hollandaise over fresh asparagus certainly can be a memorable occasion. A fresh fish fillet with a pale golden beurre blanc is something you won't easily forget either.

In addition to hollandaise and beurre blanc, emulsion sauces include mayonnaise, béarnaise, and beurre rouge. Egg-and-cream liaisons for soups and stews are also emulsion sauces, as are some salad dressings. There are many other emulsions in food: cake batter, butter, milk, pâte à choux (page 229), and so on.

An emulsion is a combination of two liquids that ordinarily do not go together, such as oil and water. In foods, an emulsion is simply liquid fat and a water-type liquid held together. Actually, three components are necessary for a stable emulsion: (1) one liquid broken into millions of tiny droplets, (2) another liquid that stays around and between all the droplets, and (3) an emulsifier that keeps the droplets from joining with each other to form big drops that form a layer and separate out. Two other factors—(4) a thick starting base and (5) temperature—are also important. Let's look at them more closely.

1. *One liquid in tiny droplets.* Mechanical action—beating or shaking—is needed to break one of the liquids into millions of droplets. In an emulsion like mayonnaise, the liquid in tiny droplets is the oil. It has to be added slowly at first. When you start to drizzle in the oil, there is a fair amount of liquid for the oil drops to move around in and avoid the wires of the whisk or the blades of the blender. It is relatively easy for the oil to avoid being broken into fine droplets. Later, when millions of oil droplets are packed tightly into the liquid, the oil that is added is held in place and cannot scoot out of the way of the blade.

2. *Another liquid between the droplets.* With the other liquid, the main concern is that there be enough. The second liquid in mayonnaise is the water in the lemon juice, the water in the egg or egg yolk, plus a little vinegar in the prepared mustard if that's used. That's not much water. I was once asked, "Why is it that whole-egg mayonnaise recipes always work, but egg-yolk recipes frequently fail?"

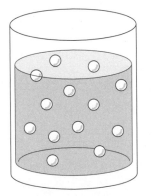

Plenty of Water Between Oil Drops

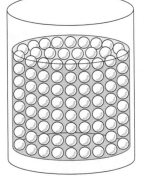

Not Enough Water Between Oil Drops

Think about the proportions in mayonnaise. Whole-egg mayonnaise contains only 3 tablespoons of water-type liquid—the liquid between the droplets—1 tablespoon from lemon juice and 2 tablespoons from egg white. Even that little bit of water can hold 1 to 2 cups of oil. If you continue to add oil, however, so many droplets will be packed in that there is no longer enough water to go between them. The oil will start to separate out of the mayonnaise.

With egg-yolk-only mayonnaise, there can be a real shortage of water to go between the droplets. Make sure that any egg-yolk mayonnaise recipe you are following has a bare minimum of 2 teaspoons of water-type liquid per cup of oil. Most mayonnaises have 1½ to 3 tablespoons of liquid per cup of oil. After part, say a third to half, of the oil is added and the mayonnaise is thick, add sufficient water-type liquid to get this up to a minimum of 2 teaspoons per cup of oil.

3. *The emulsifier.* Many foods act as emulsifiers. Egg yolk is the superemulsifier, but egg white, gelatin, and skim milk can also serve in this role. These all have molecules with a water-loving end and an oil-loving end. They are sometimes called *chemical emulsifiers.* In mayonnaise, the oil-loving ends of the emulsifiers in egg yolk dissolve in the oil droplets and coat each droplet. Chemical emulsifiers are very effective at keeping droplets apart.

    Fine powders can also act as an emulsifier. The particles coat the droplets and help keep them apart. The powders in the Poppy Seed Dressing on page 286 hold the emulsion pretty effectively. The mustard in mayonnaise, as dry mustard or prepared mustard, helps form and stabilize that emulsion. Powders are not as powerful as chemical emulsifiers, but they do help.

4. *A thick base.* For thick emulsion sauces, a thick beginning base is needed. How this thick base works is demonstrated by mayonnaise (page 289).

5. *The temperature.* Emulsion sauces all have proteins in different forms and amounts. When you get

proteins too hot, they coagulate. This is a permanent change—scrambled eggs cannot be unscrambled. Butter sauces have no egg, but do have milk proteins, which can also permanently coagulate.

With mayonnaise, any temperature in the moderate range is fine. With hollandaise, béarnaise, and the butter sauces, however, you need to exercise strict control. You need two liquids to form an emulsion. To make butter a liquid, you have to get the sauce hot enough to melt the butter but not so hot as to coagulate its proteins. Similarly, you need a temperature high enough to unwind the egg yolk proteins to provide part of the thickening, but not so high that the proteins join tightly together to form scrambled eggs. Another tightrope for the cook to walk.

Vinaigrette and other salad dressings

Vinaigrette, a simple mixture of oil and vinegar, is a short-lived emulsion that usually has to be whisked or reshaken just before being poured. Only a fine powder (like mustard) is available to act as emulsifier—no egg yolk or other chemical emulsifier. The more fine particles a vinaigrette or salad dressing contains, though, the better its chance of holding together. In Mixed Greens with Walnuts (page 314) the ground-up walnuts hold the dressing together better than a plain vinaigrette, but still the dressing has to be stirred or shaken just before being used.

Some salad dressings do contain enough fine particles to hold the emulsion together, as the following recipe shows. If the oil is drizzled in very slowly, the dry mustard, paprika, and confectioners' sugar provide enough particles to hold the emulsion together.

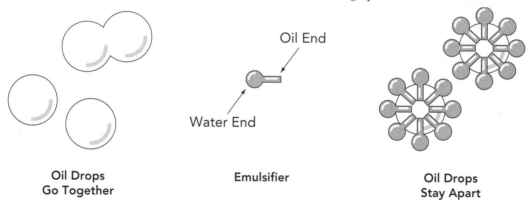

Oil End

Water End

**Oil Drops
Go Together**

Emulsifier

**Oil Drops
Stay Apart**

Fresh Fruit Flower with Poppy Seed Dressing

MAKES 6 SERVINGS

Dramatic orange, red, yellow, and cool green fresh fruit make this an eye-catching dish. Poppy seed dressings have been around for years, but some old classics are hard to beat. There is nothing like cold fruit in the summer, and this tangy, sweet dressing sets it off perfectly. The dish is a cooling accompaniment to spicy foods.

> **what this recipe shows**

Fine particles like confectioners' sugar, dry mustard, and paprika coat tiny oil drops as if they had been dusted with a powder puff, stabilizing the emulsion.

Drizzling in the oil very slowly helps this emulsion hold.

POPPY SEED DRESSING

$^1\!/_2$ cup confectioners' sugar

3 tablespoons cider vinegar

$^1\!/_2$ teaspoon salt

1 teaspoon dry mustard

$^1\!/_2$ teaspoon hot paprika

$^1\!/_8$ teaspoon cayenne

$^3\!/_4$ cup vegetable oil

1 teaspoon poppy seeds

FRUIT FLOWER

1 head Boston or Bibb lettuce

2 ripe papayas, peeled, halved, and cut crosswise into $^1\!/_4$-inch slices

2 ripe avocados, peeled, sliced lengthwise, and rubbed with $^1\!/_2$ lemon

4 oranges, sectioned

2 grapefruit, sectioned

1 pint fresh strawberries, rinsed and hulled

Sprigs of fresh mint for garnish

1. Place the sugar, vinegar, salt, dry mustard, paprika, and cayenne in a blender or food processor. With the machine running, very slowly drizzle in the oil. Stir in poppy seeds by hand. Set aside.

2. Arrange lettuce on a serving platter or on individual salad plates. Arrange an outer circle of papaya slices. Cut long avocado slices in half and arrange avocado slices in a circle, slightly overlapping the papaya circle. Arrange orange sections in a circle slightly overlapping avocado slices. Arrange grapefruit sections in a circle slightly overlapping orange sections. Pile strawberries in the center. When ready to serve, drizzle poppy seed dressing over all. Garnish with sprigs of fresh mint.

More about chemical emulsifiers

Very few recipes have enough fine powder to hold an emulsion. Most rely on chemical emulsifiers like egg yolk to stabilize them. These emulsifiers, with one end that is attracted to water and another that is attracted to oil, do two things—they coat the liquid droplets and prevent their joining together, and they change the inward pull (surface tension) of one of the liquids in the emulsion. That liquid loses its inward pull and becomes, so to speak, juicy, so that it can run between the droplets of the other liquid.

In the center of a cup of liquid, the pull on the molecules by the surrounding molecules is the same in all directions. But on the surface the liquid does not have a pull on top, and the surface molecules are pulled toward the center. This pull toward the center, called *surface tension,* makes it easy for liquids that do not readily dissolve in each other to keep to themselves.

You can picture the surface of one liquid being pulled tightly away from the surface of another liquid next to it. This is exactly what happens on the surfaces between oil and water or vinegar. The oil floating on the top of the water is being pulled upward into itself, and the water or vinegar on the surface next to the oil has a strong pull down toward the water.

When an emulsifier is added to a mixture of oil and water, a big change takes place. Take the egg yolk in mayonnaise as an example. Without it, the oil will sit on top, while the water in the lemon juice and egg white stays on the bottom. The mayonnaise will not go together, because both liquids are pulling tightly to themselves.

Egg yolk reduces the inward pull of the water, and the surface of the water changes entirely. It is no longer covered with water molecules that are pulled tightly toward the center. Instead, the emulsifier lines itself across the surface with its water-loving end dissolved in the water and its oil-loving end sticking up and touching the oil.

Since the emulsifier is dissolved in the water,

there is no longer a strong inward pull on the surface of the water. Without this inward pull, the water is

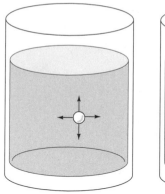 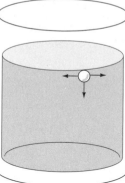

Equal Pull on Interior Molecules   **Unequal Pull on Surface Molecules**

free to mix and mingle with the oil. That is exactly what happens. Meanwhile, when the oil is broken into tiny droplets by a whisk or blender, every droplet is coated with a one-molecule-thick layer of the emulsifier. Now when two droplets of oil bump into each other, they bounce off instead of joining.

Mayonnaise—a classic example of an emulsion

To make mayonnaise, you start with an egg yolk or a whole egg and a little liquid—lemon juice, water, or prepared mustard, which contains vinegar. In the case of whole-egg mayonnaise, most of the liquid is in the egg white. The emulsifiers in the egg yolk dissolve in the liquid and drastically lower its surface tension. Then you turn on the blender or start whisking vigorously and adding the oil. The blender or whisk breaks the oil into tiny droplets. The emulsifier coats these droplets to prevent their running together, and the water with a very low surface tension is free to flow between the oil droplets. You're on your way to a good mayonnaise.

Mayonnaise, hollandaise, béarnaise, and butter sauces are thick because they have millions of droplets of one liquid packed into a tiny amount of the other liquid. The thickness of the sauce is determined by the thickness of the starting mixture and how many droplets you pack into the liquid.

For a very thick mayonnaise, start with a thick yolk mixture, then choose any or all of these options to make the yolks thicker: freezing, adding salt, or heating.

Commercial mayonnaise processors use the knowledge that frozen yolks stay thicker even after they are thawed to their advantage. In fact mayonnaise processors can get by with only about a third of a frozen yolk per cup of oil, and you can do the same at home: Whisk lemon juice into yolks (1 tablespoon per yolk) and freeze for at least 8 hours or overnight for a good mayonnaise base.

Using salt is ideal for *unheated* emulsions. To get the full benefit of the salt, whisk it into the egg or egg yolk right at the beginning. Salt on raw egg yolks that are to be heated, however, is another matter. Salt seems to speed up and encourage curdling when the yolks are heated. If you are heating yolks, salt may make it even trickier to keep the yolks from scrambling. Wait until the yolks have cooled a little before adding salt.

The risk of salmonella has made it necessary to heat egg yolks, and in spite of the dilution of the lemon juice and water, heating produces the desired thick mixture.

So, for a thick mayonnaise, start with a thick liquid (egg yolk, salt, and a little lemon juice or prepared mustard). Then, after you whisk in about a third of the oil and the emulsion is thick, whisk in some lemon juice, then continue to whisk in oil. Depending on the thickness you want, 1 1/2 to 3 tablespoons of total liquid per cup of oil makes a normal mayonnaise. Two teaspoons liquid per cup of oil is the bare minimum.

One of the larger chef's training centers teaches students to make mayonnaise by whisking together an egg yolk, salt, and Dijon mustard, then whisking in oil. Students are told to add a little lemon juice for flavor as they go. The recipe is excellent in that you start with a thick base. The salt thickens the egg yolk, and the Dijon is thick but does provide a little liquid. It is fairly easy to whisk the oil into droplets in this thick mixture and get an emulsion started.

However, some students' mayonnaises worked, and some began to separate as they whisked in more oil. For the students who added a little lemon juice shortly after they got a thick emulsion, the mayonnaise worked fine. They had enough liquid to go between the droplets. But the students who simply whisked in oil had problems with their mayonnaise separating near the end of the addition of oil. The instructor needed to point out that the lemon juice was not just for flavor; it also supplied a vital liquid component for a stable emulsion.

Vegetable oils such as corn, canola, peanut, and olive are frequent choices. Any kind of *refined* oil will do. Unrefined oils, like extra-virgin olive oil and unrefined corn oil (sold at health food stores) separate in the refrigerator the next day. Apparently something in these oils interferes with the emulsifiers in egg yolk. That something, Harold McGee reports, turned out to be the mono- and diglycerides in the unfiltered natural oils.

The following recipe should help you sidestep emulsion problems. It includes a heating procedure similar to that of the American Egg Board to kill salmonella bacteria. Lemon juice and water dilute the yolks so you can get them fairly hot without scrambling. The sugar also helps prevent scrambling. After heating, this mixture is thick enough to be a good base for the mayonnaise.

Homemade Mayonnaise

MAKES ABOUT 1½ CUPS

what this recipe shows

Diluting the egg yolks with lemon juice and water makes it easier to heat them without curdling.

Adding sugar also helps prevent curdling.

Holding for a minute at 160°F (71°C) kills salmonella.

Cooling for 5 minutes brings the temperature down and allows an emulsion to be formed.

Adding salt makes a thicker base for an easier emulsion.

Adding dry mustard and cayenne aids in keeping the oil drops apart.

Adding the oil very slowly at first allows the initial oil drops time to hit the blade and be broken up.

Avoiding unrefined oil (such as extra-virgin olive oil), which contains monoglycerides, helps prevent separation of the emulsion.

2 large egg yolks

2 tablespoons fresh lemon juice

2 tablespoons water

½ teaspoon sugar

1 teaspoon dry mustard

1 scant teaspoon salt

Pinch of cayenne (optional)

1 cup canola, peanut, vegetable, or pure olive oil, *not* extra virgin

continued

Heat the egg yolks, lemon juice, water, and sugar in a small skillet over very low heat, stirring and scraping the bottom of the pan constantly with a spatula. At the first sign of thickening, remove the pan from the heat but continue stirring. Dip the pan bottom in a large pan of cold water to stop cooking. Scrape into a blender, blend for a second or so, then let stand uncovered at least 5 minutes to cool. Add the dry mustard, salt, and cayenne if using. Cover and, with the blender running, drizzle the oil in very slowly at first, down the center hole into the egg mixture. Transfer mayonnaise to a clean container and chill immediately. This will keep for at least 7 days refrigerated.

at a glance
Mayonnaise

What to Do	Why
Have ingredients close to room temperature.	Very cold or very hot ingredients prevent an emulsion from forming.
Use a thick base.	Having a thick base makes it easier for the emulsion to form.
Add salt to the yolks or eggs to thicken them.	To get a thicker base.
Use ingredients like cayenne or mustard, which contain a fine powder.	Fine powders coat oil droplets and help keep them apart.
Beat oil in slowly at first.	When there are only a few drops, they can scoot out of the way of the blade and not get broken up.
Make sure you have enough water-type liquid.	There must be enough water to go between the oil drops and keep them apart.
Do not use unrefined oils such as extra-virgin olive oil.	Unrefined oils contain monoglycerides and cause emulsions to separate.

Modern spice- and herb-flavored mayonnaises have real punch. Honey, dill, and hot sesame oil in mayonnaise make a different sauce for iced boiled shrimp in Ring of the Sea Shrimp in the following recipe. The recipes for Curried Rice and Barley Salad (page 292) and Summer Vegetable Salad (page 293) feature mayonnaise in its conventional role as salad dressing, although both are flavored with spices.

Mayonnaise can also make a beautiful puffed browned topping. Because it contains egg, if a little starch or cheese is stirred in, mayonnaise will puff and brown when heated. The golden puffed open-face sandwich Jack, Avocado, and Sprouts on Whole Wheat (page 49) owes its puff to mayonnaise, as does the quick fish dish, Fish Fillets Under Dill Soufflé (page 294).

Ring of the Sea Shrimp with Fish Market Sauce

MAKES 10 TO 12 HORS D'OEUVRE SERVINGS

Brightly colored seashells in an ice ring on a bed of green keep shrimp cold and make a dramatic presentation, but you do need a lot of colorful seashells. If you have shell-shaped serving dishes, use them for the dipping sauces. You'll also need a disposable baby diaper to absorb liquid from the melting ice.

what this recipe shows

Condensation gives the ice ring a cloudy appearance when first placed in a warm room, but after a minute or two the surface melts so the cloudiness disappears and the bright shells are dramatic in crystal-clear frozen suspension.

Serving boiled shrimp in an ice ring keeps them wonderfully cold.

30 to 40 bright-colored 1- to 2-inch
  seashells

1½ pounds large shrimp, cooked,
  peeled, and deveined as directed on
  page 373 and chilled

1 recipe Fish Market Sauce
  (below)

1 recipe Traditional Cocktail Sauce
  (page 292)

8 to 10 leaves fresh kale

1. Boil seashells for several minutes to kill anything harmful that might be on them. Arrange shells in a large ring mold, fill three-quarters full of water, and freeze overnight.

2. Prepare shrimp and both sauces.

3. Cut a circle from the center absorbent part of a disposable diaper; it should be slightly smaller than the ring mold. Place the diaper circle on a pedestal dish, absorbent side up. Arrange kale leaves around the diaper so that as water drips onto leaves it will drain into diaper. Properly arranged, diaper will absorb liquid from melting ice.

To absorb water from a melting ice mold, cut a piece out of a disposable diaper and place it under the mold.

4. Unmold the ice ring. Bottom surface of ring may not be even. If not, chip away with an ice pick or place a small wad of aluminum foil under low side to even it when putting it in place. Center the ring on top of the kale. Place a clear glass bowl inside the hole in the ice ring to hold shrimp. Pile shrimp high in the bowl. Place dishes of dipping sauces at the base of the pedestal dish. Ice ring will hold up for about 2 hours in a room that is not too hot.

Fish Market Sauce

1 cup Homemade Mayonnaise
  (page 289), or store-bought

1 clove garlic, minced

2 teaspoons chopped fresh dill or
  1 teaspoon dried

2 teaspoons Dijon mustard

½ teaspoon hot sesame oil

1 tablespoon honey

Stir together all the ingredients in a bowl. Chill well before serving.

Traditional Cocktail Sauce

1 cup bottled chili sauce

Finely grated zest of $1/2$ lemon

1 tablespoon lemon juice

2 heaped tablespoons prepared
horseradish

$1/2$ teaspoon hot pepper sauce

Stir together all the ingredients in a bowl. Chill well before serving.

Curried Rice and Barley Salad

MAKES 12 SERVINGS

Striking colors—golden curried rice with the vivid red and green of peppers and exotic flavor blends of curry, roasted peanuts, and coconut—make this a dramatic dish to behold and to taste. A friend, Virginia Dunbar, introduced me to rice salads and pointed out what a great dish they are for large parties.

what this recipe shows

Combining barley with rice gives the salad base an unusual texture and taste.

Stirring curry powder into the mayonnaise gives a golden cast to the rice and barley as well as flavor.

4 cups Chicken Stock (page 267) or
canned chicken broth

2 cups medium-grain rice

1 large onion, chopped

2 medium jalapeños, seeded and finely
chopped

1 red bell pepper, seeded and coarsely
chopped

1 yellow or green bell pepper, seeded
and coarsely chopped

3 tablespoons mild olive oil

1 tablespoon sugar

1 cup 10-minute barley or small to
medium barley, cooked according to
package directions

1 cup golden raisins

2 jars (6 ounces each) marinated
artichoke hearts

1 cup Homemade Mayonnaise
(page 289) or store-bought

$1^{1}/_{2}$ to 3 tablespoons mild or hot curry
powder

1 tablespoon Chinese hot chili oil

2 tablespoons finely chopped fresh
ginger

$1/2$ cup flaked coconut, canned or
frozen

1 cup roasted salted peanuts, chopped

6 scallions, green parts included, sliced
into thin rings

1.   Bring the stock to a boil in a large saucepan over high heat. Stir in the rice and bring back to a boil. Turn the heat to low, cover, and simmer 18 minutes. Set aside.

2.   Sauté the onion, jalapeños, and peppers in olive oil in a large skillet over medium heat until just soft. Sprinkle with sugar and stir in well. Set aside.

3.   Stir together the cooked barley, raisins, rice, and onion-pepper mixture in a large mixing bowl. Drain the artichokes and save the liquid. Stir the drained artichokes into the mixture. In a small mixing bowl, stir together the artichoke liquid, mayonnaise, curry powder, chili oil, and ginger. Stir the mayonnaise mixture into rice-barley mixture and refrigerate for several hours or overnight for flavors to meld. Serve cold. Spoon onto a large serving platter. Sprinkle with coconut, peanuts, and scallions and serve.

Variation: *Cooked boneless chicken or cooked shrimp or scallops can be added.*

Summer Vegetable Salad

MAKES 8 TO 10 SERVINGS

Caper-tarragon mayonnaise is good on anything, especially potatoes, artichokes, and cucumbers. This salad is perfect with grilled fare.

what this recipe shows

Mixing mayonnaise with alkaline ingredients like potatoes reduces its acidity, making it a good growth medium for bacteria, so this dish should be served cold.

2 packages (10 ounces each) frozen
   green peas

1 tablespoon *and* $^3/_4$ cup olive oil
   ($^3/_4$ cup plus 1 tablespoon *total*)

$^1/_8$ teaspoon *and* $^1/_4$ teaspoon white
   pepper ($^3/_8$ teaspoon *total*)

$^1/_4$ teaspoon *and* 1 teaspoon dried
   tarragon (1$^1/_4$ teaspoons *total*)

$^1/_2$ teaspoon sugar

9 medium-size (about 3 pounds) red
   new potatoes, scrubbed

1 tablespoon *and* 1 tablespoon *and*
   $^1/_2$ teaspoon *and* $^1/_4$ teaspoon salt
   (2$^3/_4$ tablespoons *total*)

1 package (10 ounces) frozen artichoke
   hearts

3 large cucumbers

3 tablespoons fresh lemon juice

1 cup Homemade Mayonnaise
   (page 289) or store-bought

2 teaspoons juice from bottled capers

1 tablespoon *and* 1 tablespoon drained
   capers (2 tablespoons *total*)

$^1/_2$ teaspoon dry mustard

1 head Boston or leaf lettuce

6 hardcooked eggs, sliced

6 scallions, green parts included,
   chopped

3 tablespoons chopped parsley

continued

1. Thaw the peas, drain, and let stand a few minutes on a paper towel. Transfer to a bowl, add 1 tablespoon olive oil, $^1/_8$ teaspoon white pepper, $^1/_4$ teaspoon tarragon, and the sugar, and stir. Refrigerate until ready to assemble salad.

2. Slice the potatoes $^1/_4$ inch thick and cook in boiling water with 1 tablespoon salt until just tender, about 5 minutes. Drain, place in a bowl, and set aside. Cook artichoke hearts according to package directions, drain, place in a bowl, and set aside. Peel the cucumbers, halve them lengthwise, scoop out and discard the seeds. Slice $^1/_4$ inch thick. Cook the slices in rapidly boiling water with 1 tablespoon salt just until they turn pale green, about 1 minute. Drain and set aside.

3. Whisk together the lemon juice, $^3/_4$ cup olive oil, $^1/_2$ teaspoon salt, and $^1/_4$ teaspoon white pepper in a bowl. Spoon half of this dressing over the potatoes and refrigerate until ready to assemble the salad. Spoon half of the remaining dressing over the artichokes, mix, and refrigerate until ready to assemble. Combine the rest of the olive oil dressing and cucumbers in a bowl and refrigerate until ready to assemble. Stir together the mayonnaise, caper juice, 1 tablespoon capers, crushed, dry mustard, and 1 teaspoon tarragon in a bowl and refrigerate until ready to assemble. Make sure all ingredients are well chilled before serving.

4. Cover a large platter with lettuce. Arrange the potatoes on the lettuce, top potatoes with the artichokes, and place cucumbers on top of the artichokes. Coat the salad with a thin layer of the mayonnaise mixture. Press egg slices into mayonnaise around the base. Toss peas with $^1/_4$ teaspoon salt and sprinkle them on top of salad. Sprinkle 1 tablespoon capers, chopped scallions, and parsley over all. Serve immediately.

Variation: *Cook 1 pound shrimp until just cooked through as directed on page 375, plunge into ice water, and drain well; peel and devein. Mix more of the olive oil dressing that goes over the potatoes, cucumbers, and artichokes and stir into the shrimp. Chill until ready to assemble. Press shrimp into mayonnaise just above the layer of hardcooked eggs.*

Fish Fillets Under Dill Soufflé

MAKES 6 SERVINGS

This mayonnaise-cheese topping puffs to a marvelous golden brown for a dramatic dish. Use sole, flounder, orange roughy, or any other mild-tasting fish. My son fixed this in his toaster oven in the dorm to impress his dinner guests. It's instant and elegant. Sizzling Broiled Tomatoes with Herbs (page 359) and Roasted Asparagus with Lemon-Chili Oil (page 330) are ideal accompaniments.

what this recipe shows

Egg in the mayonnaise plus cheese makes a handsome puffed and browned topping.

1 tablespoon butter to grease pan

6 (about 1 to 1$^1/_2$ pounds) fish fillets, skin removed

1 clove garlic, minced

3 sprigs parsley, chopped

1 teaspoon chopped fresh dill

1 tablespoon Dijon mustard

1 cup Homemade Mayonnaise (page 289) or store-bought

$^1/_3$ cup freshly grated Parmesan

1. Preheat the oven to 400°F (204°C). Butter a 10 × 15-inch jelly-roll pan.

2. Place the fillets on the pan. Stir together the garlic, parsley, dill, mustard, mayonnaise, and Parmesan in a mixing bowl. Spread the mayonnaise mixture evenly over each fillet. Bake in the top third of the oven until the topping puffs and starts to brown, about 20 minutes.

at a *glance*
Emulsions

What to Do	Why
Start with a thick base.	A thick base will help to make a thick sauce.
Add salt or use frozen egg yolks for a thick sauce.	Salt or freezing thickens egg yolks.
Use an emulsifier.	Emulsifiers coat droplets of one of the liquids to prevent their running together, and emulsifiers lower the surface tension of the other liquid so it runs between the droplets.
Whisk or beat as the oil or butter is added.	Mechanical action is needed to break up the liquid into droplets.
Add oil or butter very slowly at first.	It is more difficult to break up the fat into droplets when it has room to scoot out of the way.
Be sure to have enough of the liquid that is not in droplets.	You must have some liquid between the drops, or they will be forced together.
Watch the temperature.	Proteins need the heat to unwind but not so much that they coagulate.

TWO KINDS OF EMULSIONS

An emulsion can be tiny droplets of oil or fat held in water, like mayonnaise or hollandaise, or the other way around—droplets of water in oil or fat, as in butter.

What determines whether an emulsion is going to be an oil-in-water or a water-in-oil emulsion? Madeleine Kamman, the noted French cooking teacher, and I went over what we knew and what we didn't know.

We knew it did not depend on the relative amounts of the two liquids. In a typical mayonnaise (1 egg, 1 tablespoon lemon juice, 1 teaspoon salt, 1 teaspoon dry mustard, and 1 cup oil), there are only about 3 tablespoons of water—one in the lemon juice and two in the egg white. A whole cup (or even more) of oil droplets is being dispersed in just 3 tablespoons of water. With such a large amount of oil and small amount of water, mayonnaise

should be a water-in-oil emulsion, but it is not. It is an oil-in-water emulsion.

If the determining factor is not the relative amounts of the two liquids, what is? This was when I took a close look into the role of emulsifiers in lowering surface tension. An emulsifier dissolves more readily in one liquid than in the other. One of the liquids keeps its surface tension and pulls itself into tiny droplets. The emulsifier dissolves more easily in the other liquid and lowers its surface tension so that it is the one that becomes juicy and runs between the droplets. Here was the answer that Madeleine and I were looking for: The liquid in which the emulsifier dissolves more easily becomes juicy and is the liquid between the drops of the other.

Liaisons

A liaison is any ingredient used to bind components and thicken a liquid to make a sauce. It may be as simple as a starch. The term also is used to refer to a mixture of egg yolk and cream or sour cream that is whisked in to quickly bind a sauce just after you remove it from the heat. This is a quick, casual emulsion with the egg yolk binding fat and liquid to thicken slightly and create a smooth sauce. These liaisons are barely stable and, if refrigerated, need to be reheated with great care like hollandaise.

Butter sauces

Once when I was teaching at L'Academie de Cuisine in Bethesda, Maryland, I did a few quick demonstrations to prove the importance of the three parts to an emulsion. I made a successful mayonnaise with correct amounts of the parts of an emulsion, then a failed mayonnaise—short of one of the necessary parts. I then got the failed mayonnaise to work by supplying its missing ingredient. I felt that I had shown how essential all three parts of an emulsion were: droplets, liquid, and emulsifier.

At that point Phyllis Richman, restaurant critic for *The Washington Post,* asked about beurre blanc, an emulsion that has only two ingredients—vinegar

and butter. "Where's the emulsifier?" asked Phyllis.

I groaned, "Oh, Phyllis, why do you do this to me? Well, let's think—vinegar and butter. That's it—butter. With only 81 percent fat, it has 19 percent water and milk solids. The milk solids are proteins and emulsifiers."

This was easy to prove. We made a large amount of reduction (shallots and vinegar boiled down) and divided it into three batches. I whisked butter into one batch and got a nice sauce. To the next batch, I added cream (more milk product—more emulsifier) and reduced it, then whisked in the butter; I had an even thicker, nicer sauce. To the last batch of reduction, I whisked in clarified butter—butter with the milk solids removed—instead of regular butter. Without any emulsifier, it would not even begin to form a sauce.

You can easily see the weak spot of butter sauce. Hollandaise and béarnaise, with the great emulsifier—egg yolk—have an abundance of emulsifier, but butter sauces are on the lean side, you might say, as far as emulsifiers go. No wonder those French chefs are always adding cream and reducing it before they whisk in the butter! They're adding a little extra emulsifier as insurance. I'm all for it.

With butter sauces, there's also the usual worry

about getting the emulsion too hot. The proteins from butter join together at an even lower temperature than egg yolk proteins do. In addition, since the butterfat droplets do not have a lot of emulsifier coating them, the butterfat crystallizes together into clumps of fat when the sauce gets cool. Not only do you have to worry about getting the sauce too hot; you have to worry about letting it get cold, too. If that happens, you can re-create the sauce by heating a small amount (¼ cup) of reduced cream, then whisking in the cold butter sauce a little at a time, just as you did the butter when you made the sauce initially.

Butter sauces are favorites with caterers. The sauces are elegant, and they are so acidic they can safely sit at room temperature without risk of bacterial growth. They are also the easiest of the emulsion sauces to make. It's simply a matter of whisking butter into a vinegar reduction and getting the temperature high enough to melt the butter into liquid form, but not so high as to coagulate the milk proteins. Instant sauce!

Good cooks have their own preferred ways of making butter sauce. Some like to whisk cold butter into the reduction on high heat. Others like to remove the hot reduction from the heat, whisk in room-temperature butter, and place the pan back on the heat for a few seconds if the butter is not melting in fairly fast. These sauces get to a coat-the-spoon thickness only and are never as thick as starch-bound sauces.

Whatever the procedure, it is crucial not to let the sauce get above 130°F (54°C), the temperature of hot tap water—not very hot. Milk proteins coagulate at 136°F (58°C), only slightly hotter, at which point the sauce will break.

Butter sauce can be made with white wine (beurre blanc) or red (beurre rouge). Either can be made with or without cream, but they are definitely thicker and more stable with reduced cream. The color of beurre rouges with added cream is quite beautiful. If made with one of the paler red wines, it can turn a lovely salmon color.

Beurre Blanc

MAKES ABOUT 1½ CUPS

what this recipe shows

Shallots add sweetness to mellow the sharpness of the vinegar.

Using many shallots and more wine makes the taste smooth and mellow rather than astringent like some butter sauces.

Using cream adds emulsifiers and helps make the sauce more stable and slightly thicker.

Whisking room-temperature butter into the hot reduction off the heat keeps the sauce warm but under 130°F (54°C), where it will separate.

5 medium shallots, chopped

⅓ cup Chenin Blanc, dry Rhine wine, or other medium-dry white wine

¼ cup white wine vinegar

½ cup heavy or whipping cream

¾ pound (3 sticks) butter, softened

¼ teaspoon salt

continued

1. Place the shallots, wine, and vinegar in a medium-size heavy saucepan over medium-high heat. Boil and reduce to about $^1/_4$ cup. Add the cream and continue to boil. Reduce to less than $^1/_4$ cup again. Remove from the heat.

2. Whisk in chunks of butter, whisking constantly, until all the butter is added. Return to low heat for a few minutes if butter is not melting. Do not let the sauce go over 130°F (54°C). If the sauce starts to break from overheating, remove from the heat and whisk until it cools a little. It will come back together. Whisk in the salt. Taste and add more salt if needed. The sauce can be held for about 30 minutes in a wide-mouth thermos that has been rinsed with hot, not boiling, water or held for hours in a 120°F (49°C) water bath.

Beurre Rouge

MAKES ABOUT 2 CUPS

Gerry Klaskala, a very imaginative chef from Atlanta, makes beurre rouge without cream. Since it is thinner than a sauce made with cream, he thickens it with pureed shallots. This is my simplified version of Gerry's sauce.

what this recipe shows

Pureed shallots add not only flavor but also thickness to the sauce.

Reducing a whole bottle of wine down to a cup intensifies the flavors.

Removing the reduction from the heat and using room-temperature butter keep the sauce warm but under 130°F (54°C), where butter sauces separate.

5 medium shallots

1 bottle (750 ml) red wine

$^3/_4$ pound (3 sticks) butter, very soft

$^1/_2$ teaspoon salt

$^1/_2$ teaspoon fresh lemon juice

Blend the shallots with 1 or 2 tablespoons of the wine in a blender until pureed. Boil the pureed shallots and the rest of the wine in a large heavy saucepan until reduced to about $^1/_3$ cup. This may take as long as 10 minutes. Remove from the heat. Whisk in the butter. Add salt and lemon juice. Taste and add more salt if needed.

Triple-Layered Fillets in Golden Beurre Blanc

MAKES 4 SERVINGS

This elegant but easy three-star entree was inspired by a similar dish from the restaurant l'Oustaù de Baumanière in Provence. Serve the dish on dinner plates with a touch of either green or pink in the border if you have them. The sauce is pale golden, and the pink salmon layer shows delicately between the white flounder layers. I serve a first course like Cocktail Puffs with Escargots (page 230), then follow this beautiful fish dish with a salad like Mixed Greens with Walnuts (page 314) and finally a lemony dessert.

The beurre blanc is a very acidic sauce and can be made ahead and safely held warm in a wide-mouth thermos. The asparagus can be blanched ahead and the fish packets prepared ahead, leaving only the poaching and draining of the packets and the saucing and arranging of the plates to do at serving time.

what this recipe shows

Adding cream provides an extra emulsifier and imparts a milder, richer taste.

4 tablespoons ($^1/_2$ stick) butter to grease foil

8 (about 1 pound) flounder or sole fillets, skinned

4 slices (about $^3/_4$ pound) salmon fillet—tail pieces are fine

Salt

Butter to dot fish

About 2 quarts Fish Stock (page 269) or water with 1 onion, quartered, and 1 lemon, quartered

1 recipe Beurre Blanc (page 297)

15 asparagus spears, steamed 2 minutes or blanched 1 minute

1 tablespoon salmon caviar

1. Preheat the oven to 450°F (232°C). Tear off five pieces of aluminum foil, each large enough to wrap a fillet. Lightly butter the shiny side of each piece.

2. With a veal pounder, gently flatten the flounder fillets slightly. Trim to a rectangular shape. Gently flatten the salmon slices and trim to the same size as the flounder fillets. Save all the trimmings. Place a flounder fillet on a piece of foil, sprinkle lightly with salt, and dot with a small amount of butter. Place a piece of salmon on top, sprinkle with salt, and dot with butter. Place another flounder on top, sprinkle with salt, and dot with butter. Fold the foil across the top and fold the ends up. Do not seal tightly. Foil is intended just to hold the three layers together, not to seal. Repeat with remaining fish. Make an extra packet with all the trimmings. The packets can be made ahead to this point and refrigerated until serving time.

3. Bring the stock to a boil in a deep baking casserole. Place the packets in the stock and place the casserole in the oven. Bake for 10 minutes. Partially open one end of each packet and stand vertically in a bowl to drain.

4. When ready to serve, prepare the beurre blanc.

5. To serve, spoon beurre blanc onto a dinner plate (see Note). Place one of the fish stacks on the sauce, near the bottom of the plate. Arrange three asparagus tips in a fan shape along the top edge of the fish. Place a small amount of salmon caviar on top of fillets. Repeat. Serve at once.

Note: *Do not heat the plates to serve. A hot plate may cause the sauce to break.*

Steamed Snapper with Beurre Rouge

MAKES 6 SERVINGS

A snow-white fish fillet accented with bright green snow pea points against a background of deep maroon beurre rouge—a rich and elegant dish. You can use cod, haddock, or any other mild thick white fish as well as snapper.

what this recipe shows

Using pureed shallots thickens the sauce when there is no cream to thicken the emulsion.

continued

2 tablespoons vegetable oil

Salt

6 (about 2 pounds) snapper fillets

1 quart water

6 snow peas

1 recipe Beurre Rouge (page 298)

Oil and salt the fillets. Steam until done, about 10 minutes per inch of fillet thickness. Bring water to a boil, add 1 tablespoon salt, and plunge the snow peas into the boiling water. Remove immediately. Cut each snow pea lengthwise at an angle to form two long pointed pieces. Sauce six plates with beurre rouge and place a fillet carefully in the center of the sauce. Garnish each fish fillet with two pieces of snow pea and serve.

Two classic emulsion sauces: hollandaise and béarnaise

Hollandaise and béarnaise are two classic French butter-and-egg emulsion sauces. Hollandaise, the sauce that so enraptured my friend, is delicately flavored with lemon juice and served warm, usually with vegetables, as in Broccoli and Cauliflower in Artichoke Cups with Hollandaise (page 303), or eggs, as in Eggs Sardou (page 203). Béarnaise, which starts with a reduction of vinegar and herbs, has a more powerful taste; it is usually served with steak but is also suitable for fish, eggs, or vegetables and may be used cold on a sandwich. The two are made in a similar fashion and share the same characteristics.

1. *Butter in droplets.* Melted butter is beaten to break it up into tiny droplets, in a blender or in a saucepan with a whisk, stirring constantly. Some recipes call for clarified butter, which is all fat and makes a very thick sauce. If you use clarified butter, be sure you have enough water-type liquid to go between the fat droplets.

2. *Liquid between the drops.* In hollandaise sauce, the liquid is primarily lemon juice; in béarnaise, it is reduced vinegar. A little water may also be included. As in any emulsion, there has to be enough liquid to go between the droplets of fat. Even if there was enough to start with, there can be some loss through evaporation if the sauce was heated very slowly or is reheated. If the sauce begins to curdle, remove from the heat and whisk in some liquid—water, lemon juice, or vinegar—right away.

3. *The emulsifier.* In both hollandaise and béarnaise, it's egg yolks—the best one there is. Yolks are heated with the lemon juice or vinegar to form a thick base.

4. *Heat.* In classic hollandaise or béarnaise, all the action takes place over heat, in a double boiler to keep it very low. Constant whisking keeps the heat distributed evenly. It's tricky. The temperature has to be high enough for the sauce to thicken, yet not so high that it curdles and separates. When this starts to happen, you can see oily melted butter beginning to accumulate around the edge of the pan and the top of the sauce beginning to glaze. If you see the sauce starting to separate like this, immediately jerk the pan from the heat and whisk vigorously. It also helps to pour the sauce into a cool container—anything to bring the temperature down fast.

The blender takes a lot, but not all, of the risk out of these sauces. In Conventional Blender Hollandaise (page 302), which can be used when and where eggs are sure to be safe from salmonella, hot melted butter is drizzled into room-temperature yolks. The mixture thickens in the blender without separating. If the eggs and lemon juice are used cold, however, the emulsion will come out too thin. The sauce can be saved by reheating it just enough to thicken the yolks. Sometimes simply pouring the sauce into the still-warm pan that was used to melt the butter and stirring are enough. If you put the pan on direct heat, though, you have to stir vigorously and watch carefully not to let the sauce get too hot. Remember, eggs scramble at 180°F (82°C), which is not very hot.

In Quick Blender Hollandaise (page 302), I use the procedure recommended by the American Egg Board for killing salmonella in egg yolks by heating them with lemon juice and water. The warm egg yolks are combined in the blender with melted, but not hot, butter. The mixture gets nice and thick without separating.

Ideally, hollandaise and béarnaise sauces should be made shortly before use. In restaurants, however, the sauces may be made during preparation time in the afternoon and held in a warm water bath until evening. The problem with holding the sauces is that you have partially cooked egg yolks being held at an ideal temperature for bacteria to grow. The acidity from the lemon juice may be high enough to control the growth; then again, it may not.

At home, you can make the sauce and hold it for a short while, say up to an hour, over hot water or in a wide-mouth thermos that was rinsed in hot water. If you plan to hold the sauce, make a fairly lemony one and make it as close to serving time as possible.

Hollandaise or béarnaise can be refrigerated and reheated over hot water in a double boiler, taking care not to let the bottom of the pot touch the water. You must stir constantly. Reheated sauces, however, do not have the stability of freshly made ones.

I have never frozen hollandaise sauce, but my friend Margaret Fogleman does it all the time. She takes the frozen hollandaise out of the freezer the night before she plans to use it and puts it in the refrigerator. Shortly before serving, she brings water in a double boiler to a boil, removes it from the heat, and puts the hollandaise over the hot water. Then she stirs frequently until the sauce is warm.

Classic Hollandaise

MAKES ABOUT 1⅓ CUPS

what this recipe shows

Once the yolk-lemon juice mixture begins to thicken, it has reached a temperature high enough to kill salmonella.

Whisking in the melted butter over hot, not boiling, water off the heat prevents the yolks from scrambling.

Adding salt to the hollandaise after the ice cubes are added and the hot water has cooled prevents the yolks from scrambling.

4 large egg yolks

3½ tablespoons fresh lemon juice
   (about 2 lemons)

1 tablespoon water

½ pound (2 sticks) unsalted butter,
   melted

¼ teaspoon salt

⅛ teaspoon cayenne

Place the egg yolks, lemon juice, and water in the top of a double boiler or in a bowl resting over the top of a medium saucepan of simmering water. It is important that the top of the water be well below the upper part of the double boiler or the bottom of the bowl. Have the melted butter ready to drizzle in. Whisk constantly. The second that the yolk mixture begins to thicken slightly, remove the top of the double boiler or the bowl from above the hot water and continue whisking. Turn off the heat. Add four ice cubes to cool the hot water a little. Put the pan or bowl of yolks back above the hot water. Whisk in the melted butter, drizzling it in very slowly. If at any time the sauce looks as if it is about to break, remove bowl and continue whisking to cool it down or whisk in 1 teaspoon cold water. With constant whisking, whisk in the salt and cayenne. When all the butter is incorporated, taste and add more salt or cayenne as needed.

Conventional Blender Hollandaise

MAKES ABOUT 1⅓ CUPS

Use this recipe only if you use pasteurized yolks or are confident that your eggs are salmonella free.

what this recipe shows

Having the egg yolks and lemon juice at room temperature and the melted butter very hot ensures that the egg yolks get hot enough to thicken the sauce.

Adding salt to the yolks thickens them slightly and makes the emulsion form more easily.

Using 3 tablespoons lemon juice supplies enough water to go between the oil droplets for a stable emulsion.

The amount of lemon juice is sufficient to make the sauce acidic and slow bacterial growth.

4 large egg yolks at room temperature

3 tablespoons fresh lemon juice at room temperature (about 1½ lemons)

⅛ teaspoon salt

Pinch of cayenne

½ pound (2 sticks) unsalted butter, melted and very hot

Place the egg yolks, lemon juice, salt, and cayenne in a blender or in a food processor with the steel knife. Turn the machine on and add hot melted butter in a fine drizzle. If the egg yolks and lemon juice are at room temperature and the butter really hot, the temperature at which eggs thicken (144°F/62°C) will be reached in the blender or processor and the sauce will thicken. If not, pour the sauce into the still-warm container the butter was melted in and whisk quickly. The heat from the pan is usually enough to thicken the sauce. If not, place the pan and hollandaise over low heat for a few seconds and whisk constantly. Remove from the heat and continue to whisk for 1 full minute. Taste and add more salt and cayenne if needed.

Variation: *To convert hollandaise to an instant béarnaise, add 1 tablespoon capers and ¼ teaspoon tarragon to the blender after the butter and blend in well.*

Quick Blender Hollandaise

MAKES ABOUT 1¼ CUPS

This recipe includes techniques for killing salmonella.

what this recipe shows

Since the yolk-lemon juice mixture is hot, the melted butter should not be very hot to prevent the sauce from exceeding 180°F (82°C) and scrambling the yolks.

4 large egg yolks

3½ tablespoons lemon juice
(about 2 lemons)

1 tablespoon water

⅛ teaspoon salt

Pinch of cayenne

½ pound (2 sticks) unsalted butter,
melted but not extremely hot

Place the egg yolks, lemon juice, and water in a small (7- to 9-inch) heavy skillet. Heat over very low heat, stirring constantly (I use a flat wooden spatula to scrape the bottom or a fork and stir with the tines lying flat on the bottom of the pan) until the mixture just begins to thicken. Continue stirring but immediately remove from the heat. Continue stirring for 1 minute. Scrape the mixture into a blender or a food processor with the steel knife and add salt and cayenne. Blend for a few seconds, then let cool 1 to 2 minutes. With the machine running, add the melted butter in a fine drizzle. Sauce will thicken before all the butter has been added. Blend in all the butter, taste for seasoning, and add more salt or cayenne as needed.

Variation: *To convert hollandaise to an instant béarnaise, add 1 tablespoon capers and ¼ teaspoon tarragon to the blender after the butter and blend in well.*

Broccoli and Cauliflower in Artichoke Cups with Hollandaise

MAKES 8 TO 10 SERVINGS

Tufts of vivid green broccoli and white cauliflower are nestled in artichoke bottoms and laced with golden hollandaise to create an elegant vegetable side dish. Excellent with meat, chicken, or fish dishes.

> **what this recipe shows**

Adding a little sugar and salt to vegetable cooking water seasons and preserves the firm texture.

Adding oil or butter to the cooking water gives a very light glaze to a vegetable as it is lifted out of the water.

2 cans (8 to 10 ounces each) artichoke
bottoms, drained (see Note)

1 cup canned chicken broth

¼ teaspoon *and* 1 teaspoon *and*
1 teaspoon sugar (2¼ teaspoons *total*)

2 teaspoons *and* ¼ teaspoon *and*
2 teaspoons *and* ¼ teaspoon salt
(4½ teaspoons *total*)

1½ pounds broccoli, 1½-inch
florets only

1 tablespoon *and* 1 tablespoon butter
(2 tablespoons *total*)

⅛ teaspoon white pepper

Pinch of cream of tartar or ½ teaspoon
vinegar (optional)

1 small head cauliflower, 1½-inch
florets only

1 recipe Quick Blender Hollandaise
(page 302)

continued

1. Drain the artichoke bottoms well and rinse several times. Place in a small saucepan with the broth and ¼ teaspoon sugar to be ready to heat at serving time.

2. Bring about 1½ quarts of water to a boil in a large saucepan. Add 1 teaspoon sugar, 2 teaspoons salt, and the broccoli florets. Boil 4 to 5 minutes. Remove from the heat. Add 1 tablespoon butter to the broccoli and water; it will melt and float on top. Using a slotted spoon, remove the broccoli to a bowl. Sprinkle with ¼ teaspoon salt and the pepper and toss gently to mix.

3. Bring about 1½ quarts of water to a boil in another large saucepan. Add 1 teaspoon sugar, 2 teaspoons salt, a pinch of cream of tartar or ½ teaspoon vinegar if water is hard, and the cauliflower florets. Boil 5 minutes. Remove from the heat. Add 1 tablespoon butter to the cauliflower and water; it will melt and float on top. Remove the cauliflower to a bowl using a slotted spoon. Sprinkle with ¼ teaspoon salt and toss gently to mix.

4. To serve, heat the artichoke bottoms in broth and drain. Place some broccoli florets in half of each artichoke bottom and cauliflower florets in the other half to create a green and white dome in each. Trim stems of florets as needed to make them fit. Arrange the filled artichoke bottoms on a serving platter or on plates as a side dish. Drizzle the hollandaise in a line between the broccoli and cauliflower, leaving some broccoli and cauliflower showing. Serve immediately.

Note: *Fresh artichoke bottoms are wonderful if you're up to the cooking and trimming, but canned are fine in this dish. Do not use artichoke hearts.*

Classic Béarnaise

MAKES ABOUT 1⅓ CUPS

what this recipe shows

When the yolk-vinegar mixture begins to thicken, it has reached a temperature high enough to kill salmonella.

Whisking in the melted butter over hot, not boiling, water off the heat prevents the yolks from scrambling.

Adding salt to the béarnaise after the ice cubes are added and the hot water has cooled prevents the yolks from scrambling.

½ pound (2 sticks) unsalted butter

4 shallots, finely chopped

2 tablespoons fresh tarragon leaves

4 white peppercorns, crushed

¼ cup white wine vinegar

⅓ cup dry white wine

4 large egg yolks

¼ teaspoon salt

Pinch of cayenne

Heat the butter in a medium saucepan over medium heat just to melt. Boil shallots, tarragon, and peppercorns in vinegar and wine in a nonreactive medium-size saucepan over medium heat until reduced to about $1/4$ cup. Strain into the top of a double boiler. Whisk in the egg yolks. Place the top over the bottom of the double boiler containing simmering water. Make sure that the top of the water is below the bottom of the upper part of the double boiler. Whisk constantly. The second that the yolk mixture begins to thicken slightly, remove the top of the double boiler from above the hot water and continue whisking. Turn off the heat. Add four ice cubes to cool the hot water a little. Put the pan of yolks back above the hot water. Whisk in the melted butter, drizzling it in very slowly. If at any time the sauce looks as if it is about to break, remove the top and continue whisking to cool it down or whisk in 1 teaspoon cold water. With constant whisking, whisk in the salt and cayenne. When all the butter is incorporated, taste and add more salt or cayenne as needed.

Pan-Seared Steak with Béarnaise

MAKES 2 SERVINGS

One of Nathalie Dupree's students was taking haute cuisine French cooking classes and had inflicted fancy creations on her family night after night. On the fifth day, her six-year-old pleaded, "Mom, I've had it with this gourmet junk. Couldn't we just have a simple steak and béarnaise tonight?"

When I can't grill, I prefer pan searing to broiling. You get a hard crust and a juicy center. Serve with a baked potato, a good salad such as Mixed Greens with Walnuts (page 314), and crusty bread such as Crusty French-Type Bread (page 24).

1 recipe Pan-Seared Steak (page 398) $1/2$ recipe Classic Béarnaise (page 304)

Prepare steaks as directed. The steaks will be medium-rare. Serve immediately with a dollop of béarnaise on each plate. Pass the rest in a bowl.

Variation: *Béarnaise-flavored Quick Blender Hollandaise (page 302) or Conventional Blender Hollandaise (page 302) may be used instead.*

the recipes

Mixed Greens with Walnuts

Spinach and Orange Salad with Pine Nuts

Fresh Fruit with Ginger

Corn on the Cob with Honey

Homemade Applesauce

Apple Wedges

Chili Oil

Roasted Asparagus with Lemon-Chili Oil

Fresh Green Bean Salad with Basil and Tomatoes

Broccoli Salad with Bacon and Sweet-Sour Dressing

Cheddar-Crusted Chicken Breasts with Grapes and Apples in Grand Marnier Sauce

Sherried Rice and Barley with Almonds

Country Summer Squash

Old-fashioned Grated Sweet Potato Pudding

Great Baked Pumpkin with Herbed Barley and Chickpeas

Golden Tomato Bake

Spicy Indian Fried Cheese

Sweet-Tart Fresh Strawberries

Roasted Sweet Onions with Balsamic Vinegar

Carrots with Raspberry-Chambord Sauce

Herbed Brussels Sprouts

Fine Cabbage Threads with Buttered Crumbs

Oven-Fried Herbed Potatoes

Shallot Mashed Potatoes with Garlic

Poached Pears with Walnut–Blue Cheese Rounds

Parmesan-Crusted Zucchini Fans

Sizzling Broiled Tomatoes with Herbs

Basil Vinaigrette

Grilled Vegetables with Chili Oil

Oven-Dried Cherry Tomatoes

Indian Vegetable Stir-fry

Artichoke Leaves with Hollandaise

treasures of the earth

A ripe, sweet peach bursting with juice is a luscious, intense pleasure. Bright green asparagus spears or vivid red tomato slices make a plate come alive. Fruits and vegetables complement other foods and add excitement to our meals. The brilliant colors, sweetness, and refreshing juiciness of fruits enhance many classic meat dishes: duck à l'orange, duck with Montmorency cherries, and chicken with grapes, to name a few.

On a practical note, we owe our very lives to plants. Plants directly or indirectly feed the world. Only the green chlorophyll in plants can take energy from the sun, combine it with carbon dioxide in the air and with water to make sugar, which plants then store as starch. Animals could not exist without food from plants. Current dietary guidelines place renewed emphasis on the role of fruits and vegetables in a healthful diet.

The cook's challenge is to preserve and highlight the bounty of color, flavor, texture, and nutrition of fruits and vegetables, nature's jewels. It would be a sin to lose the crisp, bright green of broccoli or the sweetness of fresh corn. It is important to know how to select the best, how to store to preserve quality, how to keep the colors bright and true during cooking, how to cook to enhance taste, and how to save as many nutrients as possible. Some fruits and vegetables are toxic or potentially toxic if not properly cooked or cared for. Cooks need to know about these too. Some plants are also potent medicines—from the herbs of folklore to modern prescription drugs.

Fruits and vegetables— food from plants

Plant parts

Fruits and vegetables are both plant parts. A fruit is the ripened plant ovary, which contains seeds. Vegetables may be plant leaves like lettuce or spinach, plant stems like asparagus, plant roots like turnips or carrots, tubers like potatoes, or rhizomes like ginger. Some so-called vegetables, like tomatoes, peppers, okra, squash, eggplants, cucumbers, snow peas, snap beans, and avocados, are, botanically speaking, fruits. Seeds like peas and lima beans are technically fruit parts.

Here is a table of some common vegetables classified according to which part of the plant is eaten.

COMMON VEGETABLES AS PLANT PARTS	
Parts Above Ground	
Leaf	Brussels sprouts, cabbage, chard, Chinese cabbage, chicory, chive, collards, cress, dandelion, endive, kale, lettuce, mustard greens, onion (green part), parsley, spinach, watercress
Stem	Asparagus, kohlrabi
Petiole (stem that supports leaves)	Celery, rhubarb
Flower	Broccoli, cauliflower, globe artichoke
Fruit	Avocado, cucumber, eggplant, melon, okra, pepper, snap bean, snow pea, squash, tomatillo, tomato
Seed	Chickpea, corn (sweet), dry and green peas, dry bean, lima bean
Parts Below Ground	
Root	Beet, carrot, celeriac, horseradish, parsnip, radish, rutabaga, sweet potato, turnip
Tuber (enlarged underground stem)	Jerusalem artichoke, potato
Corm (base of a stem)	Taro
Bulb	Garlic, leek, onion (white part), shallot
Rhizome (underground stem)	Ginger

Living plant parts

Picking or cutting a fruit or vegetable from the plant removes it from its source of water and nutrients, but it is still very much alive. Its cells continue to conduct their normal activities.

Don't look now, but your lettuce is breathing

Picked fruits and vegetables take in oxygen. Just as we do, their cells use it for metabolism (the breaking down of complex compounds—starches, sugars, and acids—into energy, water, and carbon dioxide, which they give off). Plant specialists call this breathing *respiration*.

I always thought that plants take in carbon dioxide, not give it off. And, in fact, plants do take in carbon dioxide when they are growing in the sun. Chlorophyll uses it to make sugar. The process of photosynthesis by which plants create food from sunlight, air, and water, however, is totally different from respiration.

The key to preserving produce quality is to slow down its metabolic breakdown—slow down its respiration. If produce is breathing fast, it's deteriorating fast. All of its metabolic changes and aging speed up, and it begins to go bad. Keeping produce cold and limiting its oxygen supply slow down these processes. Chilling and wrapping produce in plastic does just this. Produce like lettuce and celery in supermarkets is often wrapped in plastic for longer shelf life.

My husband, Arch, was skeptical about fruits and vegetables breathing. "You can't tell people that their lettuce is breathing. They'll think you're off your rocker!"

I am reasonably certain that a greater power from on high seated my husband and me next to Dr. Robert Shewfelt at an Institute of Food Techno-logists' dinner. Dr. Shewfelt is one of the world's authorities on postharvest care of fruits and vegetables. When I realized what his specialty was, I said, "Please tell my husband that lettuce is breathing." Dr. Shewfelt assured Arch that not only was lettuce breathing but so were all fruits and vegetables.

"Where are their lungs?" asked Arch innocently.

"It's not muscular breathing, but fruits and vegetables do respire. They take in oxygen and give off carbon dioxide."

Arch was not convinced. "Well," he said, "what is lettuce's respiration rate?" Rob Shewfelt, who had run into skeptics before, replied: "I don't remember lettuce's exact rate, but I was just working with green beans. They respire to produce 250 milliliters of carbon dioxide per kilogram per hour. That is why they deteriorate so quickly."

Notice in the table that potatoes, grapes, and apples have low respiration rates, so they keep well. More perishable vegetables, like lettuce and beans, have much higher rates. Green bananas breathe slowly, but ripe bananas breathe very fast.

TYPICAL POSTHARVEST RESPIRATION RATES OF FRUITS AND VEGETABLES

Milliliters of carbon dioxide given off per kilogram per hour at 59°F (15°C)

Potato	8	Pear	70
Grape	16	Strawberry	75
Lemon and orange	20	Lettuce	200
Apple	25	Banana	
Cabbage	32	Green	45
Carrot	45	Ripe	200
Peach	50	Green beans	250

American Society of Heating, Refrigerating and Air-Conditioning Engineers, 1974, as adapted in R.B.H. Wills, et al. *Postharvest: An Introduction to the Physiology and Handling of Fruit and Vegetables.* An AVI Book. New York: Van Nostrand Reinhold, 1989.

APPROXIMATE STORAGE TIMES AND BEST STORAGE TEMPERATURES FOR SELECTED FRUITS AND VEGETABLES

Fruit or Vegetable	Approximate Storage Time (weeks)	Best Temperature
Berries	1 to 2	30° to 40°F (−1° to 4°C)
Broccoli	1 to 2	30° to 40°F (−1° to 4°C)
Bananas	1 to 2	50°F (10°C)
Green beans	1 to 3	30° to 40°F (−1° to 4°C)
Lettuce	1 to 3	30° to 40°F (−1° to 4°C)
Tomatoes	1 to 3	50°F (10°C)
Asparagus	2 to 4	30° to 40°F (−1° to 4°C)
Cucumber	2 to 4	41° to 48°F (5° to 9°C)
Avocado	3 to 5	41° to 48°F (5° to 9°C)
Green pineapple	4 to 5	50°F (10°C)
Cabbage	4 to 8	30° to 40°F (−1° to 4°C)
Celery	6 to 10	30° to 40°F (−1° to 4°C)
Apples	8 to 30	30° to 40°F (−1° to 4°C)
Pears	8 to 30	30° to 40°F (−1° to 4°C)
Beets	12 to 20	30° to 40°F (−1° to 4°C)
Carrots	12 to 20	30° to 40°F (−1° to 4°C)
Lemons	12 to 20	50°F (10°C)
Onions	12 to 28	30° to 40°F (−1° to 4°C)
Potatoes	16 to 24	41° to 48°F (5° to 9°C)
Sweet potatoes	16 to 24	50°F (10°C)

E. G. Hall, *Mixed Storage of Foodstuff*s. Sidney: CSIRO: Food Research Circular No. 9, 1973, as adapted in R.B.H. Wills, et al. *Postharvest: An Introduction to the Physiology and Handling of Fruit and Vegetable*s. An AVI Book. New York: Van Nostrand Reinhold, 1989.

How to keep produce fresh longer

The real secret to slowing deterioration of produce is to limit its breathing and water loss. Chilling, limiting oxygen supply, and keeping produce in a humid atmosphere do just this.

Chilling   Chilling should be an immediate step. Cooling produce causes all its metabolic activities, including respiration, to slow down. "Quick cooling of rapidly respiring tissue such as spinach or broccoli is critical, as the tissue is 'expending' significant portions of its shelf life at higher temperatures," states Dr. Shewfelt in an article entitled "Postharvest Treatment." In fact produce cooled in the field right after being picked remains fresher and in better shape all the way through packing, shipping, and marketing.

The ideal storage temperature, however, is not the same for everything. Semitropical and tropical fruits and vegetables can be harmed by temperatures that are ideal for other produce; they suffer what is known as *chill injuries.* After injury, they don't ripen; they just rot. And, as the singing banana says, "Never put bananas in the refrigerator." Their ideal storage temperature is well above average refrigerator temperatures of 36° to 40°F (2.2° to 4.4°C). The suggested temperature for tomatoes is 50°F (10°C), and they remain more flavorful when not refrigerated. One of the critical flavor components in tomatoes, (Z)-3-dexenal, disappears when tomatoes are chilled.

Never refrigerate tomatoes. One of the critical flavor components, (Z)-3-dexenal, disappears when tomatoes are chilled.

The preceding table of storage times and best storage temperatures is arranged with the most perishable items at the top. Although 32°F (0°C) is the freezing point of pure water, storage slightly below this temperature will not freeze the liquids in the fruits and vegetables. Notice that the ideal storage temperature for a lot of fruits and vegetables is around refrigerator temperature. This is one reason many produce markets are so cold and damp. The proprietor is appropriately more concerned with keeping produce fresh and avoiding financial loss than with the comfort of the consumer. You will notice that employees are sometimes bundled up in sweaters, boots, and caps.

Limiting oxygen   Since fruits and vegetables breathe in oxygen, an obvious way to keep them fresh longer is to limit their oxygen supply to slow down their breathing. This is exactly what happens when lettuce is wrapped in plastic wrap. Shippers loosely wrap great stalks of bananas in plastic to slow their breathing and keep them from ripening before they get to the market.

Marketers are well aware that consumers don't like vegetables wrapped in plastic at the store. We want to be able to handle the produce. So most supermarkets openly display vegetables—some sprinkled intermittently with a fine mist of cold water to prevent wilting.

Keeping them moist   You know how sad wilted lettuce looks. It doesn't take much time for this to happen, particularly in a dry atmosphere. With a loss of only 5 percent moisture, lettuce wilts. Most fruits and vegetables are more than 70 percent water, with melons and lettuce on the high end at 90 to 95 percent water and sweet potatoes and corn on the low end with about 75 percent water. As food writer Harold McGee puts it, fruits and vegetables are "elegant wrappers for water."

Each plant cell is turgid—rigid and firm—because it is packed with water. Plant cells lose their firmness or turgidity when they lose water. They wilt. Semipermeable membranes in the cells of fruits and vegetables allow the passage of water but not proteins, nucleic acids, and other vital plant solutes. Fruits and vegetables on or off the plant give off water. When they are attached to the plant, this is no problem because the moisture is replaced by water absorbed through the plant roots. Once the fruit or vegetable is removed from the plant, however, it goes limp if this moisture loss is not slowed or prevented.

Produce packers and shippers pay close attention to humidity and airflow. They keep humidity as high as possible without encouraging mold growth, and they limit air movement to retard evaporation. In most supermarkets, greens are sprayed with a fine mist of cold water at regular intervals. This keeps greens crisp and turgid; however, excess moisture and surface wetting cause mold growth and rot. Sprinkling gives consumers crisp lettuce but at a price. We pay for the added water and in a few days we have slimy lettuce if we don't dry it.

This is not a problem for stores that count on rapid turnover—greens are there for only a day or two.

The produce clerk goes over the greens several times daily and discards browned leaves. Remember the old adage: One rotten apple will spoil the whole barrel. Produce with mold or rot should be discarded immediately before the mold or rot can spread.

Discard produce with mold or rot immediately or the mold will spread.

In addition to keeping humidity high and limiting airflow in storage areas, commercial packers extend nature's way of coating fruits and vegetables with a waterproof coating. Packers often spray apples, citrus, cucumbers, peppers, rutabagas, and occasionally sweet potatoes with a waxy coating that prevents moisture loss and can give the produce a shiny look. This wax coating should be washed or peeled off since it can alter the flavor of a dish. When you put a lemon wedge in hot tea, you may notice a little oily film floating on the surface. This is waxy coating that has melted. If you are using the zest, it is important to wash lemons or oranges with hot water, especially if you are using zest to make a flavored oil.

For storage, you can limit evaporation by placing fruits and vegetables in plastic bags. The plastic bag does three things: It limits moisture loss to the air; it limits airflow around product, thus limiting evaporation; and it limits oxygen, thus slowing the produce's deterioration.

Produce that has reasonably long shelf life, such as carrots, cabbage, or squash, keeps well refrigerated, uncut, and unwashed. Put it in enclosures that limit oxygen like the crisper drawer, plastic containers, or the new plastic vegetable bags with tiny holes. These bags are designed to reduce surface moisture and minimize rot and mold. They somewhat reduce the oxygen supply but do not cut it off.

NATURAL WAXY COATINGS ON PLANTS

Cooks know they can't put a vinaigrette on a salad until just before serving it since the dressing causes the greens to wilt. Many people think it's the vinegar that causes the wilting. Straight vinegar, which is reasonably acidic, can produce wilting, but diluted vinegar in salad dressings doesn't wilt the greens as much as the oil and salt. Plants are relatively waterproof, and vinegar is a water-type liquid.

If plants weren't somewhat waterproof, every time it rained the leaves would soak in water, get heavy, and pull the plant down. Leaves have a waterproof, waxy coating, called the *cuticle,* which makes raindrops roll off. It was interesting that the 5,000-year-old Alpine traveler (the "Iceman") discovered in Italy had a grass raincoat. In living plants, the cuticle limits evaporation and helps maintain water pressure in the cells.

Both the oil and the salt in a salad dressing contribute to the fast wilting of the greens. Try this: Tear a piece of lettuce into thirds, then put one third in oil, sprinkle one third with salt, and put the last third in water or a cup of water with a tablespoon of vinegar. The piece in water (or water and vinegar) will still be in good shape after 30 minutes, but in a short time the salted piece will be limp and you can see visible discoloration where oil has penetrated the piece in oil. There is a noticeable loss of crispness.

Lettuce is essentially waterproof. Oil and salt in a vinaigrette contribute as much to wilting as does vinegar—perhaps even more.

Moisture is a problem. To keep delicate produce like lettuce very crisp, you must keep it quite moist, yet any surface water can cause rot. With lettuce, which is very perishable, I like to make sure that the cells are filled with water (turgid) before storing, so I soak the lettuce in cold water for 10 to 30 minutes. For short-term storage (one day), wrap in a wet paper or cloth towel, place in a zip-top plastic bag, squeeze out the air, seal (to limit oxygen), and refrigerate.

If the lettuce is to be stored for longer than a day, after soaking it in ice water I remove as much surface water as possible by spinning in a salad spinner or a clean pillow case. Then I wrap the lettuce with one or two sheets of dry paper towel to absorb excess moisture and store it in a zip-top plastic bag. I squeeze out as much air as possible before sealing the bag. I used to simply store lettuce in a plastic bag with paper towels, but I find that squeezing out the air (cutting off its oxygen)

Lettuce will keep as long as several weeks if you squeeze the air out of its plastic bag and seal tightly after washing and drying it.

greatly improves storage life. You can keep lettuce for several weeks in this way.

Green salads With cold, crisp greens for a salad, make sure that the greens are dry before the dressing is added. If the greens are wet, the dressing washes off and ends up in a watered-down puddle in the bottom of the bowl. You will have a poorly flavored salad that tastes as if it needs more dressing.

As just discussed, the dressing will cause the greens to wilt, so it is important not to let the dressing touch the greens until just before serving. Use a small amount of dressing—less than you think you need. Toss the salad really well with clean hands or utensils until all the leaves have some dressing on them. Taste and add salt or more dressing as needed. You may be surprised by how little dressing a salad needs if the greens are dry.

I personally love cold salad plates. I believe that they keep the greens a little crisper. You can put the plates in the freezer compartment for 10 minutes before adding the salad.

PERLA MEYERS'S ONE-STEP TOSSED SALAD

Perla Meyers has a clever technique for home green salads. She prepares the dressing and puts the amount needed in the bottom of the salad bowl. Then she crosses the long-handled serving fork and spoon across the top of the dressing to prevent the greens from touching it. She then piles the prepared greens on top of the crossed uten-

sils. She covers the bowl with plastic wrap and refrigerates until serving time. Since everything is together, this procedure saves time when you are ready to serve. You do not have to look through a crowded refrigerator for the dressing. Simply reach down, pull out the utensils, toss well, taste, and serve.

Mixed Greens with Walnuts

MAKES 8 SERVINGS

Whenever I asked what was in the dressings used on the wonderful salads in France, the reply was usually "oil and vinegar." They were very delicate and not vinegary like American dressings. These are crisp greens with a mild dressing thickened and lightly flavored with ground roasted walnuts—excellent with pasta, roasted meats, or fish.

what this recipe shows

Using less vinegar than the standard one part vinegar to three parts oil produces a more delicate flavor balance and a delicious salad dressing.

Dry salad greens require a minimum of dressing.

Adding ground roasted walnuts thickens the dressing and adds flavor.

1 small head Boston lettuce	2 tablespoons white wine vinegar
1 small head Bibb lettuce	2 teaspoons sugar
1 small head romaine lettuce	1/2 teaspoon white pepper
1 1/4 cups coarsely chopped walnuts	1 teaspoon Dijon mustard
2 tablespoons butter	1 cup canola, corn, peanut, mild olive, or other vegetable oil
1/2 teaspoon *and* 1 1/4 teaspoons salt (1 3/4 teaspoons *total*)	1 pound fresh mushrooms
2 large shallots, peeled and halved	

1. Wash the lettuces and soak for 10 to 30 minutes in cold water. Spin dry, then wrap in a clean towel to dry thoroughly. Remove and place in zip-top plastic bags, squeeze out the air, seal, and refrigerate.

2. Preheat the oven to 350°F (177°C).

3. Spread the walnuts on a baking sheet and roast until lightly browned, about 10 minutes. While the walnuts are hot, stir in the butter and 1/2 teaspoon salt. With the steel knife in the workbowl, turn on the food processor and drop the shallots one at a time down the feedtube onto the spinning blade to mince. Add the vinegar, sugar, 1 1/4 teaspoons salt, pepper, and mustard. With the processor running, slowly drizzle in the oil. Add 1/4 cup of the walnuts and process a few seconds to grind. Set aside.

4. When ready to serve, slice the mushrooms thinly. Tear the lettuce into bite-size pieces and place in a large mixing bowl. Add the mushrooms and 1/2 cup dressing and toss very well to coat each piece. Taste and add more dressing or salt if needed. Place on a cold salad platter or individual salad plates. Top with remaining walnuts and serve immediately.

Spinach and Orange Salad with Pine Nuts

MAKES 6 TO 8 SERVINGS

Cold, fresh orange slices with a slightly salty honey dressing and dark green spinach with pine nuts are visually striking and an ideal complement for many dishes—Juicy Pork Tenderloins with Spicy Chinese Sauce (page 406), turkey, chicken, or fish. I use this salad over and over again.

what this recipe shows

Roasting, whether in a skillet or in the oven, enhances the flavor of nuts.

A minced shallot in the dressing adds a sweet onion flavor.

Adding a few ground nuts both thickens the dressing and enhances flavor.

1 tablespoon butter

$^1/_3$ cup pine nuts

$^1/_4$ teaspoon *and* 1 teaspoon salt ($1^1/_4$ teaspoons *total*)

1 shallot, peeled and halved

1 tablespoon cider vinegar

$^1/_2$ teaspoon white pepper

1 teaspoon Dijon mustard

1 tablespoon honey

$^1/_2$ cup canola, corn, peanut, mild olive, or other vegetable oil

2 large bunches (about 2 pounds) spinach, washed, drained well, and dried

1 small red onion, very thinly sliced

6 navel oranges, peeled and cut into $^1/_4$-inch slices

1. Melt the butter in a heavy skillet and add the pine nuts. Roast the pine nuts over low to medium heat, stirring constantly and taking care not to burn them, about 2 to 3 minutes. Sprinkle with $^1/_4$ teaspoon salt and set aside.

2. With the steel knife in the workbowl, turn on the food processor and drop the shallot down the feedtube onto the spinning blade to mince. Add the vinegar, 1 teaspoon salt, pepper, mustard, and honey. With the processor running, slowly drizzle in the oil. Add 2 tablespoons roasted pine nuts and process to puree.

3. Tear the spinach into bite-size pieces and place in a large mixing bowl. Add the onion and toss with enough dressing to coat. Taste and add salt if needed. Arrange the orange slices overlapping around the outer edge of a large cold serving platter and drizzle with any remaining dressing. Pile the spinach in the center and sprinkle with roasted pine nuts. Serve immediately.

at a *glance*
Green salads

What to Do	Why
Wash and soak greens in cold water.	To fill cells with water for crispness.
For short-term storage (one day), wrap in wet toweling, place in zip-top plastic bag, squeeze out air, seal, and refrigerate.	To limit oxygen supply.
For longer storage, spin greens dry, place in zip-top plastic bag with a dry paper towel, squeeze out air, seal, and refrigerate.	To limit oxygen supply and to keep surfaces dry for mold prevention.
Make sure salad greens are dry before dressing.	To prevent dressing from washing off.
Toss well with a smaller amount of dressing than you think you will need, taste, and add salt or dressing as needed.	Dry, well-tossed greens need less dressing.
Serve on cold plates immediately after dressing.	To preserve crispness.

Fruits: vine ripe and ripe

Plants progress through the stages of growth and development (prematuration) to maturation, ripening, and finally senescence (decline). An edible part of a plant is mature when it reaches its full size and quality. Ripening occurs near the end of maturation and continues until decline. Ripening partially overlaps both maturity and senescence.

Changes take place in both fruits and vegetables as they mature and ripen, but, remember, fruits and vegetables are different. Vegetables reach their prime and then, with age, some decay and some become tough or even woody. While changes take place in vegetables as they mature, they are nothing like the changes that occur in fruits.

The mission of fruits, plant ovaries, is to spread the plant's seeds. By becoming overwhelmingly desirable to animals, fruits get animals to carry them off, eat them, and thus spread the seeds. Fruits go from a hard, sour, inedible, almost invisible part of the plant to a sweet, juicy object with brilliant colors and enticing aromas. They change in color, taste, aroma, size, weight, texture, and nutrient content.

Changes in color The most obvious change when a fruit ripens is the color. Complex chemical changes take place, including a change in acidity that causes the green chlorophyll to break down. Some fruits, like bananas and apples, are already bright yellows, reds, or oranges under the green. The bright colors are merely hidden by the layer of green chlorophyll. Others, like tomatoes, make their red and orange carotenoid compounds as the chlorophyll starts to break down.

Raspberries accumulate anthocyanins, the compounds that give them their rich bluish red color as they ripen. Raspberries with deep red color will be riper. Cantaloupes become a paler yellow-tan as they ripen and the green disappears, but Crenshaw and Persian melons may remain light green when they are ripe.

WHEN FRUITS RIPEN

In an excellent article titled "Ripe Now," food writer Jeffrey Steingarten divides fruits into five categories:

1. Fruits that never ripen after they are picked: soft berries, cacao, cherries, grapes, citrus fruits, litchis, olives, pineapples, and watermelons
2. Fruits that ripen only after picking: avocados
3. Fruits that ripen in color, texture, and juiciness but not in flavor or sweetness after they are picked: apricots, blueberries, figs, melons other than watermelons, nectarines, passion fruit, peaches, and persimmons
4. Fruits that get sweeter after they are picked: apples, cherimoyas, kiwifruit, mangoes, papayas, pears, sapotes, and soursops
5. Fruits that ripen in every way after harvest: bananas

Changes in taste Fruits undergo major changes in taste as they ripen. They become less acidic and sweeter. Big starch molecules break down into sugars to make them sweeter and more desirable. Some fruits become sweeter as they mature by storing sugar sap from the plant. These, of course, will not get any sweeter after they are removed from the plant.

In fact a large number of fruits will never get any sweeter after they are picked. I find this depressing, knowing we depend largely on fruits that are harvested before they are sweet and ripe. Once you have had sweet, juicy fruit picked from the tree, it spoils you. Several years back, I picked apricots from a friend's tree in El Paso. Ever since, apricots have been a great disappointment. None comes close to the rich smells, the sweetness, and the juiciness of those ripe apricots.

Changes in aroma Numerous chemical changes take place in the proteins and acids of fruits and vegetables that cause them to develop sensuous, luscious-smelling volatile compounds. We have only four major taste sensations—sweet, sour, salty, and bitter—but the "taste" we experience in our minds comes from the intimate intertwining of taste and smell. The distinctive aromatic compounds of fruits and vegetables are in great part responsible for their unique characteristic taste. For example, grapes have as many as 225 volatile compounds. Some fruits have a main characteristic volatile compound like ethyl 2-methylbutyrate in apples or iso-amylacetate in bananas, but the full aroma is a subtle combination of many compounds. This makes duplicating nature's complex wonders like real fruit flavors difficult and sometimes nearly impossible.

Many times with fruits, melons in particular, smell may be the best indicator of ripeness. Sniff the blossom end—the end away from the stem.

> Scent may be the best indicator of ripeness for melons. Sniff the blossom end before you buy.

Production of ethylene As fruits ripen, in addition to enticing volatile compounds, many fruits start to produce the ethylene gas that speeds up ripening. Some fruits and vegetables are picked green to minimize handling and shipping losses. Produce shippers then use ethylene to ripen fruits like bananas and tomatoes quickly when they reach their destination. Ethylene doesn't make oranges sweeter or riper, but it does improve their color and is sometimes used on Florida orange varieties that stay green even when ripe.

When you need to ripen fruit quickly, you can use the same technique on a smaller scale. Simply place the fruit to be ripened in a closed, but not airtight, container (such as a paper bag) with something that is giving off ethylene, such as an apple. By combining this ethylene treatment with a little warmth, you will speed up ripening very effectively. When I want avocados to ripen, I warm them for 15 seconds each in the microwave and close them loosely in a paper bag overnight with several apples in a slightly warm place. I do not seal the bag. You want oxygen to get in to speed ripening and carbon dioxide to be able to escape.

The clear plastic fruit ripeners that are available do the same thing. They enclose the fruit but have some holes to let gases in and out. Placing fruit in the sun provides heat to speed up ripening but does not increase the ethylene in the surrounding atmosphere.

Changes in size, weight, and texture With many fruits and vegetables, size is an indicator of maturity. Size is used to determine harvest time for okra and strawberries. Weight can be a good indication of maturity, too. Density and weight are good ways for selecting mature cabbage. Harvest time for snap beans is determined by the seed weight and the weight-to-length ratio.

There can be enormous changes in texture with ripening. There is a breakdown of the firm substances—hemicelluloses and firm pectic substances—that hold cell walls together. As fruit ripens, more and more of these firm pectic substances convert to water-soluble pectins, and the fruit or vegetable becomes softer and softer.

The conversion of these firm pectic substances to water-soluble pectins affects the thickness of a fruit or vegetable juice. If tomatoes are pressed when cold, enzymes convert the firm pectic substances to water-soluble pectins, and the juice is very watery. If you want thick juice, first heat the tomatoes to 180°F (82°C) to kill the enzymes. When the tomatoes are pressed, the juice will have a good amount of the firmer pectic substances and be quite thick. The thick juice holds particles that would ordinarily settle out, making it even thicker.

With apples or grapes, on the other hand, the juice should be clear. These fruits are often chilled before pressing, which helps the enzymes do their work and convert the solid pectic substances to clear, water-soluble pectins. The sediment can then be removed by straining. Processors sometimes add extra enzymes to ensure this.

As fruits ripen, along with the many other changes, the seeds mature and natural waxes may develop on the skin. A natural waxiness on the surface of some melons indicates ripening.

Ripeness and nutrient content The maturity of a fruit or vegetable when it is picked has a major influence on its nutrient content. (See page 323 for more details.)

The sad truth is that many fruits and vegetables are harvested before their prime to cut shipping and storage losses. Bananas, apples, and avocados are just about the only produce that we can ripen ourselves and have them taste ripe. A few growers will hand pack and ship near-ripe produce to you (see Sources—Ripe produce, page 478, for a limited list).

You can call your state's department of agriculture and ask for a list of growers in your area who permit the public to come to the fields and pick their own products. The Georgia *Farmers and Consumers Market Bulletin* carries an extensive list of these farms during harvest season, and your state may have a similar publication.

Some changes take place in fruits and vegetables the instant that you remove them from the plant. The minute you pick corn, for instance, enzymes in the corn rush into action to convert sugars into other compounds. Corn is at its sweetest best when cooked immediately after picking. The next best alternative is to chill it instantly after harvest to slow down the actions of these enzymes. There are also some new varieties of corn that are so sweet that they are delicious even with some loss of original sweetness.

Through both crossbreeding and genetic engineering we have many superior fruits and vegetables already on the market. Crossbreeding for disease resistance and greater yields has been common practice for over a hundred years. Now genetic engineering is being used to refine plant characteristics.

We do not live in a perfect world and must make do with fruits that are not vine ripe. Adding a little sweetener can help compensate for the missing sugar that Mother Nature did not have the opportunity to impart. Here are two recipes that make up the shortfalls of fruit and vegetables that may not be at their prime. In the first, both sugar and fresh ginger enhance the fruits to give an amazingly sweet, fresh taste. In the second, a little honey in the cooking water for corn on the cob adds to nature's own sweetness for an almost field-fresh taste.

Fresh Fruit with Ginger

MAKES 10 SERVINGS

Orange slices pressed against a clear glass container alternate in rows with strawberries to create a dramatic, alluring dish. Chunks of melon and grapes fill the center, and the top is blanketed with red strawberries. Fresh ginger makes cold fruit even more refreshing. I have had guests stand by the bowl, eating and replenishing their plates several times.

> **what this recipe shows**
>
> Adding a little sugar enhances the taste of fruit that is not fully ripe.
>
> The slight bite of ginger adds a crisp, fresh taste.

$2^1/_2$- to 3-inch piece fresh ginger

1 cup boiling water

1 cup *and* 1 cup room-temperature water (2 cups *total*)

2 cups sugar (see Note)

6 navel oranges, peeled and sliced $^1/_2$ inch thick

1 ripe cantaloupe, peeled and cut into 1-inch cubes

1 ripe honeydew, peeled and cut into 1-inch cubes

1 bunch Red Flame seedless grapes (about 1 pound)

1 quart strawberries

12 slices candied ginger, finely chopped

1. Process the fresh ginger to a fine mince in a food processor with the steel knife. Place the minced ginger in a small bowl and add boiling water. Let stand for 30 minutes.

2. Heat 1 cup water and 2 cups sugar in a small saucepan on the stove or in a glass bowl in the microwave until the sugar dissolves. Stir in 1 cup water. Strain the liquid from the ginger water into the sugar water. Press the ginger against the strainer to squeeze out all flavored liquid. Discard the ginger. Place the ginger-sugar water in the refrigerator to chill well.

3. Use a tall clear glass bowl (a clear wine cooler is perfect) and press the orange slices against the glass in a row around the bottom. Put the cantaloupe and honeydew chunks and grapes in the center to hold the orange slices against the glass. Fill to the top of the orange slices. On top of the orange slices, arrange a row of strawberries against the glass. Again, place melon chunks and grapes in the center to hold the strawberries against the side. Sprinkle on about a quarter of the candied ginger. Arrange another layer of orange slices against the glass above the row of strawberries and fill the center with melon and grapes. Sprinkle with another quarter of the candied ginger. Continue layering until the container is filled. Cover the entire top with strawberries packed close to each other and sprinkle with any remaining candied ginger. Pour the cold ginger-sugar water over fruit. Cover with plastic wrap and refrigerate overnight and until ready to serve.

Note: *This can be prepared with three packets of aspartame sweetener instead of the sugar.*

Corn on the Cob with Honey

MAKES 4 TO 6 SERVINGS

My husband, whose father prided himself on growing fine sweet corn, says the only way to cook corn on the cob is to call back to the house from the cornfield, "Is the water boiling?" When the answer is yes, pull the corn and run, shucking it as you go, to get it into the pot as fast as you can. Since most of us can't do this, a little honey in the cooking water helps restore nature's sweetness. This is excellent for casual dinners and is a natural with fried chicken.

what this recipe shows

Adding honey to the cooking water adds a delicate sweetness to corn on the cob.

6 ears fresh sweet corn, such as white Silver Queen

1/3 cup honey

1/4 pound (1 stick) butter, melted

1/2 teaspoon salt

1/8 teaspoon white pepper

Shuck and silk the corn and cut away any blemished areas. Bring the water and honey to a rolling boil in a large pot. Add the corn and cook for 4 minutes after the water comes back to a boil. Stir together the melted butter, salt, and white pepper in a baking dish long enough to hold an ear. Drain the corn, roll each ear in the butter mixture, and serve immediately.

Cuts, bruises, and browning

When you cut or bruise fruit, phenolic compounds and enzymes that have been kept separate in the fruit's cells come in contact with oxygen from the air or from the airspaces between the cells and immediately cause browning. Browning can take place even without the enzymes, as in canned fruit that has been heated to kill the enzymes. Phenolic compounds and oxygen can still react at the surface of the fruit and the residual air at the top of the can.

Commercial processors use sulfur dioxide to prevent browning, but that is not practical for home cooks. Scientists thought about removing the phenolic compounds by selective breeding, but apples, bananas, cherries, peaches, and pears all contain one or more of these compounds. The Sunbeam peach and a few other varieties of fruits have no phenolic compounds, but the reality is that so many fruits contain them that their elimination is not likely.

A more practical approach is to keep oxygen away from the cut surface. Coating fruit with sugar or syrup prevents oxygen in the air from getting to the cut surface, but the oxygen between the cells is still there. Lowering the temperature also helps. Cooling slows down all chemical reactions. Blanching or brief cooking kills the enzymes that help this reaction along, so blanching fruit that is going to be frozen helps to slow browning. Plain table salt slows browning, but for it to be effective you need to use so much that it would spoil the taste. In any case, with each of these remedies the browning is simply slowed, not stopped.

The simplest procedure is to place the cut fruit in water with ascorbic acid (vitamin C). Acids slow the activity of the enzymes that cause browning. Any acid will help, preferably a mild-tasting acid like lemon juice or cream of tartar. However, ascorbic acid is the most effective since it both slows the enzymatic reaction and blocks browning of the compound that is formed and therefore stops the process quickly.

You can use Fruit Fresh, a product available at grocery stores, but it is essentially vitamin C. Instead, just crush a vitamin C tablet and dissolve it in a bowl of water. As you slice the fruit, toss it immediately into the vitamin C water. Or, since orange juice is high in vitamin C, place the sliced fruit in cold orange juice until ready to use. Many people use water with a little lemon juice, but I prefer the taste of orange juice with most fruits—especially bananas. If you want a kick to the dish, add a ground-up chili pepper, which is high in vitamin C. Putting the chili in a bowl of water will prevent browning.

Sliced raw potatoes first turn a pinkish tint, then brown, then gray when exposed to air. With potatoes, simply placing the raw slices under water usually prevents this discoloration. (More about potato discoloration on page 347.)

Fresh produce and nutrition

This is not a nutrition book, so I'll limit myself to three major points on this topic:

1. How important fruits and vegetables are in our diet

2. How widely nutrient content varies, not just between apples and oranges but from one orange to another

3. How important maturity and immediate cooling are for preserving vitamin content

The dietary importance of fruits and vegetables Your body does not make vitamin C, the B vitamins, or a number of other vitamins. You are therefore completely dependent on taking in (eating) an adequate daily supply. There is no accumulation of water-soluble vitamins like C and B from day to day, as there is of the fat-soluble vitamins A and E. Any excess of these water-soluble vitamins is simply removed from your body in urine. This means that you need C and B every single day for your body to run at its best.

Vitamin C also falls into the group of nutrients called *antioxidants,* which combine readily with other substances like cancer-causing free radicals and thus are considered cancer fighters. Beta-carotene (a precursor of vitamin A), vitamin E, and selenium are other antioxidants, and we are realizing more than ever how important it is to have an adequate daily supply of all of these nutrients.

That's why it's important to have some idea of what is in the vegetables and fruits available to us. For many people, fruits and vegetables are the only source of vitamin C, yet most of us think only of citrus fruits when we're looking for vitamin C. As the following table shows, broccoli, brussels sprouts, and kiwifruit are also good sources—in fact, ounce for ounce, they all have nearly twice the vitamin C of oranges. Chilies have five times as much as oranges, guava and black currants are also a generous source, and even strawberries have as much C as oranges. Our options for good health are abundant!

The table also gives figures for beta-carotene and for folic acid, which recently has been identified as essential to the neurological development of a fetus and as playing a role in reducing strokes and heart attacks. Notice in the table that mustard and turnip greens are real power foods in all three nutrients.

Carrots and sweet potatoes contain large amounts of beta-carotene. Spinach and broccoli have more folic acid than many other fruits and vegetables, but leafy greens, which we think of as high in folic acid, are not equally endowed—collards have less than a tenth as much as spinach. Lima beans and chickpeas are the winners in folic acid but do not have large amounts of vitamins A and C. A variety of fruits and vegetables helps ensure balanced nutrition.

VITAMIN C, BETA-CAROTENE, AND FOLIC ACID IN SELECTED RAW FRUITS AND VEGETABLES (PER 100 GRAMS)

Food	Vitamin C	Beta-Carotene (Vitamin A)	Folic Acid
	(mg)	*(IU)*	*(mcg)*
Broccoli	93.2	1,540	71
Brussels sprouts	85	883	61
Carrots	9.3	28,100	14
Chickpeas	4	67	557
Collards	23.3	3,330	11.5
Currants, black	181	230	—
Guava	184	792	—
Kale	120	8,900	29.3
Kiwifruit	98	175	—
Lemons	52	29	10.6
Lima beans, large	—	—	395
Mustard greens	70	5,300	187
Oranges	53.2	205	30.3
Peppers, hot chili	243	770	23.4
Peppers, sweet red	190	5,700	22
Pumpkin	9	1,600	16.2
Spinach	28.1	6,720	194
Strawberries	56.7	27	17.7
Sweet potatoes	22.7	20,100	13.8
Turnip greens	60	7,600	194

USDA Agriculture Handbooks Nos. 8–9 and 8–11 and NutriForm nutrition database.

Differences in nutrients Everyone realizes that different fruits and vegetables contain different amounts of nutrients, but nutrients also vary in individual samples of the same fruit or vegetable. There can, for example, be 20 times more vitamin A in one grapefruit than another! Here are a few values from the work of Ronald R. Eitenmiller et al., who published studies on the ranges of vitamin A, folic acid, vitamin C, and calcium in some fruits. You will notice that there is sometimes a difference between the findings of Eitenmiller published in 1987 and the older USDA Agriculture Handbook No. 8–9 figures, which are "averages."

RANGES OF NUTRIENT CONTENT OF SOME FRUITS IN ORDER OF DECREASING VITAMIN C							
	Vitamin C		Vitamin A		Folic Acid		Calcium
	(mg/100 g)		(IU/100 g)		(mcg/100 g)		(mg/100 g)
Fruit	Range (USDA)		Range (USDA)		Range (USDA)		Range (USDA)
Orange	49–57	(45)	349–830	(200)	18–22	(17)	14–29	(43)
Grapefruit	25–30	(37)	39–777	(259)	5.2–7.6	(9)	7–11	(15)
Cantaloupe	23–30	(42)	2290–2870	(3220)	18–30	(17)	4.9–8.7	(11)
Melon	12.5–33	(25)	119–495	(40)	4.5–7.1	(—)	2.4–5.4	(6)
Peach	6.4–9.0	(6.6)	532–755	(535)	3.3–4.2	(3.4)	2.6–5.7	(5)
Apple	1.7–6.1	(5.7)	45–415	(53)	0.6–1.1	(2.8)	3.3–5.8	(7)

Ronald Eitenmiller, et al. "Nutrient Composition of Red Delicious Apples, Peaches, Honeydew Melons, Florida Pink and Texas Ruby Red Grapefruit, and Florida Oranges." Research Report 526, University of Georgia, Agricultural Experiment Stations, December 1987, 21 pp., with USDA Agriculture Handbook No. 8–9 (1982) averages in parentheses.

Many people have suggested that supermarket produce makes a poor nutrient showing compared to farmers' market samples because the latter are assumed to be fresher and thus in possession of all of their vitamins and minerals. Freshness is not the only factor. The big difference in nutrient content from one grapefruit to another or one apple to another results from genetic differences in varieties of fruits and vegetables, growing conditions of the plant, maturity at the time of harvest, postharvest handling, storage conditions, and any processing. In 1987 and 1989 articles in the *Journal of Food Quality,* R. J. Bushway et al., compared vitamin C and alpha- and beta-carotene contents of supermarket produce and roadside stand produce. They found that supermarket and roadside produce had no practical differences in these nutrients.

Growing conditions that influence nutrient content are soil content, moisture, and sunlight. Plant products from soil that is deficient in nitrogen, for example, have lower protein content, while soil with excess nitrogen produces plant products with decreased vitamin C and fiber and with imbalances in amino acids that lower the protein quality. Similarly, excess phosphorus in the soil tends to decrease vitamin C, while excess potassium tends to increase it.

Reduced moisture produces plant products that are lower in vitamin C. Plants that get more sunlight have more vitamin C and beta-carotene. Temperature during the growing period also has an influence on vitamin content. The ideal growing temperature for higher vitamin C content is 68°F (20°C). The B vitamins accumulate best in green vegetables grown at 50° to 59°F (10° to 15°C) and best in tomatoes at 80° to 86°F (27° to 30°C). If there is any application of plant hormones, this will increase vitamin C and protein in many plants.

The quality factors: maturity and immediate cooling Probably the single most important factor in the quality of a fruit or vegetable is its maturity at harvest. Major chemical changes occur in a plant as it progresses from mature to ripe to declining. Vitamin C increases with maturation of green vegetables like snap beans but tends to decrease with advanced ripening in fruits such as peaches. Beta-carotene increases with the maturity of carrots but tends to decline in peanuts.

With such major differences caused by variety, growing conditions, harvest time, and postharvest care, there most certainly can be major differences in nutrient content of the same kind of fruit or vegetable.

In "How to Keep Produce Fresh Longer" (page 311), the vital importance of immediate chilling for quality is emphasized. Produce that is chilled in the field right after picking stays in better condition. Rapid chilling is very important for vitamin retention, too. Postharvest handling does not make significant changes in protein, fiber, or minerals in plant products, but it can make an incredible difference in vitamin content. Spinach can lose 50 percent of its vitamin C during its first 24 hours after harvest, even if it is kept at cool room temperature 68°F (20°C). Unfortunately, the consumer has no control over handling and can only observe the storage at the point of sale.

Cooking fruits and vegetables— the death of the cells

As we have seen, fresh fruits and vegetables are made of living, breathing cells. With heat from any source, plant cells die and experience dramatic changes. Regardless of the cooking method, the death of the cells brings on all these changes. Heat immediately denatures the cell membranes that control the flow of water out of the cells. The cells lose water and become limp even if they are cooked in water.

The insoluble pectic substances that provided rigid support between the cells convert into water-soluble pectins. Everything but the cellulose and lignin in the cell walls softens. Once the glue between the cell walls (the pectic substances) is converted to pectins and dissolves, the cells start to separate. They shrink as the proteins coagulate; they rupture and leak and lose water; and they separate and fall apart.

Many other changes take place with the death of the cells. As the cells soften and fall apart, the fruits or vegetables may experience changes in texture, color, taste, and starch and nutrient content.

Changes in texture: slowing down and speeding up cell breakdown

When a fruit or vegetable is heated, regardless of the method—boiling, steaming, broiling, baking, frying, or grilling—the living cells die, lose water, soften, and fall apart. These major changes in texture can be modified with certain ingredients. Sugar, calcium (in molasses, brown sugar, or hard water), and salt all have some effect.

Sugar helps fruits and vegetables retain their texture longer by slowing down the conversion of pectic substances around the cell walls. Fruit cooked with sugar remains intact when, without the sugar, it becomes mush. If you immediately put fruit into a concentrated sugar syrup to cook, the high sugar concentration on the outside of the fruit pulls water out of the fruit but maintains the firmness of the pectic substances around the cell walls. This is ideal for fragile berries that would otherwise fall apart as they cook.

Firmer fruit, though, may shrivel and get tough with too much sugar. A good procedure for cooking firm fruit such as apples is to start in water or a weak sugar solution. This tenderizes the fruit a little while not drawing out enough water to cause shriveling. Then, as the fruit cooks, you can add more and more sugar to keep the pectic substances firm.

These two recipes for Homemade Applesauce and cooked Apple Wedges illustrate this. In the first, the apples are cooked without sugar, become mushy, and are easily mashed into applesauce. In the second, the same apples cooked in the same way but with sugar remain firm.

Homemade Applesauce

MAKES 4 SERVINGS

Having not had homemade applesauce since I was a child, I was startled when I made this recipe: A dish I had dismissed as ordinary was extraordinarily good.

> **what this recipe shows**
>
> Heating causes fruit or vegetable cells to lose water and soften; pectic substances that hold cells together convert to soluble pectins and dissolve, cells fall apart, and the fruit or vegetable becomes mushy.

5 medium apples (I like Golden Delicious), peeled, cored, and cut into wedges	$^1/_2$ cup packed light brown sugar 1 teaspoon ground cinnamon $^1/_2$ teaspoon salt

Microwave the apples in a glass container covered with plastic wrap for 10 minutes on High or until very soft (see Note). Mash the apples into applesauce with a potato masher. Stir in brown sugar, cinnamon, and salt. Serve hot or cold.

Note: *The apples can be cooked on the top of the stove in a heavy covered saucepan. Add 3 tablespoons water and cook over medium-low heat until apples are very soft, about 20 minutes.*

Apple Wedges

MAKES 4 TO 6 SERVINGS

Simple apple wedges are a refreshing addition to a meat sauce and are excellent stirred into cooked vegetables like carrot slices. The cooking time here is longer to prove a point, but for most dishes I cook them for only 4 to 5 minutes so that they retain a little crispness.

> **what this recipe shows**
>
> Adding the sugar before cooking preserves the insoluble pectic substance "glue" that holds the cells together and keeps the apple wedges intact.
>
> Brown sugar, containing calcium, also helps in preserving firmness.

5 medium apples (I like Golden Delicious), peeled, cored, and cut into wedges $^1/_2$ cup packed light brown sugar 1 teaspoon ground cinnamon	$^1/_2$ teaspoon salt 2 tablespoons butter

Stir together the apples, brown sugar, and cinnamon in a glass mixing bowl. Cover with plastic wrap and microwave on High for 8 minutes (see Note). Sprinkle with salt, add butter, and stir gently. Serve hot or cold.

Note: *The apples, sugar, and cinnamon can be cooked on the top of the stove in a heavy covered saucepan. Add 3 tablespoons water and cook over medium-low heat until wedges have softened slightly, about 15 minutes.*

Adding molasses to baked beans helps retain their shape for long periods of cooking.

Calcium reacts with the pectic substances to form insoluble calcium compounds that make food firmer. Calcium is present in molasses, brown sugar, and hard water. If you add molasses early when cooking baked beans, they will take much longer to get tender. Another way to look at it is that this allows the beans to retain their texture in long-cooked dishes like Boston baked beans. Green beans and some other vegetables can become so tough when cooked in hard water that they are inedible.

On the other hand, calcium compounds are sometimes added to tomatoes during processing to keep them firmer and prevent loss of shape. For the same reason, calcium is sometimes added to fragile fruits like raspberries when they are processed. You can keep whole raspberries firmer in a sauce by adding a tablespoon or so of brown sugar.

Many recipes for vegetables instruct you to drop them into boiling salted water, and some, my grandmother's included, tell you to add a pinch of sugar as well. Having the water boiling kills enzymes and limits vitamin loss—that I understood. The sugar, my grandmother said, would make my vegetables garden sweet—in many cases this was just adding a little sugar that Mother Nature would have provided if the produce had not been picked before it was fully mature. But what about the salt? Chances are the main reason for salt in the water when you cook vegetables is flavor. However, salt does slightly elevate the boiling temperature of water, and a 1990 article by J. P. Van Buren et al. in *The Journal of Food Science* pointed out that snap beans get softer 10 percent faster in salted water than plain water. So salt in the water does make some vegetables cook a breath faster.

Changes in color: keeping them bright

Heat directly affects the color as well as the texture of many fruits and vegetables. When cooking, the trick is to retain as much of the natural color as possible without major sacrifices of texture or nutrients.

Blanching and color Blanching vegetables—dropping them briefly into boiling water (less than a minute), then plunging them into cold water to stop the cooking—causes a dramatic change in color. Caterers often blanch vegetables for raw vegetable platters to get brilliant colors and a more attractive presentation. The vegetables are still essentially raw, just much brighter in color.

Fresh vegetables have air between the cells. This film of air clouds the true bright colors of vegetables. When you drop a vegetable into boiling water, the air expands, floats to the surface, and bubbles away. The vegetable suddenly becomes much brighter in color. Broccoli and green beans turn from a pale whitish green to a brilliant vivid green. Even the change in color of bright-colored vegetables like carrots is apparent.

Keeping the greens green

All green vegetables are green because they contain two types of chlorophyll—chlorophyll a, which is bright blue-green, and chlorophyll b, which is bright yellow-green. Green vegetables have three times as much of the a-type as the b-type, which explains their bright blue-green color. This vibrant green created by the two types of chlorophyll will change, however, if the magnesium in the chlorophyll is lost from the compound. Unfortunately, the chlorophyll in the cells loses its magnesium easily when heated.

Cooking time Hydrogen, present in all acids, immediately replaces the magnesium and causes bright green chlorophyll to go to a sad, brownish yellow-green. All vegetables contain acids. When these vegetables are heated, their cells are damaged and killed so that the acids that were kept separated from the chlorophyll in the living cells now come in contact with it and the color change begins. The amount of acid varies in vegetables, so there is a slight difference in how fast this change will take place in different vegetables. Essentially, with most green vegetables, you can count on having 7 minutes of heat before there is a major color change. After 7 minutes of cooking time, the color change becomes more and more pronounced.

To preserve the bright green color, you need to cook a green vegetable for only 7 minutes, but some vegetables, like big stalks of broccoli or fat green beans, will not get tender in that time. You can cut larger vegetables into smaller pieces that will get done in 7 minutes—trim the broccoli into 1½-inch florets and slice the stems into ⅓-inch slices. This may have been the reason for French-cut green beans. If your primary desire is tenderness, you probably will have to cook longer than 7 minutes and accept the color change.

Rule of thumb: Cook green vegetables for less than 7 minutes in order to retain their bright color.

Cooking methods can have some influence on saving the green, but the 7-minute rule cannot be ignored. Cell destruction is cell destruction, regardless of the cooking method. The classic French way of cooking vegetables by plunging them into a large pot of rapidly boiling salted water and cooking uncovered until just tender is an excellent scientific approach. The large amount of water dilutes the acids as they leak out, and having the pot uncovered lets the volatile acids evaporate. This method minimizes the amount of acid in contact with the chlorophyll.

Steaming is also a good method. As the acids come out of the vegetables, they are constantly rinsed away to the bottom of the pot by the steam. Stir-frying is another excellent method. Vegetables for stir-fries are cut into small pieces to cook rapidly. The cooking is uncovered, letting volatile acids evaporate. Cooking is very fast on the high heat, which limits the time for the chlorophyll to lose its color.

Some books advise adding a pinch of soda to the cooking water to neutralize the acids and keep vegetables bright green. This does keep vegetables a beautiful green; however, alkalis like baking soda also destroy the cell walls in crisp vegetables, and the vegetables become mushy.

Here are basic directions for Steamed Broccoli and Chili Oil, a recipe for enhancing the flavor of steamed vegetables. Notice that the Chili Oil does not contain any acid.

Steamed Broccoli

Trim the florets into pieces about 1½ inches long. Peel away any tough part at the bottoms of the stalks, then slice stalks on the diagonal into ⅓-inch slices. Steam the slices for 2 minutes, then add the florets and steam an additional 5 minutes. Drain, toss with Chili Oil (recipe follows), and serve immediately.

To steam ahead, rinse the broccoli in cold water immediately after steaming to stop cooking. Drain well and store refrigerated in a tightly sealed zip-top plastic bag. When ready to serve, toss with the dressing and reheat briefly over medium heat in a large skillet. Serve immediately.

Chili Oil

ENOUGH FOR 1½ BUNCHES (ABOUT 4 STALKS) OF STEAMED BROCCOLI

My daughter, who has examined many religions, said the broccoli with chili oil was a sensation at a Buddhist breakfast. I suppose you could say this got rave reviews from vegetable experts.

> **what this recipe shows**
>
> How much of the white veinlike material that holds the seeds in the pepper you include determines the heat imparted by the pepper.
>
> This is a vinaigrette without the vinegar. Some flavor components dissolve in oil and some in water or water-type liquids like vinegar. This recipe contains both oil and water to carry more flavors. The water replaces vinegar to prevent acidity from discoloring the vegetable.

1 medium jalapeño with half of the seeds and veins removed, minced	3 tablespoons water
1 shallot, minced	½ teaspoon sugar
⅓ cup peanut, corn, or blended vegetable oil (see Notes)	1 teaspoon salt
	¼ teaspoon white pepper
	2 bunches Steamed Broccoli (page 327)

Heat the jalapeño, shallot, and oil in a small saucepan over medium heat for several minutes. Let stand for 5 to 10 minutes for flavors to meld. While the oil is still warm, stir in water, sugar, salt, and pepper. When ready to serve, spoon the flavored oil over the broccoli and toss gently.

Notes: *This oil is also excellent on other vegetables.*
Do not use strong olive oil; its flavor will interfere.

The role of cooking utensils The type of cooking utensil can also affect the color of green vegetables. Iron or tin replaces magnesium and causes chlorophylls to change to a brownish green. An iron skillet or tin-lined pan will discolor green vegetables. Old cookbooks tell readers to cook green vegetables in unlined copper pots. A beautiful vivid green copper-chlorophyll forms, but this much copper is not safe in the diet. Food processors sometimes add a chlorophyll compound that contains a harmless trace amount of copper to keep peas bright green, but this is not available to home cooks.

It is easier to preserve the green in frozen vegetables than fresh. Before they were frozen, the vegetables were blanched, then drained and flash-frozen. Some of the acids in the vegetables that cause the change to olive drab are removed in blanching. You could, of course, do this too. If you need perfect bright green vegetables and are afraid that your preparation circumstances might not be the best (as might be the case in some catering situations), blanch the vegetables and plunge them immediately into ice water to stop the cooking. Dry and refrigerate tightly sealed in a zip-top plastic bag with the air squeezed out until ready to cook. Cook the vegetables uncovered in a large quantity of boiling salted water for no longer than 5 minutes for a perfect green color.

"Acidic" dressings If the small amount of acid exuding from vegetable cells during cooking is enough to make green vegetables discolor, it's easy to guess that an acidic dressing will wreck their color. Bright green snow peas and broccoli in a pasta primavera with a vinaigrette discolor with standing. Marinated asparagus or green beans discolor after a short period with the dressing on them. Dressings with lemon juice or vinegar are such a nice taste complement to vegetables—is there anything you can do to use these and still preserve the color?

There are three possibilities for preventing color problems: avoid acid, dress the vegetable just before serving, or keep the vegetable raw.

1. *Avoid acids.* I pointed out that the Chili Oil for Steamed Broccoli is essentially a vinaigrette with water instead of vinegar. You can also use citrus zest (but not the juice) for flavor without acid, and the Roasted Asparagus with Lemon-Chili Oil (recipe follows) is a perfect example of this.

2. *Dress the vegetables just before serving.* If the dressing is mildly acidic, it takes a little time for discoloration to occur. The combination of a low-acidic dressing and pouring the dressing on the green vegetable just as it is served can give you a salad that will remain bright for about 20 minutes—long enough for the salad to be served and eaten. The Fresh Green Bean Salad with Basil and Tomatoes (page 330) illustrates this.

3. *Keep the vegetables raw.* For my husband's birthday we had a family bring-a-dish supper. My daughter-in-law, Beth McCool, an opera singer who is also a good cook, brought a broccoli salad that was unusual and very good. We all had seconds and sat around the table talking for about an hour. I expected the broccoli to turn olive-drab in the vinegary dressing, but when I fished out one of the pieces, it was still a good green—the same color as raw broccoli.

The broccoli in the salad *was* raw. I guessed that the natural protective coating of the raw vegetable had slowed or prevented the acid in the dressing from getting to the cells and causing the loss of bright green chlorophyll and decided to check that theory. Cutting a piece of raw broccoli in half, I put one half in the dressing raw and blanched the other half, which would remove its protective coating, and then put it in the dressing. When both halves went into the dressing, the blanched half was a really vivid, bright green; the raw broccoli was a good green but not as bright as the blanched. In about 30 minutes, I pulled out both pieces and rinsed off the dressing. Sure enough, the blanched piece that had had its protective coating removed was now a brownish, olive-drab green, while the raw piece had not changed color at all.

So, as you can observe in the Broccoli Salad with Bacon and Sweet-Sour Dressing (page 332), raw broccoli fares well in an acidic dressing.

Roasted Asparagus with Lemon-Chili Oil

MAKES 6 SERVINGS

I started playing with flavored oils as a fat reduction technique—a very small amount of an intense-flavored oil will give great taste to a bland low-fat dish. Then I tried Barbara Tropp's Chili-Orange Cold Noodles in her *China Moon Cookbook* and became even more hooked on flavored oils. A little lemon-chili oil and lemon zest give real fresh lemon flavor to asparagus. This is an ideal light vegetable to accompany many main courses—perfect with roast chicken (page 389).

what this recipe shows

Citrus fruit (lemon, lime, orange, or tangerine) zest can impart a fresh citrus flavor to vegetables without discoloring them as acidic fruit juices do.

Water in the dressing dissolves and carries water-soluble flavors.

1 shallot, minced

1 teaspoon dried red pepper flakes

1/4 teaspoon white or black pepper

1/4 cup peanut, corn, or blended vegetable oil

Finely grated zest of 3 lemons

1 teaspoon water

25 to 30 spears fresh asparagus, about 1 pound

1/2 teaspoon salt

1. Bring the shallot, pepper flakes, ground pepper, and oil to a simmer in a small saucepan. Simmer over very low heat for about 4 minutes. Remove from the heat and let stand for 5 minutes. Stir in the zest of two lemons only (save one to garnish) and the water. Let stand for at least an hour, then strain the oil into a small bowl.

2. Preheat the broiler or preheat the oven to 500°F (260°C).

3. Snap the tough bottoms off the asparagus spears. Arrange in an oblong heatproof dish, stir in 2 tablespoons of the prepared flavored oil, and roast the spears about 3 inches from the broiler for 5 minutes, or 6 to 7 minutes on the top shelf of the oven. Spoon the rest of the flavored oil over the asparagus. Sprinkle with salt. Taste and add more salt if needed. Sprinkle with the remaining zest and serve immediately.

Fresh Green Bean Salad with Basil and Tomatoes

MAKES 8 SERVINGS

Real summer tomatoes, ripe, deep red, and full of flavor, are true treasures of the earth. Crisp-cooked, bright green beans are piled high in the center of a white platter and encircled with overlapping vivid red tomato slices. Fresh basil and a basil dressing finish off this refreshing salad, a great summer buffet dish.

Dressing the green beans right before serving maintains their green as long as possible.

Minimizing the amount of vinegar in the dressing preserves the bright green longer.

1½ pounds fresh green beans
(see Note)

1 tablespoon sugar

1 tablespoon *and* 1 teaspoon salt
(4 teaspoons *total*)

1 recipe Garlic-Basil Dressing
(below)

5 firm ripe tomatoes, sliced

4 sprigs fresh basil for garnish

1. Leave the pointed tips on the green beans, but snap off the stem tips. Wash. Bring a large pot of water to a boil. Add the sugar and 1 tablespoon salt. Drop in the green beans and cook until crisp-tender, about 6 to 7 minutes. Drain the beans and plunge them into ice water to stop cooking. Drain well and refrigerate.

2. When ready to serve, toss the green beans with ⅓ cup of the dressing in a large mixing bowl. Taste and add more dressing or salt as needed. Pile the beans high in the center of a large white platter. Arrange the tomato slices overlapping around the edge. Sprinkle the tomatoes with salt and drizzle 3 tablespoons of the dressing on top of them. Garnish with the basil sprigs. Serve immediately.

Note: *This is one of the better ways to use canned beans if you cannot get fresh. The dressing is flavorful enough to cover any canned taste. Buy high-quality canned whole beans and rinse several times.*

Garlic-Basil Dressing

1 clove garlic

1 shallot

2 tablespoons red wine vinegar
(or water; see Note)

1½ teaspoons salt

⅛ teaspoon freshly ground black pepper

2 tablespoons Dijon mustard

1 tablespoon sugar

¾ cup canola, corn, or vegetable oil

15 fresh basil leaves

Turn on the processor with the steel knife and drop the garlic and shallot down the feedtube onto the spinning blade to mince. Add the vinegar, salt, pepper, mustard, and sugar. With the processor running, slowly drizzle in the oil. Add the basil leaves and coarsely chop with several on/off pulses.

Note: *If you need the salad to remain bright green, eliminate the vinegar altogether and substitute water.*

Broccoli Salad with Bacon and Sweet-Sour Dressing

MAKES 6 TO 8 SERVINGS

This combination of raw broccoli, sweet onion, raisins, and bacon in a sweet-sour dressing is astonishingly good and is frequently the first dish emptied on a buffet. It is quick and easy to prepare and ideal for bring-a-dish dinners.

what this recipe shows

Keeping the broccoli raw with its natural protective coating intact helps prevent acid from causing discoloration.

The high acid content of the sweet-sour dressing keeps down bacterial growth and makes this dish safer for outdoor picnics.

½ cup Homemade Mayonnaise
(page 289) or store-bought (see Note)

⅓ cup sugar

½ teaspoon salt

3 tablespoons cider vinegar

1 medium-size sweet onion, such as
Vidalia, Texas, or Walla Walla, or
1 white onion, chopped

1 bunch broccoli, 3 to 4 stalks, florets
and stems separated

⅓ cup raisins

8 pieces crisp-cooked bacon

1. About 3 to 4 hours before serving time, stir together the mayonnaise, sugar, and salt in a medium mixing bowl. Add the vinegar a tablespoon at a time, stirring in well after each addition to thin evenly and prevent mayonnaise lumps. Add the onion. Cut the broccoli florets into small bite-size pieces and cut the stems into ⅓-inch slices. Stir the broccoli into the dressing and refrigerate. Stir several times before serving.

2. When ready to serve, stir in the raisins. Crumble the bacon and stir in. Taste and add salt as needed, stir in well, and taste again. Serve cold.

Note: *This dish is good even with reduced-fat mayonnaise.*

glance

Keeping greens green

What to Do	Why
Use a large quantity of water as if you were cooking pasta.	To dilute the acids from the vegetables.
Add salt to the water.	For flavor and for slightly faster cooking.
Have the salted water at a rolling boil before you add the vegetables.	To prevent vitamin C loss.
Cook vegetables uncovered.	To allow volatile acids to evaporate from the boiling water.
Regardless of method, time cooking carefully.	Chlorophyll loses its bright green color after 7 minutes of cooking.

Keeping the reds red

Three major groups of compounds give red fruits and vegetables their colors: anthocyanins, betalains, and carotenoids. Anthocyanins and betalains are water-soluble pigments. Anthocyanins give many fruits and vegetables their beautiful red, purple, and blue colors, and betalains are responsible for the red color of beets. The carotenoids are red-orange compounds that color red tomatoes and red bell peppers but are primarily responsible for orange colors. Carotenoids dissolve in oil, not water, and are described on page 337.

Keeping nuts from turning blue, red cabbage red, and cherry muffins from having blue circles—anthocyanins   Eggplant, radishes, red-skinned potatoes, red cabbage, cherries, red grapes, blueberries, blackberries, raspberries, and strawberries all get their red and blue hues from anthocyanins. Some fruits and vegetables contain only two or three types of anthocyanins, while others have as many as twenty. Produce containing anthocyanins may also contain chlorophyll and/or carotenoids as coloring agents. Fruits that have anthocyanins in their skins but not in their flesh include some varieties of cherries, apples, plums, and peaches. Compounds that are derivatives of anthocyanins contribute to the color of some cherries, cranberries, currants, elderberries, purple figs, peaches, plums, raspberries, rhubarb, and purple turnips.

Since they are water soluble, the blue-red anthocyanin colors easily leach out of the fruit or vegetable into the sauce or cooking liquid. In the recipe for Cheddar-Crusted Chicken Breasts with Grapes and Apples in Grand Marnier Sauce (page 335), the color of the Red Flame grapes and apple skins will wash out into the sauce upon standing. Therefore, the grapes are added just before serving.

Anthocyanins change color as acidity/alkalinity changes. They remain red as long as they are acidic but turn blue as they become alkaline. Red cabbage may, for no apparent reason, turn an unappetizing blue while cooking—because some of the acids have evaporated. To bring back the red color, add a little vinegar or lemon juice (a teaspoon will probably be enough) or cook the red cabbage with something acidic like apples.

Cherry muffins may have a blue discoloration around the cherries because of the baking powder or baking soda (alkalis) in the batter. You can prevent this color change by adding something acidic or by substituting buttermilk or sour cream for milk in the recipe.

Walnuts contain some of these red-hued compounds next to the skin and may cause a blue discoloration in foods. Some walnut breads have a bluish color. If you want to prevent this color change, roast the nuts. The temperatures in roasting are high enough to cause a reaction that changes these compounds to others that do not discolor. Roasting for about 10 to 12 minutes at 350°F (177°C) is sufficient.

When anthocyanins occur with other phenolic compounds, you can get strange colors. Substances called *flavonols* or *flavones* turn yellow with alkalis and combine with the blue from anthocyanins to give odd greens and grayish greens. Apparently a similar compound that occurs in sunflower seeds turns a magnificent jade green when alkaline.

Beets—betalains   Betalains, the color compounds in beets, are also sensitive to acidity/alkalinity. Below a pH of 4 (acidic), the color shifts to violet. Above a pH of 10 (alkaline), the color changes to yellow. Beets have so much color content that they do not become pale when cooked in water even though a lot of color is leached out of the beets themselves.

at a glance
Keeping reds red

What to Do	Why
Add lemon juice or vinegar to red cabbage to prevent discoloration. Or cook with something acidic like apples.	To keep the cabbage acidic.
Substitute a little lemon juice, sour cream, or buttermilk (something acidic) for some liquid in the batter to prevent a blue discoloration around cherries and berries containing anthocyanins.	To keep the batter acidic.
Roast walnuts in a 350°F (177°C) oven for 10 to 12 minutes to prevent their discoloring baked goods or sauces.	To change anthocyanins to other color-stable compounds.
For a deep violet red or a deep red color in beets, keep beets very acidic for violet red, mildly acidic for deep red.	Betalains, the red compounds in beets, are violet red at pHs below 4 and red at pHs below 10. They can change to yellow if alkaline (above pH 10).

Cheddar-Crusted Chicken Breasts with Grapes and Apples in Grand Marnier Sauce

MAKES 8 TO 10 SERVINGS

Cheese-browned, moist chicken breasts with a rich-looking mahogany sauce are served with Sherried Rice and Barley with Almonds (recipe follows). This awesome centerpiece dish will impress even haute-cuisine friends, but the flavors are hearty and reminiscent of wonderful Sunday dinners of your childhood. As a bonus, it's low fat and full of soluble fiber—pectin in the apples plus fiber in the barley and brown rice.

what this recipe shows

Adding apples and grapes just before serving minimizes the color loss that results from the anthocyanins dissolving in the sauce upon standing.

1 medium onion, quartered

1/2 teaspoon dried sage

1 1/2 cups homemade Chicken Stock (page 267) or canned chicken broth

4 cups water

10 skinless and boneless chicken breast halves

2 cans (10 1/2 ounces each) beef consommé

2 tablespoons dark brown sugar

1/2 cup orange marmalade

1/4 cup cornstarch dissolved in 1/4 cup cold water

3 red apples (Rome or McIntosh), cored and cut into wedges

3 tablespoons lightly salted butter

Nonstick cooking spray

1/4 pound cheddar with good orange color, grated (about 1 cup)

Sherried Rice and Barley with Almonds (page 336)

1 small bunch Red Flame grapes, about 30 grapes

3 tablespoons Grand Marnier

1. Bring the onion, sage, stock, and water to a boil in a large saucepan. Add the boneless chicken breasts. Bring back to a simmer, simmer for 2 minutes, cover, and remove from heat. Let stand about 10 minutes. Remove the chicken breasts to a plate and cover to keep warm. (This double stock from cooking the chicken is excellent to cook the rice and barley in.)

2. Heat the consommé, brown sugar, orange marmalade, and cornstarch-water mixture in a large saucepan over medium-high heat until it reaches a slow boil, stirring constantly. Remove the sauce from the heat and set aside.

3. Sauté the apples in butter in a heavy skillet over medium heat until just browned, 2 to 3 minutes on each side. Set aside.

4. Spray a baking sheet with nonstick cooking spray. Dry the chicken breasts and place them on the baking sheet. Sprinkle each chicken breast with grated cheddar and slip under the broiler just to melt and brown the cheese.

continued

5. Mound up the hot Sherried Rice and Barley with Almonds in the center of a large platter. Arrange the chicken breasts against the sides of the mound. Reheat the sauce and stir in the grapes, apple wedges, and Grand Marnier. Spoon the grapes and apples and a little sauce over the top of the mound, letting the grapes and apples cascade down the side of the mound. Drizzle some sauce over each chicken breast and serve.

Sherried Rice and Barley with Almonds

MAKES 8 TO 10 SERVINGS

I think we have forgotten how good barley is. It seems that no matter how big a batch I make of this, I never have any left over. The heartiness of the brown rice and barley is tempered by the sweetness of the apples and onions, and the sherry and almonds add a touch of class.

what this recipe shows

Grains that cook in approximately the same length of time can be cooked together.

A small amount of dried fruit can impart good flavor.

2 medium onions, chopped

1 tablespoon canola, corn, peanut, or blended vegetable oil

2 tablespoons sugar

1 cup brown rice (see Note)

1 cup barley (see Note)

8 dried apple slices, chopped

4 cups water, homemade Chicken Stock (page 267), or canned chicken broth

2 tablespoons *and* 1 tablespoon butter (3 tablespoons *total*)

2 teaspoons *and* 1/4 teaspoon salt (2 1/4 teaspoons *total*)

3 tablespoons dry sherry

1 cup roasted almond slivers

1. Sauté the onions in oil in a heavy pot that has a lid over medium-high heat until soft. Sprinkle with the sugar and continue cooking for 1 minute. Add the rice, barley, apples, water, 2 tablespoons butter, 2 teaspoons salt, and sherry. Stir well and bring to a boil. Cover and cook at a low simmer for 40 minutes.

2. While the rice is cooking, preheat the oven to 350°F (177°C). Spread the almonds on a baking sheet and roast until lightly browned, about 10 minutes. Stir in 1 tablespoon butter and 1/4 teaspoon salt while the nuts are hot.

3. Heat the rice-barley mixture an additional minute with the lid off and toss gently with a fork. Stir in half of the roasted almond slivers. Sprinkle the remaining half of the almonds over the top to garnish. Serve immediately.

Note: *Both brown rice and barley are now available in 10-minute-cooking brands, which makes this a quick dish. Cook the grain and apple mixture for only 10 minutes in Step 1.*

Keeping the oranges orange—carotenoids

It's much easier to retain the color of yellow, orange, and red-orange fruits and vegetables than it is for green or red ones. Carotenoids, the yellow, orange, and red-orange pigments in fruits and vegetables, dissolve in oil, not water, and are relatively stable. Unless they are badly overcooked, they do not lose their color. Carotenoids are so stable that they are used as natural colorants for other foods. Extracts of annatto, alfalfa, carrots, paprika, and tomatoes are all used as coloring agents.

Many chemical compounds have the same basic composition but differ only in the detailed way that their atoms are arranged. Carotenoids change their structure with prolonged overcooking, and this is why their color eventually changes. As they go from the intensely colored *trans* structure to the paler *cis* structure, there is a corresponding change in the color of the vegetable.

You may have noticed that badly overcooked yellow squash becomes pale and washed-out looking. In the Country Summer Squash recipe, the squash is cooked for only about 10 minutes, so there is hardly any change in color. The dish as a whole is paler because of the onions. Carrots have such large amounts of carotene that they usually retain their color even when overcooked.

Country Summer Squash

MAKES 4 TO 6 SERVINGS

This is my mother's method of cooking squash, which my family adores. Mine never came out like hers until she explained that she mashed the squash with a potato masher. This is a versatile vegetable dish, good with pork, beef, chicken, or fish.

what this recipe shows

Yellow, orange, and red carotenoid-containing fruits and vegetables can tolerate longer cooking times without color change than can green vegetables.

6 medium or 8 small (about 2½ pounds) yellow crookneck squash, cut into ½-inch slices

1 teaspoon sugar

1 tablespoon *and* 1 teaspoon salt (4 teaspoons *total*)

3 strips bacon

2 large onions, chopped

¼ teaspoon white pepper

1. Place the squash in a large saucepan with enough water to cover. Add sugar and 1 tablespoon salt. Cook uncovered at a low boil until the squash is very tender when pierced with a fork, about 10 minutes. Drain and place the squash in a medium mixing bowl. Mash thoroughly with a potato masher. Set aside.

2. Fry the bacon in a large skillet until crisp. Remove and set aside on paper towels to drain, saving drippings. Add the onions to the drippings in the skillet and cook over medium heat until soft and just beginning to brown, about 4 minutes. Add the squash and cook for several minutes, stirring constantly. Add 1 teaspoon salt and white pepper. Crumble the bacon and stir in. Taste and add more salt or pepper if necessary. Serve hot.

Rutabagas are interesting. Raw, they contain a large amount of the pale cis structure of carotenoid pigments. When cooked, they change to the more intense trans form and go from a pale orange to a deep orange.

Since vegetables containing carotenoids retain their colors well, they are ideal candidates for longer cooking methods like baking, and here are three examples. Old-fashioned Grated Sweet Potato Pudding (recipe follows), a family recipe with little resemblance to modern sweet potato casseroles, is a refreshing change for holiday meals. The Great Baked Pumpkin with Herbed Barley and Chickpeas (page 339) is an outstanding recipe. The Golden Tomato Bake (page 340) is a modernized version of old Southern stewed tomatoes.

Old-fashioned Grated Sweet Potato Pudding

MAKES 8 SERVINGS

One of my students prepares this dish, my grandmother's recipe, every Christmas. One year she took it to a nursing home, where tears rolled down the cheeks of a patient. The elderly lady said, "I haven't had sweet potatoes like this since I was a girl." Indeed this has been a traditional sweet potato preparation since my great-grandmother's time. The sweet potatoes are grated raw and baked as a custard with ginger as the predominant flavoring.

what this recipe shows

Deep orange carotenoid-containing fruits and vegetables can be cooked for longer times than green vegetables without loss of color or nutrients.

Stirring after partially baking prevents the outer edges from overcooking before the center is done.

Nonstick cooking spray

1 pound (2 large or 3 small) sweet potatoes peeled and cut into chunks

1/2 cup packed dark brown sugar

1/2 cup packed light brown sugar

3/4 teaspoon salt

2 teaspoons ground ginger

2 tablespoons cornmeal, white or yellow

1 large egg

2 large egg yolks

1 cup heavy or whipping cream

1 tablespoon pure vanilla extract

1. Preheat the oven to 325°F (163°C). Spray a 9 × 13 × 2½-inch casserole with nonstick cooking spray.

2. Finely chop the sweet potatoes to the texture of large rice in several batches in a food processor with the steel knife, using on/off pulses. Mix the sweet potatoes, brown sugars, salt, ginger, and cornmeal in a large mixing bowl. Stir in the egg, egg yolks, cream, and vanilla.

3. Pour into the casserole and bake for 15 minutes. Stir from the outside to the middle (see Note). Continue baking and stir again after 10 minutes. Cook until lightly browned and just set, about 40 minutes in all. Serve hot or at room temperature.

Note: *The edges of a baked casserole tend to overcook and get dry before the center is done. My grandmother stirred twice during cooking before the custard set firmly. You can also use Bake-Even Strips, which are strips soaked in water that evaporate during cooking to cool the sides of the pan (see Sources—Bake-Even Strips, page 477).*

Great Baked Pumpkin with Herbed Barley and Chickpeas

MAKES 12 SERVINGS

Even a huge roasted-to-perfection turkey takes a backseat to this magnificent browned pumpkin filled with the most intriguing aromas. My daughter, Terry, and I are fond of Annemarie Colbin's recipes in both *The Book of Whole Meals* and *Natural Gourmet*. Terry prepares a stuffed pumpkin similar to Annemarie's. I tried to preserve the drama of the dish in my own version. A small- to medium-size pumpkin is ideal for this dish.

what this recipe shows

Orange carotenoid-containing fruits and vegetables retain their color and nutrients well during cooking.

Juices from the pumpkin add sweet complex flavors to the grain and bean filling.

1 fresh pumpkin, about 10 inches in diameter (see Notes)

2 large onions, chopped

1/4 cup dark sesame oil (see Notes)

2 cloves garlic, minced

3 cups quick-cooking barley

3 cups homemade Chicken Stock (page 267) or canned chicken broth

3 cups water

2 cans (16 ounces each) chickpeas, drained

2 cans (16 ounces each) navy beans, drained

1 tablespoon dried oregano

2 teaspoons dried basil

1 teaspoon dried rosemary

1/2 cup orange marmalade

2 tablespoons light brown sugar

1/2 cup chopped parsley

1/4 cup soy sauce

1 teaspoon salt

1/2 teaspoon white pepper

1 to 2 tablespoons Chinese hot chili oil

Large green leaves, for garnish

1. Cut the top from the pumpkin slightly lower than for a jack-o'-lantern. Scoop out and discard the seeds. Set aside.

2. Preheat the oven to 400°F (204°C). Cook onions in sesame oil in a large skillet over low to medium heat until the onions are just clear, not brown. Stir in the garlic and cook briefly. Cook the barley in the stock and water for 10 minutes at a low simmer in a large covered saucepan.

3. Stir together the onion mixture and the rest of the ingredients except garnish in a large mixing bowl. Spoon the mixture into the pumpkin and fit the lid back in place. Place on a small pizza pan in the lower center of the oven and bake for 1 hour. The pumpkin should be lightly browned.

4. Serve the pumpkin surrounded by large green leaves. Remove the lid and lean it against the side of the pumpkin. Serve a piece of the pumpkin with the filling.

Notes: *If a pumpkin is not available, use another large squash, such as spaghetti squash, turban squash, or another variety.*

Sesame oil should not be heated to high temperatures and ordinarily is used in small quantities as an added flavoring. The temperature is kept low here to avoid damage to the oil.

Golden Tomato Bake

MAKES 12 SERVINGS

In this modern version of my grandmother's stewed tomatoes, a crunchy, buttered crumb topping covers tomatoes, which are accented with fresh basil and lemon zest. Jane Brock, an outstanding creative cook and product developer in Jackson, Mississippi, introduced me to the simple crusty topping, which I like much better than stirring in the crumbs.

what this recipe shows

Baking tomatoes leaves them with their color but not their texture, making them good for casseroles.

Mixing the butter with some of the crumbs produces more even distribution of the butter than simply pouring it on top.

4 cans (14^1/$_2$ ounces each) diced tomatoes

4 medium-size ripe tomatoes, peeled, seeded, and chopped, or 2 cans (14^1/$_2$ ounces each) tomato wedges, drained (if using fresh tomatoes, save 1 long strip of peel)

1^1/$_2$ teaspoons salt

1 teaspoon white pepper

Finely grated zest of 1 lemon

9 fresh basil leaves, sliced thinly lengthwise

1^1/$_2$ cups *and* ^{1}/$_2$ cup fine bread crumbs (2 cups *total*)

^{1}/$_3$ cup packed light brown sugar

6 tablespoons butter, softened

2 pretty sprigs fresh basil to garnish

1. Preheat the oven to 350°F (177°C).

2. Stir together the diced tomatoes and fresh tomatoes in a large mixing bowl. Add the salt, pepper, lemon zest, and sliced basil leaves and stir together well. Taste and add salt if necessary. Spoon the tomato mixture into a large 9 × 6 × 3-inch baking casserole. Sprinkle evenly with 1^1/$_2$ cups crumbs.

3. Mix together the brown sugar, ^{1}/$_2$ cup bread crumbs, and butter in a mixing bowl. Sprinkle the butter-crumb mixture evenly over the top of the tomato mixture. Bake for 20 to 25 minutes, until the tomatoes are bubbly and the crust is lightly browned. Garnish with a tomato rose made by coiling around a long piece of tomato peel. Arrange the fresh basil sprigs with the rose. Serve hot, at room temperature, or even cold.

MORE CHAMELEONS: TANNINS AND OTHER COLOR CULPRITS

Why does cauliflower sometimes turn pink or brown when cooked? What makes some brewed coffees and teas look cloudy and muddy instead of clear? The culprits are tannins, the astringent components in persimmons or strong tea that make your mouth pucker. Foods containing tannins—including many fruits, especially apples, peaches, grapes, nuts like almonds, tea leaves, and coffee and cacao beans—undergo various color changes when cooked in alkaline or hard water. Just the tiniest trace of metal (5 parts per million) will make tannins darken. If you have hard water that contains metallic ions, you may want to use bottled water for coffee or tea or water exposed to a silver-containing filter or other system that removes metals.

Is your coffee or tea cloudy or muddy? Use bottled water—your tap water is probably hard.

Not surprisingly, the other solution is to counteract alkalinity with some acid. Icings that contain coffee and egg yolks will turn green if they become alkaline. Just add a little acid like lemon juice, cream of tartar, vinegar, or brown sugar (which is acidic) to prevent this green.

Hard water, or water that has been softened, is alkaline and can cause color changes in potatoes, rice, and onions, which contain flavonoids. Potatoes cooked in hard water will turn a cream color or even have bands of yellow. You can avoid this color change by adding about $1/2$ teaspoon cream of tartar, lemon juice, or vinegar to each gallon of water. Rice and onions can turn yellow when cooked in alkaline water. Again, you can prevent it with cream of tartar, lemon juice, or vinegar. Even cabbage and cauliflower show some yellowing when cooked in alkaline water. Some of the flavonoid compounds (in sunflower seeds, page 334) turn a beautiful jade green.

Potatoes suffer from several types of discoloration. In cooked potatoes, phenolic compounds react with iron and cause a grayish black discoloration, usually near the stem end more than other areas. This is called *stem end blackening* and appears as the potato cools. Some potatoes seem more susceptible to this than others. Climate and soil conditions are probably responsible.

To prevent this blackening, you can add $1/2$ teaspoon cream of tartar to the cooking water when potatoes are half done or add $1/2$ teaspoon per pound to potatoes that are to be mashed. I have a chef friend who always adds a little vinegar to potatoes to be mashed. All of these remedies work.

Color enhancers

Compounds that are vivid in color or have the ability to stain or dye can serve as color enhancers. Examples include spices like curry powder (which contains turmeric), saffron, and turmeric. The red-orange color of the tomatoes stained by turmeric is brilliant in the Spicy Indian Fried Cheese (recipe follows). Other ingredients can contribute color, too. In Sweet-Tart Fresh Strawberries (page 343), the berries are stained a deep, rich red by the brown balsamic vinegar and brown sugar.

Spicy Indian Fried Cheese

MAKES 4 SERVINGS

Tomatoes stained vivid red-orange by the turmeric in Indian spices contrast with startlingly green barely cooked peas and cilantro to create a brilliant dish with fascinating flavors. It is not spicy as in hot from peppers, just a forceful exotic taste.

what this recipe shows

Spices like turmeric and saffron can stain ingredients to enhance color.

³/₄ pound ricotta salata, goat's milk montasio (see Notes), or other firm heat-processed cheese, sliced ¹/₂ inch thick

¹/₂ teaspoon *and* ¹/₂ teaspoon *and* 1¹/₂ teaspoons salt (2¹/₂ teaspoons *total*)

¹/₄ teaspoon *and* ¹/₂ teaspoon *and* ¹/₄ teaspoon white or black pepper (1 teaspoon *total*)

¹/₂ cup flour

¹/₂ cup ghee (see Notes)

2 medium onions, chopped

2 cloves garlic, minced

2 tablespoons minced fresh ginger

¹/₄ cup canned chicken broth

1 teaspoon turmeric

¹/₄ teaspoon cayenne

2 teaspoons ground coriander

1 teaspoon sugar

1 teaspoon paprika, preferably recently opened

1 tablespoon Garam Masala (page 343)

4 large ripe tomatoes, peeled, seeded, and chopped, or 2 cans (14¹/₂ ounces each) diced tomatoes

1 package (10 ounces) frozen green peas

4 sprigs fresh coriander (cilantro, Chinese parsley), coarsely chopped, stems and all

1. Sprinkle cheese slices with a mixture of ¹/₂ teaspoon salt and ¹/₄ teaspoon pepper. In a plate or pie pan, stir ¹/₂ teaspoon salt and ¹/₂ teaspoon pepper into ¹/₂ cup flour. Flour seasoned cheese slices and carefully shake off excess flour.

2. In a large, heavy skillet over medium-high heat, fry the cheese slices in ghee until well browned, 1 to 2 minutes to a side. Remove to a platter. Add the onions to the skillet and sauté, stirring. Add garlic and ginger and sauté briefly, taking care not to burn. Pour in the chicken broth and add the turmeric, cayenne, coriander, sugar, paprika, Garam Masala, 1¹/₂ teaspoons salt, ¹/₄ teaspoon pepper, and a third of the tomatoes. Simmer 10 minutes, add the rest of tomatoes and the peas, and simmer briefly. Return cheese slices to hot sauce and sprinkle with chopped cilantro.

Notes: *Goat's milk montasio is available from The Mozzarella Company (see Sources—Cheese, page 477).*

Ghee is clarified butter that is cooked to boil off all the water and brown the butter and dairy solids.

Garam Masala

1¹/₂ 2- to 3-inch sticks cinnamon, broken into pieces

¹/₂ teaspoon cardamom seeds

¹/₂ tablespoon whole cloves

¹/₄ cup coriander seeds

¹/₄ cup cumin seeds

2 tablespoons whole black peppercorns

In a heavy skillet, roast spices over medium heat, stirring constantly, for several minutes, until fragrant. Grind (a small coffee grinder is excellent for this) and store in a small jar, tightly sealed.

Sweet-Tart Fresh Strawberries

MAKES 12 SERVINGS

Since Marcella Hazan introduced balsamic vinegar to Americans, many people put it on strawberries. My friend Terry Ford, owner of the *Enterprise* newspaper in Ripley, Tennessee, adds brown sugar, which finishes off the sweetness and gives an even deeper red color to the berries. Some recipes call for fresh cracked black pepper. I like to add a little salt. They are so eye-catching that you need to serve them in clear glass—I use a tall, moderately narrow vase. These are ideal at parties or as a light dessert.

what this recipe shows

Both brown sugar and balsamic vinegar stain berries for a deeper red color.

The lightly acidic balsamic and the sweet brown sugar add refreshing sweet tangy flavor notes to enhance the berry flavor.

2 quarts fresh strawberries, hulled

1 cup packed light brown sugar

¹/₄ teaspoon salt

¹/₂ cup balsamic vinegar

About 1 to 2 hours before serving time, toss the strawberries with the brown sugar, salt, and vinegar in a large mixing bowl. Refrigerate for about 30 minutes, then toss again. Refrigerate and toss again. Drain and serve cold in a clear glass container.

Changes in taste: for better or worse

In cooking, many compounds break down, others combine, and some evaporate off. Many of these changes affect taste. Two families of vegetables that are altered dramatically with cooking are onions and members of the genus *Brassica* (cabbage, cauliflower, broccoli, for example). When onions cook, some of their strong sulfur compounds dissolve and break down, and many evaporate. New compounds are formed. Some of the compounds that form in onions are even sweeter than sugars. Roasted Sweet Onions with Balsamic Vinegar have a caramel sweet mellowness.

Roasted Sweet Onions with Balsamic Vinegar

MAKES 6 TO 8 SERVINGS

Sweet onions—such as Vidalia, Maui, and Walla Walla—caramelized to a deep brown are basted with balsamic vinegar and a little butter. I loved Carlo Middione's whole roasted onions at Vivande restaurant in San Francisco so much that I rushed home to make my own version. I cut the onions in half and place the cut sides down so that a large area ends up caramelized. I like to prepare them in an iron or other heavy skillet, which makes it very easy to deglaze and boil down the drippings. These are an ideal complement for roasted meats.

what this recipe shows

Onions become sweeter when cooked.

4 medium-size sweet onions, cut in half horizontally and peeled

3 tablespoons good olive oil, such as Berio, Berillo, or Plaginol

Salt

White pepper

1 to 2 tablespoons water

$1/3$ cup balsamic vinegar

2 tablespoons light brown sugar

2 tablespoons butter

1. Preheat the oven to 350°F (177°C).

2. Rub the onion halves with olive oil. Sprinkle the halves all over with salt and white pepper. Place the onions, cut side down, in a large ovenproof skillet. (You can cover a wooden skillet handle with foil to protect it but should not place such a skillet in an oven over 375°F/191°C.) Bake until the onions are soft, about an hour.

3. Scoop up the onions with a spatula and place them in a serving dish with the browned sides up. Stir together the water, balsamic vinegar, brown sugar, and butter in the skillet that the onions were roasted in. Scrape the skillet to dissolve the baked-on brown juices from the onions. Heat to a boil. Boil for several minutes to reduce, then spoon some over each onion. Taste a piece of onion and add salt if necessary. Serve hot or at room temperature.

As we have seen, cooking causes mass destruction of the cells and many chemical changes. Compounds with low boiling points boil off. Some compounds break down, and some combine to create new substances with different tastes.

Many vegetables like carrots, sweet peppers, and fennel seem to become sweeter when cooked. How can this be when the actual weight or percent of sugar in raw carrots is the same as in cooked carrots? One possible answer is that some of the double sugars break down into the single sugars glucose and fructose, which could cause a taste change. Another is that the cell destruction causes increased sweetness: In raw carrots the cell structure is so firm and rigid that even chewing does not liberate a lot of the sugars, but when cooking destroys the cell structure the sugars become available. A third possibility is that evaporation or chemical changes during cooking remove some acidic or non-pleasant-tasting compounds, leaving the carrots with a sweeter taste. All three factors could be at work. In any case, when carrots are cooked with a little brown sugar they became so sweet that the slight sharpness of raspberries is ideal with them.

Carrots with Raspberry-Chambord Sauce

MAKES 8 SERVINGS

Intensely orange carrot slices surrounded by a moat of vivid red make this dish visually sensational. The concept of carrots and raspberries came from Glorious Foods of Dallas, a marvelous caterer that prepares pureed carrots with raspberry sauce. This is a great side dish for pork, chicken, or fish.

> **what this recipe shows**
>
> **Carrots increase in perceived sweetness when cooked.**

2 pounds carrots, peeled and cut into 1/4-inch slices

1 tablespoon *and* 1/4 cup packed brown sugar (5 tablespoons *total*)

1 tablespoon *and* 1 teaspoon salt (4 teaspoons *total*)

2 packages (10 ounces each) frozen raspberries, thawed

2 tablespoons cornstarch

2 tablespoons Chambord, Grand Marnier, or other raspberry or orange liqueur

4 tablespoons (1/2 stick) lightly salted butter

1/4 teaspoon white pepper

Finely grated zest of 1 lemon

1. Boil the carrots in water to cover with 1 tablespoon brown sugar and 1 tablespoon salt in a large pot until fork-tender, about 10 minutes. Drain well.

2. To prepare the sauce, drain the juice from the thawed raspberries into a medium saucepan. Stir in cornstarch and heat over medium heat, stirring constantly, until the juice thickens. Stir in the Chambord and gently fold in the raspberries.

3. When ready to serve, reheat the carrots by sautéing in butter in a large skillet over medium-high heat for about 2 minutes. Sprinkle with 1/4 cup brown sugar, 1 teaspoon salt, white pepper, and lemon zest. Mound the carrots attractively on a platter, leaving a border around the edge. Spoon the bright red raspberry sauce into the border around the carrot mound and serve hot.

Unlike onions, members of the genus *Brassica* (cabbage, cauliflower, broccoli, brussels sprouts, kale, mustard, rutabagas, collards, turnips, and others), do not fare well flavor-wise with extended cooking. The longer you cook these vegetables, the stronger and more unpleasant they taste. Just 2 minutes matters. Increasing the cooking time of cabbage from 5 minutes to 7 minutes *doubles* the amount of strong-smelling hydrogen sulfide gas (the smell of rotten eggs) that the cabbage produces.

Broccoli, brussels sprouts, and cauliflower can smell even worse than cabbage if overcooked. Cooking these vegetables for a short time and uncovered as in the Herbed Brussels Sprouts helps maintain their green color and also their mild flavor by preventing many of the smelly compounds from forming and allowing those that are formed to escape.

As with other green vegetables, it's best to cut the *Brassicas* into small even-size pieces so that they cook very fast. Follow the trimming and cooking directions in Steamed Broccoli (page 327) for beautifully cooked broccoli. For cabbage, try the Fine Cabbage Threads with Buttered Crumbs (page 347)—a quick, easy dish that shows how delicious this vegetable can be when cooked briefly.

> **Increasing the cooking time from 5 to 7 minutes *doubles* the amount of strong-smelling hydrogen sulfide gas produced by cabbage.**

Herbed Brussels Sprouts

MAKES 6 SERVINGS

Barbara Peterson, a caterer and cooking teacher in Baton Rouge, prepares brussels sprouts in a classic manner with herb-seasoned crumbs—ideal for quickly cooked sprouts. This is an excellent side dish for meats.

what this recipe shows

Short cooking time keeps the sprouts bright green, mild, and sweet in taste.

6 cups water

2 teaspoons *and* 1 teaspoon salt (1 tablespoon *total*)

1 pound very small brussels sprouts

4 tablespoons ($^1/_2$ stick) butter

$^1/_2$ teaspoon sugar

$^1/_2$ cup Italian-seasoned dry bread crumbs

$^1/_8$ teaspoon white pepper

1. Bring the water with 2 teaspoons salt to a rolling boil in a large pot. Add the brussels sprouts. If some sprouts are larger, cut an X in the stem end for more even cooking. Cook 6 minutes. Drain.

2. Melt the butter in a heavy skillet and add the brussels sprouts. Stir to coat. Add 1 teaspoon salt, sugar, crumbs, and white pepper. Toss to coat the sprouts with crumbs. Serve hot.

Fine Cabbage Threads with Buttered Crumbs

MAKES 4 SERVINGS

Cabbage cooked just to a pale green and tossed with seasoned crumbs and caraway is a quick, simple preparation that highlights the sweet, fresh taste of quickly cooked cabbage.

> **what this recipe shows**
>
> Cooking members of the genus *Brassica* (cabbage, cauliflower, broccoli, brussels sprouts, and others) for a very short time preserves their mild, sweet taste.

1 small head cabbage

4 tablespoons (½ stick) butter

2 teaspoons caraway seeds, crushed (see Notes)

½ cup dry bread crumbs

½ teaspoon *and* 1 tablespoon salt (3½ teaspoons *total*)

2 teaspoons sugar

1. Cut the cabbage in half and place it, cut side down, on a cutting board. Slice the cabbage into thin shreds like delicatessen slaw.

2. Melt the butter in a large skillet. Stir in the caraway seeds, bread crumbs, and ½ teaspoon salt and cook, stirring constantly, until lightly browned. Set aside.

3. Bring a large pot of water with the sugar and 1 tablespoon salt to a boil. Add the cabbage. Cook for 4 minutes. Drain well. Stir the cabbage into the skillet with the crumbs. Reheat for just a minute and pour onto a serving dish. Taste and add salt if necessary. Serve hot.

Notes: *Crush the caraway seeds in a freezer-type zip-top plastic bag by hitting with a hammer or a meat pounder.*

A mixed herb and spice seasoning called Spike contains oregano, basil, marjoram, rosemary, and thyme, and more than a dozen spices and is available at the spice counter in many grocery stores. It is excellent on cabbage. Substitute 2 teaspoons for the caraway seeds.

Persnickety potatoes: changes in starch

Starch is affected not only by cooking but by chilling as well. This presents advantages and disadvantages. High-starch plant parts like potatoes and rice demand attention to a variety of factors.

Changes with chilling When potatoes are refrigerated for several hours, some of the starches in them convert to sugars, creating French-fry nightmares for restaurants that slice potatoes, place them in cold water to prevent discoloration, and refrigerate them overnight to prepare ahead for a rush. The next day, because of the higher sugar content, the potatoes get very brown when fried before they are done in the center.

Fortunately, you can use this change to your advantage if you like to bake potato sticks for low-fat fake French fries. Placing the potatoes unpeeled in the refrigerator overnight or for a couple of days to increase their sugar will encourage them to brown more than they usually do in the oven, as the following recipe shows.

You might also notice that potatoes harvested early in the season have high starch content, which limits browning. Early potatoes may have starch on the surface, which covers the sugars that brown faster. Rinsing the potatoes several times in water can eliminate this problem.

Oven-Fried Herbed Potatoes

MAKES 4 TO 6 SERVINGS

I have made "oven fries" for years, but it was Maggie Waldron's recipe in *Cold Spaghetti at Midnight* that taught me to chill the potatoes to get them browner. These are perfect with grilled meats.

what this recipe shows

Standing in the refrigerator for 24 hours or more makes some of the starch in potatoes convert to sugar, and this makes them brown better.

2 large (1½ to 2 pounds) Russet Burbank (Idaho) baking potatoes

Nonstick cooking spray

1 tablespoon olive oil

2 teaspoons salt

¼ teaspoon cayenne

2 teaspoons finely minced fresh rosemary

¼ cup freshly grated Parmesan

1. Place the potatoes, unpeeled, in the refrigerator for a day or two. When ready to cook, scrub, but do not peel, and cut into fat French fry sticks. Rinse well under running water.

2. Preheat the oven to 450°F (232°C). Spray a baking sheet with nonstick cooking spray.

3. Steam the potato sticks for 8 minutes. Pat dry. Stir together the olive oil, salt, cayenne, and rosemary in a medium bowl. Add the potato sticks and toss well to coat. Arrange in a single layer on the baking sheet. Bake until lightly browned, about 1 hour.

4. Sprinkle the Parmesan over the potatoes and return to the oven just long enough to melt the cheese, about 4 minutes. Taste and add salt as needed. Serve immediately.

Changes with cooking The starch content of a potato is determined by variety and growing conditions and is a good indicator of how the potato will cook. The size of the starch granules varies with different varieties, just as the size and shape of starch granules vary in different plants. When potatoes are heated, their starch granules absorb water from their surroundings and swell. Starch granules of Russet Burbank potatoes are larger and swell to a larger size than the granules of new red potatoes. That's how they get their characteristic dry, mealy texture—they absorb more water from around the cells.

So, when you cook high-starch potatoes like the Russet Burbanks (also known as Idaho Russets or just Idahos), their cells separate and become dry and mealy. These are excellent for baking; they break open and separate and easily absorb butter or sour cream. Yukon Golds are a little lower in starch content but still high enough to be good baking potatoes. They have a golden yellow flesh and a buttery taste.

Lower-starch varieties like red potatoes (Red La Soda, Superior Red Pontiac) and some round white varieties (such as Katahdin and White Rose) absorb less water when cooked, swell less, are firm and waxy, and their cells do not separate. These potatoes are perfect for boiling to make scalloped potatoes or potato salad.

Russet Burbanks also produce better fried potatoes because they do not turn limp or absorb as much grease as lower-starch potatoes. When potatoes are fried, the starch granules on the surface expand immediately with the heat and dry the surface as they absorb water from the interior of the potato. With this expansion, they partially dry the interior and seal the surface. That's why French fry recipes specify well-dried potatoes: If starch granules get water from the surface moisture, they won't dry and seal the potatoes and will therefore absorb more grease. Fries made from lower-starch/higher-moisture potatoes (red potatoes) get limp with even a short standing period because of all the steam trapped under the surface.

High-starch potatoes, in contrast, make crisp, nongreasy fries—but only when the cooking time and temperature are right. As discussed in Chapter 2, when cooked just seconds too long, the potatoes run out of internal moisture. Without this moisture to turn to steam, which pushes outward and helps prevent grease from being absorbed, fries become greasy. Because high-starch potatoes are lower in moisture than waxy potatoes, correct cooking time is crucial.

So is temperature. For very crisp fries, the Idaho Potato Commission suggests chilling the potatoes for up to 2 hours just before frying and then precooking them at 350°F (177°C) until the fries just begin to color. After this first frying, they should be drained and placed in a single layer on a baking sheet and held at room temperature or refrigerated until completely cool. You can hold them overnight in the refrigerator. When you're ready to serve, fry again at 375°F (191°C) until browned and crisp. Some chefs like to precook at the slightly lower temperature of 325°F (163°C), and some like the final frying temperature slightly higher at 400°F (204°C).

Choose Russet Burbanks (Idahos) for baking and frying and red potatoes for boiling. If you want to estimate the starch content of your potatoes, make a brine of 1 cup salt and 11 cups water. Potatoes that float in this brine are less dense, which means they contain less starch.

> *To estimate starch content of potatoes:* Put them in a brine of 1 cup salt and 11 cups water. Those that float contain less starch.

Mashed potatoes When potatoes are boiled, their starch granules swell. If you vigorously break those swollen granules, perhaps by processing cooked potatoes in a food processor with the steel knife, you liberate massive amounts of starch and essentially have wallpaper paste. Anyone who has tried to make mashed potatoes in the food processor has encountered this glue. How can you make nongluey mashed potatoes?

For mashed potatoes, the Idaho Potato Commission's literature suggests precooking Idahos for 20 minutes at 140°F (60°C), well below a simmer. Then cool them and, when ready to mash, bring to a boil and cook until tender. Heat milk almost to a boil. Drain the potatoes well, mash, then beat in butter, salt, and hot milk. Serve immediately. The commission claims this precooking allows the starch to retrograde (firm with cooling—see "Retrogradation," page 275), which makes less gluey mashed potatoes. After starch has been swollen (gelatinized) and retrograded, it loses its ability to dissolve in water. Even when you break the cells during mashing, the retrograded starch does not give you a gooey mess.

> Are your mashed potatoes gluey? Precooking potatoes allows the starch to firm with cooling.

Jeffrey Steingarten came to a similar conclusion. In an article in the January 1989 issue of *HG* magazine, he explained that he got the finest mashed potatoes by precooking the potato slices at 160°F (71°C) for 20 to 30 minutes, draining, and then cooling by running cold tap water over them. This cooking at 160°F (71°C) swells and gelatinizes the starch. Then, when cooled, the swollen starch becomes firm. You have probably noticed that cooked rice becomes hard when cold—the same retrogradation process.

Steingarten explains that this 160°F (71°C) cooking is not too difficult. Get the water to 175°F (79°C). Add your peeled, washed potato slices. The potatoes cool the water down to about 160°F (71°C). If you want to be precise, attach a cooking thermometer to the pot and maintain this temperature by adding a little cold water if they get too hot or by turning up the heat a little if they get too cool. After cooking and cooling and a 30-minute stand, go on to the final cooking by either steaming or simmering. Cook until tender, drain well, and reheat briefly to dry out thoroughly.

Steingarten recommends pushing the cooked potato slices once through a ricer while they are still very hot. This is less damaging and ruptures fewer cells than the old-fashioned potato mashers, which repeatedly rupture cells.

How much butter you add is your business, but you should add it first to melt and then stir in the hot milk or cream. Personally, I go with less butter but do stir in some hot cream that is seasoned with salt and white pepper.

Cook's Illustrated recommends boiling potatoes in their skins so that they absorb less water for better potatoes for mashing or potato salad. I have not personally compared boiled in skin to peeled and sliced, but I have been satisfied with the peeled and sliced for mashed potatoes.

To simplify all of this—for less gummy mashed potatoes, precook potato slices for 20 minutes below a simmer, then cool. When ready to serve, simmer to reheat, drain *well,* and mash, adding butter, salt, pepper, and hot milk or cream as desired.

Shallot Mashed Potatoes with Garlic

MAKES 4 TO 6 SERVINGS

Two of our Atlanta restaurants serve unusual mashed potatoes. The Buckhead Diner concocts an excellent combination of pureed celery root and mashed potatoes, and the Georgia Grille serves shallot mashed potatoes, which I adore. I started with Georgia Grille's recipe, added precooking, used more shallots and garlic, and increased the cream in place of some of the butter—all the things I like in mashed potatoes—and then I tried to simplify it all. Good mashed potatoes are perfect with roast chicken (page 389), London Broil (page 378), or Marinated Grilled Skirt Steak (page 386) or whenever you need some warm loving comfort food.

what this recipe shows

Precooking potatoes at 140°F (60°C) to 160°F (71°C) and then cooling swells (gelatinizes) and recrystallizes the starch, making it no longer soluble in water.

A little vinegar in the water adds a hint of flavor and helps prevent discoloration caused by hard water or stem end blackening.

4 to 5 medium Burbank Russet (Idaho)
 or Yukon Gold potatoes, peeled and
 sliced $1/3$ inch thick

3 medium shallots, crushed

2 cloves garlic, crushed

2 teaspoons *and* 2 teaspoons salt
 (4 teaspoons *total*)

$1/2$ teaspoon cider vinegar

5 shallots, finely minced

2 cloves garlic, finely minced

3 tablespoons butter

$1/4$ teaspoon white pepper

1 cup heavy or whipping cream

$1/2$ to 1 cup half-and-half or milk

1. Precook the potatoes by heating enough water (about 6 cups) to cover them in a large pot over high heat until hot but well below a simmer. Turn heat down to low and add the potato slices, crushed shallots and garlic, 2 teaspoons salt, and vinegar. The potatoes will cool the water down to about 160°F (71°C). Cook at about this temperature, well below a simmer, for about 20 minutes. Drain and run cold water over the potato slices to cool them. Let them stand in water in the pot until you're ready to finish cooking, or drain, cover, and refrigerate.

2. When ready to mash, sauté the minced shallots and garlic in butter over medium heat until just soft in a heavy, medium saucepan. Stir in 2 teaspoons salt and the white pepper. Add the cream and half-and-half and heat gently. Set aside.

3. To finish cooking, add water to the cool potatoes if necessary, bring to a boil, and simmer until fork-tender, about 5 minutes at a simmer. Drain, remove the crushed shallots and garlic, put the potatoes back in the pot, and heat for about a minute to dry them out. To mash, push the potatoes once through a ricer or large-holed strainer into a medium mixing bowl. Bring the shallot and cream mixture to a boil and stir into the hot mashed potatoes. Taste and add more salt or pepper as needed. Serve immediately.

Rice—more starch cookery challenges

Whole books have been written on rice alone. It is one of the major grains of the world and the staple food in many countries. There are thousands of different varieties with fascinating characteristics, from rices with intense hot popcorn aromas to sticky gooey sweet rices. I have to limit rice coverage in this book to basic rice cookery and problems.

In addition to the many varietal differences, there are processing differences. "White rices" have the hull and bran removed and about 90 percent of American processed rices are enriched—thiamin, niacin, iron, calcium, riboflavin, or vitamin D added. Brown rices have the hull, but not the bran, removed; thus they have almost double the fiber of white rices

and they are also higher in vitamin E and magnesium. Converted rices are parboiled or pressure steamed *before* milling, which transfers the nutrients in the bran into the endosperm, making it higher in nutrients than other "white rices."

Rices are divided into three major grain types, long-grain (the length is three or more times the width), medium-grain (the length is about or a little less than two times the width), and short grain (the length is less than two times the width).

Long-grain varieties (like the aromatic rices basmati and Jasmine, to name a couple) are frequently but not necessarily grown near the equator (many Texas and Carolina rices are long grain). They are higher in amylose, the long, straight-chained starch.

They cook in separate fluffy or dry grains. Just as cooling sets the starch in cooked potatoes, cooling makes high amylose rice harden. After the amylose has been heated above its gelation temperature (around boiling), then chilled, the long starch chains link tightly together. This rejoining of amylose is even more dramatic when there is not much water present like in long-grain rice. When the heated amylose cools and dries it actually forms hard crystals. Think about how hard refrigerated, leftover Chinese take-out rice is the next day.

When rice is reheated, these crystals melt and the rice becomes soft again. In dishes where rice is cooked, chilled, and not reheated—as in rice salad—this can be a disaster. If you stir the rice together with other moist ingredients while it is hot, you can avoid the low moisture cooling and drying that produces such hardness. Alternatively, you can use medium-grain rice that is lower in amylose and does not harden when cooled.

If you make rice pudding using leftover rice that has cooled and hardened, the rice will soften while the pudding is hot but, if you chill it, this rice that has been allowed to harden and dry earlier will harden again. As you ordinarily make rice pudding, even if you cook the rice first, it is combined while hot with the sugar, milk, and eggs and has never experienced the drying and cooling that causes the hardening of the amylose.

In rice puddings, the rice can settle to the bottom and form a dense layer with the custard sitting on top. This is the same situation that you have when fruit or nuts settle to the bottom in muffins and cakes. The batter is simply not thick enough to hold up the ingredients. In a rice pudding, the rice has not been cooked long enough to exude starch to thicken the custard enough to hold up the rice. Or, possibly, a long-grain rice that does not exude much starch was used. In any event, you need more starch to thicken the custard. If you are not willing to cook the rice 30 minutes or more for it to get starchy, simply sprinkle a little cornstarch over the cooked rice before stirring it with the other ingredients.

Medium-grain rice is grown in Italy, Spain, California, and many other moderate climate locations. It is higher in amylopectin, the smaller branched starch and is lower in amylose than long-grain rice. Thus it does not harden when cooked and cooled like long-grain rice does. It's a good choice for rice salads or rice pudding.

The Italian arborio rice strains are fat varieties of medium-grain rice. They characteristically have a small area of not fully developed starch in the center which combined with the traditional method of cooking keeps just the center hard while the rest of the grain is soft and creamy, giving risotto its incredible creaminess with just a little bite in the center.

Short-grain rice, sometimes called sushi rice, is high in amylopectin and cooks soft and sticky. It is ideal for sushi. Glutinous rice, waxy varieties, sometimes called sticky or sweet rice, is even higher in amylopectin and cooks very sticky. It is used in East Asian dishes and desserts.

There are many different ways to cook rice from steaming to simply boiling in a large amount of water and draining. Most package directions suggest cooking rice with $1^3/_4$ to 2 volumes water, salt, and butter on low heat for 15 to 18 minutes. Brown rice requires about 2 volumes water and 45 to 50 minutes cooking time. I'm a great believer in going with whatever method works best for you. Many Asian cooks rinse the rice, once or many times, or even soak rice. This is probably a good idea with rice that may contain many impurities, but American rice which is of very high quality doesn't require this and you are rinsing away the nutrients that were added to the rice.

Dried legumes

Once again, whole books are written about legumes. In fact, there are even many books on just tofu, a product of soybeans. I have confined my coverage to basic cooking problems.

The most common problem with dried beans, peas, lentils, etc., is the "hard-to-cook" phenomenon. Sometimes you can have a dried legume that you have soaked overnight and literally cooked all day and the peas or beans stay hard. Storage conditions can be the cause. Dried legumes stored cool, around or below 40°F (4.4°C) and 50 percent humidity, cook nicely. However, those stored at high temperatures around 100°F (38°C) and high humidity (80 percent) increase in hardness and decrease in digestibility. Soaking in salt water before cooking can eliminate these problems. If you have a container of legumes that was difficult to cook, soak the next batch in salt water (about 1 tablespoon per gallon) for a couple of hours.

To reduce long cooking times, dried legumes are usually soaked first. Initially water seeps into the seed only through the hilum and micropyle, the small area where the seed was attached to the plant. After some water has been absorbed by the seed, water does start to soak through the seed coat so seeds with thinner coatings do soften faster. This is still a slow process and beans are usually soaked overnight. Soaking time can be reduced to 1 hour if the beans are boiled first in an abundance of water for 2 minutes.

Ordinarily in cooking, the firm pectic substances between the cells convert to soluble pectins and dissolve so that the cells fall apart and the vegetable softens. However, sugar or calcium prevents this conversion and softening. Calcium forms a firm compound with the pectic substances and makes the vegetable hard even after long cooking. Different varieties of dried legumes can contain small amounts of calcium, which, as you would guess, can make them harder to cook. Fortunately, dried legumes also contain phytic acid or other phytate compounds that combine with the calcium and prevent hardening. Some varieties have higher amounts of these phytate compounds, which aid in tenderization during cooking.

Cooks can use calcium's hardening effect to their advantage. Recipes for Boston baked beans include molasses or brown sugar. Both of these contain sugar and calcium, which allow the beans to retain their shape even when cooked for long periods of time.

One problem associated with beans is flatulence. Big sugars in legumes called oligosaccharides pass through the upper intestines undigested. Then, in the lower intestines, the bacteria love these sugars but produce gases when they devour them. The whole subject is not thoroughly understood and other parts of legumes possibly contribute to gas, too. Some legumes may even have a compound that reduces gas production. At any rate, some varieties are more gas producing than others. Pouring off the soaking water and rinsing a time or two may reduce these sugars a little since they do dissolve easily in water.

The unexpected stability of nutrients

You would expect different cooking methods to result in different losses of nutrients, and there are differences—more so with vitamins than with minerals—but not as much as you might think. The following table shows the variation of losses in minerals and several vitamins for selected raw vegetables and cooking methods. The mineral loss is relatively small (5 to 10 percent) no matter how the vegetable is cooked. On the other hand, the loss in vitamins varies somewhat, but many vegetables lose only 20 percent or less no matter what the cooking method. This is not to say, however, that there isn't great variability in the nutrient content, because there are very large differences in nutrient content before cooking (see page 322).

All cooking methods are relatively equal: Many vegetables lose only 20 percent or less of their vitamins no matter how they're cooked.

APPROXIMATE LOSS IN MINERALS AND VITAMINS ACCORDING TO COOKING METHOD FOR SELECTED RAW VEGETABLES, IN PERCENT

Vegetable	Minerals[1]	Vitamin A	Vitamin B$_1$	Vitamin B$_2$	Vitamin B$_6$	Vitamin C
Potatoes: baked in skin	0	—	15	5	5	20
Potatoes: boiled in skin	5–10	—	20	5	5	25
Potatoes: hashed-brown[2]	5–10	—	60	15	—	75
Sweet potatoes: baked in skin	0	10	15	5	5	20
Sweet potatoes: boiled in skin	5–10	15	20	5	5	25
Tomatoes: baked, boiled, stewed	0	5	5	5	5	5
Greens[3,4]	5–10	5	15	5	10	40
Roots and bulbs[3,5]	5–10	10	15	5	5	30
Other vegetables[6]	5–10	10	15	5	10	20

USDA Agriculture Handbook No. 8–11, U.S. Government Printing Office, 1984.

[1.] Calcium, iron, magnesium, phosphorus, potassium, sodium, zinc, copper, and manganese.

[2.] Potatoes were pared, boiled, and held overnight before being hash-browned.

[3.] Cooked in small or moderate amount of water until tender, drained.

[4.] Dark, leafy greens such as collards, mustard greens, turnip greens, spinach, Swiss chard, beet greens, Chinese cabbage, and wild greens.

[5.] And other vegetables of high starch and/or sugar content, such as beets, carrots, onions, parsnips, rutabagas, salsify, turnips, summer and winter squash, green peas, lima beans, and other immature seeds of legumes.

[6.] Such as asparagus, bean sprouts, broccoli, brussels sprouts, cabbage, cauliflower, eggplant, kohlrabi, okra, and sweet peppers.

Minimizing nutrient loss in cooking Vegetables lose nutrients in cooking primarily in two ways: water-soluble compounds (like minerals, vitamin C, and some of the B vitamins) dissolve in the cooking water, and heat destroys or makes unavailable certain nutrients (vitamin C and thiamin). The simplest way to avoid loss of water-soluble nutrients is to not cook the vegetable in water or to cook it in such a manner that the cooking liquid remains part of the dish, as it does in a casserole. You can cook vegetables whole in their skin and avoid a little nutrient loss. When you consider that you can lose as much as 50 percent of the vitamin C in some produce in 24 hours simply by not keeping it refrigerated, losses during cooking take on a new perspective.

Some of these cooking losses can be minimized. To preserve vitamin C in boiling, it is very important that the cooking water be at a boil *before* the vegetable is added. In warm water, the enzymes in vegetables become very active before the water gets hot enough to kill them. Some of these enzymes cause great destruction (up to 20 percent loss) of vitamin C in the first 2 minutes of cooking. Having the water boiling before you add the vegetable can prevent this vitamin C loss.

Some produce loses 50 percent of its vitamin C within a day if not refrigerated.

There is also loss of more water-soluble vitamins when you cut up vegetables because there is more surface from which vitamins can dissolve into the water. At the same time, however, you reduce cooking time, which saves vitamins.

Some cooking methods—cooking vegetables in a closed container with a minimum of water—will slightly improve the water-soluble vitamin content, but green vegetables will be an unappetizing olive drab, and those from the genus *Brassica* will be very strong tasting. It doesn't matter how many vitamins you preserved if you can't get anyone to eat them! I think a realistic approach is needed in this area. Don't strain a gnat and swallow a fly.

Cooking has little effect on the nutritive value of carotenoids. Even with the color change from one structure to another, the nutrient value does not change. Most of the carotenoids in yellow or orange vegetables are precursors of vitamin A. This is not true, however, for red vegetables that get their color from carotenoids.

Carotenoids do deteriorate with lengthy exposure to oxygen. Dried carrot chips, for example, lose their beta-carotene if packaged in air but not if nitrogen packaged.

Cooking and bioavailability of nutrients You have seen that the comparative nutritional profile of fruits and vegetables is complex even before cooking. One grapefruit can have 20 times more vitamin A than another (page 323). Spinach can lose 50 percent of its vitamin C in 24 hours if it is not refrigerated.

The effects of cooking on fruits and vegetables are no less complex. Cooking can remove some toxins and harmful compounds from foods. As discussed later (page 365), cooking removes cyanide compounds from lima beans and others. Cooking destroys enzymes in soybeans and other beans that inhibit our digestion of proteins (page 365). Cooking also converts some nutrients that are present in foods in unusable forms to forms that the body can absorb.

Corn is the classic example of cooking to make essential nutrients in a food available for use in the body. Corn contains lysine, one of the essential amino acids, but in a form that our bodies cannot use. Cooking corn with a strong alkali converts the lysine to a form that is bioavailable. This makes corn a food that can be the staple or primary food of a culture.

The Indians of South and Central America have known this for centuries. How on earth did these early tribes know to cook corn with an alkali—ashes or burned shells or bits of limestone—to make it more nutritious? They probably didn't.

What they may have learned accidentally was that cooking corn with an alkali softened and puffed the kernels and made the corn much easier to grind. Food preference researcher Dr. Paul Rozin tells of asking South American Indian women in remote villages why they cooked the corn with ashes? They explained that it will not grind right and you cannot make tortillas from the meal if you don't do it that way.

Early cultures that learned to raise corn but did not learn to cook it with an alkali suffered severe dietary deficiency diseases such as pellagra.

Cooking methods

No matter how fruits and vegetables are heated, the same dramatic events occur—cell walls shrink, cells lose fluid, firm pectic substances between cells convert to pectin and dissolve, and the cells fall apart.

As we have seen in the preceding sections, this destruction of the cells brings about changes in texture, color, taste, and nutrients regardless of the cooking method. Although the differences produced by various cooking methods are much smaller with plant foods than with meats, there are practical reasons for choosing one method over another in many situations.

Boiling, blanching, and poaching

Cooking fruits or vegetables in water does remove some of the water-soluble vitamins; however, as already mentioned, this loss can be minimized by having the water boiling before plunging the vegetables in. This immediately kills enzymes that are destructive to vitamins.

Cooking in water does not produce the crust or

sweet surface compounds that you get from plunging foods into hot fats. It does offer controlled temperature since the water, and therefore the fruit or vegetable, can never get over 212°F (100°C). However, this is not a great advantage. Even with the water well below a simmer you get all the results of the death of the cells—changes in color, texture, and taste.

You do have control over some of these changes with the cooking time. For example, you can keep green vegetables green by not cooking them longer than 7 minutes, keep *Brassicas* sweet and pleasant tasting by not cooking them for more than 5 minutes, and enhance the taste of onions and carrots with longer cooking.

When cooking in water, you also have the ability to limit or control how much flavor is extracted from a food. For example, if you want to extract flavor, as in stock making or sauce making, you can put the food into cold water, bring the temperature up slowly, and simmer for a long time. If you have ever tasted the onions or celery left after straining stock, you know they are almost devoid of flavor.

On the other hand, if you want the flavor to remain in the fruit or vegetable, you should bring the liquid to a boil, then add the produce and cook for a limited time. In most of the recipes in this chapter I wanted to preserve flavor in the fruit or vegetable. Here is a recipe in which I wanted to extract part of flavoring into the sauce. So, in this case, I started poaching the pears in cold liquid.

With fruit, ripeness determines how much sweetener is needed and how much cooking time. If you have ripe fruit, you will need very little sugar, though this is always a judgment call. As for cooking time, when asked "How long should you poach pears?" Jacques Pépin answered, "From a minute to an hour." Only the ripeness will tell.

Poached Pears with Walnut–Blue Cheese Rounds

MAKES 6 SERVINGS

Poached pear halves spread into a fan and presented simply with blue cheese rounds are impressive and delicious. Master Pastry Chef Chris Northmore of the Cherokee Town and Country Club in Atlanta serves a blue cheese mousse with whipped cream and the outline of a chocolate pear on each serving plate with his poached pears. I have included the directions for Chris's elegant presentation here with my recipe.

what this recipe shows

Beginning the cooking of the pear halves in room-temperature, not hot, liquid extracts some flavor into the sauce.

Adding sugar to the poaching water helps preserve the shape of the pears.

1 cup walnut halves or pieces

2 tablespoons butter

$1/8$ teaspoon salt

1 package (3 ounces) cream cheese, softened

4 tablespoons ($1/2$ stick) butter, slightly softened

5 ounces Saga blue or Très Bleu cheese

$1/4$ cup heavy or whipping cream

3 large ripe pears

$1^{1}/_{2}$ cups water

$2/3$ cup sugar

$1^{1}/_{2}$ cups sauterne or Rhine wine

1 teaspoon cornstarch dissolved in 2 tablespoons water

6 ounces semisweet chocolate, melted

6 sprigs fresh mint

1. Preheat the oven to 350°F (177°C).

2. Roast the walnuts on a medium baking sheet for 10 to 12 minutes. While the nuts are hot, stir in 2 tablespoons butter and sprinkle with salt. Separate out $1/4$ cup of the smaller walnut pieces. Set aside.

3. Process the cream cheese, softened butter, and blue cheese in a food processor with the steel knife until smooth. Add the cream and blend in. Add the smaller walnut pieces and mix in with two or three quick on/off pulses. Place the mixture in the refrigerator for 10 minutes so that it is easier to work with. When chilled, shape the cheese mixture into six mounds on wax paper, flatten each into a 1-inch-tall disk, and line the sides of each cheese disk with toasted walnuts. Return the disks to the refrigerator.

4. Peel, core, and halve the pears lengthwise. Place in room-temperature water with the sugar and sauterne in a medium saucepan. Poach over medium heat until just tender. Remove the pear halves with a slotted spoon to a dish and chill. Boil the poaching liquid to reduce it to about 1 cup. Stir the cornstarch mixture into the reduced liquid and heat, stirring constantly, until slightly thickened. Chill.

5. Make a piping cone of parchment or use a pastry bag with a very small round tip. Pipe the outline of a large pear in melted chocolate onto each plate, slightly to the left of the center of the plate. Use fresh mint leaves as leaves at the top. Cut four slits in each pear half up to 1 inch from the stem end, but not all the way to the top of the pear. Fan one pear half out on each plate, half inside and half outside the right border of the chocolate pear.

6. Spoon 2 tablespoons thickened poaching liquid into the open space inside the chocolate pear. Arrange a cheese disk inside the chocolate pear in the thickened liquid. Chill until ready to serve. These can be prepared ahead, covered with plastic wrap, and refrigerated.

Steaming

Steaming has some of the advantages of boiling. (In fact you can substitute steaming for the brief uncovered boiling called for in recipes.) As long as you steam *Brassicas* no longer than 5 minutes, their flavor will be mild and pleasant. Steaming green vegetables for less than 7 minutes will maintain their bright green chlorophyll.

Steaming is also advantageous for large quantities of food. It is much easier to steam large trays of vegetables in a commercial steamer than to deal with huge pots of boiling water. Most restaurants, hospitals, and schools steam their vegetables. For home steaming, you do not need a lot of special equipment. For small amounts, use a bamboo steamer or the inexpensive fold-up steamer basket that fits inside a saucepan.

Parmesan-Crusted Zucchini Fans

MAKES 6 TO 8 SERVINGS

These browned and bubbly cheese-crusted fans are a delicious and good-looking way to serve zucchini. They are excellent with light fish dishes and a lovely accompaniment for dishes with a tomato sauce.

> **what this recipe shows**
>
> Steaming softens zucchini and extracts a little of its liquid so it can be shaped.

Nonstick cooking spray	1/2 teaspoon salt
8 small zucchini, scrubbed	1/4 teaspoon freshly grated nutmeg
3 tablespoons butter, melted	1 cup freshly grated Parmesan

1. Preheat the broiler. Spray a baking sheet with nonstick cooking spray.

2. Cut four or five slits lengthwise in each zucchini to within 1/2 inch of the stem end. Steam the zucchini in a steamer or a large pot with a steamer basket until just soft enough to bend without breaking, about 4 minutes (see Note).

3. Spread each zucchini out on the baking sheet like a fan (see Variation). Brush lightly with butter and sprinkle with salt and nutmeg. Generously cover each fan with Parmesan. Slide under the broiler until the cheese melts and is lightly browned. Serve hot.

Note: *Zucchini can be microwaved to soften. Arrange in one layer on a glass dish and cover with a glass lid or microwave-safe plastic wrap. Microwave 5 minutes on High, then go to Step 3.*

Variation: *Fans are perfect for individual servings, but slabs are more practical for a large buffet platter. Trim the stem ends from four large zucchini and cut them crosswise into pieces 2 to 3 inches long. Then slice each piece horizontally to make slabs about 1/2 inch thick and 2 to 3 inches long. Steam the slabs for 4 minutes. Place the slabs in a single layer on a baking sheet and follow Step 3 for broiling. To serve, arrange the slabs in overlapping circles to form a big flower pattern.*

Braising and stewing

Braising, cooking tightly covered in a small amount of water, is often used for tenderizing tough cuts of meat. Frequently, fruits are added to impart sweetness and flavor to braised meats. Although braising is not usually required for vegetables, recipes for braising celery, which is fairly tough, are common. Endive, radicchio, and fennel are also braised. Braising in a flavored liquid will help cut the bitterness of endive and radicchio.

Stewing consists of simmering in water for a long time. Tomatoes, which hold their color well, are often stewed. Many consider the traditional Southern way of cooking vegetables like green beans and collards to be stewing. Both are simmered for over an hour with chunks of salt pork and sometimes a hot chili pepper.

Oven cookery—roasting, broiling, and baking

In roasting, the liquid that comes out of the damaged vegetable cells is immediately evaporated, and this drying concentrates flavor. Long roasting of vegetables like onions and root vegetables caramelizes the exterior to give complex sweet tastes (see Roasted Sweet Onions with Balsamic Vinegar, page 344). For roasting, vegetables are oiled or buttered, salted, and peppered before cooking. Sometimes herbs or onions are added for flavor. Roasting does not work well with most green vegetables, which can lose their bright color in 7 minutes. Green beans shrivel and dry up before they get tender. But a few tender green vegetables can be quick-roasted or broiled in a very hot oven. An example is the Roasted Asparagus with Lemon-Chili Oil on page 330. Properly timed, roasted vegetables have wonderful texture, slightly firmer than when cooked in water, concentrated intense flavors, and great color from the high temperature.

Broiling is an ideal way to heat and form a crust or to melt cheese on a surface without any real cook-ing. This is just right for something that will lose its shape with much heat, like a tomato half (see Sizzling Broiled Tomatoes with Herbs, below).

Yellow, orange, and orange-red vegetables retain their colors well and can be cooked for long times. In casseroles, where you are not concerned with shape, vegetables like squash, carrots, sweet potatoes, red peppers, and tomatoes are ideal candidates for longer baking. Examples are the Old-fashioned Grated Sweet Potato Pudding (page 338) and Golden Tomato Bake (page 340).

Firm carotenoid-containing vegetables like pumpkins and acorn, turban, and other hard squash also make them perfect for baking. The firm shell holds a filling in the Great Baked Pumpkin with Herbed Barley and Chickpeas (page 339).

Sizzling Broiled Tomatoes with Herbs

MAKES 8 SERVINGS

Simple broiled tomatoes are an ideal accompaniment for cheese dishes and many meat dishes, so I serve them often, especially considering how quick they are to prepare. Basil Vinaigrette and buttered-herbed crumbs accent the fresh tomato halves here.

> **what this recipe shows**
>
> **Because of the stability of the carotenoid compounds, tomatoes easily retain their bright orange-red color but must be cooked a limited time to hold their shape.**

4 medium to large almost-ripe tomatoes, cut in half horizontally
1 teaspoon salt
$1/2$ teaspoon white pepper
$1/2$ cup Basil Vinaigrette (page 360)

1 cup herbed stuffing mix
4 tablespoons ($1/2$ stick) butter
4 or 5 sprigs fresh basil for garnish

1. Preheat the broiler.

2. Arrange the tomato halves cut side up on a baking sheet. Sprinkle with salt and white pepper. Spoon 1 tablespoon vinaigrette over each half.

3. Place the stuffing mix in a zip-top plastic bag and roll over it with a rolling pin to crush into finer crumbs. Sauté the crumb mix in butter in a medium skillet just to melt the butter and coat the crumbs; do not brown. Divide the crumbs over the tomato halves. Broil 4 to 5 inches from the heat for several minutes to heat through and brown the crumbs. Serve hot, at room temperature, or cold. Garnish with the fresh basil.

Basil Vinaigrette

MAKES ABOUT 1 CUP

You can use any vinaigrette for the preceding tomato recipe, but here is a simple, flavorful one.

what this recipe shows

Using less vinegar than the traditional 1 part to 3 parts oil provides an excellent, less acidic flavor balance.

Adding sugar enhances the taste, and balances the acidity.

1 clove garlic

2 shallots

1 teaspoon salt

1 teaspoon sugar

$^1/_4$ teaspoon white pepper

2 teaspoons Dijon or honey mustard

2 tablespoons balsamic vinegar

$^3/_4$ cup mild olive oil (Berio, Berillo, or Paginol)

10 fresh basil leaves or $^1/_2$ teaspoon dried

With the steel knife in the workbowl, turn on the food processor and drop the garlic and shallots one at a time down the feedtube onto the spinning blade to mince. Add the salt, sugar, and white pepper. Scrape the sides of the bowl with a spatula and process to blend. Add the mustard and vinegar. Turn the processor on and slowly drizzle in the oil with the processor running. Add the basil leaves and process with two or three quick on/off pulses.

Grilling

Over high heat, a hot grill makes deep brown, sweet, caramelized streaks, making vivid-colored vegetables look magnificent. The cooking time is short, producing crisp-cooked, juicy vegetables. Vegetable slices or wedges can be oiled, grilled, and served hot or at room temperature. They are also excellent with a light dressing like Chili Oil, as in the following recipe.

Grilled Vegetables with Chili Oil

MAKES 6 SERVINGS

Vivid-colored vegetables with dramatic caramelized deep brown grill marks can be prepared along with grilled meats or ahead, held at room temperature, and tossed with heated Chili Oil at the last minute to warm them slightly. They are great for casual dinners.

what this recipe shows

Oiled vegetable slices grilled over high heat are sweet from the caramelized grill marks yet can still have a medium texture if cooked fast enough.

4 onions, quartered by cutting down
through the stem end

1 red bell pepper, quartered and seeded

1 green bell pepper, quartered and
seeded

1 yellow bell pepper, quartered and
seeded, if available

1 small globe or Japanese eggplant,
unpeeled, sliced lengthwise ¹/₂ inch
thick

10 large mushrooms, halved

Olive oil for coating vegetables

Salt

White or black pepper

1 recipe Chili Oil (page 328), reheated
or hot from preparation

8 sundried tomatoes, coarsely chopped,
or Oven-Dried Cherry Tomatoes
(below), cut into quarters

1 cup oil-cured black olives, pitted

1. Rub all the vegetable pieces except the tomatoes and olives with olive oil and sprinkle with salt and pepper. Grill over high heat until the grill marks are dark brown but the vegetables still have some crispness. Grill the onion quarters on the cut sides so that if they break apart all the pieces will have browned edges. Place the grilled vegetables in a large mixing bowl.

2. When ready to serve, add the Chili Oil and toss to coat well. Taste and add salt if needed. Arrange the vegetables on a large serving platter. Sprinkle with the tomatoes and olives. Serve hot or at room temperature.

Drying

Drying, removing the water in fruits or vegetables, intensifies both flavor and color. These Oven-Dried Cherry Tomatoes are a cross in color and intensity between fresh tomatoes and the very strong commercial sundried tomatoes.

Oven-Dried Cherry Tomatoes

MAKES 8 SERVINGS

These cherry tomato halves shrink to about three-fourths their original size, become a deep red in color, and have an intense real-tomato taste. They are a much brighter red than commercial sundried tomatoes and are an excellent accompaniment to meats. They can be used as a flavorful topping for pizza, in a salad, or simply as a flavorful color accent to a dish (see Note).

what this recipe shows

Drying intensifies both color and flavor.

Covering the baking sheet with foil prevents its discoloration from acids in the tomatoes.

About 35 cherry tomatoes, 2 to 3 pints

1 to 2 tablespoons salt

continued

1. Preheat the oven to 200°F (93°C).

2. Line a large baking sheet with aluminum foil. Cut the tomatoes in half horizontally and arrange them cut side up on the baking sheet, touching each other. They will shrink to about three-quarters of their original size. Sprinkle well with salt. Bake in the center of the oven for 2 to 3 hours. Serve hot, at room temperature, or cold.

Note: *For an impressive buffet dish, arrange the slabs in the variation of Parmesan-Crusted Zucchini Fans (page 358) in overlapping circles to resemble a flower. Then pile the deep red oven-dried tomatoes in the center.*

Deep-fat frying and stir-frying

Deep-fat frying and pan-frying are covered in detail in Chapter 2. Many fried vegetables are battered. As indicated on pages 162 to 163, batters should be an appropriate thickness for the food and usually should not contain sugar, which hastens browning and can brown the crust before the vegetable is cooked. To keep batters from absorbing fat, they should contain little or no fat themselves.

High-starch vegetables like potatoes, and starch coatings present special problems. At high temperatures the surface starch swells and seals the food. If a lot of moisture is trapped inside, the food becomes soggy and limp. In the classic French fry method (page 349), the potatoes are precooked at a lower frying temperature—325° to 350°F (163° to 177°C)—to cook and remove water from the interior. When ready to serve, they are fried at a high temperature—375° to 400°F (191° to 204°C)—to seal the surface and crisp the fries. Pages 348 to 353 give details about changes in starch in vegetables during cooking.

In selecting a vegetable to fry, take into account the texture. Frying something that instantly turns soft and mushy like ripe tomato slices can be a mess, while firm green tomatoes fry beautifully. Okra is a very successful firm vegetable to fry.

Stir-frying, which is very fast, is an ideal technique for keeping green vegetables bright. Stir-frying is not just for Chinese dishes. You can use Italian, Indian, Greek, or any other seasoning. In the following recipe, Indian spices give the stir-fried green beans and potatoes both interesting color and taste.

Indian Vegetable Stir-fry

MAKES 4 SERVINGS

Cumin, coriander, mustard, turmeric, and sesame seeds add an Indian touch to this unusual stir-fry. The colors are vivid and dramatic, and so is the taste. I usually serve this with chicken, but it would also be excellent with lamb.

what this recipe shows

Quick cooking keeps green vegetables a brilliant green.

1 teaspoon ground cumin	$1/2$ teaspoon turmeric
2 teaspoons ground coriander	2 teaspoons sesame seeds
1 teaspoon dry mustard	1 teaspoon salt

$^1/_8$ teaspoon cayenne

$^1/_8$ teaspoon white pepper

1 tablespoon dark sesame oil

1 medium potato, baked, peeled, and
cut into $^1/_2$-inch cubes

3 tablespoons peanut or vegetable oil
for stir-frying

1 medium onion, chopped

1 red bell pepper, seeded and cut into
$^1/_2$-inch strips

2 tablespoons dry sherry

$^3/_4$ pound green beans, stems removed
and cut into 1-inch pieces

$^1/_3$ bunch spinach, about 10 leaves,
cut into $^1/_4$-inch strips

4 scallions, green parts included, sliced
into thin rings

6 sprigs cilantro (fresh coriander),
chopped

1. Stir together the cumin, coriander, mustard, turmeric, sesame seeds, salt, cayenne, and white pepper in a small bowl as a spice mixture. Pour the sesame oil over the potato cubes in a medium bowl. Toss the potatoes to coat. Sprinkle with half of the spice mixture. Stir and set aside.

2. Heat an empty wok until the top edge is hot to the touch. Add the peanut oil, onion, and bell pepper and stir-fry for about 30 seconds to soften. Add half of the sherry, quickly cover the wok, and let steam for 20 seconds. Add the green beans and stir-fry until bright green, about 1 minute. Cover and steam for 10 seconds. Add the potato cubes and spinach strips and stir-fry for several seconds, until the spinach turns bright green. Remove from the heat. Taste and add additional salt or spice mixture. Spoon onto a serving platter. Garnish with the scallion slices and chopped cilantro and serve immediately.

Microwaving

Microwaving vibrates the water molecules in the food itself to create friction and heat. This can give you a completely different texture than traditional cooking methods.

Jeffrey Steingarten points out that fewer volatile compounds are created when microwaving. Since smell and taste are intimately entwined, you can get different tastes with microwaving from those created by traditional cooking methods.

Whole vegetables cooked in their skins in the microwave can be excellent. A microwaved whole artichoke (recipe follows) or an unpeeled onion, turnip, or beet is delicious. Since you are not cooking in water, vegetables retain their flavor and water-soluble vitamins. There is also good color retention if a small amount of vegetables is microwaved at a time, allowing very short cooking times. I prefer to microwave, rather than cook in water, the two apple recipes, Homemade Applesauce (page 325) and Apple Wedges (page 325), that demonstrate the role of sugar in cooking fruits and vegetables.

Microwaving is very fast for small to medium amounts of food, but the more food, the longer the cooking time. In addition, unless you have a microwave with a rotating turntable, the process requires time and attention (turning, stirring, rotating, and standing) to ensure even cooking.

Foods cook more evenly in round dishes than in square or rectangular containers. Microwave energy from two sides overlaps in the corners of a square or rectangular dish and cooks the corners twice as fast as the rest.

The thickest portion requires the longest cooking time and should be turned to the outside to absorb the most energy. With broccoli, this involves arranging the broccoli stems out for cooking, then rearranging them more attractively for serving.

Some fruit and vegetable dishes are definitely not suited to microwaving. For example, with stuffed tomatoes the shell will cook to a soft mush by the time the filling gets hot.

Artichoke Leaves with Hollandaise

MAKES 3 TO 4 HORS D'OEUVRE SERVINGS

This old-time classic hors d'oeuvre is hard to beat.

Microwaving is a quick, simple way to prepare an artichoke.

2 large artichokes, rinsed and stems cut off close to the base, sharp leaf tips trimmed if desired

1 recipe hollandaise (page 301 or 302)

Wrap each artichoke in microwave-safe plastic wrap. Microwave one at a time for 6 to 7 minutes on High. Let stand 5 minutes. Push the leaves down to spread out and make them easier to remove. Serve warm, at room temperature, or cold with hollandaise for dipping and a plate for the leaves, which are discarded after the edible portion has been eaten.

Fruits and vegetables: beautiful, alluring, and sometimes deadly

You may think that plants lead a glamorous life. They bask in the sun. They feel the breezes through their leaves. Gentle rain caresses them. They're fed by nutrient-rich soil. Ah, but their life isn't that easy. Plants have a big problem—they can't move. They can sway in the breezes, but they have to stay right where they are with their roots anchored in the earth.

Their immobility is a major handicap in both reproduction and survival. If a plant drops its seeds right on the ground under it, many of the seeds will not have a chance to survive. They will be in the shade of the mother plant, and all the seedlings will be crowded together in the same place, depriving each other of space and nutrients.

If a plant could just get its seeds farther away by themselves so that each had space and light, they could really grow. This is the role of those sweet, enticing fruits. Their job is to lure animals (including human beings) to pick them, move away with them, eat them, then spread the seeds, hopefully in a nice area with rich soil and sunlight, where the seedlings can have space and grow.

Finding a good home for its babies is just one of a plant's problems. A plant is in constant danger of being eaten by insects, animals, or humans. Animals run and hide from their predators, but plants have to remain where they are. What can plants do?

They resort to chemical warfare. Mushrooms sometimes poison on the spot anything that tries to eat them; onions and chili peppers give off fumes that temporarily deter or even blind their attacker. When you cut an onion, enzymes that have been isolated in one area of the plant's cells come in contact with other fluids in the cells and produce those well-known fumes that bring tears to your eyes. Sometimes plants contain chemicals that are poisonous to their primary predators and/or to man.

For instance, lima beans and cassava (also called *manioc* or *yuca* with one *c*) contain cyanide. In fact, certain varieties of beans are not permitted to be grown commercially in the United States because of their high cyanide content. Some Asian varieties contain thirty to fifty times as much cyanide as varieties grown here. When lima beans are cooked, we not only make them tender and digestible; we also convert the cyanide to a small amount of hydrogen cyanide gas, which floats harmlessly away into the air with the steam. In this way we make this potentially poisonous plant perfectly safe to eat.

Cassava, the source of tapioca, is a vegetable similar to potatoes and is indigenous to the Americas. It is now a major food in many parts of the world. The traditional method of preparation is to chop or mash the cassava and allow it to stand. This gives the enzymes in the cassava time to convert the cyanide to harmless compounds.

Many fruit seeds also contain cyanide—apple seeds, peach pits, apricot pits. Fortunately, they are rarely eaten. There have been a few reported cases of **Fifteen apricot pits can kill a child.** poisoning from people who roasted and ate apple seeds or of small children who ate apple seeds or apricot pits. Fifteen apricot pits can kill a child.

Raw soybeans and certain other legumes contain enzyme inhibitors that prevent digestion of proteins. They create chemical changes in the linings of the intestine that make it impossible to absorb nutrients. Regular consumption of raw soybeans and other varieties of raw beans can result in death. Once cooked, these protease inhibitors are eliminated and the beans are safe and nutritious.

Rhubarb leaves are deadly poisonous. Under no circumstances should they be eaten. Many grocery stores now remove the leaves from rhubarb stalks before putting them on the counters. Although oxalates have been blamed for human deaths from eating rhubarb leaves, experts believe that their toxins are anthraquinone glycosides. Spinach, which is harmless, contains about the same amount of oxalate as rhubarb leaves, supporting this assumption.

Many of the poisons in foods we eat are present in such small amounts that they are completely benign. For example, potatoes, the most popular vegetable in the United States, contain a poison called *solanine*. An average potato contains about 5 milligrams per 100 grams (3.5 ounces) of potato, which is harmless. However, if the potato is left in sunlight and allowed to turn green, the amount can increase twenty times and become dangerous. Fortunately, solanine is located right under the green area of the skin. If you peel a green potato to remove about $1/16$ inch under the green, you remove the poison. Sprouts on a potato also contain high amounts of solanine. You should cut out and discard all sprouts. **A green layer under the skin of a potato is toxic; peel to remove about $1/16$ inch under the green for safety.**

Some poisonous vegetables, herbs, and spices are eaten in such small amounts that they are harmless. A generous sprinkle of nutmeg on your eggnog is quite safe, but if you eat two whole nutmegs you may hallucinate wildly.

Parsley contains two known toxins, myristicin and apiole. But since parsley is used primarily as a garnish, we usually eat small, harmless amounts. Some dishes, such as tabbouleh, contain large amounts of parsley. It is not advisable to eat large quantities of parsley on a regular basis.

Fruits can contain poisons, too. Mangoes contain a poison just under their skins that is similar to poison sumac. Some people are more sensitive to it than others and get a red, itchy rash on any part of the skin exposed to this area of the mango.

Most of our food plants have been bred for generations to improve their size, taste, and nutrient content and to minimize their toxins. Cooking and many of our traditional methods of preparing foods remove toxins. Still, it doesn't hurt to have a healthy respect for plants. The fact that something is "natural" does not mean it is good for you.

Plants as medicine

Just as some plants are potent poisons, some are strong medicine. Plants have been used for centuries for their medicinal qualities. Many herbal remedies and folk medicines have been proved by scientists as valid medications. Aspirin, digitalis, quinine, and morphine all originally came from plants. Cranberry juice is usually part of the treatment for urinary tract infections. The vapors from hot chicken soup aid in clearing clogged nasal passages.

In my own family, my aunt recounts how eighty years ago my grandmother nursed her tiny premature twins who had pneumonia. The doctor said the boy had a chance but that the little girl's breathing was so bad she would not make it through the night. My grandmother got the two oldest girls to take turns holding and rocking Uncle Lewis while she put an onion poultice on Aunt Lou's chest and held her close through the night. The strong irritating fumes enabled my aunt to breathe, and she survived, too.

There are many books on foods as medicines and much fascinating current research on compounds or components of plants that aid in maintaining good health.

Food allergies

Allergies develop just like immunity to disease. After an earlier exposure, a person's body thinks something relatively harmless like dust, pollen, or a certain food is a dangerous invader and produces antibodies to fight it. When the person contacts the dust, pollen, or food again, these antibodies cause the release of chemicals such as histamines from cell tissue.

This released histamine can have dramatic effects—blood vessels can become larger, and the smooth involuntary muscles of the internal organs and breathing passages can contract. Changes in the sizes of cells cause fluids to ooze through cell walls, producing inflammation and swelling like puffy eyes and a stopped-up nose. The contraction of involuntary

muscles can cause cramps at the least. Contraction of air passages brings on an asthma attack. Severe allergic reactions can trigger anaphylactic shock and can quickly lead to death.

Heredity plays a role in some allergies like asthma, but not an apparent role in many others like food allergies.

Food allergies are not to be taken lightly. The medical literature documents many fatalities and near fatalities from anaphylactic shock caused by foods from peanuts to pine nuts to shrimp, foods that are nutritious and completely harmless to most people. According to a 1976 paper by F. Speer in *American Family Physician,* the ten foods that people are most commonly allergic to are cow's milk, chocolate and cola (the kola nut family), corn, eggs, the pea family (primarily peanuts), citrus fruits, tomatoes, wheat and other small grains, cinnamon, and artificial food colors.

While researchers agree in general on lists of the major food allergies, they differ in details. Susan L. Hefle, co-director of the Food Allergy Research and Resource Program at the University of Nebraska, published a 1996 paper in *Food Technology* on food allergens. She listed the most common sources for adults as peanuts, tree nuts, soybeans, fish, and shellfish. For children, they are cow's milk, eggs, soybeans, peanuts, wheat, and tree nuts.

At the 1995 annual conference of the International Association of Culinary Professionals, Daryl Altman, M.D., and Dr. Steve Taylor, food scientist, spoke on food allergies. Although only 1 or 2 percent of the population have true allergies, as many as 20 percent think that they have them. While this is frustrating to restaurant personnel, chances should not be taken. For adults, they said, the most common offenders are tree nuts, peanuts, fish, shellfish, and eggs. For children, they are eggs, peanuts, milk, soy, and fish. The Food Allergy Network can provide helpful information to consumers and professionals (see Sources—Food allergies, page 479).

Persons identified as suffering from a specific food allergy are advised to scrupulously avoid that food. Allergic reactions can be treated with epinephrine and antihistamines, and some allergies might be helped with a desensitization program. When you are asked if a certain ingredient is in a dish you have prepared, don't just assume the person is trying to guess your secret recipe. Someone's well-being may depend on your knowledge and honesty.

When a reaction occurs, immediate medical help is essential. Some symptoms are a very flushed face and difficulty in breathing. Call 911 immediately, because the person can die within minutes!

People with severe allergies are advised to carry an epinephrine kit and antihistamines such as Benadryl at all times. Ask them to self-medicate and insist on getting them to a hospital. Even though they may get better with self-medication, there are frequent relapses.

In addition to life-threatening allergies, there are other food allergies and sensitivities. One of these is a sensitivity to wheat gluten. The Gluten-Free Pantry is a source of special products for those with this condition (see Sources—Gluten-free, page 478).

the recipes

fine fare from land, sea, and air

CHAPTER

The challenges of meat, seafood, and fowl cookery

A s an old cook's tale goes, the best chefs of the land were preparing a great feast. The most famous chef kept his dish for the royal occasion a secret. His food was the finest, and anticipation ran high. Everyone wondered and speculated about what magnificent marvel he was preparing. All were stunned when the dish the great chef presented was *perfectly roasted chickens.*

The chef's message was clear and his appraisal correct: This apparently simple culinary feat—roasting a chicken—is incredibly difficult to do exactly right. You must cook the dark thigh and leg meat to an internal temperature above 160°F (71°C) to lose its raw metallic taste and to melt its stringy gristle, yet the broad, exposed breast will be dry and tough if you get it much over 150°F (66°C). Here is a challenge, indeed!

Whole books have been written on how to cook beef, lamb, fish, chicken—even on the individual methods of grilling, roasting, and steaming. Nevertheless, meat (animal muscle) cookery shares many common problems whether the meat is beef, veal, pork, lamb, fowl, fish, or shellfish.

369

What is meat? Why is some light and some dark? How do proteins cook? What happens when meats are heated? Why does chicken, meat, or even fish shrink and dry out? Why are some fish or meats tender and some tough? How do you select a quality piece of fish or chicken or steak, and then how should you cook it?

What is meat?

Animal cells of similar kind join together to form tissue of four primary types: muscle, nerve, connective, and epithelial. There are three different kinds of muscles: skeletal, cardiac, and smooth. Skeletal muscles comprise about 40 percent of body weight and are the main tissues used as meat. Skeletal muscle fibers are about 50 micrometers in diameter and extremely long. These long muscle fibers are held in bundles by connective tissue. It is these bundles of fibers or muscles that we generally term *meat.*

White meat and dark meat

Muscle cells contain different types of the basic muscle protein called *myosin.* A single cell contains several types of myosin side by side, but frequently it contains much more of one type than another. Some muscles are also much higher in *myoglobin,* the oxygen-holding compound that gives meat its red color. It all depends on how the muscles are used.

Think about how differently the muscles of a land animal must work compared to those of a fish. Land animals must constantly pull against gravity just to stand up. Fish, on the other hand, don't have to fight gravity; they control their buoyancy and remain effortlessly suspended in water. With one little flick of its tail, a fish can glide along for a long period without having to move many muscles; however, when a bigger fish is after it, the fish has to move fast to get away. A fish's muscles are designed for periods of little or no work but capable of a sudden burst of great speed when needed. These "fast" muscle cells burn glycogen, a carbohydrate in the cell fluids ready for instant use and speed.

In contrast, most land animals depend on steady, constantly working muscles just to move around and keep from falling down. These slow-working muscle cells burn fat, a process that requires oxygen. So these cells need a ready supply of oxygen. Myoglobin is the compound in the muscle that receives oxygen from the blood and holds it in the cell, ready for use. Cells that contain myoglobin have a red color. Cells that use more oxygen than others contain more myoglobin and are a darker red.

Chicken legs and thighs, which move constantly and need oxygen, are dark meat. In domestic chickens, which don't fly, the breast muscles are not used, so they are white. Wild game birds like ducks, which do fly, have dark breast meat.

Hogs are usually confined to small pens and are generally less active than cattle, so some cuts of pork are quite light compared to beef, especially cuts from little used muscles such as the chops and tenderloin.

The muscles of some whales provide a striking example of this phenomenon. Since whales are mammals, they do not get a constant supply of oxygen from the water, as fish do through their gills. Cells in some whale muscles must hold lots of oxygen for long periods under water. As a result, they contain large amounts of myoglobin and are almost black.

How proteins cook

Natural proteins are separate and individual large coiled or wadded-up molecules with bonds across the coils or loops to hold the protein together in a single unit. When the protein is heated, exposed to acid, or even air, some of these bonds break and the protein pops loose and unwinds. It is now somewhat straightened out with some of its bonds exposed and protruding. It has changed from its natural form, so these unwound proteins are said to be *denatured*. Almost immediately this unwound (denatured) protein with bonds sticking out bumps into another unwound protein with exposed bonds, and they bond together (cook or coagulate).

You see this happening when you fry an egg. You can see through the raw egg white in the pan because the natural proteins are separate, individual, bound units. There is plenty of room for light to go between them. When the egg is heated and the individual proteins unwind (denature) and join together (coagulate), there is no longer room for light to go between. The egg white becomes an opaque, solid white. The same thing happens when meats cook. You know how glassy raw chicken, fish, or shrimp looks—some light can come through between the individual proteins. Then, when the proteins join together (cook), there is no longer room for light to get through. The chicken, fish, or shrimp becomes solid white.

In addition to changes in appearance as proteins join together they take up less space or shrink. Loosely joined proteins contain a lot of moisture. There is liquid that was trapped between proteins as they joined, and there may be water adhering to the proteins themselves. These barely cooked proteins are juicy and tender.

As heating or exposure to acid or air continues, more natural bonds may break and join, existing bonds may tighten—both pulling the proteins more tightly together. Moisture trapped between proteins is squeezed out. The proteins shrink and become drier and tougher.

How muscles cook

Taking beef muscles as an example, as the proteins in the muscles unwind (denature) and join together (cook), they shrink, becoming drier and tougher just as any proteins do. Up to 120°F (49°C), these bundles of protein strands shrink in diameter only, not in length, and a minimum amount of moisture is lost. As the tem-

**Natural
Proteins**

**Denatured
Proteins**

**Coagulated
Proteins**

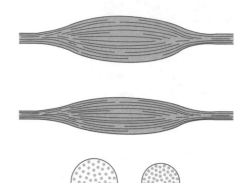

Up to 120°F (49°C)
Diameter Shrinks
Small Water Loss

Above 120°F (49°C)
Length Shrinks, Too
Big Water Loss

perature of the muscle increases above 120°F (49°C), lengthwise shrinkage and greater water loss begin.

So it should not be surprising that careful control of heat is the secret of many perfectly cooked, tender, juicy pieces of meat, fish, or fowl. Even seafood can become dry and tough when overcooked. Overcooked shrimp or lobster can be so tough that you need a steak knife to cut it. As Paul Prudhomme says, you don't cook shrimp; you just *heat them through,* enough for the proteins to bond together. The following recipes for Salty-Sea Perfect Tender Shrimp and Spicy, Garlicky New Orleans Herbed Shrimp (page 374) both depend on gentle cooking for tender shrimp.

A friend from the Gulf Coast bitterly objects to the common practice of rinsing cooked shrimp in cold water or plunging them into ice water to stop the cooking, saying that you wash away all the flavor of your seasonings in the cooking water. So, for the best of all worlds, I cool the shrimp in ice water that also has been seasoned.

> Shrimp cooking liquid makes an excellent stock.

Shrimp are best cooked in their shells, and the cooking liquid makes an excellent stock for another dish.

Salty-Sea Perfect Tender Shrimp

MAKES 6 SERVINGS

Plump, tender shrimp cooked and cooled in seasoned herb water have just a hint of salty sea water taste. Anita Kidd-Humphreys, an executive dining room chef in Atlanta who cooks perfect shrimp, swears by 5 minutes of cooking time for 1 1/2 pounds of "large" (26 to 35 count) shrimp in their shells. Her Floridian friend, Dan Fisk, taught her to cook them exactly this long and then to plunge them instantly into ice water. I have essentially gone with Anita and Dan's timing, but I like to start taking the shrimp out a few seconds early. They are a party hit served cold as in Ring of the Sea Shrimp with Fish Market Sauce (page 291) or as part of another dish (Lemon-Burst Shrimp with Caviar, page 373).

Seasoning both the cooking and the cooling water adds flavor.

Timing is vital to perfect tenderness.

6 cups *and* 6 cups water (3 quarts *total*)	1 onion, quartered
1 large seafood seasoning boil bag like Old Bay or Zatarain's (see Note)	1 lemon, quartered
	30 ice cubes
2 tablespoons salt (preferably sea salt)	1½ pounds large (26 to 35 count) shrimp in shells

1. Bring 6 cups water with the seafood seasoning bag, salt, onion, and lemon in it to a boil in a large pot. Simmer over low heat for about 4 minutes. Scoop out 2 cups of this flavored liquid and place in a large mixing bowl. Add ice cubes to the bowl, stir, and let cool.

2. Add 6 cups water to the pot with hot liquid and seasoning bag and bring back to a full boil. Add the shrimp. When the water comes back to a simmer, immediately turn heat down to medium-low. Simmer the shrimp 4½ minutes, then start scooping them out with a large slotted spoon or strainer. Place the shrimp instantly in the prepared bowl of flavored water with ice cubes. Stir shrimp to cool evenly and add more ice cubes if all melt. Let shrimp stand about 5 minutes in the flavored ice water. Drain, peel, and devein. Serve cold.

Note: *If you prefer, you can tie your own blend of spices in a cheesecloth bag.*

Lemon-Burst Shrimp with Caviar

MAKES 12 TO 15 SERVINGS

Lemon, caviar, shallot, cream cheese, and shrimp are a fabulous flavor blend, and the colors are just as enticing: the dark green of the cucumber skin, the pale pink shrimp, a burst of yellow from the lemon wedge, and a dash of startling red or black from the caviar. When you bite into the hors d'oeuvre, the juices of the lemon, brilliantly fresh with the zest, blend with the salty caviar. These can be a single elegant hors d'oeuvre served before dinner or a mainstay for a cocktail buffet. Prepare them ahead without the caviar, then add very cold caviar just before serving.

Cooking and cooling shrimp in seasoned water adds to flavor.

Carefully timing the cooking produces tender shrimp.

Chilling caviar for several hours before assembling gives it a crisp, clean, salty taste—no hint of fishiness.

continued

1 pound large (26 to 35 count) shrimp
in shells

2 medium shallots

1 package (8 ounces) cream cheese

½ teaspoon salt

2 large long seedless cucumbers or
3 regular cucumbers

3 thin slices thin-skinned lemon

1 small jar (2 ounces) red or black
caviar, well chilled

1. Cook, peel, and devein the shrimp according to directions in Salty-Sea Perfect Tender Shrimp (page 373).

2. With the steel knife in the workbowl, turn the food processor on and drop the shallots down the feedtube onto the spinning blade to mince. Add cream cheese and salt and process to mix. Slice each shrimp in half horizontally so that each half still looks like a shrimp. Slice cucumber into about ¼-inch slices. Cut each lemon slice (peel on) into quarters, then cut each quarter into three small wedges.

3. Generously spread cucumber slices with 1 tablespoon of the cream cheese–shallot mixture. Place a shrimp half, cut side down, on each cucumber slice. Stand a small lemon wedge, peel up, in the curl of the shrimp. Chill. Just before serving, spoon a small mound of caviar inside curl of shrimp at the base of the lemon wedge.

Spicy, Garlicky New Orleans Herbed Shrimp

MAKES 6 SERVINGS

Herbs, garlic, and black and red pepper give this dish exciting, complex tastes and tantalizing aromas—an inviting hors d'oeuvre served at room temperature without the rice, with toothpicks. As a hot main course, spread on white rice and scattered with bright scallions and parsley, it's a rich-looking platter. With a salad such as Mixed Greens and Oranges with Brandied Dressing (page 186) or Mixed Greens with Walnuts (page 314) and Crusty French-Type Bread (page 24), this is a marvelous meal.

Dried spices and herbs, which have been open for months and not tightly reclosed, turn brown, look like straw, and lose flavor, so be sure to use only recently opened jars to give the dish its full flavor.

what this recipe shows

Using a quick stock from the shrimp shells adds flavor and complexity.

Using both dried and fresh herbs imparts intense and diverse flavors.

1½ pounds large (26 to 35 count)
    shrimp (see Note)

4 medium shallots or ½ medium onion,
    chopped

3 tablespoons mild olive oil

3 cloves garlic, chopped

¼ pound (1 stick) butter

1 large bay leaf

1 teaspoon dried basil leaves

1 teaspoon dried oregano leaves

1 teaspoon dried rosemary leaves

1 teaspoon salt

½ teaspoon cayenne pepper, preferably
    recently opened

½ tablespoon paprika, preferably
    recently opened

1 teaspoon ground black pepper

½ medium-size thin-skinned lemon,
    unpeeled, seeded, and coarsely
    chopped

1 tablespoon fresh oregano leaves

1 teaspoon fresh thyme leaves

¼ cup *and* ½ cup chopped parsley
    sprigs (¾ cup *total*)

4 cups hot cooked long-grain white or
    brown rice

4 scallions, green parts included,
    chopped, for garnish

1. Peel the shrimp and set aside. Place the shrimp shells in a large saucepan, add water just to cover shells, and simmer over medium heat for 4 minutes. Strain liquid stock from shells into a medium bowl and discard shells. Pour the stock back into the saucepan and return to medium heat to boil down to about ½ cup.

2. Meanwhile, in a large skillet, sauté shallots in olive oil over medium heat just to soften, about 2 minutes. Add the garlic, butter, dried herbs, salt, cayenne, paprika, black pepper, and lemon. Simmer over low heat 4 to 5 minutes to blend flavors. Add the stock and continue to simmer 5 minutes. Remove and discard the bay leaf. Add fresh oregano and thyme, ¼ cup parsley, and shrimp.

3. Cook the shrimp over medium heat, stirring gently, until just cooked through, 3 to 4 minutes. Remove from the heat and let stand 1 minute. Cut a large shrimp in half to make certain they are cooked through. Cook a minute or two more if not done. Spoon immediately over rice and garnish with scallions and remaining parsley.

Note: *This can be transformed into a very fast dish by simply sautéing the garlic, shallots, and herbs in oil and butter and then adding peeled shrimp for a quick sauté. However, the quick flavorful stock from shrimp shells does add depth and richness of flavor.*

Variation: *This dish can also be prepared with scallops, fish fillets, chicken breasts, or even snails. (Eliminate the step for making shrimp-shell stock.)*

Carefully cooked, boneless chicken breasts can be juicy and tender. It takes just 7 minutes to steam a medium boneless breast. In Hot Thai Curried Chicken with Coconut Milk and Avocados, slices of cold raw chicken breast are added to a boiling sauce.

The sauce is then brought back to a simmer, simmered for 2 minutes, and removed from the heat to stand for 4 minutes for tender, juicy chicken slices. This typical Thai curry dish can be quite hot, depending on the amount of curry paste.

Hot Thai Curried Chicken with Coconut Milk and Avocados

MAKES 6 TO 8 SERVINGS

Gentle coconut milk calms the fiery curry paste in this thick sauce that one restaurant critic described as coating the chicken and avocados "like a rich brocade." This dish appears on the menus of many Thai restaurants, sometimes as Masaman Chicken. I fell in love with it at the Thai House, an excellent restaurant in Houston.

Here is my Americanized version, which my daughter, who has little time to cook, favors whenever she needs to carry a dish somewhere. Make a trip to an Asian market and open a few cans, and—Voilà!—an intriguing dish. Cold fruit like the Fresh Fruit with Ginger (page 319) is a cooling companion.

> **what this recipe shows**
>
> Slicing the chicken breasts across the grain prevents the slices from shrinking and curling when cooked.
>
> Peeling and slicing the avocados at the last minute prevent browning.

2 teaspoons oil

1 can Masaman curry paste (see Notes)

4 cans (13$^1/_2$ ounces each) unsweetened coconut milk (see Notes)

1$^1/_2$ sticks red sugar (see Notes) or $^1/_3$ cup packed light brown sugar

$^1/_2$ cup homemade Chicken Stock (page 267) or canned chicken broth

8 boneless chicken breast halves

2 avocados

1. Mash oil and curry paste together in a large wok or large skillet over medium heat. Add a few tablespoons coconut milk and continue stirring to thin curry paste. Add more coconut milk and stir together well. When curry paste is well blended in, add rest of coconut milk and turn heat up to medium-high to bring to a boil. Add red sugar and chicken stock. Boil vigorously for a minute, then turn down slightly to a gentle, steady boil to reduce. Boil gently, stirring frequently, and reduce to about 3 cups of noticeably thicker sauce. Even in a large wok, this reduction will take about 30 minutes—longer if you do not keep it at a slow, steady boil.

2. Cut chicken breasts across the grain at a slight angle into $^1/_2$-inch slices. (The grain runs lengthwise.) When sauce is reduced, add chicken breast slices and bring back to a boil. Turn heat to medium, simmer 2 minutes only, and remove from the heat. Let stand uncovered in the hot wok about 4 minutes to finish cooking.

3. Meanwhile, peel and slice the avocados. Arrange the slices over the bottom of a serving platter. Pour the chicken and sauce over the avocados. Serve hot. This also can be served over rice or noodles.

Notes: *Masaman curry paste is available in Asian markets (particularly Thai) and comes in a can the size of a tuna can with a green or white label with green lettering. Canned coconut milk and red sugar are also available in Asian markets.*

This is quite hot with one can of curry paste. If you prefer a gentler dish, use $^1/_4$ to $^1/_2$ can.

Tender and tough cuts of meat

Butcher Merle Ellis's statement in *Cutting-up in the Kitchen* tells it like it is: "There are two kinds of meat: tender and tough." In general, meat is tough if it is a much-used muscle and tender if it is a little-used muscle. That means muscles like the legs, hips, shoulders, and neck are tough, and muscles in the center of the back of the animal—the rib and loin area—are tender. Tougher muscles that are used a lot may contain more myoglobin and may be a deeper red in color. They are also very flavorful. The same principles apply to pork, lamb, and fowl. Since veal is from a young animal, it is usually tender throughout.

Packers and butchers cut beef into quarters—two front quarters and two hind quarters. The front quarter is divided further into the shoulder area (the chuck), which is very flavorful but tough, and the ribs (prime ribs, rib-eye steaks), which are tender. The hind quarter is divided into three sections: the short loin just behind the ribs (club steaks, T-bone steaks, porterhouse steaks, and filets), which is tender; the sirloin (sirloin steaks), which is also tender; and the hip area (the round), which is tough.

Knowing about cuts of meat can help you get better value (though cuts and their names vary—see pages 380–381). Sometimes the best buys in meats are those cuts that lie just at the edge of the tender area but that the butcher sells as cheaper tough cuts. In the front quarter, this would be the chuck blade roast, which lies right next to the tender rib area. You can identify this chuck roast by the bone, which is thin and does not have a protrusion in the middle, as do the bones in cuts of chuck nearer the front of the animal (seven-bone chuck roast). This relatively inexpensive roast contains a beautiful rib-eye steak below the blade bone, as well as meat just below and above the blade bone that is perfect for stir-fries. So, for about half what you would pay for two rib-eye steaks, you can have enough rib-eye to cut into two steaks and an abundance of meat for a second big stir-fry dinner.

On the back edge of the tender area of the beef, right next to the sirloin, is the top round. This is the inside of the leg and does less work than most leg muscles. The butcher may label steaks from the round as London broil or top round, but there is so little standardization that a butcher two doors down may call the same cut something different. The steaks toward the middle or tougher end of the round are rectangular and quite tough. Look for meat for this London broil recipe that is kidney bean shaped, with an indentation in the center area. This is the tenderest cut of the top round and, properly cooked to no more than medium-rare and thinly sliced, it is an excellent steak and a great buy. I could not have gotten our five teenagers through high school without London broil and scalloped potatoes!

An advantage of the tougher cuts of meat is their excellent flavor. The cuts on the "fringes" of the tender cuts can be very flavorful, not too tough, and relatively inexpensive.

Look for cheaper cuts on the edge of the tender area, such as the chuck blade roast. For about half the price of two rib-eye steaks, you get a rib-eye for two plus meat for a big stir-fry.

Chuck Blade Roast Showing Rib-eye

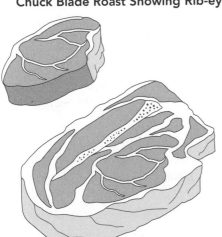

London Broil

MAKES 6 SERVINGS

London broil is affordable, wonderfully flavorful steak—all meat with no waste. Cooked rare to medium-rare and thinly sliced at an angle, it will thrill the steak lovers in your home and is one of the leaner, better-for-you cuts. Serve with baked, scalloped, or mashed potatoes (Shallot Mashed Potatoes with Garlic, page 350), a good salad such as Mixed Greens with Walnuts (page 314), and Crusty French-Type Bread (page 24).

This is an excellent two-meal dish. Buy a steak large enough so that 60 percent will feed your family the first meal. For the second meal, use the rest sliced thinly with all the remaining sauce. Add a large carton of sour cream and ¹/₂ pound sliced mushrooms for a marvelous stroganoff.

what this recipe shows

The tenderest London broil comes from the kidney bean–shaped cut with an indentation in the center.

1¹/₂ to 2 pounds London broil, about
   1¹/₄ inches thick

1 large clove garlic, minced

3 tablespoons beef concentrate
   (I like B-V; see Notes)

¹/₄ cup vegetable oil

Juice of 1 large lemon
   (2 to 3 tablespoons)

¹/₄ cup water

2 to 3 tablespoons butter

1. Preheat the oven to 425°F (218°C)—see Notes.

2. In a large freezer-type zip-top plastic bag, marinate the meat in a mixture of garlic, beef concentrate, oil, and lemon juice for about 1 hour at cool room temperature. Save the marinade.

3. To roast 1¹/₄-inch-thick London broil, place the meat on a roasting rack or cooling rack over an inch-deep pan and place on a shelf located in the upper quarter of the oven. Roast about 7 minutes, turn over, and roast 7 more minutes. For 1-inch-thick meat, roast about 6 minutes to a side.

4. Let the meat stand for 7 minutes before slicing. Drain marinade into a small saucepan. Add water and boil several minutes. Add butter. Transfer the meat to a cutting board and very thinly slice at an angle enough to serve everyone. Place uncut London Broil and slices on a serving platter. Reheat sauce and pour over all the meat.

Notes: *For this dish, a restaurant would use concentrated beef stock (glace de viande) or veal stock, but home cooks rarely have such treasures on hand. I have tried many store-bought concentrates and like B-V "The Beefer-Upper" best by far. Publix and Winn Dixie carry it in the South. (See Sources—B-V, page 477, which lists the processor to call for the name of a distributor* *near you.) If B-V is out of the question in your area, make your own beef concentrate by placing 1 cup beef consommé in a medium saucepan. Stir in 1 teaspoon instant beef bouillon and 1 teaspoon light or dark brown sugar. Boil down to reduce to about ½ cup. Use all to replace 3 tablespoons concentrate.*

This can be broiled, grilled, or microwaved as well as roasted. No matter what the method, it is important to cook no more than medium-rare.

- To broil, place meat on a rack over a broiling pan 4 inches from the heat and cook 5 minutes to a side for 1 inch thick. For 1½ inches thick, broil 5 minutes to a side, turn broiler off, and turn oven down to 350°F (177°C). Leave in oven 3 minutes more.

- To grill, proceed according to your best estimate for your grill, somewhere around 4 to 5 minutes to a side.

- To microwave, place meat and marinade in a heatproof 8-inch by 12-inch dish. Cover with wax paper and microwave on High 2 minutes. Turn meat over and microwave at 70 percent for 6 minutes.

If roasting, broiling, or grilling, drain marinade from bag into a medium saucepan. Add ¼ cup water to bag to rinse and add to marinade. Boil marinade gently for at least 4 minutes, whisk in butter, and remove from heat. In microwaving, stir water and butter into hot drippings in dish.

Cuts of meat—what's in a name?

Butchers have several names for the same cut of meat, and these names can also vary from place to place—this can be very confusing. Also, butchers break down quarters of beef in different ways depending on demand. For example, the butcher can cut the loin into steaks (club steaks, T-bone steaks, and porterhouse steaks) or cut out the whole filet for filet mignon or tenderloin, remove the bone, and cut New York strip steaks, also called *Kansas City strip steak* or *Delmonico steak*, from the other side of the bone. Butchers in different places may cut a section of a quarter quite differently, too. If you buy your meat at a supermarket and have a question about where a cut comes from, ask the butcher exactly which cut it is and where it's from.

In addition, some cuts of meat have misleading names. "Mock" tenders or chuck tenders are anything but tender. They are lean and pure meat shaped like a filet mignon, but they are from a muscle that is heavily used and very tough. Eye of round looks like a tenderloin but, again, is tough. It is, in fact, the toughest part of the round.

On the next two pages is a chart listing some of the names used for cuts of meat from the different sections of each quarter. I have used names from the National Live Stock and Meat Board and from Merle Ellis's book, *Cutting-up in the Kitchen*. For more in-depth information on the various cuts, I highly recommend Merle's book.

at a glance
Some common names for cuts of meat

Beef	Pork	Lamb
Short Loin	**Loin**	**Loin**
Club steak	Chops (all types)	Chops
T-bone steak	Pork tenderloin	Double loin chop
Porterhouse steak	Canadian bacon	(English chop)
New York strip steak	Top loin roast	
(Kansas City strip steak,	Crown roast	Saddle of lamb
Delmonico steak, boneless	Blade roast	
club steak, shell steak)	Rib roast	
Filet mignon	Country-style ribs	
Tournedos	Sirloin roast	
Chateaubriand		
Medallions		
Sirloin		
Pin bone sirloin		
(next to the short loin)		
Flat bone sirloin		
Round bone sirloin		
Wedge bone sirloin		
Top sirloin		
Rib	**Rib**	**Rib**
Standing rib	Spareribs	Rack of lamb
Prime rib	Bacon	Rib chops
Rib-eye steaks		Crown roast

Beef	Pork	Lamb
Chuck	**Shoulder**	**Shoulder**
Pot roast	Boston butt	Square-cut shoulder
Blade chuck roast	Picnic/shoulder ham	Arm chop
Boneless shoulder roast		Blade chop
Center-cut chuck roast		Neck slice
Seven-bone steak chuck		
Swiss steak		
"Mock"/chuck tenders		
Ground chuck		
Round	**Leg**	**Leg**
Top round (London broil)	Whole ham	Leg sirloin chops (leg chops, lamb steaks, sirloin steaks, loin end steaks)
Bottom round	Ham (shank portion)	
Eye of round (breakfast steaks, wafer steaks)	Center-cut ham slices	Whole leg
	Top-leg (inside) roast	Half leg
Round steak		Short leg
Rump roast		American leg
Denver pot roast		
Sirloin tip		
Ground round		
Shank		**Shank**
Brisket (all types)		Shank
Flank		
Flank steak		
Rolled flank steak		

Veal

I did not include veal in this table because veal is very young beef and veal cuts are similar to beef. The term *milk-fed veal* refers to the calf that is still on milk as opposed to grass. The meat of this young calf will be pale pink in color since it has a low iron intake. This used to be a criterion for fine veal—if it was not very pale, it was not good. There are now, however, farms that raise fine veal that is a rich, deep pink, and perhaps tint is not the best indicator of quality today. One young farmer who produces this excellent veal said that when he got skeptical chefs to compare his deeper pink veal to the lighter that they considered fine veal, they conceded that his product was superior—superb in flavor and very tender—and now buy all their veal from him.

Veal does not contain the fat marbling that aged beef does, so it demands great care in cooking—overcook veal just the slightest bit, and it will be dry and tough. Since veal is not as rich in flavor as beef, some cuts are traditionally served with flavorful sauces to enhance the meat. The Lemon Veal with Cream (page 263) is a typical example of a classic veal dish.

Tough or tender—other considerations

Although in general a piece of meat is tough if it is a much-used muscle and tender if it is a little-used muscle, this is not the whole story. Scientists can measure the force required to cut a piece of meat, but this does not necessarily correspond to what people consider tender. The juiciness and the amount of fat marbling, as well as how difficult it is to bite through the meat, influence our judgment. The nature of the muscle matters greatly—whether it is an often-used muscle and tough, or a relatively little-used muscle and tender; whether it contains streaks of fat (marbling) to melt, flavor, and tenderize by creating liquid between fibers; and whether it contains collagens that melt at relatively low temperatures to provide juiciness. All contribute to our concept of tenderness.

Proteins hold moisture on their surfaces and trap moisture when they join loosely. This water-holding ability of proteins is influenced by acidity. This means that changes in the acidity of meat that occur before, during, and after rigor may influence meat tenderness. (See "Rigor, Aging, and Tenderness," page 383, for details.) Even muscles that should be tender can be made tough if they are not allowed to go through rigor and relax again under proper conditions. Although you have no control over processing, you should recognize that processing does affect tenderness and buy from reliable suppliers who process meats correctly.

USDA beef grades

Meat grading is a voluntary service provided by the Consumer and Marketing Service of the USDA. The components of beef grading are:

- the age of the animal at the time of slaughter—older cattle have darker meat with a coarser texture
- texture—smooth, finely textured muscle is more tender
- appearance—consumers prefer firm, cherry-red meat
- marbling—intramuscular fat with flecks of fat in the lean

There are eight USDA quality grades, and all older cattle are graded in the three lowest: Commercial, Utility, and Cutter and Canner. For young cattle, the emphasis is heavily on the degree of marbling, which ranges from "abundant" to "devoid."

The top grade of Prime is juicy, tender, and flavorful and has a high degree of marbling. Choice has a little less marbling but is still of very high quality. If the marbling is slight, the grade is Select. Standard grade has a high proportion of lean meat with very little fat and is dry unless cooked by moist methods.

Most of the meat in the supermarkets is Choice, but there is a wide variation within this grade. You can select the higher end of this quality by looking for the pieces with the most marbling. The major market for Prime is the hotel and restaurant trade, but you can find it in specialty gourmet markets or locally run meat markets.

The "new" meats

Pork breeders have been working for years to breed leaner meats. As a result, we now have lower-fat products, especially ham and pork tenderloin. According to the National Pork Producers Council, the average fat content of pork has decreased 31 percent since 1983. Lean cuts of pork are very low in fat and contain more nutrients than poultry. Chops and pork tenderloin can be the basis of quick, low-fat meals and are ideal for today's hurried lifestyle. The pork council does extensive advertising and offers excellent customer service providing information and recipes for pork, "the other white meat" as they call it.

Lower-fat meats need care with both cooking time and temperature. They do not have the fat that can melt and provide a juicy mouthfeel, so they can get very dry when cooked. They're not the least bit forgiving, as are the higher-fat meats.

The beef industry is now breeding for lower-fat meats, too, and like the pork producers, beef producers are anxious to make consumers understand that low-fat meats should not be overcooked. Properly cooked, low-fat meats can be tender and juicy.

I realize that the beef industry will probably forge rapidly ahead in production of leaner beef and that it is for our better health—but, I must admit, I hope there always will be a tiny bit of high-priced prime beef left for that every-once-in-a-while when you long for a richly marbled, incredibly juicy steak.

Rigor, aging, and tenderness

I was present when someone asked a noted seafood expert and chef how he selected his fish. He had said that he, personally, went every morning to the dock and selected fish for the restaurant for the day. I expected him to emphasize things like "firm flesh, shiny and transparent eyes, dark pink to bright red gills," but instead he said he would not buy a fish unless it was "stiff as a board" (still in rigor).

All animal muscles have their own storage supply of glycogen or carbohydrates, which they burn for energy. This process creates lactic acid as a waste product. When circulation stops, blood no longer flows through the muscles, and this lactic acid is not removed. In addition to the protein myosin, muscles contain *actin*, which normally slides past the myosin during muscle movement. When lactic acid builds up in the muscles, the actin and myosin that normally slide past each other react chemically to form actomyosin. This causes muscles to contract tightly and become, as the seafood expert said, "stiff as a board."

Eventually, though, the muscles relax again. The temperature and the species are the two major determinants of how long it takes muscles to go into rigor and how long they stay in that state. Rigor takes about a day to pass in beef, for example, but only about six hours in pork and chicken. Fish can take a few hours to a few days, depending on the species.

Going through rigor improves meat in several ways. First, it greatly improves the texture; meat that does not go through rigor has a gummy texture. Second, the increase in acidity slows spoilage of meat and also enhances the water-holding ability of the meat proteins. Meat that has been allowed to go through rigor properly is juicier than meat that has not.

One thing that will prevent proper rigor is the animal's being under stress or working very hard just before slaughter. When that does happen, the animal's muscles will have used up their supply of glycogen or carbohydrates, and lactic acid will not have formed. The resulting meat will therefore be less acidic, darker in color, and more easily spoiled, with a poorer texture.

Meat is hung at moderate temperatures (61°F/16°C) for 16 to 20 hours postmortem to enhance its tenderness and ensure its proceeding into rigor. The hanging is done to stretch out the muscles before rigor sets in. If the muscles are allowed to contract as rigor starts, the bonding of the actin and myosin is stronger and the meat is tougher. The temperature is kept moderate because rapid chilling of meat before it goes into rigor will toughen it.

So a dark, poor-textured piece of meat or one that should be tender but is very tough may have been processed poorly. Even fish can be tough if not allowed to go through rigor properly. One seafood authority tells of a floating fish-processing factory that was freezing fish fillets within minutes of the time the fish were pulled from the water. This sounds worthwhile, but the resulting fish were so tough that the procedure had to be altered.

Meat benefits greatly from aging after rigor—in both texture and flavor. Aged meat has a buttery texture and a more intense meaty flavor. Meat should be held at 34° to 38°F (1° to 3.3°C) to control bacterial growth. An optimal aging period for beef is 11 days. Aging can be hastened by holding meat at a warmer temperature (70°F/21°C) for 2 days under high humidity (85 to 90 percent) to prevent moisture loss and with ultraviolet lights to control bacteria. The meat found in most markets is fast aged and gets a little additional aging during shipping and in the store.

Meat can be aged wet (sealed in Cryovac) or dry (exposed to the air). Nearly all meat today is wet aged, sealed in Cryovac. Because of the moisture and weight loss of costly meat, dry-aged prime beef is extremely expensive. Even if you are willing to pay the price, it is difficult to find—obtainable only from a handful of specialty suppliers. You can age meat in your refrigerator at home. Place the unwrapped meat on a rack over a dish that is lined with a paper towel and leave uncovered 2 to 7 days in your refrigerator, which is between 36° and 40°F (2.2° and 4.4°C). It will turn dark, and the surface will dry out. When you are ready to cook, cut away the dried surface area.

Methods of tenderizing

Cooks have both physical and chemical ways to tenderize meats. We can grind, chop, or pound meat. We can even tenderize by the way that we slice meat. Although we may not have thought of them as chemical tenderizers, cooks have long used marinades and the natural enzyme tenderizers in pineapple and papaya.

Physical methods

Most physical methods of tenderizing cut or tear long meat fibers across the grain. Large supermarkets now offer a wide selection of ground meats, turkey, pork, veal, and several cuts of beef—ground chuck, round, and sirloin—with different fat contents. Some stores even offer a coarse grind called a *chili grind*. With a

food processor or meat grinder attachment it is easy enough to grind your own so that you can control tenderness, flavor, and fat content.

Flattening with a meat pounder is not only an excellent way to thin and tenderize meat but is also ideal for evening pieces of meat of different thickness such as a boneless chicken breast. Pounded to a uniform thickness, a chicken breast will cook faster and more evenly. A meat pounder or even the bottom of a heavy saucepan can be used. To prevent tearing, put the meat between two sheets of wax paper or plastic wrap before pounding.

Cookware shops carry meat tenderizers that have thin spikes or blades that partially cut fibers. Depending on the device, you either pound the meat or roll it across the meat. Swiss steak and country fried steak are two dishes for which such tenderizers are typically used.

Slicing and carving

Muscle fibers can be as long as a foot. When meat is sliced with the grain (that is, parallel to these fibers), you have a long, tough, uninterrupted fiber that may be unchewable. The same piece of meat thinly sliced across the grain has only a tiny length of fiber (the thickness of the slice) and is easy to chew. Even tough cuts of meat can seem tender when thinly sliced across the grain. Think how tender deli corned beef can be when it is almost shaved (very thinly sliced) even though corned beef is a relatively tough cut. On the other hand, a moderately tender piece of meat can be nearly inedible if sliced with the grain.

> Even tough cuts can seem tender when thinly sliced across the grain.

Slicing across the grain is also important when the meat is raw. When a piece of raw chicken breast or flank steak is sliced lengthwise with the grain, when it cooks it will shrink and shrivel up into a tough wad. The same cut, sliced across the grain, cooks into flat, tender slices.

In some pieces of meat, such as flank steak, it is very easy to see the grain. In others, it is more difficult. You may need to look closely or even slice a thin piece in two different directions and observe its tenderness. In chicken or turkey breasts, the grain runs along the length of the breast.

The question of the grain is less important with fish. Fish muscles naturally have very short fibers. This is why fish flakes, nearly falling apart, and is so tender.

Slicing influences flavor, too. Shaved ham seems more flavorful than the same ham cut thicker. There is much more surface exposed to oxygen and more surface in the mouth to influence taste.

Skirt steak, which is the diaphragm muscle and the cut often used for fajitas, is a challenge to cut

Flank Steak

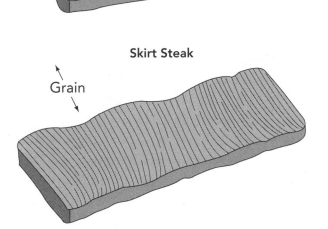

Skirt Steak

across the grain. Skirt steak is nicely marbled with fat and, when cooked no more that medium-rare and thinly sliced across the grain, is much more tender and flavorful than flank steak. Notice that the grain runs up and down the short side of the skirt steak, not along the length as in the flank steak.

Chemical tenderizers

Many plants like papayas, pineapple, and figs, to name a few, contain enzymes that can break down muscle fiber and/or collagen to tenderize meats. Powdered meat tenderizers (like Adolph's) containing one or more of these enzymes from plants are available at grocery stores. Marinades containing mild acids, which can break apart large proteins, are another form of chemical tenderizer.

Enzymes The enzymes papain and chymopapain from papayas and ficin from figs attack muscle proteins, and there is a fungal enzyme, Rhozyme P-11, that attacks meat. The fresh ginger rhizome contains a single enzyme that attacks both muscle and collagen. Bromelain from pineapple is more active toward collagen than muscle. Even honeydew contains enzymes that attack meat proteins.

These tenderizers work only when they are in direct contact with the meat, so unless they are injected into the meat, just the surface can be tenderized. To me, these tenderizers produce an undesirable mushy texture. The meat is mushy on the surface while the interior remains tough and leathery. I prefer to use marinades.

Marinated Grilled Skirt Steak

MAKES 4 SERVINGS

Long before fajitas rose to fame, I cooked these little richly marbled, flavorful steaks rare, sliced them across the grain at an angle to get a wider slice (1 inch), and served them as steak. My family loves them with scalloped or mashed potatoes (Shallot Mashed Potatoes with Garlic, page 350) and Fresh Green Bean Salad with Basil and Tomatoes (page 330).

what this recipe shows

Marinating the meat with a mildly acidic vinaigrette tenderizes and adds flavor.

Cooking the meat to rare or medium-rare limits lengthwise shrinking of the muscle fibers and produces tender, juicy steak.

Slicing thinly across the grain gives tender slices.

2 medium shallots, minced	2 tablespoons balsamic vinegar
1 tablespoon dried thyme leaves	1/3 cup vegetable oil
3 tablespoons (packed) dark brown sugar	1 1/2 pounds skirt steak
1/4 cup soy sauce	1/4 cup water
3 tablespoons Worcestershire sauce, preferably Lea & Perrins	2 tablespoons butter

1. Combine the shallots, thyme, brown sugar, soy sauce, Worcestershire sauce, vinegar, and oil in a freezer-type zip-top plastic bag. Add the steaks. Squeeze out as much air as possible and seal the bag. Place the bag in a medium bowl in the refrigerator to marinate overnight, turning bag several times for even coating.

2. Prepare a charcoal grill.

3. Drain the steaks, reserving the marinade. There will not be much. Pour the marinade into a small saucepan. Rinse the bag with 1/4 cup water and add to marinade. Simmer 4 to 5 minutes and stir in the butter.

4. Grill the steaks to medium-rare, about 2 minutes on each side. Timing will vary according to the heat of the coals and the distance of the meat from the coals.

5. Let stand 4 to 5 minutes before slicing. Thinly slice at an angle across the grain, which runs up and down the short side of the skirt steak. Bring the marinade-butter mixture back to a boil, pour over the meat, and serve immediately.

GELATIN

Gelatin is animal protein, made by processing collagen from animals. The same enzymes that break apart fiber and collagen to tenderize meat affect gelatin. In fact, some of these attack collagen (gelatin) more effectively than muscle. Most cooks are aware that raw pineapple contains a gelatin-wrecking enzyme. Figs, papayas, ginger, kiwifruit, and honeydew also contain enzymes that will ruin a gel. If you wish to use these fruits or fresh ginger in gelatin, first heat them to a temperature high enough to kill these enzymes. (Most enzymes are killed by 170°F [77°C] and all by 185°F [85°C].)

Just as acidic marinades tenderize meat, acids can interfere with gels. Avoid using large amounts of acidic ingredients in gels.

Gelatin customer service representatives claim that you can boil gelatin; however, this is not the whole truth. Yes, you can boil gelatin, but you reduce its thickening ability when you do. Gelatin should be sprinkled on room-temperature liquid to soften and expand, then should be heated just to melt thoroughly.

Another important fact to realize is that if you double the amount of gelatin in a recipe you are doubling you may end up with a rubbery mess. Some trial and error is necessary to arrive at the ideal amount of gelatin when doubling. I usually start with $1^{1}/_{4}$ times to $1^{1}/_{3}$ times the original amount of gelatin. Gelatin used with care can be a wonderful, unobtrusive thickener. I love to use it sparingly so that it is not really detectable.

Acid and marinades

Acid-based marinades both tenderize and add flavor. Acid causes the denaturation or unwinding of long proteins. In fact, you can actually "cook" (denature and coagulate) proteins with acids. In seviche, fish and scallops are "cooked" this way with lime or citrus juice.

You can use anything acidic as a marinade—wine, fruit juice, buttermilk, milk, yogurt. Marinating can be as simple as soaking chicken in buttermilk. Vinaigrettes are effective marinades that penetrate the meat fibers surprisingly well. Marinated Grilled Skirt Steak (page 386) is a good example of a marinade that both tenderizes and adds flavor.

I once experimented with different marinades by soaking meat samples for equal lengths of time in each marinade, then slicing the meat to see how deeply the marinade had penetrated. I found that oils that contained emulsifiers (mono- and diglycerides) penetrated deeper and faster, thus making better marinades. You can check the label to see if an oil contains emulsifiers. Some brands formerly contained emulsifiers, but because consumers complained about additives they were removed. Extra-virgin olive oil contains monoglycerides (which, contrary to popular opinion, are quite natural) and some of the new flavored oils contain emulsifiers like Polysorbate 80. Any of these oils will penetrate deeper and faster than pure olive, pure canola, pure corn, or pure soybean oil.

For fish, which is already tender, marinating may be brief, primarily for flavor, or a very mildly acidic marinade may be used, as in Best-Ever Marinated Shrimp (recipe follows). Tender seafood can actually be made tough by marinating it in too strong an acid for too long a period—you "overcook" it. I have had marinated shrimp that were quite tough for this reason.

Dry marinades or rubs are used for flavor. Mixtures of herbs and spices, sometimes moistened with a little oil, are rubbed on the meat before cooking. They are used with poultry, meat, or fish that is to be grilled, broiled, or even pan-fried (Prudhomme's famous blackened fish).

Liquid marinades have been on possibly contaminated raw meat and in spite of their acidity may be contaminated. *Never* use a marinade directly as a sauce. Before using, boil it for several minutes to kill bacteria that it might have picked up from the raw meat.

Oils with mono- and diglycerides (emulsifiers) make better marinades; try extra-virgin olive oil or one of the flavored oils that contains Polysorbate 80.

Best-Ever Marinated Shrimp

MAKES 8 TO 10 SERVINGS

For a big party, these are a luxury that will truly impress your guests. They are also addictive—even though you are embarrassed to take another one, you do. I started with Anita Kidd-Humphrey's excellent marinated shrimp recipe, reduced the vinegar, and added more capers, which I love.

> **what this recipe shows**
>
> A mildly acidic marinade slightly tenderizes but does not overcook delicate shrimp.

1½ pounds shrimp, cooked and peeled (page 372)

1 large onion, thinly sliced

7 bay leaves

1 cup good olive oil

2 tablespoons white vinegar

2 tablespoons capers, drained

2 tablespoons caper juice

2 teaspoons Worcestershire sauce, preferably Lea & Perrins

1 teaspoon salt

4 dashes hot pepper sauce

In a large mixing bowl, alternate layers of shrimp, onion, and bay leaves. In a small bowl, whisk together remaining ingredients. Pour over shrimp, cover, and refrigerate overnight. Drain the vinaigrette, remove some of the onions (for aesthetics), and serve cold.

Brining

Water can flow in and out of cells, and normally it flows through cell walls toward the most strongly concentrated solution in the cells or tissues. When you rub salt on the surface of meat, the salt dissolves in the meat juices to produce a concentrated solution that draws moisture out of the meat cells.

To dry meat, thin strips are rubbed with salt and placed in the sun. The moisture that the salt draws from the meat is evaporated as it is drawn out, and the salt continues to draw more moisture from the meat and thus dry it.

However, if the water were not evaporating from the surface, the salt water on the surface would become diluted by the moisture. Soon the concentration outside the cell walls would be less concentrated than the liquid in the meat cell, so the salt water would flow into the cell. Meat cells contain a lot of water, but it is water that is bound and held by the proteins. Actually, the free liquid in meat cells is very concentrated with dissolved substances.

This means that even concentrated solutions of salt water or salt and sugar water will be less concentrated than the liquid inside the meat cells. So, when meat is soaked in a salt or sugar solution, some of the liquid will go through the cell walls into the cells. Brining is a way to increase the amount of liquid inside the meat cells—a way to make meat juicier.

Brining makes meat juicier by increasing the amount of liquid inside the meat cells.

Juicy Roast Chicken

MAKES 6 SERVINGS

This recipe is a combination of my brining, food science writer Harold McGee's roasting technique, and my herbs and basting mixture. I gave this recipe out on a radio talk show, and listeners called in to report that they had never had such good chicken.

---

what this recipe shows

Absorbing water while soaking in the brine makes chicken meat juicier.

Basting the chicken with a corn syrup and butter mixture adds sugar and protein for good browning.

---

1 6-pound roasting hen	1 small orange, quartered
1 cup salt	1 bay leaf
2 onions (*total*), quartered	3 tablespoons dark corn syrup
3 leaf ends of celery stalks (*total*)	4 tablespoons ($^1/_2$ stick) butter, melted
10 sage leaves (*total*)	3 tablespoons cornstarch
10 sprigs fresh thyme (*total*)	$^1/_4$ cup cold water
1 quart *and* $^1/_4$ cup cold water ($4^1/_4$ cups *total*)	

1. Remove the neck and giblet package from the chicken and pull off the flap of fat if there is one. Rub hen inside and out with salt. Place the chicken in a pot that is just large enough to hold it and tall enough to completely cover it with water. Add ice water just to cover the chicken and stir briefly. Place in the refrigerator for 3 hours. Remove the bird and rinse for several minutes under cold running water to remove all salt inside and out.

2. While the chicken is in the refrigerator, make a simple stock by simmering the neck, giblets (no liver), 1 onion quartered, 1 leaf end of celery, 2 sage leaves, 2 thyme sprigs, and 1 quart water in a large saucepan over low heat for 1 to 2 hours. Strain and save the stock. Pull the meat from neck and add to the stock. Cut up the giblets and add to the stock.

3. Preheat the oven to 475°F (246°C).

4. Arrange a V-rack for holding the bird over a pan to catch juices. Pour $1^1/_2$ cups of the stock into the pan, reserve the rest. Place the remaining onion, leaf ends of celery, sage leaves, and thyme sprigs, plus the orange and bay leaf, in the chicken cavity.

5. In a small bowl, stir together corn syrup and butter. Brush the chicken lightly with this mixture and place in a V-rack, breast side down. Roast 20 minutes, basting with drippings in pan every 7 minutes. Remove the chicken from oven and turn so that one leg and thigh are up. Baste with the butter mixture and return to the oven for 10 minutes. Remove from the oven again, turn completely over so that the other leg is up, baste well with the butter mixture, and return to the oven for another 10 minutes.

continued

6. Remove from the oven again and turn the oven down to 325°F (163°C). Leave the oven door partially open so oven temperature will come down quickly. Turn the bird breast side up, brush well with butter mixture and return to the oven. Baste every 5 minutes. In 5 minutes, check breast meat temperature using an instant-read thermometer. The ideal temperature to remove it is 150° to 154°F (66° to 68°C). If necessary, place back in oven an additional 5 to 10 minutes until it reaches this temperature. Breast meat temperature rises about 2°F (1°C) every minute in the oven at this temperature.

7. Add as much of the pan drippings to the remaining stock as you can without getting the sauce too salty. Stir cornstarch into ¼ cup cold water and stir it into the stock-drippings mixture in a large saucepan. Heat over medium heat, stirring constantly, until the sauce thickens.

8. Allow chicken to stand at least 15 minutes after removing from the oven. Carve or partially carve the chicken and arrange back in bird form on the serving platter. Pass hot gravy in a separate bowl.

Brined Turkey

The fowl that is most familiar to Americans, after chicken, is certainly turkey. Once I tried brining chickens to enhance juiciness, turkey was just a step away. Pam Anderson of *Cook's Illustrated* magazine has written several articles on brining turkey and strongly recommends it. I basically follow the preceding Juicy Roast Chicken recipe, but I let the turkey, about 16 to 18 pounds, remain in the brine overnight, refrigerated. I preheat the oven to 475°F (246°C) and start the turkey breast side down. After 20 minutes, I turn the oven down to 300°F (149°C). After 1½ hours, I turn the turkey breast side up and continue roasting at 300°F (149°C), basting frequently with the corn syrup–butter mixture until the breast temperature reaches 153°F (67°C). The temperature will rise a few degrees with standing.

Cooking time: the crucial factor

How much should meat be cooked—to what end temperature? Is the end temperature that is right for one kind of meat right for another? How can you tell when meat is done?

Rare—medium—well

Review your priorities. If you prize juiciness and tenderness, you should cook meats no more than medium-rare. Remember the drawing in "How Muscles Cook" (page 372). Muscles start to shrink in length and start losing moisture when heated above 120°F (49°C), and above 140°F (60°C) major shrinkage and water loss take place. A piece of loin meat that should be magnificent (with little-used, tender, well-marbled muscles), properly processed, with excellent water-holding ability, can become dry, leathery, and unchewable when totally ruined by overcooking.

Of course what one person may consider perfectly cooked another may find totally unacceptable. But, in any case, keep in mind that you can't have your cake and eat it too. In meat cooked to medium or beyond, a lot of juiciness and tenderness are sacrificed. Bearing this in mind, decide which internal temperature you personally define as "done" for fish, chicken, veal, pork, lamb, and beef.

There is considerable difference in what various cookbooks define for beef as rare, medium-rare, medium, medium-well, and well-done. For example, the internal temperature for rare runs from 120° to 130°F (49° to 54°C). Here is a table for beef with frequently used descriptions of doneness.

BEEF: DONENESS, DESCRIPTION, AND APPROXIMATE INTERNAL TEMPERATURE			
Doneness	Color	Feel	Temperature
Very rare*	raw, unchanged	soft, flabby, warm center	<120°F (49°C)
Rare*	red center	soft, yielding, warm center	120° to 125°F (49° to 52°C)
Medium-rare	pink center	soft but resilient, dents but resumes shape immediately, warm center	125° to 140°F (52° to 60°C)
Medium	gray, almost complete loss of pink	firm, very warm center	145° to 155°F (63° to 68°C)
Well-done	gray and dry	very firm/rigid, hot center	>160°F (71°C)

*Some descriptions of very rare and rare beef call the center "cool." At a temperature of 120°F (49°C), the center would be warm to the touch. A truly "cool" center would be below this temperature. Color and texture of rare beef are matters of personal choice, no matter what a thermometer tells you.

Different meats, different ideal temperatures

Juiciness and tenderness are my top priorities, and my personal preferences for doneness temperatures are as follows:

Beef: 130° to 135°F (54° to 57°C)

Dark-meat chicken: 165° to 175°F (74° to 79°C)

White-meat chicken: 149° to 152°F (65° to 67°C)

Fish: about 145° to 150°F (63° to 66°C)

Meat thermometers   Two general types of meat thermometer are widely available in cookware shops. One type has a dial and a fat shaft that is inserted into the meat before it goes into the oven. It stays in the meat during cooking. The other type has a thin metal shaft that is inserted briefly for a quick check and gives you an "instant" reading on a dial—actually in about a minute. Some modern "instant-read" thermometers have a digital display instead of a dial. Instant-read thermometers are not designed to be left in the oven during cooking.

Both of these thermometers give you an average reading over the portion of the shaft that is in the meat. This is not necessarily the accurate temperature in the center of the meat that is needed, and many of these thermometers are off by over 10°F (about 5°C).

Despite their shortcomings, I believe a good thermometer is an indispensable aid to good cookery, not just for meats but for everything from yeast breads to candy. You can always check your thermometer's accuracy (page 392), and you can buy a high-quality thermometer. Scientific supply houses (see Sources—Thermometers, page 477) have very accurate digital thermometers available for between $30 and $100, depending on your specific needs. I firmly believe that a good thermometer is a worthwhile investment.

CHECKING YOUR THERMOMETER FOR ACCURACY

Any thermometer can be off. I check mine occasionally by measuring the temperature of crushed ice in water. If my thermometer says about 32°F (0°C), I know that it is close to right. Some people like to check the boiling point—212°F (100°C); however, if you are not near sea level, remember that the boiling point changes with altitude. The freezing point varies slightly, but the boiling point decreases approximately 1.8°F (1°C) for every 1,000-foot (305-meter) increase in altitude.

The effect of beginning temperature The temperature of the meat when it begins to cook has an influence on both the cooking time and the evenness of cooking. If you have a large piece of meat that is cold or near-frozen and cook it by ordinary directions, the outside may be well browned and done before the center has even gotten warm.

You can use this as an advantage with pieces of meat or cuts like filets, which are relatively small in circumference and roast on all sides at once. Starting with the meat frozen and cooking at a high temperature can give you the desired well-browned exterior on this thin piece of meat without cooking the center beyond medium-rare. This classic frozen filet recipe is a good example.

Whole Roasted Tenderloin

MAKES 6 TO 8 SERVINGS

This is my rendition of an excellent recipe from Fritz Blank, chef-owner of Deux Cheminées restaurant in Philadelphia. He based it on a roasting method of Chef Louis Szathmáry.

> **what this recipe shows**
>
> A cold or even frozen starting temperature of the meat can be used to produce a well-browned exterior without overcooking the interior.

1 whole beef tenderloin (3 to 3½ pounds), well trimmed

Salt, preferably sea salt

3 tablespoons beef concentrate, B-V (see Note)

3 tablespoons peanut or corn oil

2 tablespoons unflavored gelatin

2 cups Burgundy or other full-bodied red wine

1. Cut off the narrow tip and trim the bull nose so the tenderloin is more uniform in shape. Lightly salt on all sides and rub in. Stir together the beef concentrate and oil and rub the meat well with the mixture. Push the narrow end in and press the fat end out. Double-wrap tightly with heavy-duty plastic freezer wrap such as Saran Wrap. Place the meat on a pan and freeze solid. Allow 2 days in the freezer.

2. Preheat the oven to 425°F (218°C).

3. Heat an unlined heavy iron grill smoking hot on top of the stove. Remove the hard-frozen tenderloin from the freezer, unwrap, and sear on all four sides. Sprinkle the gelatin over the Burgundy in a 3-cup measuring cup.

4. Place the meat on a double thickness of heavy-duty aluminum foil in a jelly-roll pan and form the foil into a "canoe." Pour the wine-gelatin mixture over the meat. Loosely close the top of the canoe and roast 20 to 25 minutes depending on size. Remove from the oven and leave at room temperature or refrigerated until ready to finish cooking. May be refrigerated, ready to heat, for up to 4 days.

5. When ready for final cooking, preheat the oven to 450°F (232°C). Open the foil to expose the top of the meat and roast 15 to 25 minutes. Use an instant-read thermometer to check the temperature in the middle of the roast for desired doneness. I like 130°F (54°C) for medium-rare.

6. Allow to stand 15 to 20 minutes before serving. Slice into 1/2-inch slices, fan attractively on a serving platter, and serve with béarnaise (page 304) or your choice of sauce.

Note: *Chef Blank uses Kitchen Bouquet, but I really love B-V, which is not as widely distributed as Kitchen Bouquet. See Note on page 378 for information on locating a source of B-V.*

Cooking methods for meats

After Merle Ellis summarizes the two kinds of meat as tender and tough, he says, "There are two ways to cook meat—wet and dry. Cook tough meat wet and tender meat dry." This is a clear principle and words to cling to as we delve into the complexities of cooking meat.

Deciding how to cook some cuts of meat is easy. A naturally tender cut from the beef loin such as a thick, richly marbled porterhouse steak is a snap to cook. You have no connective tissue to tenderize, so long, slow, wet cooking is not necessary. You want the internal temperature to remain low to keep this meat juicy and tender. So, whether you grill, broil, or pan-broil, if you remove the steak from the heat rare to medium-rare, you should have juicy, tender meat.

With a shoulder roast that is tough and full of connective tissue, there is no question that you must cook it long enough and to a high enough temperature to melt some of the connective tissue to get it tender enough to chew; it will not even begin to get tender enough to chew until it is well done, above 160°F (71°C).

For some cuts the decision will be more difficult. For example, a chuck roast, depending on the cut, contains sections of three to four different muscles, each with a different tenderness. If you cook it to a low internal temperature, the meat will be tender and juicy, but it is difficult to chew because it contains so much connective tissue. If you cook it to a high enough temperature to melt the collagen in the connective tissue, the meat can be tough and overdry. Alas! You can make a pot roast of it and use the gravy to compensate for the dryness of the meat or, if it is a blade chuck roast, cut the rib-eye out and cook dry—grill or broil, slice the other sections thinly across the grain, and stir-fry.

Here is a table of different cuts of beef with my personal opinion on the cooking methods best suited for each. It was inspired by, but is different from, one by Jack Ubaldi and Elizabeth Crossman in the January 1993 issue of *Food & Wine*. These are also good guidelines for lamb and pork cuts shown in the table of similar cuts on pages 380–381. Notice that while some methods are rarely or never recommended, you may find recipes for them or personally choose them.

glance

Recommended methods of cooking different cuts of beef

	Dry			Wet			Hot Fat		
	Roasting (Baking)	*Grilling Broiling*	*Pan Broiling*	*Boiling (Poaching)*	*Steaming*	*Stewing Braising*	*Deep Frying*	*Pan-frying*	*Stir-frying*
Loin									
T-bone & porterhouse steaks		x	x						
Strip, club & Delmonico steaks		x	x						
Filet mignon		x	x						
Whole tenderloin	x	x							
Sirloin steak		x	x						
Flank steak	x	x				x			x
Hind Quarter									
Top round	x	x	x						
Top round roast	x					x			
Bottom round	x					x			
Rump roast	x					x			
Eye of round						x			
Sirloin tip	x								
Ribs									
Prime rib	x								
Rib-eye steak		x	x						
Skirt steak		x	x						x
Hanger steak		x	x						x
Short ribs	x	x							
Chuck									
Blade chuck roast	x								x
7-bone chuck roast						x			
"Mock" tenders						x			
Chuck-eye roast						x			
Brisket	x					x			
Shank	x								

Dry methods of cooking

Cuts of meat that are reasonably tender are ideal for dry cooking methods. Smaller, thin cuts cook perfectly by grilling, broiling, or pan broiling. Large, relatively tender cuts, on the other hand, are ideal for roasting (baking). If you broil or grill these thick, large cuts, the surface will overcook before you get the center even warm.

Browning—the Maillard reactions

A richly browned roast or a steak with dark grill marks has marvelous complex, sweet, meaty flavors. Well-browned meat is a taste treat. Meat whose red color has given way to gray can be dull and unappealing. How can mere color make such a difference?

Browning adds so much because it's *not* just color. We are all familiar with sugar's browning (caramelizing) and the sweet rich taste of caramel. When table sugar (sucrose) reaches very high temperatures (over 300°F/149°C), it melts, then starts to decompose—different sugars are formed, sugars break apart, and some of these rejoin. At any given moment between clear melted sugar and black compounds, there is a different mixture of sugars. More than 128 different sugars have been identified. Many of these are brown and have the wonderful taste that we associate with caramel.

The Maillard reactions are a series of complex reactions between certain sugars and proteins that produce similar sweet caramel-tasting brown products at much lower temperatures. They are named after Dr. L. C. Maillard, who first described them. Everything from baked goods to fried foods and roasts gets this rich-tasting brown coating from these complex reactions of sugars and proteins caused by heat.

Three conditions are necessary for this lower-temperature browning: proteins, reducing sugars, and a nonacidic environment. Acids prevent browning. Little if any browning occurs at a pH below 6. But with low acidity (ideally a pH between 7.8 and 9.2), the more of certain sugars and proteins in a product, the browner it gets.

Some sugars like glucose (corn syrup), a reducing sugar, brown at lower temperatures than others. Most natural compounds and doughs or batters (which contain sugars from starch that has broken down) contain mixtures of sugars. Not all sugars take part in the Maillard reactions. Reducing sugars (those with a specific configuration) like glucose are most reactive with an exposed amino group from proteins. If you substitute as little as 1 tablespoon of corn syrup for 1 tablespoon of sugar in a cookie recipe, you will get cookies that are browner and have a crisper surface.

> Substituting as little as 1 tablespoon corn syrup for 1 tablespoon sugar will make your cookies crisper and browner.

A turkey or roast chicken that is basted with a mixture containing sugars and proteins will brown faster than one basted with oil. My favorite basting combination is corn syrup (for the fast-browning glucose), melted butter (which contains both sugars and proteins from milk), and consommé (clarified beef stock and gelatin). Both the stock and the gelatin contain proteins to aid in browning.

In most recipes meat is browned before roasting, either on top of the stove in a skillet or in a high-temperature oven (frequently both in a skillet and in the oven), then the oven is turned down and the meat is roasted at a much lower temperature. This initial browning both kills surface bacteria and provides a flavorful crust. In my slow roasting procedures, I like to sear the meat in a skillet to kill surface bacteria, then cook in a low-temperature oven for juiciness, tenderness, and even cooking, then finally turn the oven up to 500°F (260°C) for a quick, thorough surface browning.

I prefer the oven browning at the end, not the beginning of roasting. During roasting, juices containing proteins and sugars have come to the surface and evaporated. So when I brown the meat at the end of the cooking time, there is a considerably greater concentration of surface sugars and proteins. With the elevated temperature, browning occurs rapidly.

Grilling and broiling—ideal cooking methods for thin, tender cuts

Grilling over coals or an open flame or broiling—in the oven or in a very hot pan with no fat—can give you the best of both worlds: incredible flavors from quick browning and mouthwatering tenderness from quick cooking (not overcooking). A good grilled or broiled steak or lamb or pork chop combines a rich, crusty-brown, fabulously flavorful surface with a juicy, tender interior.

Intense heat gets the surface of the meat to high temperatures fast. Proteins and sugars react to give us the browning reaction at its crusty, rich-flavored best. The cooking time is short, which keeps the interior of the meat at a low enough temperature to prevent toughening and excessive shrinkage and moisture loss. If you get the interior of the meat no hotter than 140°F (60°C), medium-rare, lengthwise shrinkage of the muscle and major water loss are minimized.

Of course, there are differences between grilling and broiling. For one thing, the temperature over a very hot bed of coals or in a fiery preheated skillet is higher than under a home oven broiler. For this reason, when grilling you normally set the grilling surface 3 to 5 inches from the hot coals, while broiling for thin steaks is done 1 to 3 inches below the oven broiler. For thicker, larger pieces of meat like chicken thighs the broiler distance would be greater. Nevertheless, grilling, broiling, and pan broiling all provide fast, dry, high heat, which can produce a flavorful browned surface and a juicy interior.

These methods work only on thinner cuts. With the very high heat and a short cooking time, there is not time for heat to be conducted to the center of a thick cut, and it will remain cold and raw. The meat should be no thicker than about 2½ inches. You will find that most steaks run ¾ to 1½ inches in thickness.

Pork, lamb, and chicken can be cooked nicely by grilling or broiling. In addition to cuts of beef from the ribs, loin, and sirloin, you can successfully grill or broil the cuts on the edge of the loin like the top round (London broil), which lies just behind the sirloin on the back end of the beef; a rib-eye cut from a blade chuck roast, which lies just in front of the ribs; the flank steak, which lies just below the loin; skirt steaks (the diaphragm muscle); and hanging tenders. All of these cuts have excellent flavor like the tougher cuts. When you get them from beef that is nicely marbled with fat, cook them rare to medium-rare, and slice them thinly across the grain, they are outstanding in flavor and tender.

You can also marinate these cuts to improve tenderness. (See Marinated Grilled Skirt Steak, page 386.) You want to use the cuts that are nearest the sirloin in the back (top London broil) and the ribs in the front (blade chuck roast). (See pages 380–381 to learn how to identify these cuts.)

There are many variables in grilling—the temperature of the coals, whether the meat is directly over the coals or to the side, and so on—and there can be considerable differences in oven broilers. Only trial and error and experience will give you the expertise to get the steak on your grill or under your broiler exactly as you want it. There are hundreds of books on outdoor cooking. Recent books on grilling that give good guidance include *The Thrill of the Grill* by Chris Schlesinger and John Willoughby, *Cooking with Fire and Smoke* by Phillip S. Schulz, *The Joy of Grilling* by Joe Famularo, and *The Grilling Encyclopedia* by A. Cort Sinnes.

While grilling or broiling is a good method for meat, fish, or fowl, remember that fish and white-meat chicken can overcook *quickly*. Take care in both how long the meat is on the grill and how intense the heat is. You can marinate or simply brush fish or chicken with a basting sauce. In the Thai Grilled Chicken with Honey-Chili Glaze (recipe follows), the chicken is marinated in a ginger, lemon grass, and coconut milk mixture and then grilled and coated with a honey-chili glaze.

Thai Grilled Chicken with Honey-Chili Glaze

MAKES 8 SERVINGS

Chicken is marinated in coconut milk with ginger and lemon grass for subtle exotic flavors. Peanuts, chilies, coconut, and cilantro top this off to complete the Far East medley. Serve with cold fruit like Fresh Fruit with Ginger (page 319) and Corn on the Cob with Honey (page 320).

what this recipe shows

Chicken thighs are better for grilling than breasts because the collagen in the thighs melts and keeps the meat moist.

Fresh ginger contains an enzyme that breaks down gelatin and tenderizes meat.

Vinegar, honey, and sugar in the glaze tame the heat of the peppers.

4 quarter-size slices of ginger $1/4$ inch thick, minced

1 can ($13^1/2$ ounces) coconut milk

$1^1/2$ teaspoons salt

1 teaspoon freshly ground pepper

2 2-inch pieces lemon grass

8 large chicken thighs, skin removed

1 recipe Honey-Chili Glaze (below)

$1/2$ cup flaked coconut, canned or frozen

1 bunch fresh cilantro, stems included, chopped

6 scallions, green parts included, chopped

1. In a medium mixing bowl, whisk together ginger, coconut milk, salt, and pepper. Add lemon grass and stir.

2. In a freezer-type zip-top plastic bag, marinate chicken in ginger–coconut milk mixture overnight in the refrigerator.

3. Prepare charcoal grill or preheat the broiler.

4. Drain the chicken and cook about 8 to 10 minutes on each side, depending on the size of the thighs and the heat. Take care not to overcook.

5. Remove to a bowl and pour Honey-Chili Glaze over the hot chicken. Toss to coat all pieces well. Place on a serving platter and sprinkle with coconut, cilantro, and scallions. It's delicious hot, at room temperature, or even cold.

Honey-Chili Glaze

MAKES ABOUT 1 CUP

2 small hot serrano chilies or 1 jalapeño, seeds included

2 tablespoons (packed) dark brown sugar

$1/2$ teaspoon rice or cider vinegar

$1/2$ teaspoon salt

1 cup honey

$1/3$ cup salted roasted peanuts, finely chopped

In a blender or chopper, chop together the chilies and brown sugar. Scrape into a small saucepan. Add the vinegar, salt, and honey and bring to a boil. Simmer over very low heat for about 2 minutes. Stir in the peanuts.

Pan broiling

Pan (skillet) broiling is a handy and quick method of cooking that does not require firing up a grill, though you do need good ventilation or you will set off your smoke alarm. The Pan-Seared Steak that follows gives you typical pan broiling directions, but even seafood can be pan broiled. Paul Prudhomme's famous blackened redfish is a widely known pan-seared delicacy. Large sea scallops are browned quickly in a very hot skillet and served on a bed of honey-chili glazed papaya, avocado, and tomato salsa in the Seared Scallops with Avocado and Papaya (page 399).

Pan-Seared Steak

MAKES 2 SERVINGS

When I can't grill, I prefer pan searing to broiling. You can get a hard, dark brown, sweet-flavored crust and a pink-red, juicy center. For those who cook with limited facilities, a good pan-seared steak is a godsend. You just need a burner and a heavy, unlined skillet such as cast iron that can take high temperatures without warping or damaging. Serve with a baked potato with a little blue cheese, a good salad such as Mixed Greens with Walnuts (page 314), and Crusty French-Type Bread (page 24).

> **what this recipe shows**
>
> Having the surface of the pan very hot before adding the meat helps prevent sticking.
>
> Salt draws juice from the meat surface, adding proteins and sugars to the surface and reducing its moisture content for better browning.

Salt

2 1-inch-thick rib-eye, sirloin strip, or
    other steaks of your choice

Heat an empty, heavy unlined skillet almost smoking hot (see Note). Sprinkle salt over the bottom to cover thinly. Drop the steaks into the hot pan. Sear over high heat for 3 minutes, turn steaks over, and sear until you see beads of moisture (juices) coming through the steak crust (about 3 minutes). This time will give you a medium-rare steak. Serve immediately.

Note: *Be sure to use an unlined skillet since the high heat will damage a lining. Fat and juices from the steaks really smoke when they hit the hot pan. Have a ventilator fan on or a window open, or you'll set off your smoke alarm. You could, of course, grill or broil the steaks.*

Seared Scallops with Avocado and Papaya

MAKES 4 SERVINGS

Fresh and refreshing—well-browned, big, tender sea scallops on a bed of ripe orange papayas, bright red tomatoes, and pale green avocado chunks. Sharon Shipley of Mon Cheri Caterer serves this honey-chili dressing with fruit for spectacular salads. I like the sweet-hot dressing so much that I put it on all kinds of dishes—from grilled chicken to scrambled eggs. You can serve this as a colorful first course or as a light luncheon dish or convert it to a large platter for a summer buffet.

what this recipe shows

Honey tones down the chili for a mild sweet-hot blend.

1 recipe Honey-Chili Dressing
  (page 400)

2 ripe tomatoes, peeled, seeded, and cut
  into $1/4$-inch cubes (see Note)

$1/2$ ripe papaya, peeled and cut into
  $1/4$-inch cubes

$1/2$ avocado, peeled and cut into $1/4$-inch
  cubes

Juice of $1/2$ lemon (about 1 tablespoon)

12 large sea scallops

4 tablespoons ($1/2$ stick) butter, melted

6 scallions, green parts included, sliced
  into thin rings

1. Have serving plates ready, Honey-Chili Dressing prepared, tomatoes and papaya cubed. Place the avocado in lemon juice as you cube it to prevent discoloration.

2. Cover the center of each plate with a pool of dressing and sprinkle tomato and papaya cubes into the dressing on each plate. Drain avocado cubes and add to tomato and papaya cubes on each plate.

3. Heat a nonstick skillet quite hot over high heat (but not hot enough to damage the lining). Coat the flat sides of six scallops with melted butter. Place the scallops flat side down in the hot skillet. When browned, turn over and brown other side. Place three scallops on each plate in the center of the tomato-papaya-avocado mixture. Repeat with the remaining six scallops. Sprinkle each serving with a few scallion rings.

Canned tomato wedges can be good quality— red and firm; try Del Monte.

Note: *If really ripe, red tomatoes are out of the question, use canned Del Monte Tomato Wedges. They are not cooked soft like regular canned tomatoes but are a good red and still firm.*

Honey-Chili Dressing

MAKES A LITTLE OVER 1 CUP

2 medium shallots

3 tablespoons fresh lemon juice

1$^{1}/_{2}$ teaspoons salt

$^{1}/_{4}$ teaspoon white pepper

2 teaspoons chili powder, preferably
recently opened

1 tablespoon confectioners' sugar

2 tablespoons honey

$^{3}/_{4}$ cup oil

$^{1}/_{2}$ small jalapeño, seeded and finely
chopped (optional)

In a blender or a food processor with the steel knife, turn the machine on and drop the shallots down the feedtube onto the spinning blade to mince. Add lemon juice, salt, pepper, chili powder, confectioners' sugar, and honey. Turn the processor on and drizzle the oil slowly down the feedtube. Stir in jalapeño if you want a hotter dressing.

Roasting

There are many ways to roast meat. Low-temperature slow roasting produces very even cooking and wonderful juiciness and tenderness in meats cooked no more than medium-rare. On the other hand, high-temperature roasting creates marvelous flavorful browning and is relatively fast; however, it does overcook and dry a portion of the meat. There is no hard-and-fast right or wrong. Examine the results of different methods as described here and decide for yourself which you prefer.

To sear or not to sear? Since the mid-1800s, cooks have seared meat "to seal in the juices." No worry, cooks should keep right on searing, but the reason for searing is flavor, *not* juiciness. Since the early 1930s, research by meat scientists and home economists has pointed out that there is less fluid loss in meat cooked at a constant moderate temperature than in meat that is first seared at a high temperature, then roasted at a moderate temperature.

Searing seals in juices, right? Wrong. The reason for searing is flavor.

Cooks have been aware of this for a long time. As early as 1936, *The Joy of Cooking* pointed out that the "modern" method recommended by national packers was roasting at a constant relatively low temperature (325°F/163°C). So, we know we lose more juices when we sear. It is just hard to give up this concept.

You can clearly see that a hard crust forms on the surface of seared meat, so it makes sense to say that the crust is sealing in the juices. In *The Curious Cook*, Harold McGee points out that, whether we are willing to admit it or not, we know that the crust leaks. As evidence, there is the abundant juice that accumulates on the platter from a steak that is seared well on both sides.

Searing does create sweet, incomparable flavors in the rich, brown crust. At the higher temperatures on the surface of the meat, the proteins and sugars in the meat will form complex mixtures of flavorful, sweet, brown compounds (see "Browning—The Maillard Reactions," page 395).

Slow roasting In a November/December 1995 *Cook's Illustrated* article comparing various time and temperature combinations for roasting, Pam Anderson concluded that briefly browning the outside of the meat in a skillet, then slow roasting the meat at 200°F (93°C) produced a magnificent tender, juicy prime rib—"perfect restaurant prime rib." Five different cooks have told me, "I will never cook prime rib any other way." I consider that astounding praise.

Pam's conclusions bear out exactly what we know from the science point of view. Even when pieces of meat of approximately the same size and shape are cooked so that the center reaches the same end temperature of about 130°F (54°C), they will vary greatly with cooking temperature in shrinkage, water loss, and overall doneness.

Pieces cooked at the higher temperatures, 400° to 500°F (204° to 260°C), will have a very small red, medium-rare area of 1 to 1½ inches in the center, but the outer two-thirds of the meat will be dry, overcooked, and gray. Meats cooked at more moderate temperatures, 300° to 375°F (149° to 191°C), will have a larger medium-rare area in the center but will still have (as the temperature increases) a considerable outer portion of dry, overcooked meat. Meats cooked below 250°F (121°C) to the same end temperature of 130°F (54°C) come out a beautiful medium-rare almost to the edge.

In the section "How Muscles Cook" (page 371), we saw that up to 120°F (49°C) muscles shrink in diameter only, not in length, so a minimum amount of moisture is lost. As the temperature of the muscle increases above 120°F (49°C), lengthwise shrinkage and greater water loss begin. At high cooking temperatures the outer portions of the meat are well into these temperatures and suffer great water loss before the heat can be conducted into the center. With slower, lower-temperature cooking the heat has time to penetrate and cook the center before the outside reaches such high temperatures that it shrinks and dries.

Even after the meat is removed from the oven, the temperature continues to rise. Meats cooked at higher temperatures may rise as much as 15°F (8°C), overcooking even the center. The internal temperature of meats cooked at lower temperatures rises only a degree or two.

Commercial meat processors would never think of using any method other than low-temperature, slow cooking—the weight loss is simply too great. Chris Kimball, publisher of *Cook's Illustrated,* has quoted the weight loss with a 6-pound roast cooked at 500°F (260°C) to be over 2 pounds!

So, if we have such great juiciness and tenderness cooking at low temperatures, what are we sacrificing? A major disadvantage is that we will not have good browning with all its marvelous flavors. To have the best of both worlds, quickly sear a roast on the outside on top of the stove. This will kill surface bacteria and provide some light browning. Slow-roast the meat to about 20°F (11°C) below the desired temperature, then turn the oven up to 500°F (260°C) for a quick surface browning while the desired temperature is being reached. Turning the oven up after most of the cooking will give you a deep brown crust very fast. Juices containing proteins and sugars have come to the surface during cooking, evaporated, and left a high concentration of proteins and sugars for fast browning.

Prime Rib

For the best ribs for prime rib or a standing rib roast, ask the butcher for the ribs near the small end of the rib cage. These lie next to the very tender loin area and are more tender than ribs from the large end of the rib cage, which are next to the chuck (shoulder).

This recipe uses a combination of Pam Anderson's procedure and the high heat at the end of roasting that Chris Kimball and I like to do.

Arrange a rack in a low position in the oven and preheat to 200°F (93°C). Heat a large, heavy iron (unlined) skillet until very hot. Sear the meat on all sides. Lift the roast and place a round wire rack in the skillet. Rub the meat with salt and pepper and arrange on the rack. Roast in the oven until the internal temperature reaches 110°F (43°C). Turn the oven up to 500°F (260°C) and continue cooking until the internal temperature gets to 130°F (54°C)—about 10 to 15 minutes more. Let the roast stand for 30 minutes before carving.

Fast roasting The crusty brown surface that you get by fast roasting at high heat provides a wealth of taste thrills. So what if the meat next to the outside is dried out? Most of the meat is moist and juicy and delicious beyond belief.

While I prefer slow roasting for beef, I think a whole fowl is ideal for fast roasting. I love a fast-roasted chicken with the wondrous sweet meaty flavors created by the browning reaction that permeates the whole bird.

One obvious joy of fast roasting, in addition to the marvelous crust, is speed. You need only about half the normal cooking time when fast roasting a chicken. When you need an excellent meal in a hurry, roasting a chicken or two is perfect. As Barbara Kafka says in her book *Roasting*, "When in doubt, roast a chicken. When hurried, roast a chicken. Seeking simple pleasure? Roast a chicken." I agree wholeheartedly with Barbara's method for fast-roasting chicken, and my recipe is almost identical.

Fast Roast Chicken

If possible, leave the chicken at room temperature for an hour before roasting. This is safe since the bird is not being cooked rare. Arrange a shelf slightly below the center of the oven and preheat to 500°F (260°C). Remove the neck, giblets, and liver as well as the fat from the tail area. Put a few of your favorite aromatics (sage leaves, a quarter onion, and a lemon) in the cavity. Place the bird, breast side up, in a small heavy roasting pan (I have a small Le Creuset pan that is ideal). Place in the oven, legs first. After 5 to 10 minutes, run a spatula under the bird to loosen it. Roast until the juices run clear, 40 to 50 minutes more. Remove the chicken to a serving platter. Spoon out most of the fat from the drippings and discard. Add ½ cup water or chicken broth to the pan and place it on a burner over medium heat. Scrape the bottom well to get up all the stuck bits, boil to reduce to about ⅓ cup, and pass separately.

Roasting other fowl—birds on the wing You can run into some interesting differences in cooking ducks, geese, and other game fowl as compared to chickens and turkeys. For instance, ducks have what seems to me an almost inconceivable amount of fat. After many years, I still have a vivid recollection of the day that I came to this realization.

It was the first Thanksgiving that I was away from home and family. My new husband and I had a tiny apartment in Far Rockaway, New York, a narrow strip of land between Jamaica Bay and the Atlantic Ocean. In the summer, this was a resort area, but in the winter it was a wind-swept, cold, desolate place. I had limited experience in the kitchen, but I was determined

that we were going to have a special Thanksgiving. I bought a duck and pulled out *The Joy of Cooking*.

I made a batch of dressing and stuffed it into the duck. It disappeared. The duck looked three-quarters empty inside. I decided this would not do. The dressing is one of my favorite things, and I wanted plenty. So I made another batch of dressing and put it inside the duck, too. The duck still looked quite empty. I ended up stuffing four batches of dressing into that poor duck.

After the duck had been roasting for a while, we heard this great *rrrrrrrmmmmmp*—a strange, muffled explosion from inside the oven. I jerked the oven door open. There to my wondering eyes appeared a huge mountain of dressing with just a few duck bones sticking out of it. The dressing had absorbed fat and absorbed fat and absorbed fat and expanded and expanded until the poor duck literally exploded. We could find only an occasional piece of duck meat or bone amid all the dressing.

Some of my friends have had a different type of disaster with the great amount of fat from ducks, geese, and lamb—oven fires! Be sure to drain fat off frequently when roasting these high-fat meats.

In addition to having much more fat than domestic chickens, ducks have less meat and a slightly different anatomy. For example, the thigh is connected to the body in just enough of a different location (farther back) to make it difficult to find this joint. Half a roasted duck is an appropriate entree serving per person.

Some years after my exploding duck experience, I was assigned the job of preparing the ducks in a restaurant where I sometimes worked for experience. I would stuff each duck with orange wedges, onion wedges, and celery leaves, roast them until the thighs wiggled "like they were done," then set them on a rack to cool. I poured off great quantities of fat from the duck drippings. (We saved the fat since duck fat is a very flavorful cooking fat.) While the ducks cooled, I made whatever the sauce of the evening was—sometimes with oranges and Grand Marnier, sometimes with cherries and Chambord.

When the ducks were cool enough to handle, I cut them with kitchen shears or a large knife in half down the center of the breastbone and backbone. Then I pulled the whole rib cage of bones from each duck half, leaving just the thigh and leg bone. This made the half duck easy to eat and a good plate presentation. We kept these partially boned duck halves warm in a covered area by the stove. When an order for duck came in, I grabbed a duck half, placed it in a small pan in the hot oven just to warm while I heated a serving portion of the sauce in a skillet. Then I removed the warm duck from the oven, poured the hot sauce over it, garnished it, and sent it out.

This is an excellent way to prepare ducks for a dinner party. Roast and partially bone the ducks and prepare the sauce ahead. At serving time you have just to reheat the ducks and the sauce and serve. Here is a detailed recipe for family or restaurant, Roast Duck with Caramel Grand Marnier Sauce.

Roast Duck with Caramel Grand Marnier Sauce

MAKES 4 SERVINGS

This is the restaurant technique for preparing ducks. Cook the ducks ahead, cut them in half lengthwise, and pull out all the bones except the leg and the thigh. Sauce these halves with a deep brown caramel Grand Marnier sauce, and you have a work of art. This makes beautiful individual servings (half duck per person) that are easy to eat. The caramel sauce, when browned on the duck halves, is out of this world! I like to serve duck with Spinach and Orange Salad with Pine Nuts (page 315) and Sherried Rice and Barley with Almonds (page 336).

what this recipe shows

Removing the rib cage before serving makes the duck easy to eat.

Adding corn syrup to the sugar prevents crystallization during caramelizing.

Consommé and caramel give the sauce a rich brown color.

continued

2 4- to 5-pound ducks (see Notes)

2 tablespoons *and* 2 tablespoons Grand Marnier (¹/₄ cup *total*)

Salt and pepper

2 small to medium oranges (*total*), quartered

2 small onions (*total*), quartered

2 ribs celery with leaves (*total*), cut into 1-inch pieces

2 cups water

Caramel Grand Marnier Sauce (page 405)

4 sprigs fresh thyme

4 orange slices, twisted

1. Preheat the oven to 500°F (260°C).

2. Rub each duck with 2 tablespoons Grand Marnier. Rub salt and pepper inside ducks well. Fill each duck with quarters of 1 orange, 1 onion, and 1 rib celery.

3. Place a large cooling rack on top of a large roasting pan. Pour 2 cups water into the roasting pan. Place the ducks breast side down on the rack on top of the pan. Place in the lower third of the oven and immediately turn the oven down to 350°F (177°C). Roast for 45 minutes.

4. Carefully pull the pan from the oven. You will have great quantities of hot duck fat (see Notes). Carefully drain most of this hot fat into a metal bowl. Replace rack, drain duck into roasting pan, and turn breast side up. Roast 45 minutes to an hour more, until the leg joints wiggle easily when pulled. Check the temperature of the thigh meat with an instant-read thermometer. Center of thigh meat should be at least 165°F (74°C).

5. While the ducks are roasting, prepare the Caramel Grand Marnier Sauce. About 10 to 15 minutes before the ducks are done, brush with the sauce.

6. When the ducks are cooked, let them get cool enough to handle. Then, using good poultry shears, cut the ducks in half lengthwise straight down beside the breastbone and backbone. Discard orange-onion-celery filling. With your fingers, loosen the rib cage from the breast meat. When you have the entire rib section loose, cut the joint between the thigh and the rib cage with a knife or shears. Remove all the bones but the thigh and leg bone. This gives you an easy-to-eat half duck, mostly boned. Do this with remaining duck halves. Cover the roasted duck halves with foil and keep in a warm place until near serving time.

7. When ready to serve, heat the oven to 350°F (177°C) and place the duck halves in the oven for 8 to 10 minutes to warm. Bring the remaining Caramel Grand Marnier Sauce to a boil. Arrange a duck half, leg and thigh up, on each plate. Coat well with sauce. Pass remaining sauce in a gravy boat. Garnish plates with a sprig of fresh thyme and a twisted orange slice.

Notes: *Frozen ducks are the only ones available in many areas. Allow time to defrost and, after defrosting, be sure to remove neck and other parts tucked inside duck.*

Ducks have great quantities of fat. Be sure to use a deep pan to catch this fat and drain it frequently during roasting. Many kitchen fires have been caused by duck fat. Be very careful. I have two friends who are professional cooks who have had major oven fires with duck fat!

Caramel Grand Marnier Sauce

MAKES ABOUT 1½ CUPS

3 tablespoons light corn syrup

1 to 2 tablespoons water

1 cup sugar

1 cup beef consommé (canned is fine)
   at a boil

3 tablespoons cornstarch stirred into
   ½ cup orange juice

Finely grated zest of 1 orange

¼ cup Grand Marnier

1. To make caramel in the microwave, stir together the corn syrup, water, and sugar in a heat-proof 1-quart glass measuring cup. Make sure all the sugar is damp—no dry patches. Microwave on High until the sugar begins to darken. When the bubbles are piled high—bubbles on top of bubbles on top of bubbles—the sugar has melted and is about to color. Watch carefully and remove from the microwave when the sugar begins to color. (The caramel will continue to darken after it is removed from the microwave. Once the color starts to change, it goes very fast from pale brown to charred black.) This can be done on the stove. Follow directions on pages 423 and 424.

2. Bring the consommé to a boil in a small saucepan. Carefully, all at once, pour the hot consommé into the hot caramel. This will spatter and burn you if you put only a little liquid in, so place the rim of the saucepan on the edge of the glass cup and dump the consommé in quickly. Stir to dissolve the caramel in hot consommé. Add the cornstarch–orange juice mixture and heat in the microwave or in a saucepan, stirring until the sauce thickens. Stir in the orange zest and Grand Marnier.

Many markets now carry frozen ducks and geese, but squab and game birds, although popular in many countries, are not widely used in the United States. In Southern markets, we do get fresh as well as frozen quail. Some gourmet markets are beginning to carry a wide variety of fowl, and there are also wild bird and game ranches that will ship a variety of birds overnight in cooler packs. These arrive cold, cleaned, and dressed, just ready for your aromatics and the oven.

Wild birds have all dark meat and a gamey flavor, while farm-raised birds will be lighter in both color and flavor. Most fowl can be roasted or braised, and tender fowl like smaller young chickens can be fried. Really old, tough fowl is best braised. Here is a table of general fowl sizes and the most frequently used cooking methods for each.

FOWL SIZES AND COOKING METHODS

		Roasting	Braising	Frying
Squab, Quail, Dove, Small Game Birds	¾ to 1 lb.	x		
Rock Cornish game hens	1 to 2 lbs.	x		
Fryers/Broilers	2 to 3 lbs.	x		x
Stewing/Roasting hens	3 to 7 lbs.	x	x	
Capons	6 to 9 lbs.	x	x	
Ducks	4 to 5 lbs.	x		
Geese	6 to 14 lbs.	x		
Turkeys	5 to 25 lbs.	x	x	

In Louisiana deep frying small turkeys outdoors in a large pot over big burners is in vogue. Technically, you can deep-fry even whole birds if you have a large heat source and vat for frying.

Roasting pork tenderloins Pork tenderloins (not the *loin* but the *tenderloin,* comparable to the beef tenderloin from which filets are cut) are a wonderful resource for cooks. They are low-fat, lean meat without any waste, they have an excellent nutritional profile, and, if you do not overcook them, they are also juicy, tender, and delicious. In fact, pork tenderloins are so delicate in flavor that they can pass as veal when prepared in a manner or with a sauce typically associated with veal. Some of my Austrian friends slice the tenderloin fairly thick (½ to ¾ inch) on an angle to get larger pieces, then pound these slices out flat so that they have thin, nice-size pieces of lean meat and use them to prepare Wiener Schnitzel. They swear that most people can't tell the difference between the pork tenderloin and veal.

You can let your mind go wild thinking of wonderful glazes and sauces for these tender jewels. I have taught at least five different versions of roasted pork tenderloins through the years. I've used honey mustard and rosemary or currant jelly with hot chili peppers, roasted pecans, and bourbon. Here is my favorite done with hoisin (Chinese barbecue sauce).

Austrian cooks know that pounded pork tenderloin tastes a lot like veal in Wiener Schnitzel.

Juicy Pork Tenderloins with Spicy Chinese Sauce

MAKES 6 SERVINGS

The young lady who prepared this for my recipe testing is a business executive who is on the road every week and rarely cooks. She told me that after the testing she served this for small dinner parties several times! It is easy, fast, and impressive—pieces of tender, lean meat with dark, rich-flavored sauce. The Spinach and Orange Salad with Pine Nuts (page 315) and this dish complement each other beautifully.

what this recipe shows

Small-diameter pieces of meat will cook through quickly and evenly at a high temperature.

¼ cup soy sauce

¼ cup hoisin sauce (see Note)

2 tablespoons vegetable, peanut, or similar oil

3 tablespoons sugar

2 pork *tender*loins, *not* loins, about ¾ to 1 pound each

Nonstick cooking spray

½ cup water

4 tablespoons butter

6 cups cooked rice, long-grain white or brown

6 scallions, green parts included, sliced into thin rings, for garnish

1. In a freezer-type zip-top plastic bag, mix soy, hoisin, oil, and sugar. Add the pork tenderloins and marinate (refrigerated) at least an hour or overnight.

2. Preheat the oven to 500°F (260°C).

3. Line a small roasting pan with foil, spray a cake-cooling rack or roasting rack with nonstick spray, place tenderloins on the rack, and roast for 9 to 10 minutes. Turn the tenderloins over and roast 9 to 10 minutes more. Check the temperature in the thickest portion. The center internal temperature should be above 148°F (64°C).

4. Pour the marinade from the plastic bag into a small saucepan, add water, and bring to a boil over low heat. Boil gently for several minutes, then add the butter and bring back to a boil.

5. Let the cooked tenderloins stand at least 8 minutes before slicing. Slice thinly at an angle across the grain. Arrange slices overlapping attractively on a bed of rice. Pour the hot marinade-butter mixture over the slices and garnish with chopped scallions.

Note: *This dish is highly dependent on the taste of a good hoisin sauce (Chinese barbecue sauce). Unfortunately, hoisin sauces vary greatly, and some are not that good. It is well worth a trip to an Asian market to get Koon Chun, an outstanding brand. It comes in jars or cans with a blue and yellow label.*

Roasting lamb   Lamb is particularly suitable for roasting because it's high in fat. Its fat content, however, is also the reason I prefer to roast a lamb at 350°F (177°C). When I slow-roasted it, it was almost too tender—on the verge of mushy. I like to cook it medium-rare, to about 130°F (54°C), with garlic slivers pressed into the meat and rosemary and lemon slices covering the top. Get the butcher to remove the hip bone for easier carving.

Roasting seafood   Seafood can be roasted or baked, too. Mosca's famous New Orleans restaurant (with a touch of Italian) is noted for its Parmesan-baked oysters.

Cajun-Italian Oysters

MAKES 4 SERVINGS

Mosca's, an old restaurant in the suburbs of New Orleans, is famous for its Cajun-Italian Oysters Mosca and Shrimp Mosca—tin pie pans filled with herb-crumb-and-Parmesan-crusted oysters or shrimp. This is my version. Even if you're not an oyster lover, you may become addicted to this complex blend of flavors and textures. Cold, cold beer, crisp greens with olives, fresh crabmeat, and a light olive oil dressing, and grilled crusty slices of country bread coated in garlic olive oil will complete this rustic feast.

continued

Garlic is sautéed only briefly to avoid developing strong flavors. Reducing the poaching liquid adds outstanding flavors.

1 medium-large onion, finely chopped

4 tablespoons (¹/₂ stick) butter

2 to 3 cloves garlic, minced

2 dozen oysters and oyster liquor (see Note)

¹/₂ teaspoon *and* ¹/₄ teaspoon dried oregano leaves (³/₄ teaspoon *total*), preferably recently opened

¹/₄ teaspoon dried thyme leaves, preferably recently opened

¹/₈ teaspoon cayenne

¹/₈ teaspoon black pepper

¹/₂ teaspoon salt

³/₄ cup fine fresh or dried bread crumbs

¹/₂ teaspoon dried basil, preferably recently opened

¹/₂ teaspoon dried or fresh rosemary, finely chopped

1 to 2 ounces Parmesan, grated

4 scallions, green parts included, chopped, for garnish

4 sprigs flat-leaf parsley, chopped, for garnish

1. Preheat the oven to 375°F (191°C).

2. In a large skillet over medium-low heat, sauté the onion in butter until soft, about 2 minutes. Add the garlic and sauté another minute. Add oysters, oyster liquor, ¹/₂ teaspoon oregano, the thyme, cayenne, pepper, and salt. Simmer until the oysters just begin to curl. With a slotted spoon, remove the oysters to a pie pan or baking dish. Boil down the oyster liquor and spices a few minutes longer.

3. In a medium mixing bowl, stir together the bread crumbs, basil, ¹/₄ teaspoon oregano, and the rosemary. Pour the reduced oyster liquor and spices over the oysters. Sprinkle bread crumb–spice mixture on top and toss well to coat oysters. Sprinkle with Parmesan. Bake for 15 to 20 minutes, until browned. Sprinkle with chopped scallions and parsley.

Note: *This dish is also excellent using shrimp. Prepare as for oysters but sauté the shrimp briefly and bake for 5 minutes only.*

Wet methods of cooking

Wet methods of cooking include poaching (boiling), steaming, braising, and stewing. Poaching, particularly at lower-than-simmer temperatures (simmer is about 185°F/85°C), is ideal for delicate seafood or chicken. Steaming is also a good method for delicate light meats. Braising and stewing, on the other hand, are long, slow methods, just right to melt connective tissue into soluble collagen in tough pieces of red meat or in dark-meat poultry.

Poaching/Boiling

Poaching is a time-honored method for cooking seafood and light-meat poultry. Except in a pressure cooker, boiling water can never exceed 212°F (100°C), a fact that provides a built-in temperature control. You can cook protein dishes in water very delicately so that they are very tender.

The tender poached eggs that I described on page 201 were cooked by placing the eggs in simmering water in a large, heavy pot, putting the lid on, and removing the pot from the heat. In Flounder Florentine (page 261), you poach the fish in a similar manner: Place the fish fillets in simmering liquid, cover with a piece of buttered foil, then put them in a warm oven to finish cooking. In each case, the addition of the cold raw ingredient (eggs or fish) lowers the temperature of the water so the protein cooks at a temperature below a simmer and remains tender.

At a boil, or even a simmer, you can make any meat quite tough. Even at a simmer (around 185°F/85°C), the temperature is hot enough to overcook any meat. Protein begins to lose moisture at a little over 100°F (38°C), loses considerable moisture between 140° and 150°F (60° to 66°C), and *really* loses moisture between 150° and 160°F (66° to 71°C). These temperatures are all well below the simmer. Perhaps you have had some tough boiled meat that verifies these facts.

For these delicate, tender proteins, you need both to cook at a temperature below the simmer and to watch the time carefully. Even with not-so-delicate proteins like a pork chop, timing is very important. Some of our traditional rules for cooking fish, like "10 minutes to the inch (thickness)," are correct for some thicknesses but not accurate in all cases, particularly for thin pieces. Cooks, even very good cooks, are guilty of writing recipes that call for too long a time for poaching and, in some cases, braising.

Alan Davidson in his books, *North Atlantic Seafood* and *Mediterranean Seafood,* states that the time necessary to heat any object to a given temperature is proportional not just to its thickness but to the square of its thickness. This means that if it takes 5 minutes to get a 1-inch-thick piece of meat to a certain temperature, to get a piece that is 2 inches thick to that temperature will take not 2 times 5 minutes = 10 minutes, but will take 2 squared (2×2) times 5 minutes, or 20 minutes.

Harold McGee, in doing research for his book *The Curious Cook,* found that this formula came from noted Oxford physicist Nicholas Kurti. The rule holds true for large pieces of meat like a sirloin steak but not for small pieces of meat like a pork chop, which gets enough heat from all sides as well as the top and bottom to affect the center temperature. McGee, working from formulas that he derived from these laws of physics, developed invaluable tables of cooking times for large and small pieces of meat of different thicknesses.

Here are a few time and temperature suggestions from McGee's tables both at a high simmer (210°F/99°C) and at a slightly lower temperature (180°F/82°C). Notice that at the high simmer you sometimes have only $\frac{1}{2}$ to 1 minute before the center of a piece of meat goes up 10°F (5.5°C). It is not realistic for the average cook to try to get a dish off the heat with less than 30 seconds for error. For this reason, McGee recommends cooking at the lower temperature so you have better control. You have several minutes before foods overcook at these temperatures.

The times in these tables are *total* cooking times, starting from 50°F (10°C). (This is the approximate temperature that the meat would reach after coming out of the refrigerator at a temperature of about 36° to 40°F/2.2° to 4.4°C and standing at room temperature for about 15 minutes.) If you brown 1-inch pork chops for 2 minutes to a side, that would be 4 minutes gone. When you add wine or stock and braise, for a 1-inch-thick chop (which requires only 10 minutes at a high simmer to reach 150°F/66°C in the center—juicy and tender), your remaining braising time will be 6 minutes.

EXCERPTS FROM HAROLD MCGEE'S TABLES FOR COOKING MEAT IN LIQUID		

Starting temperature of meat 50°F (10°C)

Cooking times to reach an internal temperature of:

	140° F (60° C) *fish, pink veal*	*150° F (66° C)* *chicken breast,* *barely pink veal* *& pork*	*160° F (71° C)* *chicken thigh,* *well-done veal* *& pork*
Cooking Times at a High Simmer (210°F/99°C)			
1/2" steak or chop	2 min.	2.5 min.	3 min.
1" steak or chop	9 min.	10 min.	12 min.
Cooking Times Below a Simmer (180°F/82°C)			
1" chop	11 min.	14 min.	16 min.
1" steak	12 min.	15 min.	18 min.

Harold McGee. *The Curious Cook: More Kitchen Science and Lore.* San Francisco: North Point Press, 1990.

When you compare some of your meat poaching and braising recipes to these charts, you may find the times to be longer than those recommended by McGee. Try McGee's times—you may be surprised at how much juicier your meats will be.

Steaming

Steaming is a method that you may not think about for meats, but it is an excellent quick way to cook boneless chicken breast or fish. Boneless, medium chicken breast halves steam perfectly in 7 minutes. When you let them stand 2 to 3 more minutes, the center of the meat will turn from pink to just white and the meat will be quite juicy and tender. Chicken cooked simply in this way is a perfect complement to unusual sauces and makes an elegant, low-calorie main dish. Fire and Ice—Spicy Grilled Chicken Fingers (page 258) are grilled, but they can be steamed and sauced with a spicy, sweet-hot jalapeño sauce.

Using even an inexpensive fold-out steamer basket (with a removable center stem), you can steam directly on a dinner plate ready to serve. I have a large-diameter soup pot that my steamer and dinner plate will fit into with room around the sides to lift the plate out. A simple steamed fish is the perfect base for hundreds of sauces. The fat-free Milanese Gremolata that follows is a flavorful topping for low-fat dishes. A small amount of even a high-fat sauce on a medium to large fillet will still give you a reasonably low-fat dish, like the very simple steamed Fresh Fish Fillets with Macadamia Butter (page 411), which are elegant and delicious.

Gremolata

This is a fat-free topping traditionally served on osso buco but also excellent on steamed fish or chicken.

2 cloves garlic

Finely grated zest of 3 lemons

1 cup parsley leaves

1/2 teaspoon salt

With the steel knife in the workbowl, turn the food processor on and drop the garlic down the feedtube onto the spinning blade to mince. Add the lemon zest and parsley. Process to mince. Sprinkle the cooked fillets with salt and sprinkle heavily with gremolata.

Fresh Fish Fillets with Macadamia Butter

MAKES 6 SERVINGS

A mild fish like sole, flounder, or orange roughy is a perfect match for this delicate topping. A real expert with fish, Susan Jones from Santa Clara Beach, Florida, and Hawaii, taught me the joy of macadamia nuts on fish.

what this recipe shows

Delicate macadamia nuts are a perfect complement to mild fish.

Roasting enhances the flavor of nuts.

6 (about 1½ pounds) medium-size
   mild fish fillets (such as sole,
   flounder, orange roughy)

1 tablespoon oil (canola, safflower,
   corn, peanut)

Salt

4 tablespoons (½ stick) butter

⅓ cup finely chopped macadamia nuts

4 sprigs parsley, finely chopped,
   for garnish

1 lemon, sliced, for garnish

5 sprigs parsley, for garnish

1. Lightly rub fillets with oil and sprinkle with salt. Steam or fry (see Snapper Fingers with Smoked Pepper Tartar Sauce, page 165, or Spicy Indian Fried Cheese, page 342, for frying directions). Remove to a warm serving platter.

2. Melt the butter in a large skillet. Add the chopped macadamia nuts and cook over low heat until lightly browned, about 2 minutes. Pour the macadamia nut and butter sauce over the fish fillets. Garnish the fillets with chopped parsley. Garnish the platter with overlapping lemon slices and parsley sprigs.

Poaching and steaming are favorite methods for cooking fish. Of course, species of fish differ widely in texture and flavor, and often the fish you had in mind is not available. So here is a chart to help you pick appropriate substitutes. The chart was inspired by one that *Simply Seafood* magazine carries in its issues. If, for example, you wanted a fish with a mild flavor and medium-firm texture, such as haddock, you might consider substituting tilapia based on price and availability.

at a*glance*
Substitutions of fish species

TEXTURE	Mild	FLAVOR Moderate	Full
Delicate	Flounder/Sole	Pink Salmon	Herring/Sardine
	Orange Roughy	Whiting	Smelt
	Farm-Raised Catfish		Butterfish
	Skate		
	Alaska Pollock		
Medium-Firm	Cod	Ocean Perch/Rockfish	Atlantic Salmon
	Haddock	Striped Bass	King Salmon
	Halibut	Chum Salmon	Mackerel
	Tilefish	Drums	Yellowtail
	Grouper	Amberjack	Sockeye Salmon
	Snapper	Rainbow Trout	Sablefish
	Tilapia	Mahi-Mahi	Bluefish
		Sea Bass	Carp
		Atlantic Pollock	
		Perch	
		Pompano	
Firm	Monkfish	Shark	Tuna
		Sturgeon	Marlin
			Swordfish

Bones in fish My daughter was once disappointed to find that the trout fillets she had ordered in a restaurant had bones since fillets are usually boneless. I explained that all trout, salmon, and shad have a double rib cage and that when the fish are filleted the flesh is cut off one rib cage but the bones of the smaller, inner rib cage are simply cut from the backbone and left in the flesh. These bones that are left in the fillet are called *pin bones*. With salmon, the pin bones are large and you can pull them out easily with a pair of needlenose pliers. They are easy to find since they are in a row (they were a rib cage) and are big enough to spot easily.

In trout, the second rib cage bones are much finer and more difficult to see and remove. In fine restaurants that serve whole trout, the waiters are trained to very carefully remove both rib cages from the cooked trout. With practice and some care, you will find that you can carefully pull the backbone, intact, from cooked trout with both rib cages still connected to it.

It is quite a different matter with raw trout. A group of chefs and I sat around all afternoon one Sunday with a big bucket of trout, trying to come up with an easy way to remove both rib cages to get boneless fillets. It was a frustrating afternoon. We could do it, but it was really time-consuming.

A large trout supplier I asked about the problem explained that most of the trout fillets supplied to restaurants did have the pin bones left in. He supplied a few restaurants with boneless trout fillets, but they were more expensive because they had to be filleted twice, with a portion of the flesh cut away. First the trout were filleted in the normal fashion (with the pin bones left in), then the fillets were chilled until very cold. Next they used a special tool that they had come up with to slice off a layer with the bones in it, as in slicing bacon. Some of the fillet was lost in this second slicing.

So, when working with trout, salmon, or shad you need to be aware of the difficulty of getting a true boneless fillet.

Braising and stewing

Braises, stews, fricassées, blanquettes, estouffades, étouffées, ragoûts, and daubes are all intimately related. If you take a single big piece of meat, sear it, and then cook it, tightly covered, in a liquid that comes halfway or less up the side of the meat, you will produce a braise. If the pieces of meat are small and are seared and then cooked while covered or nearly covered in liquid, you'll end up with a stew. Fricassées and blanquettes, on the other hand, are "white" stews, or stews in which the meat is not browned before it is simmered. In modern usage, chicken prepared in this way is called a *fricassée,* while veal is called a *blanquette.*

The 1961 edition of *Larousse Gastronomique* describes étouffées as meats cooked tightly sealed with little or no liquid, while estouffades are dishes whose ingredients are stewed slowly. Most daubes are stews in which the pot lid is literally sealed on with a paste of flour and water.

A true braise is made like this, according to Madeleine Kamman in *The Making of a Cook:* The cook sears and browns the meat, then places it on a layer of aromatic vegetables (onions, carrots, celery leaves, and herbs) in a heavy pot as close to the size of the meat as possible. Ideally there should be no space between the top of the meat and the lid. Since this is frequently not possible, the cook should cover the meat with a piece of aluminum foil pressed tightly against the meat. The foil should go to the edges of the pot and be curved upward to catch any condensation from the lid and prevent its dripping back into the sauce.

The cook then adds enough liquid to come about halfway up the side of the meat, covers the meat with the foil and the lid, and cooks in a medium to low (325°F/163°C) oven until a skewer inserted into the meat comes out freely. Madeleine likes to use a piece of pork rind with the fat removed under the aromatics to provide gelatin and flavor richness for the sauce. When you baste the meat with this gelatin-containing sauce, it will give a nice glaze to the meat.

Madeleine speculates that, in a braise, with the meat bound on all sides and possibly with some pressure built up in a heavy, closed pot, the juices that normally flow from the meat are held in the meat by pressure and the sides of the container. These juices in the meat pry fibers apart and tenderize. Whatever the physics in a braise, we know that you can get very tender, juicy meats in this way.

Modern cooking bags can be used for a bastardized braise—though Madeleine may chastise me for associating the word *braise* with the bags. They do offer a enclosed environment for the meat, even if the condensation on the top of the bag does drain into the sauce. I like braising large turkey breasts in these bags. Many people still cook the whole bird only on holidays, but turkey parts, which make an easy family or party meal, are now available year-round in large supermarkets.

Turkey Breast with Fresh Sage and Marmalade

MAKES 8 MAIN-COURSE SERVINGS OR 20 TO 30 HORS D'OEUVRES

This is a big, deep golden brown, juicy turkey breast. The great joy is that the preparation is easy. I once took this to a bring-a-dish party of food professionals because I was very busy and needed something that I could shove in the oven and forget. I was a little embarrassed to take such a simple dish to compete with the quail eggs, pâtés, and other fancy dishes that I knew would be there but rationalized that a hearty meat dish might be needed among all the elegant nibbles. To my surprise, four fellow food professionals later called to ask how I had cooked the turkey!

what this recipe shows

Turkey breast stays tender and juicy when the internal temperature stays just under 160°F (71°C).

Regular roasting times cannot be used as a gauge because meats cook faster in the bag.

1 6- to 8-pound turkey breast	About 20 to 30 fresh sage leaves
4 tablespoons *and* 3 tablespoons Grand Marnier (7 tablespoons *total*)	2 tablespoons cornstarch stirred into ¼ cup beef consommé (canned is fine)
Salt and pepper	½ cup beef consommé (canned is fine)
1 large turkey cooking bag (for up to 12 pounds)	Fresh sage, kale, or parsley for garnish
½ cup orange marmalade	

1. Preheat the oven to 400°F (204°C). The bags specify not using them at over 400°F (204°C). If you think your oven runs hot, set it at 375°F (191°C).

2. Rinse and pat the turkey breast dry with paper towels. Rub the meat with 4 tablespoons Grand Marnier. Rub salt and pepper into the bone side of the breast. Place the breast in the cooking bag with the meat side down. With a spoon or your hand, reach inside the bag and pile orange marmalade on top of breast. A lot of it slips off, but some will stay. Put sage leaves in the bag around the sides of the breast.

3. Gather the opening of the bag facing up so that the bag will hold the juices. Close loosely with the tie so that a little steam can escape to prevent the bag from exploding.

4. Arrange a cake-cooling rack across the top of a medium roasting pan. Pour about ½ to 1 inch of water into the pan (this will prevent any drippings that may leak from burning). Place the turkey in the bag on top of the rack (breast meat down, bone side up). Roast for 30 minutes. Remove the pan from the oven and carefully (with hot pads) turn the turkey over so that the breast is now up and bone side down. Try to avoid losing drippings. Rearrange the bag opening so the bag will hold juices.

5. Turn the oven down to 300°F (148°C). Leave the door partially open for 4 minutes to cool the oven down fast. Leave the turkey on the counter during this cooling period. Place the turkey breast back in the oven. After 45 minutes, check the temperature with an instant-read thermometer. Remove the turkey breast from the oven and insert the thermometer near but not touching the ribs. (You can insert the thermometer through the bag; just remember later that you have made a small hole.) Strive for a temperature around 155°F (68°C). If the temperature is lower than 152°F (67°C), place the turkey breast back in the oven to get it up to 155° to 160°F (68° to 71°C). Depending on your oven, this usually requires about 1 hour and 30 minutes total cooking time.

6. When the correct temperature is reached, carefully remove the turkey from the bag. Pour the drippings into a medium saucepan. Add the cornstarch-consommé mixture plus plain consommé to drippings. Bring to a gentle boil, stirring constantly. Stir in 3 tablespoons Grand Marnier.

7. Place the turkey breast on a medium serving platter. Let the turkey stand 10 minutes before slicing. The grain of the turkey breast runs from front to back. I like to slice the meat without removing it from the bone. It does not look sliced, but each slice lifts right off (see Note). Pour some of the hot Grand Marnier sauce (made from drippings) over the turkey breast. I actually pry slices apart and spoon a little in between the slices and then push slices back together so that they are still in place.

8. Garnish the platter with some kind of greens—fresh sage leaves, kale, or parsley.

Note: *To slice so the breast looks undisturbed, make the slices straight in, all the way to the breastbone. Slice one whole side of breast this way. Hold the skin so you don't tear it too much. Slice one side all the way down but still connected to the breastbone. Do the same on the other side. Now hold the breast slices on one side of the turkey in place with one hand and carefully cut them free from the breastbone by slicing along the bone. Do the same thing on the other side. The turkey breast is now completely sliced, but all the slices are resting perfectly in place so the turkey breast does not look disturbed. It is very easy for buffet guests to serve themselves.*

When Rose Beranbaum and I were traveling in Europe, we had an incredible stewed chicken in yellow wine with morels and cèpes. I was so wild about it that even at the risk of being removed from the two-star restaurant I scraped the tableside cooking pot when the waiter wasn't looking. I developed my own version, using a Mâcon-Villages for the wine, since in this country we cannot get the *vin jaune* (yellow wine) that is available in the Jura region of France. (It is a white wine that is sealed in casks and aged for seven years to reach a glorious yellow color.) At the end of cooking this chicken stew, cream is added and reduced. The dish is rich and wonderful—if you find it a bit too rich, you can stir a little cornstarch into yogurt and substitute it for the cream. (See page 183 for starch and yogurt information.)

Chicken with Wild Dried Mushrooms and Wine

MAKES 6 TO 8 SERVINGS

When you crave an old-time, rich French restaurant sauce seething with deep, complex flavors, this is the dish for you. The intensity of dried mushrooms and wine, mellowed with rich cream and chicken flavors, makes a memorable dish. This dish is made up of few ingredients simply prepared, but, as the old saying goes, "the secret of French cooking is reduce, reduce, reduce." Since this is a rich, sauced dish, it is best served with a light salad, fresh asparagus or fresh thin green beans, a good bread, and ripe figs or pears with a little cheese.

continued

Dark-meat chicken stays moist and juicy during simmering.

Having the butcher cut large chicken thighs in half across the bone adds rich flavors to the sauce from the marrow.

Seasoning both the chicken and the flour imparts full flavor.

Deglazing the pan with wine dissolves some flavor components that water cannot.

Boiling off the wine mellows any sharp alcoholic taste and reduces the possibility of the alcohol's curdling the cream.

1 ounce mixed dried morels and dried cèpes (porcini)	Flour
½ cup warm water	4 tablespoons (½ stick) butter
8 chicken thighs, cut in half across the bone	2 tablespoons olive oil
	1 cup dry white Burgundy (Mâcon-Villages)
Salt and pepper	2 cups heavy or whipping cream

1. Soak the dried morels and cèpes in warm water in a small bowl for at least 20 minutes.

2. Sprinkle the chicken pieces with salt and pepper. Season the flour with salt and pepper and dredge the chicken in the seasoned flour. Brown chicken pieces in butter and olive oil in a large, heavy skillet over medium heat until deeply browned on all sides, about 8 minutes. Cover the skillet and cook below a simmer until the chicken is quite tender, about 10 minutes. Remove the chicken and excess fat.

3. Pour in the wine and scrape the pan well to loosen any stuck browned bits. Add morels and cèpes. Add soaking water, taking care to pour the water off the top and leave some water so that any sand from the mushrooms is left behind. Boil 4 minutes to reduce and boil off some of the alcohol.

4. Add the cream and reduce over medium heat until the consistency of a medium cream sauce, about 4 minutes. Return the chicken and any drippings to the skillet and simmer about 4 minutes more.

Giuliano Bugialli's Fenneled Chicken with Almonds in *The Fine Art of Italian Cooking* has long been a favorite of mine. Onions, Italian bacon (pancetta), and well-browned chicken thighs are cooked in stock with whole almonds. The only seasonings are salt, pepper, and fennel. The result is a thick, rich, brown sauce with delicious, tender chicken and almonds. I make an Americanized version that I'm sure Giuliano would want nothing to do with.

Smothered Chicken with Ham and Almonds

MAKES 8 SERVINGS

This is an ideal dish for crowds. You can prepare it ahead, and it is even better reheated. My daughter Sherry and her fiancé, Joft, stayed up all night preparing batch after batch for my other daughter's (her twin sister's) wedding. A friend with a large catering service says that her clients request this dish time after time. Serve with pasta or rice, a salad like Spinach and Orange Salad with Pine Nuts (page 315), and a good bread like Crusty French-Type Bread (page 24).

> **what this recipe shows**
>
> **Dark-meat chicken stays moist and juicy during simmering.**
>
> **Cutting the thighs across the bone adds rich flavors to the sauce from the marrow.**

1 ounce salt pork

1 pound smoked ham

1 large red onion, quartered

2 tablespoons *and* 3 tablespoons olive oil ($^1/_3$ cup *total*)

8 to 10 large chicken thighs, each cut in half across the bone by the butcher

$^1/_2$ cup *and* 2 tablespoons flour ($^2/_3$ cup *total*)

$^1/_4$ teaspoon white pepper

2 cups homemade Chicken Stock (page 267) or canned chicken broth

$^1/_4$ pound whole almonds

1 teaspoon fennel seeds, crushed (see Note)

Salt if needed

1. Coarsely chop the salt pork, ham, and onion by hand or in a food processor. (If using the processor, process in two batches using half the onion and half the meat in each batch.)

2. Sauté the salt pork, ham, and onion in 2 tablespoons olive oil in a large, heavy frying pan or a large metal casserole over medium heat until the onion is soft and lightly browned, about 4 minutes. Remove the mixture to a bowl. Coat the chicken pieces in $^1/_2$ cup flour and the white pepper. Shake to remove excess. Add 3 tablespoons olive oil to the frying pan and the chicken and sauté over medium heat until the chicken is deeply browned on both sides, about 8 minutes.

3. Meanwhile, bring the chicken stock and almonds to a boil in a medium saucepan, then turn down to low.

4. When the chicken is browned, spoon off any excess fat, sprinkle 2 tablespoons flour into the drippings or on the chicken, and stir in the crushed fennel seeds. Cook several minutes, until the flour is well browned. Add the ham and onion mixture back to the skillet. Cook a minute, then add the hot stock and almonds. Stir well. Stir occasionally and simmer over low heat for 10 minutes, until the chicken is very tender and the sauce is a deep brown. If the sauce is still thin, remove the chicken to a bowl. Boil the sauce gently, stirring, until it reaches the desired thickness, then add the chicken. Taste and add salt if necessary. Serve hot. Can be prepared ahead (leave the sauce thin) and reheated over medium-low heat.

Note: *Fennel, like coriander, has a flavor that people feel strongly about. If you know your guests like fennel, use up to $1^1/_2$ tablespoons.*

Hot-fat cooking—deep frying and pan-frying

Hot fat is an excellent way to quickly heat and brown the surface of fish, chicken, or meat. Cooking techniques for hot fat are in Chapter 2—deep-fat frying (page 154), pan-frying (page 168), and stir-frying (page 166).

Storing meat, poultry, and seafood at home

Meat must be refrigerated or frozen until ready for use and is optimally used shortly after purchase. The ideal temperature for meat storage is 28° to 32°F (−2.2° to 0°C), which is colder than most home refrigerators (36° to 40°F/2.2° to 4.4°C). Store meat in the coldest part of the refrigerator. Many refrigerators have meat compartments that are set to be several degrees colder than other parts of the refrigerator.

Meat can be stored for a day or so in the grocery store wrap, but the plastic film on fresh meat is permeable, allowing oxygen in to maintain the bright red color (oxymyoglobin). This wrap will not prevent meat from drying out. For freezing, meat should be wrapped in nonpermeable plastic wrap like Saran or in products clearly labeled "freezer" or well wrapped in freezer paper, which has a plastic or wax coating. Foil provides a good barrier, but it does tear easily.

Large solid pieces of meat keep better than items like ground beef, which have tremendous surface area that can contain bacteria. The National Live Stock and Meat Board recommends keeping ground beef, veal, pork, or lamb no longer than 1 to 2 days in the refrigerator (36° to 40°F/2.2° to 4.4°C), while other beef cuts like steaks and roast can be kept for 3 to 4 days.

Meat can develop an off flavor even while frozen because fat continues to oxidize even while frozen. The Meat Board recommends storing ground beef, veal, or lamb no longer than 3 to 4 months in the freezer and other beef cuts no longer than 6 to 12 months. Beef, because of its higher saturated fat content, keeps better while frozen than meats that have a higher unsaturated fat content, such as seafood, chicken, and pork. The Meat Board's recommendations for ground pork are 1 to 3 months and 6 months maximum freezer storage for other cuts of pork.

Seafood is highly perishable because its enzymes are accustomed to working in cold water, a low-temperature environment. Therefore, home refrigerator temperatures do not slow down spoilage from enzyme activity as well in seafood as they do in meat. Seafood should be used immediately.

When I have to hold seafood for a day or two, I spread the seafood out over the bottom of a large strainer and pile ice on top. Place the strainer over a bowl deep enough so that any liquid is well below the bottom of the strainer. A setup like this keeps the seafood very cold, and as the ice melts the liquid rinses bacteria from the seafood. Since this wash is away from the seafood, it cannot recontaminate it as it can if seafood is simply packed in ice.

Foodborne disease

Many different types of microorganisms can grow on meat—yeasts, molds, bacteria, and viruses. Viruses are in only raw or undercooked shellfish. A few parasites also invade meat (*Trichinella spiralis*), but by far the most prevalent foodborne disease problems in meats come from bacteria.

Bacteria

In addition to concerns with killing the bacteria itself—killing salmonella and *Listeria monocytogenes* bacteria—there are toxin-producing bacteria like *Clostridium botulinum, Escherichia coli (E. coli), Staphylococcus aureus,* and *Clostridium perfringens.* With these rascals, cooks also have to worry about not allowing conditions that can produce toxins.

Instant-kill and time-temperature kill Cooks have been cautioned to minimize the time that food is in the temperature zone that enhances bacterial growth—40° to 140°F (4° to 40°C). Meat Board literature suggests that you can remember the beginning of the danger zone by "life begins at 40." Many of us are also familiar with the instant-kill temperature for most bacteria, 160°F (71°C). Many meats cooked to 160°F (71°C) are tough and dry and have lost all sensory appeal.

Fortunately, bacteria can be killed just as dead by holding the meat at lower temperatures for a time period. In Chapter 3 we saw that salmonella was killed in pasteurization of eggs by holding them at 140°F (40°C) for 3½ minutes. The FDA 1995 Food Code (available in hard copy or on computer disk, see Sources—FDA Food Code, page 479) now specifies flexible time/temperature combinations to kill bacteria. For example, safe ground beef can be prepared by heating to 155°F (68°C) for 15 seconds or to 145°F (63°C) for 3 minutes. Roast beef can even be safe rare, but the internal temperature must be held to at least 130°F(54°C) for 121 minutes. Unfortunately, home cooks do not yet have equipment other than a slow cooker to hold meat at low cooking temperatures over a long time. A number of pieces of restaurant equipment obviously can do this, as evidenced by the marvelous rare prime rib that restaurants can produce.

Home slow-roasting procedures like the one on page 400 are within reason safety-wise. The meat is seared to kill surface bacteria, roasted at 200° to 250°F (93° to 121°C), and then browned at 500°F (260°C). If the piece of meat is small enough so that the roasting time is under 4 hours, there should be no opportunity for major internal bacterial growth. The meat has not remained in the temperature zone of maximum bacterial growth—40°F to 140°F/4° to 40°C—long enough for extensive bacterial growth. There should be little live bacteria in such a piece of meat.

The USDA maintains a Meat and Poultry Hotline (see Sources—Meat and Poultry Hotline, page 479) that will advise you on conservative procedures for cooking meat. The FDA has a seafood hotline (see Sources—Seafood Hotline, page 479).

Cross-contamination   Even in spotless kitchens cross-contamination can occur. Immediately wash anything that comes in contact with raw meat. Dish cloths and sponges can be a problem. If you use a dish towel to wipe a knife that you have just used to cut raw chicken, and then use it later to dry a plate, you have just spread bacteria from the raw food to the plate.

Some restaurants use pink dish rags for work in areas where raw meat is handled and white in areas where cooked food is handled. Home cooks should be aware of the problem and handle raw meat in a confined area such as in the sink. Then, wash everything that has touched the raw meat.

People who are at risk—anyone with low immunity—and expectant mothers should not eat undercooked meat. Toxoplasmosis is transmitted in undercooked red meat. To be on the safe side, obstetricians recommend that expectant mothers not eat any raw meat, and some recommend that their patients not even eat rare meat.

Common sense along with proper storage and preparation techniques can go a long way toward minimizing the possibility of illness from foodborne disease. We take risks every time we walk across a street. Even the FDA 1995 Food Code advises restaurants that if they do not serve an at-risk population (like a nursing home), they may serve customers (who are informed that there are risks) ordering rare or even raw meat (such as steak tartare).

the recipes

Crème Brûlée in Flaky Pastry

Old-fashioned Burnt Sugar Icing

Orange Slices with Cinnamon Candied Peel

Macadamia-Almond Brittle

Classic Fudge

"You Can't Eat Just One" Creamy Fudge

Sinfully Easy Fudge

Creamy Pralines

Real Naw'llins Prawleens

Sugar-Crusted Almond Slivers with Chocolate

Roasted Walnuts with Brown Sugar Caramel

Espresso Granita

Silky Smooth Ginger Ice Cream

Creamy Lemon Chip Ice Cream

Rich Cappuccino Ice Cream

Smoothest-Ever Truffles

White Chocolate Mousse with Raspberry Sauce

The Secret Marquise

Chocolate Stonehenge Slabs with Cappuccino Mousse

Chocolate Walnut Ruffle Cake

sweet thoughts and chocolate dreams

Here are the makings of great desserts—sugar and candies, ices and ice creams, whipped cream, and chocolate. Much sugar work, ices and ice creams, and some chocolate work all involve crystals and crystallization. So, an understanding of crystals and how they form will help with many problems in these areas.

Whipped cream is a foam held up by fat particles sticking together—a totally different type of foam from that formed by beating egg whites. Fortunately for cooks, there are techniques that are helpful in both whipping cream and combining whipped cream with other ingredients.

421

Crystals in foods

When you think of crystals, you may think of lacy snowflakes on a frozen windowpane or of pointed icicles hanging from bare, black branches. Actually, many everyday substances contain crystals, including many foods. Sugar and salt crystals are familiar parts of our lives, and ice crystals are the solid form of water. Crystals are the solid form of many substances. Fine sugar crystals make fudge become firm, tiny ice crystals make ice cream thicken, cocoa butter crystals make melted chocolate solidify.

When molecules of the same compound or element get cold and slow down, they join together in a precise formal pattern peculiar to that substance. This is a crystal—molecules of a pure substance joined together in the unique pattern that this particular substance forms. A crystal is very precise, and each molecule must fit exactly into its place for the crystal to form.

What is needed to form crystals

The purity of a substance, the amount or concentration, the temperature, the presence of seed crystals, and the agitation all influence the number and size of the crystals that form.

Purity Purity is the first thing you need for crystals to form. I learned this the hard way. As a research biochemist, I sometimes spent weeks getting a specific sugar isolated from the body clean enough and pure enough to crystallize. The major problem was to get rid of all the other sugars and impurities that were preventing my unknown sugar from crystallizing.

You can use the necessity for purity when you want to prevent crystals from forming: Add a substance that is similar but not exactly the same. When the solution becomes concentrated and the molecules are packed next to each other, they try to form crystals, but the molecules of this similar substance get in the way. When you are caramelizing sugar, for example, and you want to inhibit crystallization, throw in a little corn syrup, which is primarily glucose and interferes nicely with the crystallization of table sugar (sucrose).

Many substances can interfere with crystallization. Some are large molecules that simply get in the way when crystals try to form. Other substances, like the fats and egg whites used in candy making, keep crystals small by coating tiny crystals to prevent them from collecting more molecules and enlarging into big crystals.

Concentration For crystals to form in a solution, the mixture must contain a sufficient amount of the substance to be crystallized. The amount of liquid in which the substance is dissolved must be reduced, concentrating the solution until the molecules are packed against each other and forced to form crystals.

Temperature Temperature also influences crystal formation. Warm liquids dissolve and hold more sugar than cold liquids. In candy making, you boil down the sugar syrup to concentrate it; a small amount of water (in the syrup) at a high temperature holds a great deal of dissolved sugar. When the solution cools down, the sugar molecules are packed in there just waiting for the least agitation to crystallize. Any little shake or touch of a spoon, and the crystals will fall out of the solution.

Cooling is important in determining the size of the crystals. If you stir a concentrated syrup while hot, it will crystallize into a few crystals. As the syrup cools further, these crystals will pick up more and more molecules and grow and grow, and you will end up with large crystals. In a fudge or fondant, these big crystals will be gritty on your tongue.

If the syrup cools undisturbed before stirring, you will have a more saturated solution. Then when you stir, many more crystals form instantly. So, instead of getting a few crystals that grow large, you get thousands of tiny crystals all at once. These crystals do not grow, because the syrup is already cool. With rapid stirring, the cool, supersaturated syrup crystallizes into thousands of baby crystals.

As just explained, the three major factors in forming crystals are purity, concentration, and temperature. Agitation and the presence of an object on which crystals can form play roles too.

Sugar

In cooking, sometimes you want crystals and sometimes you don't. When you're caramelizing sugar or making hard candies or brittles, preventing crystals is your goal and can be a big problem. With other candies you want crystals, but of different sizes. Sometimes you want baby-fine, tiny crystals that are not perceived by the tongue, like the tiny sugar crystals in a creamy fondant or fudge. Other times you want big, impressive crystals, as in rock candy. If you know how to control the process, you can produce exactly the kind of crystals that you want or you can prevent crystals from forming when you don't want them.

Caramel

Caramel is melted, broken-down sugar. This sounds simple enough, but it is a classic case of having to prevent crystals from forming. To make caramel, sugar is heated to such a high temperature that it melts and actually comes apart chemically. When melted table sugar gets very hot (over 338°F/170°C), it breaks down to form smaller sugars, some of which join together again to form different sugars, and these may break down again to form different smaller sugars. In other words, new sugars and similar compounds are constantly being formed during the caramel-making process. The process starts with one sugar—table sugar (sucrose)—but by the time you have dark, almost black caramel, more than 128 different sugars and related compounds have been formed.

Each one of these sugars has a different color, smell, taste, and chemical formula; many of them have a beautiful amber to deep brown color and the marvelous flavor that you associate with caramel. At any given time, you have a collection of sugars with different smells and taste. A light caramel tastes very different from a dark caramel.

Light caramel and dark caramel also behave differently for "sugar work" like caramel cages or pulled sugar. Melted sugar or sugar caramelized only to a pale straw color will harden into a very hard, glasslike sheet. The darker the caramel, the softer it will be when it hardens. With the pale melted sugar, most of the sucrose is still left to harden. In the dark caramel, a lot of the sucrose has been destroyed (converted to other compounds), so not as much is left to harden.

There are two classic methods of caramelizing sugar—the dry method and the wet method. Each has advantages and disadvantages. Then there is a modern wet method—microwave caramel.

The darker the caramel, the softer it will be when it hardens.

Dry method for caramel In the dry method of making caramel, table sugar (sucrose) is heated so hot that it melts and breaks apart. These are very high temperatures (sucrose melts at 320°F/160°C and begins to caramelize at 338°F/170°C), so it is important to use a pan that can take the heat. The lining of a lined copper pot or a nonstick coating on a pan may melt, so a heavy iron skillet or a heavy unlined pot is necessary.

A real French caramel pot is an unlined copper pot with a funnel-shaped hollow copper handle. The reason for this odd design is that the pot gets so hot during caramel making that a wooden handle may catch on fire. In fact, the pot gets so hot that if you try to lift it with hot pads, they may catch on fire, too. To move a French caramel pot, stick a foot-long length of broomstick into the hollow handle. Since the handle is funnel-shaped, the wood touches the hot metal only in a small section so the wood near your hand does not get too hot. You need to select your stirring utensils with care, too. A plastic spoon certainly will not do, and a metal spoon will get so hot you cannot handle it. A wooden spatula is ideal.

Now, armed with an iron skillet or a heavy unlined pot and wooden spatula, all you have to do is put the sugar in the pot, put it on the stove, and melt it. Unfortunately, this basically simple process is not as easy as it sounds.

If you are using a small, deep pot and melting a fair amount of sugar, you will find that even if you stir constantly the sugar on the bottom will melt and become a dark, bitter-tasting caramel before you can get the sugar on the top down to the bottom to melt. The melted sugar on the bottom blackens and burns before the rest of the sugar melts.

The iron skillet gives you a large, open surface that makes it easy to stir well. Some cooks like to sprinkle sugar evenly over the bottom of the skillet, then heat on low to medium heat, stirring constantly, until they get a medium caramel. I like to sprinkle in the sugar a little at a time; as it melts, I add more and stir. Working in this manner, you can keep the caramel from getting too dark. The only problem with the iron skillet is that the dark background makes it difficult to tell exactly what color the caramel is.

Some candy makers sprinkle an even coating of sugar in a large skillet and slip it under the broiler to melt and caramelize. If you do this, you need to keep your eyes right on the sugar the whole time—when it melts, the caramel can darken in a flash, and you have to pull it out of the broiler fast.

Whether on the stove or under the broiler, it is just short of impossible to make a medium to light caramel with the dry method. For the lighter caramels, you need to use the wet method, which gives you more control.

Wet method for caramel This method is a little more complicated, but it gives you the control that you need to make a light or medium caramel. In this technique, add enough water to dissolve the sugar, then heat and stir until the sugar is dissolved. Before the water gets too warm, some chefs dip their fingers in and stir around the sides and bottom of the pot to make sure there are no undissolved crystals.

Then boil gently, without stirring or shaking, until the water boils off and the sugar melts. Trouble can come while you are boiling off the water—if you shake or stir before all the water has boiled off, the sugar will crystallize out, and you can end up with a grainy mess. It does not take much to start this crystallization. Some of the sugar water that has splattered on the pan during boiling can evaporate and leave sugar crystals clinging to the pan. If one of these crystals falls into the concentrated sugar solution, it may cause the sugar to crystallize.

You can wash down the sides of the pan with a pastry brush dipped in cold water to dissolve any crystals that are clinging there or oil the sides of the pot to prevent crystals from sticking. Oiling the sides also helps prevent the syrup from boiling over. A third way is to cover the pot with a lid for 2 to 3 minutes while boiling so that the water that condenses on the lid drains back down the sides, rinsing them.

Oiling the sides of the pot helps prevent syrup from boiling over.

There is an easy way out, one that I've already mentioned. Remember, for crystals to form a substance must be pure. All you have to do is to add a sugar that is similar to, but not exactly the same as, table sugar. It will interfere with the crystallization process and make caramelizing nearly fail-safe. Sucrose is a double sugar, consisting of two simple sugars, glucose and fructose, joined together. So, to make caramel making easy, just add a little corn syrup (mostly glucose) to the mixture. Although glucose is one of the two sugars in table sugar, it is not exactly the same. Glucose acts as an "impurity"; when crystals try to form, glucose molecules get in the way.

Another trick is to add a few drops of lemon juice, vinegar, or cream of tartar. Acids break down some of the sucrose molecules into the two sugars that they are made of—glucose and fructose. What a good idea! Now you have three different sugars. Both glucose and fructose act as impurities and prevent sucrose's crystallization.

Certainly the easy way out in making caramel is to add a little corn syrup and/or lemon juice or vinegar right at the beginning. This takes the worry out of caramel. You don't have to wash down the sides of the pan. You can simply keep a watchful eye and get the caramel to the darkness you want.

Once the caramel gets hot, the chemical changes in sugar take place quickly, and caramel goes from light to medium to dark very fast. You can tell when the water has boiled off and it is time to watch the caramel carefully. The bubbles take on the look of "bubbles on top of bubbles on top of bubbles."

You will love making caramel in the microwave. It is a simple, quick, and nearly fail-safe method using corn syrup and/or a few drops of lemon juice or vinegar.

Microwave Caramel

Place ¹/₂ cup sugar in a 2-cup glass measuring cup, add ¹/₄ cup corn syrup and 4 or 5 drops of lemon juice. Stir until no white patches of dry sugar are left. Add 1 tablespoon water if necessary.

Place the dampened sugar in the microwave and microwave on High until bubbles start piling up on top of each other. The time depends on the amount of sugar and water and the power of the microwave. Start watching very carefully. The second you see a color change take place—the melted sugar turning to pale tan—be on the alert. Permit the color to darken only slightly, then remove the caramel and let it stand. It will continue to darken and may get as dark as you want. If not, microwave it for about 10 seconds longer and, if needed, in additional 10-second increments until you get a color slightly lighter than the color you want. It will continue to darken after you remove the caramel from the microwave.

Caramel in action   You can make wonderful desserts with caramel, such as flan, crème brûlée, burnt sugar icing, and many more.

Caramel is the most difficult part of flan—it's nothing to stir together flavored milk, sugar, and eggs for the custard—but making caramel in minutes in the microwave takes all the trouble from this dish. I think you will be surprised at how easy it is to make Mesmerizingly Smooth Flan (page 220).

Crème brûlée is another custard with caramel, in this case as a thin crust on top of the custard. For this dish it is important to chill the custard well before making the caramel. You need a very cold starting temperature to protect the fragile custard from overheating. When the custard is well chilled, sprinkle sugar or light brown sugar evenly over the top and slip it under the broiler just long enough to melt and brown the sugar. You can also use a small propane torch with a flame spreader. Chill your custard immediately. Crème brûlée is ordinarily made in heatproof custard cups, but it can be surrounded by a flaky crust as in this recipe.

Crème Brûlée in Flaky Pastry

MAKES 6 SERVINGS

This is an elegant dessert—a delicate, satiny custard with a crunchy thin layer of clear amber caramel in a buttery crust. Since this is a classic rich dessert, it is best served with lighter meals that can stand a rich finish.

what this recipe shows

Egg yolks give the custard a satiny smooth texture.

Chilling the custard well before melting the sugar on top saves it from curdling.

1 recipe Simple Flaky Crust (page 110)	⅓ cup sugar
Nonstick cooking spray	Salt
1 large egg white, beaten	5 large egg yolks, beaten
3 cups heavy or whipping cream	½ cup light brown sugar (not packed)

1. Preheat the oven to 400°F (204°C).

2. Make 6 pastry shells in 3-inch round pans or use a muffin tin for 6 large muffins. Spray the pans or muffin tin with nonstick cooking spray. Do not make the pastry shells more than 1 inch high. After arranging pastry in the pan, place it in the freezer for 10 minutes, cover with parchment, weigh down with pennies or beans, and bake for 18 minutes. Remove the parchment and weights. Return the crusts to the oven for about 8 minutes to dry out well. Glaze with the beaten egg white and return to the oven for 3 minutes to set and seal crust. Allow to cool on a rack.

3. Bring the cream, sugar, and a tiny pinch of salt just to a boil in a heavy saucepan. Remove from the heat. Ladle ¼ cup hot cream into the egg yolks and whisk to blend. Then whisk egg yolks into the saucepan with the hot cream. Stir constantly and reheat over very low heat just until mixture coats the spoon. Do not overheat. If the mixture gets over 180°F (82°C), the egg yolks will curdle. Place custard in refrigerator to chill well.

4. Preheat the broiler. Remove shells from muffin tin and place on a baking sheet. Spoon chilled custard into pastry shells. Custard should come almost to the brim of the shells. Place in freezer for 5 minutes. Sift brown sugar through a large strainer over each custard for an even coating. Slide custards under the broiler 2 to 3 inches from the flame until the sugar melts and bubbles only. You can also use a propane torch; I use one with a flame spreader. Touch the sugar coating with the flame just to melt and caramelize it.

Variations: *For a quicker version, you can omit the pastry in Step 2, pour the hot custard into small ovenproof dishes, chill well, and proceed as in Step 4.*

For White Chocolate Crème Brûlée, in Step 3, after custard has thickened and cooled slightly, stir in 3 ounces grated white chocolate.

Old-fashioned Burnt Sugar Icing

MAKES ENOUGH FOR A 9-INCH THREE-LAYER CAKE

Once you've had old-fashioned burnt sugar icing, you can never forget how good it is. I remember sitting on the grass under a tree at a country church dinner-on-the-grounds when I was really little, eating the caramel icing off my piece of cake, then leaving the bare cake and going back to get another piece for the icing.

This recipe is a real old-time recipe from Margaret Fogleman, an outstanding home and semiprofessional cook who prepares meals for over a hundred at her church in Marion, Arkansas. When Margaret told me she occasionally had trouble with the icing curdling, I asked whether she used milk or cream. She said she used half-and-half or milk and, sure enough, the last time she made it with milk, it had curdled.

what this recipe shows

With fewer proteins than milk and a higher fat content to protect the proteins from the acidic conditions and heat, cream or half-and-half will not curdle as plain milk can.

Permitting the icing to cool before beating produces finer crystals and a smoother icing.

2½ cups *and* ½ cup sugar
  (3 cups *total*)

1¼ cups heavy or whipping cream or
  half-and-half

¼ pound (1 stick) butter

Salt

1 teaspoon pure vanilla extract

1. Combine 2½ cups sugar, cream, butter, and a pinch of salt in a large (at least 2½-quart) heavy saucepan and bring to a boil over low to medium heat. Stir to dissolve sugar. Remove from heat.

2. Heat ½ cup sugar in a heavy unlined saucepan (see Note) until it melts and just turns to a good brown color. Use a wooden spoon or spatula to stir. Once the sugar starts to color, the color change goes very fast. Take care not to burn the caramel.

3. Return cream mixture to heat. Add hot caramel in a fine stream. Stir in well, then stop stirring and heat to soft-ball stage (234° to 240°F/112° to 116°C). Remove from heat. Let cool 45 minutes without stirring or shaking. Pour into a mixer bowl. Add the vanilla and beat on low speed until icing is glossy but still spreadable (about 10 minutes). Ice cake immediately.

Note: *Be sure to use an unlined saucepan; the high heat will damage a lining. You can make the caramel easily in the microwave (page 425), if desired.*

Orange Slices with Cinnamon Candied Peel

MAKES 6 SERVINGS

I adore Marcella Hazan's simple, hearty, wonderful food, like the orange slices with candied orange peels that were among the original recipes she developed for a restaurant in Atlanta called Veni Vidi Vici. Here's my version—cold orange slices and the faint flavor of caramel and Grand Marnier with crunchy bits of golden candied peel, an elegant dessert that is simplicity itself.

what this recipe shows

Boiling orange zest and pouring off the water twice removes some bitter taste from the peel.

Lemon juice breaks down some sucrose into glucose and fructose and adding corn syrup (glucose) provides different sugars that interfere with sucrose's crystallization.

6 navel oranges	$1/3$ cup corn syrup
1 cup sugar	Nonstick cooking spray
$3/4$ cup water	$1/4$ cup boiling water
2 (2- to 3-inch) cinnamon sticks	3 tablespoons Grand Marnier or other
$1/4$ teaspoon fresh lemon juice	orange liqueur

1. Peel strips of zest $1/8$ to $1/4$ inch wide and 1 to 2 inches long from oranges, leaving white pith behind. Bring the strips of zest to a boil in a medium-size heavy unlined saucepan of water (see Note), then pour off the water. Add fresh water and bring to a boil again. Pour off again.

2. Add sugar, $3/4$ cup water, cinnamon sticks, lemon juice, and corn syrup and boil without stirring or shaking until sugar melts and just begins to caramelize. When bubbles begin to pile up on top of bubbles, sugar is close to caramelizing. At the first sign of color, remove from the heat.

3. Spray two forks with nonstick cooking spray. Spray a piece of foil. Lift out pieces of peel with a fork, letting peel drain a few seconds over the pot, then place on the foil. Use the other fork to separate and spread out peel. Try to keep clumps no larger than three or four pieces of peel; draining over the pot helps. Do not touch hot peel with fingers—a burn from molten sugar can be severe.

4. When most of the peel has been removed, carefully drizzle the hot syrup into $1/4$ cup boiling water in a small saucepan. Keep your hands out of the way; the water can steam up. With a wooden spoon, stir well to blend. Heat for a minute or two if necessary to get it to blend. When blended, remove from heat. Remove the cinnamon sticks and stir in Grand Marnier.

5. Peel the white pith from the oranges and slice them crosswise $1/2$ inch thick. Chill well. Arrange orange slices on individual serving plates or a large platter, drizzle Grand Marnier syrup over them, top with the pieces of candied peel, and serve.

Note: *Be sure to use an unlined saucepan; the high heat will damage a lining. The mixture can be caramelized easily in the microwave if desired. Stir together sugar, cinnamon sticks, lemon juice, and $1/4$ cup only corn syrup in a 4-cup glass measuring cup and microwave on High until sugar turns a golden tan, about 5 minutes, depending on the microwave oven. Remove from microwave, drop in orange zest strips, and proceed with Step 3.*

Spun caramel Those elegant desserts you've seen nestled in golden threads of spun caramel are not as difficult as they look. The secret is the temperature. As caramel begins to cool, it gets to a stage at which thin threads form when pulled.

You can make these golden threads easily with caramel that you have made in the microwave. When the caramel has cooled a little, dip the tines of a fork into the caramel, touch the fork to the rim of the measuring cup to anchor the threads, then raise the fork into the air. (Pastry chefs like a utensil with many tines so that they can make many threads at once. Some cut off a whisk not far from the handle so that they have many points to touch to the caramel. Rose Levy Beranbaum uses an angel food cake cutter with the tines bent alternately in different directions.) Fine gold threads will run from the fork or other instrument to the measuring cup. If the caramel is not quite cool enough, you will not get threads. In that case, wait a minute and try again.

Caramel cages Fancy desserts are sometimes served with caramel cages over them. You can make them at home too. The big problem is sticking. You can drizzle lines of caramel in a crisscross cage pattern over a curved surface easily enough with a little practice, but it's not so easy to pull the cage free when it cools. To avoid sticking, use the old double-grease technique— butter, chill, and oil the backs of large, round metal dippers or small metal bowls or spray them heavily with nonstick cooking spray. Taping a Teflon sheet (see Sources—Nonstick sheets, page 477) over the bowl or dipper and then spraying with nonstick cooking spray makes an excellent nonstick mold.

Pulled sugar Pulled sugar is the work of master sugar artisans. Fragile, iridescent roses, delicate flowers that look like priceless china, and other forms can be made by pulling sugar.

When I first saw Roland Mesnier, the White House pastry chef, pull sugar into a pale shimmering woven basket with fragile flowers of the softest salmony pink on the handle, I thought it was the most wonderful thing I had ever seen. I whispered to Rose Beranbaum, who was sitting next to me, "We've got to teach this right away. These are the most beautiful flowers I've ever seen." Rose answered, "You've never tried this, have you?" Roland made it look so easy.

I was so excited about pulled sugar and determined to pull mine perfectly. My pulled sugar was the finest in class, the palest, most delicate shade. You could pull it thinner than you can imagine into see-through iridescent petals of incredible beauty.

The whole class abandoned their efforts at pulled sugar and used mine. Alas, I was not able to make a single petal. In my determination to make perfect pulled sugar in spite of the terrible pain of handling the hot sugar, I had burned my fingers so badly that I could do nothing more.

To make pulled sugar, you first prepare melted sugar or the palest caramel, then cool it a little, and while it is still very hot, pull and fold, pull and fold, much as you would pull taffy.

This is an art that you need to watch a master perform. Recipes are not as important as technique. Old recipes call for 4 cups sugar, 1 cup water, 2 tablespoons corn syrup (glucose), and ¼ teaspoon cream of tartar (acid). Chef Roland likes to go with less. Some recipes suggest cooking the syrup to 312°F (156°C). He just melts the sugar or lets it go to the palest possible caramel. As you might suspect, you need to employ every good technique for candy making (page 433) when boiling down the syrup.

Once you pour out and fold the syrup, start pulling it. (Pastry chefs often touch their burning fingers to their chef's jacket to dispel some of the heat.) Pull and fold the hot mass of sugar again and again.

You cannot wear rubber gloves. The heat would melt them instantly. Some old-time candy makers suggest white cotton gloves like those Southern ladies used to wear everywhere. Gradually, the sugar takes on a sheen and becomes whiter and shiny. You can work tiny amounts of oil-based color into the sugar as you are pulling, but keep the color pale for the most beautiful flowers.

Use a lamp with a large bulb or a heat lamp to keep your mass of pulled sugar warm while you are working with a portion. If the work you are shaping hardens too fast, you can hold it under the lamp.

Caramel for Pulled Sugar

Use a heavy, unlined saucepan that is smaller than your burner. Stir together 2½ cups sugar, ⅔ cup water, and a little lemon juice or cream of tartar, then heat gently to dissolve, making sure you dissolve all the crystals. You can check for undissolved crystals by stirring with your hand and running your fingers against the sides and bottom of the pan *before* the liquid gets hot. Once the sugar is dissolved, stir in 4½ tablespoons glucose or corn syrup well.

Bring the sugar-water to a boil, then cover it for 3 minutes to wash down the sides. At this point you may clamp on a sugar thermometer and boil without shaking or stirring until the syrup reaches 312°F (156°C) and the sugar is just beginning to caramelize. It is important that you catch the syrup when it is at the palest straw color. If the caramel gets too dark, the sugar will be too soft to pull properly.

Dip the bottom of the saucepan into a pan of cold water to stop the syrup's cooking, then pour the syrup out on a marble slab. As the edges cool, lift them toward the center with a metal spatula. Start to pull the sugar as soon as it cools enough to hold it for seconds at a time.

Sugar syrups

When sugar is caramelized, all the water is boiled off and the sugar melts and breaks down. There is no need to know how concentrated the syrup is. For many candies, though—hard candies, some brittles, taffies, caramels, divinities, fondants, fudges, and pralines, to name a few—you can't control the final texture of the candy without knowing how concentrated the sugar syrup is.

The more concentrated the syrup is, the harder the candy will be. You want lollipops hard, divinity firm but not rock-hard, and caramels soft, so you need to know how much to boil the syrup to get the texture you want.

Candy makers solved this problem in a very simple way a long time ago by dropping a tiny bit of hot syrup into ice water. If the boiling syrup forms an inch or two of thread in the water, it is said to be at the *thread stage*. At this point the syrup will cool to a nice thick syrup but will not harden. If the boiling syrup forms a *soft ball,* it will harden into a soft but firm candy, like fudge or fondant. If it forms a *firm ball,* it will be perfect for caramels. If it forms a *hard ball,* it is just right for divinity, taffy, and nougat. At a somewhat higher temperature, syrup separates into threads that are hard but pliable; this is called *soft crack* and is perfect for butterscotch and taffies. At an even higher temperature, the threads become brittle; this is called *hard crack,* and the syrup is just right for brittles.

Instead of checking the syrup in ice water, you can use a candy thermometer. The less water there is in a syrup, the hotter it can get and the harder the candy will be. Temperature is a perfect indicator. Many good candy recipes give both the temperature and the description (such as "245°F/118°C, firm ball"). The table lists the various stages and their temperatures.

SUGAR SYRUP TEMPERATURES			
Temperature	Term	Description	Use
230–234°F / 110–112°C	Thread	Forms 2-inch threads	Syrup
234–240°F / 112–115°C	Soft Ball	Ball that flattens	Fondant, fudge, pralines
244–248°F / 118–120°C	Firm Ball	Ball that holds its shape	Caramels
250–266°F / 121–130°C	Hard Ball	Ball is hard and firm	Divinity, nougat
270–290°F / 132–143°C	Soft Crack	Hard, pliable threads	Taffy
300–310°F / 149–154°C	Hard Crack	Hard, brittle threads	Brittles, lollipops
320–350°F / 160–177°C	Caramel	Syrup from tan to brown	Flan, caramel cages

Basic candy formulas—how much interference do you need? In many candies you want crystals, but controlled crystals. As just explained, you can control candy texture (hardness) through the concentration of the sugar syrup. Another way of controlling size and number of crystals is with ingredients. Some ingredients like corn syrup and acids help prevent crystallization. Others like fats and egg whites coat crystals and prevent their growth.

In making caramel, brittles, or hard candies, you want to prevent crystallization altogether. You need enough interfering substances like corn syrup or acids to do this. Notice in my Microwave Caramel (page 425) I use both corn syrup and acid (lemon juice) to make it nearly fail-safe. The Macadamia-Almond Brittle recipe on page 434 contains ½ cup corn syrup plus brown sugar, which is an acidic ingredient. You will observe that most hard candy recipes like lollipops and lemon drops have a significant amount of corn syrup.

Basic candy formulas vary considerably. On page 432 is a table of general formulas for candies. The amount of interfering sugar (corn syrup, in particular) varies. Several factors influence how much corn syrup is needed—whether there are other ingredients in the recipe like fat, egg white, or milk proteins that also interfere with crystallization; whether you are stirring the candy cold or hot; whether the final product is hard and crisp or soft and chewy.

Let's look at an example of each of these factors. Toffee has a large amount of fat, which interferes with crystal growth, so it does not need as much corn syrup as a hard candy, brittle, or taffy that is dependent on corn syrup alone to prevent crystallization. For divinity, a concentrated sugar syrup is beaten into egg whites while the syrup is very hot. Stirring a concentrated syrup while it is hot encourages big crystals (page 436). You need twice as much corn syrup to keep the crystals fine in divinity as you need in fondant or fudge, which is not stirred until cool. Soft and rich candies like caramels contain a lot of both corn syrup and fat.

BASIC CANDY FORMULAS

Crystalline

Fondant	Fudge
1 cup sugar	1 cup sugar
1 tbsp corn syrup (or $^1/_{16}$ tsp cream of tartar)	1 tbsp corn syrup
	1 tbsp butter
$^1/_2$ cup water	$^1/_2$ cup milk
237°F/114°C	234°F/112°C
Soft ball	Soft ball

Amorphous or Noncrystalline

Caramels	Taffy	Toffee	Lollipops
1 cup sugar	1 cup sugar	1 cup sugar	1 cup sugar
1 cup corn syrup	$^1/_4$ cup corn syrup	1 tbsp corn syrup	$^1/_3$ cup corn syrup
$^1/_4$ cup butter		$^3/_4$ cup butter	
1 cup cream (or evaporated milk)	$^1/_3$ cup water	$^1/_4$ cup water	$^1/_2$ cup water
248°F/120°C	261°F/127°C	300°F/149°C	310°F/154°C
Firm ball	Hard ball (or 275°F/135°C, Soft crack)	Hard crack	Hard crack

Special Textures

Divinity	Marshmallows
1 cup sugar	1 cup sugar
2 tbsp corn syrup	1 tbsp corn syrup
1 egg white	1 tbsp gelatin
$^1/_4$ cup water	$^1/_4$ cup water
252°F/122°C	248°F/120°C
Hard ball	Firm ball

Helen Charley. *Food Science*, 2nd ed. New York: John Wiley & Sons, 1982.

Sugar syrup candies For boiled sugar syrup candies you need all the skills you have seen in caramel making, plus a few more. You will need a large (2- to 3-quart) heavy unlined saucepan and a candy thermometer. Sugar syrups should be boiled down on a burner larger than the saucepan so that the sides are well heated. To prevent crystals that may form on the pan sides from dropping into the syrup, you can oil the inner sides of the saucepan, rinse the sides down with a pastry brush dipped in water, or place a lid on the syrup for 2 to 3 minutes while boiling. Be sure that all the sugar is dissolved initially, clamp on the candy thermometer, and heat without stirring or shaking until the recipe temperature is reached. If the recipe calls for a cooling period before stirring, let stand without touching or shaking. Pour candy from the pan and do not scrape the pan.

at a glance
Candy making

What to Do	Why
Use heavy unlined metal saucepans, preferably of a metal that conducts heat well.	A heavy pan made of a good conductor will prevent scorching and burning. The temperatures reached in candy making are very high and will melt or damage some pan linings.
Use a saucepan larger than you think you need.	Sugar syrups tend to boil over, especially those containing molasses.
Oil the sides of the saucepan.	Oiling helps prevent boil-over and helps keep crystals from sticking to the sides.
Cook syrup on a burner larger than your saucepan.	A large burner keeps the sides of the pan hot; this helps prevent crystals from sticking to the sides.
Heat sugar, water, corn syrup, and other ingredients over low heat, stirring constantly until all the sugar is dissolved.	All sugar crystals must be dissolved so that they will not cause crystallization.
Brush down sides of saucepan with a pastry brush dipped in cold water, wipe sides with a damp paper towel, or cover mixture for 2 to 3 minutes over medium heat.	Sugar crystals must not be left on the sides of the pan or they will cause crystallization.
Clamp on candy thermometer and boil over medium heat without shaking or stirring until desired temperature is reached.	Agitation can cause crystallization.
Never dip a spoon that has been dipped in syrup back in the syrup without washing.	Crystals clinging to the spoon can cause crystallization.
Remove candy thermometer from pan and pour boiled syrup or candy from pan. Do not scrape pan.	Scraping or stirring can cause crystallization.
With candies that call for cooling before stirring, cool to suggested temperature before stirring.	Cool syrup produces finer crystals.

Brittles Brittles hover on the edge between melted sugar (caramel) and a very concentrated boiled syrup. Sometimes a brittle is nothing more than caramelized sugar poured over nuts. Such a recipe might read: "Cover the bottom of a buttered shallow pan with warm roasted peanuts. Heat 2 cups of sugar in a heavy iron skillet over low heat, constantly stirring, until melted and light brown. Pour over nuts. When hard, break into irregular pieces." Other brittle recipes do not caramelize the sugar but instead make a concentrated syrup containing brown sugar, molasses, or honey. Some have vanilla extract stirred in at the end. A good brittle recipe contains corn syrup or an acid or both to interfere with crystallization. The nuts should always be warm; cool nuts make the syrup cool down too fast, and it will not spread thin enough.

Some brittle recipes contain baking soda, which foams when you add it. A small amount just neutralizes acidity. Some specialty brittles on the market are very flaky and airy. These are made using acid and a larger amount of baking soda. The acid and the baking soda form carbon dioxide gas, which puffs up the candy.

Brittles can change amazingly in flavor. Right after they are made, the taste may be disappointing, but a day or two later the flavors are wonderful.

> **A good brittle recipe contains corn syrup or an acid or both to interfere with crystallization.**

Macadamia-Almond Brittle

MAKES ABOUT 2½ POUNDS

Delicate-flavored roasted macadamias and almonds in super buttery, crunchy caramel brittle are *so* good—the Rolls-Royce of brittles. This brittle improves with age and becomes even more buttery tasting after a day or two.

what this recipe shows

Corn syrup aids in preventing crystallization.

Brown sugar, vanilla, and roasted nuts add flavor.

The small amount of soda in the recipe just neutralizes the acidity. This is not enough to make the brittle flaky.

Popcorn salt, which has fine crystals, sticks to the nuts better than regular salt.

1 cup almond slivers

2 cups roasted macadamia nuts

1 tablespoon *and* ¼ pound (1 stick) butter (9 tablespoons *total*), cut into tablespoon-size pieces

¼ teaspoon popcorn salt

1 cup packed light brown sugar

1½ cups granulated sugar

½ cup light corn syrup

½ cup water

1 teaspoon baking soda

½ teaspoon pure vanilla extract

1 teaspoon cold water

Nonstick cooking spray

1. Preheat the oven to 350°F (177°C).

2. Spread the almond slivers on a baking sheet and roast until very lightly browned, 7 minutes. Add the macadamia nuts and continue to roast until all nuts are lightly browned, 4 to 5 minutes. Remove from the oven and turn it down to 150°F (66°C), leaving the door open to cool the oven off. While nuts are still hot, stir in 1 tablespoon butter and sprinkle with popcorn salt. When the oven has had about 4 to 5 minutes to cool down, place the nuts back in the oven to keep warm.

3. Butter two 10 × 15 × 1-inch nonstick baking sheets or jelly-roll pans with 1 tablespoon butter. Combine the brown sugar, granulated sugar, corn syrup, and water in a 3-quart heavy unlined saucepan over medium-high heat. Heat and stir with a wooden spoon until sugar is dissolved. You should no longer be able to feel any grains of sugar against the bottom of the pan when you stir. Move pan off the heat and, with a wet pastry brush or wet paper towel, wipe any grains of sugar from the sides of the pan above the liquid level. Place pan back on heat. Clip on candy thermometer and bring syrup to a boil. Do not stir or shake.

4. Boil syrup to hard-crack stage (300° to 310°F/149° to 154°C). Depending on the heat, this may take about 10 minutes. Stir in ¼ pound butter. The temperature will drop. Continue heating until temperature comes back up to soft crack (270° to 290°F/132° to 143°C). Remove from heat. Stir together baking soda, vanilla, and cold water in a small bowl. Stir in soda-vanilla mixture and continue to stir for 30 seconds or so. Candy will foam. Stir in nuts and continue stirring until all nuts are coated.

5. Pour onto the baking sheets. Spray two forks with nonstick cooking spray and use to pull the hot candy out as thin as possible. With a large spatula, turn candy over. When cool enough to handle, stretch with your fingers. Allow to cool completely, then break into pieces. Store in an airtight tin with wax paper separating layers of candy.

Variation: *This brittle may be made with peanuts. Warm 3 cups canned roasted peanuts in a 150°F (66°C) oven and substitute for the almonds and macadamia nuts in Step 4. An unusual and delicious brittle may be made using 3 cups smoked almonds (about 3 small cans).*

Lollipops, lemon drops, and other hard candies Lollipops, lemon drops, and other hard candies are all made from a sugar syrup boiled down until it is very concentrated, with flavoring and coloring added at the end. Then the candy is left to harden. Review "Candy Making at a Glance" (page 433). Specifically, old-time candy makers recommend that you grease the sides of the pot in which you boil down the syrup to prevent boiling over. Also, be sure that you are using a pot that is large enough.

Sugar's ability to attract and hold moisture from the atmosphere is a big disadvantage in making hard candy. You've probably heard or read the warning, "Don't make this dish on a rainy day." This is generally said of recipes with a high sugar content.

You can boil a syrup to a low enough water content that it will become hard when cooled, but if it cools in a very moist room it may pick up so much water that it remains gooey. This can happen both during candy making and with already prepared candy. A piece of hard candy sitting uncovered in a warm, moist room can become quite sticky.

Some sugars attract moisture more than others. Fructose, one of the two sugars that make up table sugar, attracts and holds moisture more than sucrose or glucose, the other sugar in table sugar. A loaf of bread with honey in it keeps better and stays moister than one made with sugar because honey is about 42 percent fructose. When making hard candies, therefore, it is important not to use much fructose.

You may think you're *not* using any fructose, but if you add an acid to table sugar you instantly form *glucose* and *fructose* because acids break down sucrose into its two component sugars. For a good

> Hard candy that cools in a very moist room may pick up so much moisture that it remains gooey.

hard candy, then, you need a little bit of a different sugar to interfere with crystallization, but you want to stay away from fructose and any acid that will form fructose. Most hard candy recipes contain sugar, water, and corn syrup (glucose) as an interfering agent. To finish the candy, a flavoring oil and color are added with a minimum of stirring.

Don't use fructose in hard candies; it retains moisture.

Another problem with adding acid to boiled syrup candies is that the longer you boil a syrup containing acid, the more fructose you form. In other words, both how much acid you add and how long you cook the syrup influence the amount of fructose formed. This makes it very difficult to arrive at just the right amount of acid to use in a recipe.

Lemon drops that contain fresh lemon juice pose a tricky fructose problem, but candy makers came up with a solution long ago: Roll the drops in granulated sugar after they're formed to coat the water-absorbing fructose with a protective layer of sucrose and thus keep it from the air.

Taffies What is taffy? You might think of it as a hard candy like a lollipop that was pulled and twisted over and over again to incorporate tiny air bubbles while it was hot. Those thousands of tiny air bubbles soften and lighten taffy and make it opaque, not clear like hard candy. Taffy recipes usually contain sugar, a fair amount of corn syrup (¼ to ⅓ cup per cup of sugar), water, and flavoring. Taffy recipes with molasses are notorious for boiling over. If you're making taffy with molasses, be sure to use a large pot and to oil the sides.

Molasses will make taffy boil over, so if you use molasses, use a large pot and oil the sides.

Caramels Caramels contain milk, cream, or sweetened condensed milk—sometimes all three. The sugar in milk (lactose) caramelizes at a lower temperature than table sugar (sucrose), so the presence of milk in any form lowers the temperature at which you get browning and caramel flavors. Caramels are normally cooked to firm-ball stage (about 246°F/119°C).

When ready, the caramels are poured out to cool in a lightly buttered pan or on lightly buttered foil between confectioners' bars (heavy bars that can be pulled away for easy cutting when the candy cools).

Caramels can be flavored with vanilla, and some recipes contain honey, brown sugar, or molasses for additional flavor. Since acidic ingredients like brown sugar or molasses can cause milk to curdle, such recipes also contain a little baking soda to reduce the acidity and prevent the milk proteins from curdling. Caramel recipes always contain corn syrup to interfere with crystallization.

Big crystals, little crystals

When you crystallize sugar, you can make baby-fine crystals hardly detectable on the tongue (smooth fondant) or huge crystals like those in the rock candy that you may have had as a child.

If you want to make baby-fine crystals, you have your job cut out for you. First you must let your concentrated sugar solution cool undisturbed. If crystals form while the syrup is hot, they will grow as it cools and the candy will be grainy. But if you let the solution cool undisturbed, millions of baby crystals will form all at once when stirred.

Big crystals—rock candy Rock candy is a perfect example of growing giant crystals. You place a damp string with a few crystals of sugar on it in very concentrated sugar syrup, and crystals start to grow on it. Once the crystals have begun to form, they continue to grow larger and larger in undisturbed syrup.

Little crystals—fondants Fondant is a concentrated sugar syrup that is cooled and then worked to form a thick creamy paste of fine crystals. It is flavored and cut into different shapes and used as candy centers, or it can be rolled out into a ¼-inch-thick sheet and draped over cake layers for a perfectly smooth coating. You will notice that in all fondant recipes the syrup is worked vigorously at the crucial time when the crystals are forming to get fine crystals to form all at once.

There are many different recipes for fondant. Helen Fletcher in her book *The New Pastry Cook* lets the processor do all the hard work.

Food Processor Fondant

Heat sugar, water, and about 1½ tablespoons of corn syrup per cup of sugar to the soft-ball stage (238°F/114°C). Pour into the food processor with the steel knife. Wash the candy thermometer well and reinsert into the syrup. Let the syrup cool undisturbed in the workbowl to 140°F (60°C), about 30 minutes. Remove the thermometer. Add desired flavoring or coloring and process 2 to 3 minutes, until syrup completely converts from a glassy syrup to an opaque paste. When thoroughly cooled, store sealed at room temperature for 24 hours. Use or refrigerate for later use.

Since major effort is required for fondant making, many professionals purchase it ready made from suppliers like Maid of Scandinavia or Wilton (see Sources—Fondant, page 478). Refrigerated, fondant keeps for months.

Fudge Fudge is a sugar syrup enriched with butter or cream and flavored, frequently with chocolate. The syrup is heated to the soft-ball stage (236°F/113°C), cooled without disturbing to 110°F (43°C), then, like fondant, beaten vigorously until opaque. It is then poured into pans, allowed to set, and cut into desired pieces.

Because fudge contains fat and protein from milk products and cocoa particles, it does not need as many sugar crystals to thicken it as do fondant and some other candies. Fudge recipe temperatures are usually slightly lower than for fondants—around 236°F (113°C) for fudge, closer to 240°F (116°C) for fondant.

Classic Fudge

MAKES ABOUT THIRTY 1½-INCH SQUARES—1¼ POUNDS

Stories go that beaten fudge was originated at Smith College, and it is sometimes called *college fudge*. Classic fudge is smooth textured and soft enough to cut without breaking.

what this recipe shows

Using half-and-half reduces the risk of curdling that exists with milk. (The acidity of the chocolate can curdle milk proteins.)

Stirring steadily during the first part of cooking helps prevent curdling.

Using a little corn syrup slows crystallization and produces a creamier fudge.

Letting the fudge cool undisturbed to 110°F (43°C), then beating it vigorously, creates millions of very fine crystals all at once for smooth texture.

continued

1 tablespoon butter for greasing pan

2 ounces unsweetened baking
   chocolate, chopped

2 cups sugar

$^2/_3$ cup half-and-half

2 tablespoons light corn syrup

2 tablespoons butter

1 teaspoon pure vanilla extract

Pinch of salt

1. Grease an $8 \times 8 \times 2$-inch pan well with butter.

2. Heat chocolate, sugar, half-and-half, and corn syrup in a heavy saucepan over medium-low heat, scraping the bottom of the pan gently with a wooden spatula. Clamp a candy thermometer on the side of the pan and boil the mixture gently, frequently scraping to prevent burning until the temperature is 236°F (113°C). Do not stir or shake. Turn off the heat.

3. Drop 2 tablespoons butter on top, but do not stir or shake. Allow to cool to 110°F (43°C). Remove the thermometer. Add vanilla and salt and beat vigorously until fudge begins to lose its shine and a small amount dropped from a spoon holds its shape. Pour or knead and press (if necessary) into greased pan.

4. Let cool, then cut into $1^1/_2$-inch squares. Store and serve at room temperature.

Fudge recipes using marshmallows, or marshmallow cream, take advantage of the fact that marshmallows are made with a sugar syrup boiled to the firm-ball stage and also contain egg whites and a little gelatin. The egg whites and gelatin in the marshmallows coat the sugar crystals as they form to keep them small and produce an unusually creamy fudge. Because it is easy to make and very creamy, I love fudge made with marshmallows.

You can also make excellent fudge using the microwave. My friend Rosanne Greene, who ran a wonderful cooking school in Dallas for many years, makes instant fudge in the microwave by melting together sweetened condensed milk, chocolate, and butter. With the condensed milk, you already have a boiled-down, concentrated sugar syrup, and when you add the acidic melted chocolate, it causes the proteins in the milk to set. The combination of the proteins tightening and the chocolate becoming solid at room temperature results in a set fudge of soft consistency. Made with a good-quality chocolate, this is incredibly delicious.

"You Can't Eat Just One" Creamy Fudge

MAKES ABOUT SIXTY 1½-INCH SQUARES

This is my version of my daughter-in-law Janice Hecht's marshmallow fudge. Different versions of this tried-and-true fudge recipe have been around for a long time. You will see it called *million-dollar fudge, Sees fudge,* and so forth. It's hard to beat.

> **what this recipe shows**
>
> Roasting enhances the flavor of walnuts.
>
> Egg whites and gelatin in marshmallows and the butter coat the sugar crystals as they form, keeping them small and producing a creamy fudge.

2 to 3 tablespoons butter for greasing pan

1½ cups walnut or pecan pieces

2 tablespoons *and* 4 tablespoons butter (6 tablespoons *total*)

¼ teaspoon salt

3 cups sugar

1 can (5 ounces) evaporated milk (⅔ cup)

12 ounces semisweet chocolate chips or good-quality semisweet chocolate, such as Lindt, Tobler, or Hershey's Special Dark, chopped into ¼-inch pieces

2 teaspoons pure vanilla extract

1 jar (7 ounces) marshmallow cream

1. Preheat the oven to 350°F (177°C).

2. Line bottoms and sides of two 8 × 8 × 2-inch pans with foil and grease well with butter.

3. Spread nuts on a medium baking sheet and roast until lightly browned, 10 to 12 minutes. While nuts are hot, stir in 2 tablespoons butter and salt. Pour into a large mixing bowl.

4. Bring the sugar, evaporated milk, and 4 tablespoons butter to a low boil in a large heavy saucepan, stirring constantly. Stop stirring. Clamp on a candy thermometer and keep at a low steady boil over medium-low to medium heat until the mixture reaches 238°F (114°C), 4 to 5 minutes.

5. While sugar-milk mixture is boiling, place the chocolate chips, vanilla, and the marshmallow cream into the large mixing bowl containing the roasted nuts. Have everything in the bowl before the cooking time is up.

6. When the sugar-milk mixture has reached 238°F (114°C), pour it immediately over the nut mixture. Stir quickly to blend everything. Pour into the pans. Cool until firm. Remove from pans and peel off foil. Cut into 1½-inch squares.

Sinfully Easy Fudge

MAKES ABOUT THIRTY 1½-INCH SQUARES

This is my version of microwave fudge. Rosanne Greene uses a whole stick of butter in hers, but with milk chocolate this makes the fudge soft unless kept in the refrigerator.

what this recipe shows

Roasting enhances the flavor of walnuts.

Condensed milk provides boiled-down, concentrated sugar syrup.

Chocolate, which is acidic, aids in firming condensed milk just as lemon juice does in a condensed milk lemon pie.

1 tablespoon butter for greasing pan

1 cup pecan or walnut pieces

1 tablespoon butter

1/8 teaspoon salt

1 can (14 ounces) sweetened condensed milk

7 ounces milk chocolate chips, chopped

11 ounces semisweet chocolate chips or good-quality semisweet chocolate such as Lindt, Tobler, or Hershey's Special Dark, chopped into 1/4-inch pieces

1 teaspoon pure vanilla extract

1. Preheat the oven to 350°F (177°C).

2. Line the bottom and sides of an 8 × 8 × 2-inch pan with foil and grease well with butter.

3. Spread nuts on a medium baking sheet and roast until lightly browned, 10 to 12 minutes. While nuts are hot, stir in 1 tablespoon butter and sprinkle with salt.

4. Combine the condensed milk and all the chocolate in a large glass bowl or 4-cup measuring cup and melt in the microwave at 50 percent power about 3 minutes, stirring frequently (see Note). As soon as the chocolate is melted, remove from the microwave and stir in vanilla and roasted nuts. Pour into the prepared pan. Refrigerate to cool and set. Remove from pan, peel off foil, and cut into squares.

Note: *To make on the stovetop, simply heat chocolate and condensed milk over low to medium heat in a heavy saucepan, stirring constantly, until the chocolate melts. Then proceed as directed at the end of Step 4.*

Pralines   With classic New Orleans pralines, you want the sugar to crystallize, but in fine crystals. Notice that at the crucial point in the following recipes you need to stir vigorously to get fine crystals to form; then you must instantly spoon the candies into individual-size pieces before the crystals grow larger.

Real Naw'llins Prawleens is a small recipe that cools fast so the candy stands only 3 to 4 minutes before beating. The first pralines spooned up have fine crystals, since the small candy pieces cool very quickly and the crystals do not have a chance to collect more molecules and grow. But the crystals are growing by the second in the candy still in the hot pot. There is an enormous difference between the first candy that is spooned up and the last. The early ones are very smooth, and the last one grainy. It's helpful if you have two people to spoon out the pralines at the crucial moment.

I once had some very interesting "chewy" pralines from Texas. How would you keep the pralines from crystallizing and make them chewy? Well, how about the old similar-but-not-the-same-sugar trick? Sure enough, when I read the label on the package, the first or second ingredient was corn syrup. To make them, I took the classic praline recipe but substituted 3/4 cup corn syrup for part of the sugar. I got chewy caramels that ran together before they set, so I backed off on the corn syrup. With these you have plenty of time to spoon them up. In fact, if you don't cool the mixture in the pot until thick throughout, they will run together. These two recipes, Creamy Pralines and Real Naw'llins Prawleens, really illustrate how corn syrup slows crystallization.

Creamy Pralines

MAKES ABOUT TWENTY-FOUR 2-INCH PRALINES

Instead of hard, granular pralines, these have a much smoother, fine-crystal texture from the corn syrup.

what this recipe shows

Roasting enhances the flavor of nuts.

Because of the corn syrup, which slows crystallization, these must be boiled to a slightly higher temperature than the Real Naw'llins Prawleens to ensure a firm set.

Corn syrup slows crystallization so much the setting is slow and these will run together if you do not wait until the mixture has cooled and is quite thick before spooning out.

Butter for greasing foil

1 1/2 cups pecan or walnut pieces

3/4 cup almond slivers (optional)

2 tablespoons *and* 2 tablespoons butter
   (4 tablespoons/ 1/2 stick *total*)

1/4 teaspoon salt

1 cup packed light brown sugar

3/4 cup granulated sugar

1/3 cup light corn syrup

1/2 cup canned evaporated milk

1 teaspoon pure vanilla extract or good
   rum like Myers's

continued

1. Preheat the oven to 350°F (177°C). Generously grease a large piece of foil with butter. Set aside.

2. Spread the nuts on a baking sheet and roast until lightly browned, 10 to 12 minutes. If using almonds too, put them on the baking sheet near the oven door so that you can watch them. While the nuts are hot, stir in 2 tablespoons butter and sprinkle with salt. Set aside.

3. Boil the brown sugar, granulated sugar, corn syrup, evaporated milk, 2 tablespoons butter, and roasted nuts until the temperature reaches high soft-ball stage (238° to 240°F/114° to 115°C). Let stand, undisturbed, for 5 minutes to cool. Add vanilla. Then start beating with a flat-ended wooden spatula. Scrape the thicker portion on the bottom loose and stir in. Stir until the mixture thickens noticeably in the center, not just the edges. This will take several minutes. Spoon heaped-tablespoon-size pralines as fast as possible onto buttered foil to cool and set. Store in an airtight container.

Real Naw'llins Prawleens

MAKES ABOUT THIRTY 2-INCH PRALINES (SEE NOTE)

This is based on a recipe from Lee Barnes, which she gave me permission to use and to teach on the condition that I pronounce *pralines* correctly—*praw* like claw, not *pray*—"prawleens." Lee owned a large cooking school in New Orleans, and you would be hard-pressed to find anyone who has been in the food business there for a while who was not helped by Lee in some way. I was fortunate to be with her and her family for one or two weeks a year for many years when I taught there, and we had many adventures together. She died much too young and I miss her.

Lee always said, "Add more nuts for your favorite people." I frequently make these pralines with two different kinds of nuts. Many of my students make batch after batch of them for the holidays. They are the best that I have ever eaten, and many others say the same.

what this recipe shows

Roasting enhances the flavor of nuts.

Concentrating the sugar by boiling ingredients to the soft-ball stage (236°F/113°C) ensures firm crystallization of sugar and firm hardening.

When these pralines start to thicken, hardening (full crystallization) is immediate since they contain no corn syrup to slow the process.

Butter for greasing foil

1¼ cups pecan or walnut pieces

½ cup almond slivers (optional)

2 tablespoons *and* 2 tablespoons butter
   (4 tablespoons/½ stick *total*)

¼ teaspoon salt

1 cup packed light brown sugar

1 cup granulated sugar

½ cup canned evaporated milk

1 teaspoon pure vanilla extract or good
   rum like Myers's

1. Preheat the oven to 350°F (177°C). Generously grease a large piece of foil with butter. Set aside.

2. Spread nuts on a baking sheet and roast until lightly browned, 10 to 12 minutes. If using almonds too, put them on the baking sheet near the oven door so that you can watch them. While the nuts are hot, stir in 2 tablespoons butter and sprinkle with salt. Set aside.

3. In a heavy medium saucepan, stir together well and boil the brown sugar, granulated sugar, evaporated milk, 2 tablespoons butter, and roasted nuts until mixture reaches the soft-ball stage (236°F/113°C). Let stand, undisturbed, for 4 minutes to cool. Add vanilla. Then start beating with a large wooden spoon. The second you feel the mixture begin to thicken, start spooning up the candies. Spoon heaped-tablespoon-size pralines as fast as possible onto the buttered foil. Let stand to cool and set. Store in an airtight container.

Note: *If you want to make a double batch, add a tablespoon of corn syrup to the ingredients. This will slow crystallization just enough to allow you to spoon up the larger batch.*

Divinity   Divinity is nothing more than a concentrated sugar syrup beaten into an egg white foam and flavored. It makes use of the fact that the egg white coats the sugar crystals and keeps them small to form creamy candies. Drizzle a hard-ball stage (250° to 266°F/121° to 130°C) sugar syrup into beaten egg whites and beat constantly until thick. Beat in the flavoring and spoon out the divinity onto buttered foil or wax paper to set. Divinity may be flavored by replacing some of the sugar with brown sugar or honey. Nuts or candied fruit can be added.

The big problem with divinity is getting the proper concentration in the sugar syrup. When you drizzle hot concentrated syrup into cold beaten egg whites, the cold egg whites immediately firm up the syrup. You can get a hard mass of sugar and egg whites that will soften as you add more hot syrup. When you first add the syrup, it is better if it is not too concentrated; this avoids the suddenly set sugar–egg white mass. If, on the other hand, the syrup is not concentrated enough, with all the water in the egg whites the sugar won't crystallize—the divinity will not become firm and will fail.

There are many fail-safe divinity recipes that take these two problems into account. In one of my favorite recipes, a thinner syrup cooked to soft-ball stage (240°F/116°C) is added at first so that you won't have a solid mass of hard sugar and egg white. Then, at the end, a more concentrated hard-ball syrup (265°F/129°C) is added to ensure that the candy will have enough sugar to set it.

There are two ways of doing this. One is to prepare two separate syrups—one cooked to soft ball and one cooked to hard ball. Drizzle in the soft-ball syrup first and have the other syrup heating so that it will get to hard ball at just about the time you finish drizzling in the first syrup. The mixer has to be running constantly. Unfortunately, this method soils two sugar syrup pots.

The other way is to prepare a syrup cooked to soft ball (240°F/116°C), drizzle in half, keep the mixer running while you return the remaining syrup to the heat and heat it to hard ball (265°F/129°C), then drizzle it in. This is a somewhat simpler method and leaves you only one syrup pot to wash.

Sugared nuts   A popular cooking teacher here in Atlanta, Ursula Knaeusel, tells her students, "You're the boss in the kitchen." Do it the way that you want to. If you want sugar crystals on nuts, you can do it. If you want a caramel coating, you can do that too. The next two recipes are beautiful examples of the do-it-the-way-you-want-it principle.

Sugar-Crusted Almond Slivers with Chocolate

MAKES ABOUT 30 CLUSTERS

Paula Murphy in Fort Walton Beach, Florida, boils raw peanuts in concentrated sugar water with constant stirring until she gets the sugar to crystallize in fine crystals on the nuts. Then she places them in the oven to melt the crystals just a little for wonderful crunchy coated nuts. Paula prepares large batches of these nuts for holiday guests. The only complaint is that no matter how many batches she makes, they seem to disappear instantly.

I have had other nuts that I analyzed as simply a more elegant version of Paula's recipe. They were almond clusters made with a crystallized sugar coating and then dipped in chocolate. I created my own version here—roasted nuts with a crunchy sugar coating topped with smooth chocolate.

what this recipe shows

Stirring the nuts as the sugar syrup becomes concentrated causes the sugar to crystallize on them for a slightly grainy, crunchy crystalline surface.

Nonstick cooking spray

2 cups almond slivers

1 cup sugar

$1/2$ cup water

12 ounces good-quality semisweet chocolate such as Lindt, Tobler, or Hershey's Special Dark, chopped into $1/4$-inch pieces

1. Preheat the oven to 350°F (177°C). Spray a baking sheet well with nonstick cooking spray.

2. Boil the almond slivers, sugar, and water in a heavy unlined skillet over low to medium heat, stirring constantly, until all the water has evaporated and the sugar crystallizes on the almond slivers. You want to keep the heat low to medium and stir constantly so the sugar will crystallize and not melt to caramel.

3. Spoon clusters of almond slivers onto the baking sheet. Do not touch hot sugar with your hands. Spray two forks well with nonstick cooking spray and use them to separate the clusters into bite-size clumps if necessary. Bake until the sugar is melted and lightly colored, about 15 minutes. Watch carefully to avoid burning. Remove when sugar is just melted. The sugar will continue to darken after being removed from the oven.

4. For a shiny chocolate coating, temper the chocolate as described on page 468. For a quick but still delicious candy, melt the chocolate in a heavy saucepan over low heat, stirring constantly. When almonds have cooled completely, dip into melted chocolate. Lift out with a fork. Drain and let excess chocolate drip back into saucepan. Place coated clusters on bare foil to set.

5. If the crystallized sugar does not hold the nuts together, stir loose almond slivers into the melted chocolate. With a fork, lift out five or six slivers at a time, drain, then drop as a cluster onto foil.

Roasted Walnuts
with Brown Sugar Caramel

MAKES 1 POUND

Terry Ford, cook, teacher, and writer extraordinaire of Ripley, Tennessee, makes marvelous cocktail nuts using a technique that is completely different from Paula Murphy's in the preceding recipe. He melts the sugar directly on the nuts, with no water out of which the sugar can crystallize, so the nuts end up with a paper-thin caramel coating. These are a perfect cocktail buffet sweet or holiday nibble. They are instant and deadly delicious; you just can't stop eating them.

what this recipe shows

The lack of water allows the sugar to melt and form a caramel glaze on the nuts.

Nonstick cooking spray

3 tablespoons butter

1 pound walnut or pecan halves

$^1/_2$ teaspoon salt

$^1/_4$ cup packed brown sugar

1. Spray a large sheet of aluminum foil with nonstick cooking spray. Set aside.

2. Spray a large heavy skillet with nonstick cooking spray. Melt the butter in the skillet and add the nuts. Cook over low heat, stirring constantly, until nuts are roasted, about 4 minutes. Sprinkle with salt and brown sugar. Continue stirring and cooking until sugar melts, about 3 more minutes. Spread out on foil to cool. Serve immediately or when completely cool. Store in a tightly sealed container. Nuts keep at room temperature for several weeks.

Ice crystals—ices, ice creams, and sherbets

As water temperature falls below 32°F (0°C), ice crystals begin to form. However, many ingredients that you may add to this water affect both the temperature at which the ice crystals form and the size of the crystals. In candy making, stirring during crystallization and adding various ingredients strongly affect the texture of the candy. Stirring and ingredients have a major effect on ice cream too. Fine ice cream has a soft, light consistency, full body—not runny or spongy—and a creamy smoothness in the mouth. Both ingredients and techniques affect the final result, whether it's an ice (granita or sorbet), ice cream, or sherbet.

Ices have only sugar and lemon juice to control their texture, while dairy products play a big role in the texture of ice cream or sherbet. Techniques like time and amount of stirring influence ices. Stirring fine air bubbles into ice creams and sherbets as they freeze lightens and softens their texture. With mousses, the light, soft texture is created by beating air into one or more of the ingredients before freezing.

Ices

Modern fruit ices (also called *water ices*) fall essentially into two categories—those that are stirred very little (granitas) and those that are stirred frequently (sorbets). All fruit ices contain four simple ingredients: fruit juice or puree, sugar, water, and lemon juice. Even with so few ingredients, getting fine ice crystals and at the same time leaving enough syrup unfrozen between the crystals for a smooth, light texture isn't easy.

The number-one priority is the sugar concentration. An excellent fruit ice has tantalizing, refreshing fruit flavor and consists of fine ice crystals packed in concentrated fruit syrup. How much syrup is left unfrozen when the ice is removed from the freezer makes the difference between a successful and an unsuccessful ice. With no syrup between the crystals the ice freezes solid, and you can attack it only with an ice pick. In fact, even if there is some syrup between the crystals, but not enough, the ice feels like a solid block when you try to scoop it. On the other hand, if there is too much syrup, you have mush.

When a mixture of fruit juice or puree, sugar, and flavoring freezes, some of the water in the fruit-sugar syrup freezes into pure ice crystals. The fruit-sugar syrup left behind becomes more and more concentrated. Since sugar lowers the freezing point of a liquid, as the syrup becomes more concentrated its freezing point gets lower and lower, so that you have some syrup left unfrozen at freezer temperature. So if you want to be able to scoop the ice when it is removed from the freezer, the initial mixture has to have enough sugar to leave some syrup at freezer temperature. The sugar concentration in the mixture is the key to everything.

While sugar is the key ingredient that controls the texture of an ice, other ingredients contribute to texture and taste. Lemon juice and water play a major role in taste. A fruit with a high pectin content or a jam made with pectin will make a softer ice. Wine, liqueur, and honey all lower the freezing point so less sugar is needed if these ingredients are used.

You must have a high sugar concentration for an ice to be scoopable. Without a little lemon juice to cut this sweetness, the taste of the ice can be cloyingly sweet. Water adds subtlety and delicacy to the taste. An ice made with a straight cantaloupe puree, for example, has a cucumberlike vegetable taste, while one diluted with water has a light, fruity taste.

The recipe

You would think that coming up with a successful recipe for an ice would be a snap since there are so few ingredients. Actually, parts of the recipe are simple. Amounts of lemon juice, water, pectin, and wine or liqueur are relatively simple. It is the amount of the vital ingredient, sugar, that is the problem.

Sugar The exact amount of sugar in the recipe is crucial for a scoopable ice. Some recipes for ices direct you to make a "concentrated" sugar syrup, then add enough of it to the fruit mixture to reach a certain specific gravity. **Sugar is the key ingredient in the texture of an ice.** Unfortunately, different people recommend different specific gravities. Madeleine Kamman's recommendation for specific gravities between 1.1335 and 1.1425 is perfect for a scoopable ice. Many other recommendations are not. However, for this method you need a saccharometer, an instrument most home cooks don't have.

Food science writer Harold McGee has a method for calculating the correct amount of sugar, but this requires facts and figures about your fruit as well as some math. He simplified workable ice recipes by reducing everything to percentage of sugar by weight in a mixture. If the total weight of a mixture is 100 grams and it has 20 grams of sugar, it is said to be a 20 percent solution. McGee figured out that you need between 25 and 35 percent sugar in a mixture to have a scoopable ice. Below 25 percent sugar an ice from the freezer is too hard to scoop, and above 35 percent it starts to get mushy with puddles.

It would be great if all you had to do was to add so many grams of sugar for so many grams of mix. Unfortunately, ices are not that simple. Not only do you have the sugar that you are adding; you've got sugars in the fruit, too—fruits like cherries and grapes have twice the percentage of sugar that fruits like raspberries or strawberries have. To complicate matters further, a little lemon juice is usually added to an ice to cut some of the sweetness. Just as fruits vary greatly in sugar content, fruits vary in acidity, too. So, unfortunately, some fruit ices need more sugar than others, and some need more lemon juice than others.

The single best reference that I have found on ices is a chapter in McGee's *The Curious Cook*. He took into account the sweetness and acidity of each fruit and figured out the amounts of sugar and lemon juice for scoopable ices for several different types of ices for thirty different fruits. Any given fruit will have different sweetness according to its ripeness and variety, but McGee's tables will give you a close estimate of the correct amounts of sugar and lemon juice for a scoopable ice for all these different fruits. If you have fought with rock-hard ices, you know what a treasure this table is.

Before I had these tables, I worked with McGee's recommendation of a mixture with a sugar content between 25 and 35 percent of the total weight. I aimed for 30 percent. For example, to design a 2-cup sorbet the total weight is the weight of 2 cups of liquid (16 ounces) plus the weight of the sugar. The weight of the sugar needs to be 30 percent of the total weight, so my formula is:

$$x \text{ (weight of the sugar)} = 30 \text{ percent of the total weight}$$
$$(16 + x)$$
$$x = 0.3\,(16 + x) = 4.8 + 0.3\,x$$
$$0.7\,x = 4.8, \text{ so}$$
$$x = \text{approximately 7 ounces}$$

Since a cup of sugar weighs about 7 ounces, this would be 1 cup of sugar.

The appropriate amount of sugar for granitas, which you want to be a firm crushed ice, is between 15 and 20 percent. This is approximately half the amount used for a sorbet. For the Espresso Granita (page 449), I started with $1/2$ cup sugar for 2 cups espresso and worked from there. I ended up reducing the sugar since I used Kahlúa, whose alcohol also lowers the freezing point.

Lemon juice The amount of lemon juice is partly a matter of personal preference, but here are some guidelines. The tartness (acidity) of the fruit itself makes it easy to select a good-tasting amount of lemon juice. Very tart fruits like lemons, limes, cranberries, and sometimes grapefruit, kiwifruit, and passion fruit need diluting with water. Tart fruits like berries need no lemon juice at all. Moderately acidic fruits like oranges and strawberries need about 1 tablespoon lemon juice per cup of fruit puree, while fruits with low acid content like apples, avocados, bananas, cherries, figs, grapes, guava, and pears need up to 2 tablespoons lemon juice per cup of puree.

Another ingredient that can add tartness is con-centrated pomegranate juice, sometimes called *pomegranate molasses*. This is available as a syrup or paste in many Middle Eastern grocery stores (see Sources—Pomegranate molasses, page 478). Diane Postro Mastro, a restaurateur who serves many different fruit granitas, says this dark red ingredient adds color, flavor, and tartness and is ideal for many fruit granitas.

Pectin Pectins are huge molecules that grab water and form soft gels. They occur naturally in fruits, and commercial pectin concentrates are available in grocery stores. Pectin, whether from fruit alone or from concentrate, is what makes jellies and jams set. Pectin from the fruit itself or from fruit preserves produces a wonderful, smooth, creamy ice with fine crystals.

Pectin keeps ice crystals small in two ways. First, just because of the large size of its molecules, pectin gets in the way when ice crystals try to form and enlarge. Second, if a lot of pectin is present when small ice crystals melt, the pectin simply holds the water in a soft gel. The small crystals melt when an ice is exposed to slightly higher temperatures (when you open the freezer door, for example). Ordinarily, when the ice gets cold again, this water freezes onto bigger crystals and enlarges them. This thawing and refreezing process creates the big crystals sometimes found on the surface of ice or ice cream in the freezer. But beware when using pectin concentrate. With too much pectin, the ice or ice cream will set like jelly and become a nonmelting solid. This destroys part of the sensuousness of ice cream—the sensation of cold melting to nothingness.

The addition of preserves is a great idea. They contain good concentrated flavor plus a small amount of pectin for creaminess. Preserves can be swapped tablespoon for tablespoon with sugar.

Honey Honey, like sugar, lowers the freezing point of water so it can be used in place of some of the sugar. Because honey consists of sugars with smaller molecules than table sugar, it is over twice as effective at lowering the freezing point as table sugar. Try substituting 1 tablespoon honey for $2^1/2$ tablespoons sugar.

Wine and liqueur   Alcohol lowers the freezing point of water, just as sugar does. Using alcohol gives you the advantage of being able to use a little less sugar, or you can leave the sugar the same and get a softer ice. Remember, at freezing temperature you will get very little flavor from wine. Try 3 or 4 tablespoons liqueur or spirits (about 35 percent alcohol) per pint of mix for a softer ice. For a straight wine ice, use 8 to 10 tablespoons of sugar per cup of wine.

Ice-making techniques

You saw in candy making how important it is to stir at the right time—if you stir fudge or fondant while it is hot, you get a grainy mess with big crystals. The point at which you begin stirring and how much and how fast you chill the mixture are just as important for getting fine crystals in ices and ice cream.

When making ices and ice creams, you need to chill the mix well before freezing begins. This way you get a lot of fine ice crystals all at once instead of a few that grow bigger and bigger as the mixture gets colder. Some experts recommend chilling for several hours to get a uniformly cold mixture. Some even recommend chilling ice cream mixes with complex flavors overnight for flavor as well as for uniform chilling.

Types of ice cream makers

Many different types of ice cream makers are on the market, starting with the old-fashioned hand-cranked model with a wooden or plastic tub, a metal container, and a wooden or plastic dasher. The dasher scrapes the frozen mixture from the sides and beats in air. Electric models are also available. The drawback of these machines is that you have to make a fairly large batch (sometimes several quarts) and deal with ice and salt.

With these freezers, the temperature is controlled by the ratio of salt to ice. Generally speaking, the higher the proportion of salt the lower the temperature. Helen Charley, author of *Food Science,* recommends one part salt to eight parts ice by weight, which gives you a brine that freezes reasonably fast.

Mable and Gar Hoffman in their book *Ice Cream* recommend 1 cup rock salt with 6 pounds crushed ice or 1½ cups table salt with 6 pounds whole ice cubes. It has been my experience that you get a smoother texture with slower cooling.

Next are the small sealed-coolant ice cream makers, both hand-cranked and electric. With these, you put the container in the freezer overnight, then fill it with a cold mix and crank by hand or automatically. Finally, there are the more expensive refrigerated units. With these, you pour in the chilled mix, turn the switch, and the machine cranks and chills automatically. The great advantages of these two types are that you can make small batches of excellent-quality ice cream and do not have to fuss with ice and salt.

Granitas   Granitas are coarse ices, frequently made with fruit, wine, or coffee. They are prepared with infrequent stirring during the freezing process, producing coarse ice crystals. They are usually made simply by placing the mixture in an ice tray, pan, or plastic container in the freezer compartment of a refrigerator and stirring occasionally. Stir several times as ice begins to form around the edges (about every 15 minutes) to prevent separation and ensure good distribution of flavor. If you wait too late to stir, the ingredients can separate so that you get some very large ice crystals. After you have crushed the ice by stirring several times and it is completely frozen in coarse crystals, if you stir and leave it in the freezer for an hour longer the crystals seem to dry for a lovely flaky texture.

Some cooks like to make granitas by freezing the mixture in large flat plastic freezer containers without stirring, then serve by scraping an ice cream scoop or heavy spoon across the surface to shave off the flavored ice.

Granitas can also be made by completely freezing the mixture in ice cube trays and then crushing the cubes in an ice crusher for a coarse texture or processing them in a food processor for a finer texture.

Espresso Granita

MAKES 4 SERVINGS

Imagine a cool breeze and you are in the warm sun in Venice on the magnificent Piazza San Marco at one of the ancient caffès that you read about, Florian's or Quadri. Large crystals of glistening brown ice melt in your mouth. There is the exquisite lingering taste of fine coffee. This granita has a little Kahlúa for a more intense coffee taste and a slightly softer texture. It is excellent served with lightly sweetened whipped cream.

> **what this recipe shows**
>
> To have a mixture that is 15 percent sugar, a good percentage for a firm crushed ice, I would need ½ cup sugar, but since I am adding a little alcohol in the form of Kahlúa, I use ⅓ cup.

2 cups cool brewed espresso or strong coffee

⅓ cup *and* 2 tablespoons sugar (about ½ cup *total*)

2 tablespoons Kahlúa or other coffee liqueur

¼ teaspoon finely grated lemon zest

¼ teaspoon finely grated orange zest

1 cup heavy or whipping cream

1. Stir together espresso, ⅓ cup sugar, Kahlúa, and zest. Refrigerate at least an hour to chill well. Pour into a metal or plastic pan or bowl and place in the freezer. When ice crystals begin to form around the edge, stir well every 15 minutes until completely frozen. Fluff crystals lightly with a fork and leave in freezer to dry for about an hour before serving.

2. In a cold bowl with cold beaters, whip cream to firm peaks. Whisk in 2 tablespoons sugar. Spoon granita into sherbet glasses, top with a generous dollop of whipped cream, and serve immediately (see Note).

Note: *Granitas and sorbets are best made and served the same day.*

Sorbets Sorbets, like granitas, can be simple fruit ices. They make light, refreshing desserts or intermezzos between courses. Sorbets vary in sweetness from very sweet to very light and may contain herbs, wine or liqueur, pectin, egg white, or meringue. They do not contain milk or cream as sherbets do. Sorbets are made in an ice cream maker that stirs constantly to produce fine ice crystals and beat in air to create a smooth, light texture—very different from the coarse granitas. A small amount of jelly or preserves in a sorbet can produce a lovely texture.

Sorbets, like granitas, can be made by freezing the mixture to a slush in the freezer compartment of the refrigerator, then processing in a food processor with two egg whites and returning to the freezer. As soon as the sorbet is frozen again, process one more time and serve immediately or return to the freezer just long enough to firm again. Such a sorbet will not have nearly the fine, light texture of one made in an ice cream maker.

Sorbets are at their very best when they highlight the true fresh flavors of fruits.

Ice creams and sherbets

Ice cream can be as simple as that—frozen sweetened cream. Ice creams and sherbets differ from granitas and sorbets in that they contain cream and/or milk or related dairy products such as evaporated milk, condensed milk, dry powdered milk, or sour cream. So we are going to introduce new variables in our challenge to get perfect light texture and smooth creaminess.

Sorbets and granitas contain primarily fruit, sugar, and water. With so few ingredients, we saw how crucial the sugar content was to get the desired texture. Achieving good texture is easier with the addition of cream or other dairy products that control crystal size for smoothness and incorporate and hold air for lightness.

Typically, sherbets are made with fruit. You can think of them as a sorbet with dairy products added. Like sorbets, they may contain wine or liqueurs; pectin, gums, or gelatin; or egg whites or meringue. They usually have more sugar but less cream or dairy products than ice cream.

Ice creams fall into two major categories. Philadelphia (also called *New York, American,* or *plain*) ice cream contains no eggs and is simply sweetened, flavored cream lightened with milk and frozen. The Silky Smooth Ginger Ice Cream (page 452) is an example of the Philadelphia style. Ice creams made with eggs or yolks, usually in the form of a cooked custard base that may or may not have a little starch, are called *French, custard,* or *French custard ice cream.* Creamy Lemon Chip Ice Cream (page 453) and Rich Cappuccino Ice Cream (page 454) are examples of the French style.

The official FDA categories of frozen desserts are (1) ice cream; (2) frozen custard, French ice cream, or French custard ice cream; (3) ice milk; (4) fruit sherbet; and (5) water ices. These match my categories except for ice milk, which simply contains less butterfat than true ice cream, and the water ices, which include granitas and sorbets.

There are many terms for frozen desserts, both historical and current. In particular, there are terms associated with various nations such as *glace* for a rich ice cream in France, *gelato* and *sorbetto* in Italy for a rich ice cream and an ice (*sorbet* in French), respectively. Even in modern usage, terms and definitions are not entirely consistent, so I have stuck closely to those used by the FDA.

Ingredients

All dairy products have several components that affect ice cream and sherbet. Milk proteins trap air when stirred, and the tiny air bubbles lighten the product. Milk solids get in the way when ice crystals are trying to grow, effectively keeping them small. The fats in dairy products coat ice crystals and prevent them from enlarging. Fats also act as a lubricant, making even ice cream with larger crystals feel smooth on the tongue. Milk and dairy products also contain emulsifiers that hold fats and liquids together, giving ice cream additional smoothness and creaminess.

Other components of dairy products lower the freezing point and thus help maintain the necessary unfrozen syrup between ice crystals, just as sugar does. These dairy products include calcium salts and other salts as well as lactose, the sugar in milk. However, be aware that lactose is not as soluble as sucrose (table sugar); if there is too much lactose in cold ice cream, it will crystallize and cause the ice cream to feel sandy with gritty crystals. Different dairy products have varying amounts of these effective ingredients. For example, cream has much more fat and less lactose than evaporated or condensed milk.

Cream is a great multipurpose ingredient in ice cream. Cream is thicker than milk and therefore much better at trapping and holding air when stirred. The high fat content of cream makes it very effective at limiting ice crystal size; it also acts as a lubricant between crystals. Cream has some of the milk solids that limit crystal size, and it has only a limited amount of the lactose (milk sugar) that can cause grittiness.

Milk, on the other hand, has proteins to trap air when stirred but does not have as much fat as cream and cannot hold air permanently as cream can. Milk does not have as much fat as cream to help limit crystal size, but it does have more milk solids, and these milk solids are even more effective than fat at controlling crystal size. Milk, with its lower fat content, does not give ice cream as smooth a mouthfeel as cream does, but it does contribute a lightness that you do not get with cream alone.

Condensed milk, evaporated milk, and powdered dry milk are major contributors of milk solids to help limit crystal size. However, all have a high concentration of lactose; if too much is used, lactose crystals form and the result is an ice cream with a sandy texture.

Ice cream can contain ingredients that we saw in granitas and sorbets such as pectin, gums, gelatin, or liqueurs. Flavoring ingredients run a wide gamut from chocolate and nuts to fruits.

The recipe

The recipe for ice cream can be only sweetened cream, lightened with a little milk. Ice cream and sherbet recipes are not as exacting as those for granitas and sorbets since the dairy products are so effective at limiting crystal size and incorporating air. The milk proteins and cream in ice cream and sherbet trap air bubbles as you stir them, and the fat, proteins, and calcium salts from the dairy products help the sugar control the number and size of the ice crystals.

A small amount of salt (about $1/8$ teaspoon per recipe) is a vital contributor to flavor. It enhances both the perception of sweetness and the flavor of the cream. In French ice creams egg yolks, the great emulsifiers, contribute silky smoothness.

Whether you are making Philadelphia or French ice cream, one step is essential for optimum smoothness if using any milk or half-and-half in the recipe. The milk or half-and-half should be heated to 175°F (79°C), just below scalding. I do not know the exact nature of the changes that this heating causes—perhaps denaturing or partial coagulation of some of the proteins. Whatever it is, the effect is a noticeably smoother texture in the ice cream. It is not necessary to heat heavy or whipping cream, which has very little protein.

Aging

After heating any form of milk (or half-and-half) in the recipe and combining the ingredients, you allow the mixture to "age" for 4 to 12 hours at a temperature between 32° and 40°F (0° and 4.4°C). The usual refrigerator temperature is between 36° and 40°F (2.2° and 4.4°C). This aging before freezing improves the body and texture of the ice cream. W. S. Arbuckle, an authority on commercial ice cream, strongly recommends aging 12 hours for rich ice creams.

Freezing

The temperature of the mix when it goes into the ice cream maker and the temperature of the freezing container itself determine how fast freezing takes place. With modern ice cream makers that have their own refrigeration units, and even with those in which you freeze the coolant overnight, the only control that you have is the temperature of the mix when it goes into the ice cream maker. A starting temperature between 27° and 35°F (−2.8° and 1.7°C) works well. Below the lower temperature the mixture will freeze immediately, and you will get large pieces of ice. Above 45°F (7°C), you run the risk of the cream's being churned to butter.

With ice cream making using ice and salt for cooling, you must take care that you have the correct ratio of salt to ice (see page 448 for recommended amounts). With too much salt, the mixture can get too cold.

In addition to the correct temperature, with a hand-cranked ice cream maker you are concerned with the cranking speed. Initially, maintaining the mixture in steady, constant motion keeps the temperature the same throughout. The minute freezing begins, faster motion is needed to keep the crystals small and to beat air into the mixture.

Taking all this into consideration, if you are using a hand-cranked machine, whether with ice and salt or with a frozen cylinder, turn slowly at first to keep the temperature even throughout the mixture as it cools. Then, when it starts to thicken or freeze, turn the crank rapidly. Eventually the ice cream becomes too thick to turn. At this point, remove the dasher and either return the cylinder to the refrigerator freezer or pack the sides and top of the ice cream maker with ice and salt. In either case, let the ice cream stand and harden.

at a *glance*
Ice cream making

What to Do	Why
Include a small amount of salt in the recipe.	Salt enhances the flavor of other ingredients.
If ice cream is sandy or gritty on the tongue, reduce amount of evaporated milk, dry powdered milk, or condensed milk in the recipe.	All have high lactose content, and lactose crystals form.
Heat any milk or half-and-half to 175°F (79°C).	For a smoother-textured ice cream.
Let ice cream mixture "age" for 4 to 12 hours at 32° to 40°F (0° to 4°C) before freezing.	To improve body and texture of ice cream.
Have the mixture between 27° and 35°F (−2.8° and 1.7°C) when it goes into the ice cream maker.	To prevent mixture from freezing too fast, which will produce large ice crystals and not incorporate air.
With hand-cranked ice cream makers, crank slowly at first.	To keep temperature the same throughout.
With hand-cranked ice cream makers, crank faster when thickening and freezing begin.	To keep crystals small and beat air into the mixture.

Silky Smooth Ginger Ice Cream

MAKES ABOUT 1 QUART

Susan Mack and Andy Armstrong heroically tested over thirty ice cream recipes for me. In searching for "something different," Susan came up with ginger preserves, which give this typical Philadelphia-style ice cream a delightful freshness.

what this recipe shows

Half-and-half heated above 175°F (79°C) makes a smoother-textured ice cream.

A small amount of salt enhances the flavors of the ingredients.

Preserves add a little pectin for improved texture.

Chilling and "aging" the mixture for 4 hours between 27° and 35°F (− 2.8° and 1.7°C) enable it to freeze at the proper rate for smooth ice cream and provide better body and texture.

2 cups half-and-half

1 cup heavy or whipping cream

1 cup sugar

1 teaspoon pure vanilla extract

$1/8$ teaspoon salt

3 tablespoons ginger preserves or

$1/4$ cup finely chopped candied ginger

Heat the half-and-half and cream just below a boil in a medium-size heavy saucepan. Remove from the heat. Stir in the sugar, vanilla, salt, and preserves. Refrigerate for 4 to 12 hours, then place in the refrigerator freezer for 5 minutes. Freeze in an ice cream maker following the manufacturer's directions.

Creamy Lemon Chip Ice Cream

MAKES ABOUT 1½ QUARTS

Starting with one of my basic French-type ice cream recipes, Andy Armstrong created this creamy concoction with sparkling bits of crushed lemon drops and grated fresh lemon zest.

what this recipe shows

Heating half-and-half above 175°F (79°C) produces a smoother-textured ice cream.

Emulsifiers in egg yolks give the ice cream a smooth, creamy mouthfeel.

A small amount of salt enhances the flavors of the ingredients.

Chilling and "aging" the mixture for 4 hours between 27° and 35°F (–2.8° and 1.7°C) enable it to freeze at the proper rate for smooth ice cream and provide better body and texture.

2 cups sugar

1 cup water

Finely grated zest of 1 lemon

$3/4$ cup fresh lemon juice (4 lemons)

$1/8$ teaspoon salt

1 cup half-and-half

1 cup heavy or whipping cream

3 large egg yolks

$1/2$ cup lemon drop candies, crushed

(see Note)

1. Boil the sugar and water together, stirring constantly, for 4 minutes to get a syrup. Stir in the lemon zest, lemon juice, and salt. Set aside.

2. Heat the half-and-half just below a boil in a heavy medium saucepan. Combine half-and-half, flavored syrup, cream, and egg yolks in the top of a double boiler. Cook, stirring constantly, over but not touching simmering water until the mixture is thick enough to coat a spoon, about 4 minutes.

3. Refrigerate for 4 to 12 hours, then place in the refrigerator freezer for 5 minutes. Stir in the crushed lemon drops. Freeze in an ice cream maker following the manufacturer's directions.

Note: *The lemon drops can be crushed by putting them in a freezer-type zip-top plastic bag and hitting them with a hammer or a meat pounder. They should be pounded into fairly fine pieces.*

Rich Cappuccino Ice Cream

MAKES ABOUT 1½ QUARTS

I loved the late Bill Neal's rich espresso ice cream at La Residence restaurant in Chapel Hill, North Carolina, and patterned this ice cream after his.

what this recipe shows

Heating half-and-half above 175°F (79°C) makes a smoother-textured ice cream.

Emulsifiers in egg yolks give the ice cream a smooth, creamy mouthfeel.

A small amount of salt enhances the flavors of the ingredients.

Chilling and "aging" the mixture for 4 hours between 27° and 35°F (−2.8° and 1.7°C) enable it to freeze at the proper rate for smooth ice cream and provide better body and texture.

1½ cups half-and-half	2 cups heavy or whipping cream
2 tablespoons instant cappuccino powder	1 tablespoon pure vanilla extract
8 large egg yolks	⅛ teaspoon salt
1 tablespoon water	Finely grated zest of ½ lemon
¾ cup sugar	¼ cup coarsely cracked espresso coffee beans

1. Bring the half-and-half and instant cappuccino nearly to a boil in a medium-size heavy saucepan. Remove from the heat and set aside.

2. Whisk together the egg yolks, water, and sugar in a medium saucepan. Whisk the hot half-and-half mixture into the egg yolk–sugar mixture. Return the mixture to medium heat and heat, stirring constantly, until fine bubbles appear around the edge or a wisp of steam comes from the mixture. Remove from the heat and stir in the remaining ingredients.

3. Refrigerate for 4 to 12 hours, then place in the refrigerator freezer for 5 minutes. Freeze in an ice cream maker following the manufacturer's directions.

Low-fat and fat-free ice creams The new commercial fat substitutes and compounds like maltodextrins are very effective in producing a rich, creamy mouthfeel in cold products. The home cook has limited access to these compounds by using ingredients that contain them. Nonfat sour cream, for example, can give you a sensationally creamy ice cream like the Low-Fat Lemon-Ginger Ice Cream (page 184), with literally no fat. When you taste this homemade nonfat ice cream, you will believe that you are eating one of the high-fat, super-premium brands.

Whipping cream

In the making of ice cream, cream is a major ingredient that coats crystals to keep them small, contributing smoothness to the texture. Cream also aids in holding air. Whipping cream does not involve crystals, but it is an integral part of many desserts and is used in recipes in this chapter.

When air is beaten into cold heavy cream, the air bubbles formed are coated with a water film filled with fat droplets. As more and more air bubbles are whipped into the cream, the film around the bubbles becomes thinner and the fat droplets, robbed of their film and coating, touch each other and stick together. The whipped cream gets firmer. Eventually, enough air is beaten into the cream that all the air bubbles are lined with fat droplets stuck to each other—the cream is firm and ready to use.

Cold fat droplets that are stuck to each other make whipped cream firm. What happens when the fat droplets get warm? They melt, which is exactly what happens to whipped cream when it gets warm.

The big factor in whipping cream is temperature. The cream should start out cold and stay cold, which you accomplish by chilling the bowl and beaters. Always place the beater and bowl in the freezer for 15 minutes before whipping cream.

Even if you have a cold bowl, beaters, and cream, if you beat 90°F (32°C) air from a hot kitchen into your cream, it may fail to whip. This is why pastry chefs have a different kitchen or try to work during different hours from regular commercial kitchen operations. Their cream needs a cold kitchen to whip, their pastry and puff pastry depend on cold layers of unmelted fat for flakiness, and their chocolate work also requires a cold room. So make sure your kitchen is at least cool when you're whipping cream.

Another major factor in producing firmly whipped cream is the amount of fat in the cream. The fat droplets lining air bubbles are what hold whipped cream up, so you need a good supply of fat. Half-and-half has 10 to 18 percent fat, light (coffee) cream has 18 to 30 percent, whipping cream 30 to 36 percent, and heavy cream has a minimum of 36 percent. For whipping you need at least 30 percent fat. Heavy cream will whip much faster and to a firmer consistency than any other cream.

Three other factors affect whipped cream. First, there is an enzyme in milk that encourages fat globules to clump together. Unfortunately, heat destroys this enzyme, so pasteurized and ultrapasteurized creams do not whip as readily as raw cream. Raw cream is rarely available; most cream in the United States today is ultrapasteurized so whipping time is longer than when raw cream could be had. Second, large fat globules clump together more easily than small fat globules; therefore homogenized milk or cream whose fat has been broken into tiny globules does not whip as well as unhomogenized. For this reason, whipping cream is not homogenized. Finally, the breed of cow matters. Jerseys and Guernseys produce milk with large fat globules, and Holsteins produce milk with smaller fat globules—not that we can exercise much choice over that.

Fats absorb odors easily, so don't store whipped cream in a refrigerator with strong-smelling foods. Cover whipped cream tightly to protect it from odors when storing it in the refrigerator.

Liquid drains out of whipped cream as it stands, but you can whisk the cream back together with a few strokes. If you want to keep the cream very firm, place a piece of cheesecloth over a strainer, pile in the whipped cream, then cover the strainer and place it over a clean bowl. Refrigerate the cream until ready to use and simply discard the liquid that has dripped out.

Whipped cream and gelatin

Packaged stabilizers for whipped cream (Whipit is one trade name) are available in some grocery stores with the baking supplies. Most are some form of modified starch or gum. Or you can dissolve a tiny amount of gelatin in water and add it to the cream near the end of whipping to get firmer whipped cream that holds up well in hot weather.

To stabilize whipped cream, look for packaged modified starch, gums, or gelatins: Whipit is one trade name.

There is a problem, however. To dissolve the gelatin, whether in water or cream, you must heat it, and adding something hot to whipped cream can be a disaster. The solution is to sprinkle the gelatin on cold liquid and let it stand until the gelatin has softened, then heat the gelatin just enough to dissolve it (boiling gelatin reduces its ability to gel) and let the gelatin cool to about body temperature before whisking it into the cream.

You can safely incorporate a small amount of slightly warm liquid into a large amount of cold whipped cream without deflating it. If you let the gelatin cool too much, it sets the second it hits the cold cream—before you have a chance to whisk it in. This gives you a splatter-shaped piece of gelatin that you can peel from the top of the cream! Body temperature or slightly above is ideal.

A sneaky way to add a tiny amount of gelatin to whipped cream is to incorporate a near-melted marshmallow into the cream at the end of whipping. One large marshmallow per cup of cream will hold whipped cream well. Cut large marshmallows into quarters (easiest with greased scissors), place

> A sneaky way to get a tiny amount of gelatin into whipped cream is to add a near-melted marshmallow at the end of whipping.

them in a warm spot (warm toaster oven) or microwave for a few seconds until they are quite soft and almost melted, then whisk them into your cold whipped cream. If you try to whisk a whole, cold marshmallow into whipped cream, the beater may sling the cream-covered marshmallow right into your face!

One final precaution about incorporating whipped cream into other ingredients: If the whipped cream is too warm or is overwhipped, it will turn into butter and a thin liquid. I remember well my first lobster mousse. I was trying to do everything just right, so I whipped my cream nice and firm, then began to whisk in my pureed lobster. I noticed what looked like a tiny yellow streak. Before my eyes, my lovely firm white whipped cream was churned into a thin liquid with small yellow blobs of butter.

Whipped cream that is to be blended with other ingredients should be whipped to form soft to medium peaks—not firm. You must be able to continue to whisk it a little more to thoroughly incorporate the other ingredients.

at a *glance*
Whipped cream

What to Do	Why
Have cream, bowl, and beaters cold and whip cream in a cool room. Keep whipped cream cold at all times.	Fat droplets need to stay cold and firm so that they will stick to each other around the air bubbles.
Whisk only a small amount of a warm ingredient into whipped cream.	Cream has to stay cold so that fat droplets around air bubbles do not melt.
Whip cream that is to be combined with other ingredients to medium-firm peaks only.	The cream needs to be whisked more to combine it with other ingredients. If it is already firm, additional beating will turn it to butter.
Keep whipped cream covered in the refrigerator.	Fats are great holders and carriers of flavor. Uncovered whipped cream will pick up tastes and odors from other items.

Chocolate

In *Theobroma cacao,* the name of the tropical cocoa bean tree, *Theobroma* means "food of the gods." Chocolate is just that to many. Some claim it is an aphrodisiac, some think it is addictive, some find it a stimulant. It does contain small amounts of potent compounds like caffeine and theobromine. Whatever its components, most people think it's pretty wonderful. I remember clearly the day I discovered chocolate ice cream.

This magical elixir started out as a bitter drink in South America. Indian runners carried chocolate to sustain them on long hazardous journeys through the jungle. In more recent times, when Lindbergh flew the Atlantic he took only a few chocolate bars. Chocolate is a concentrated, intensely satisfying food source—something substantial to have with you in the face of great danger.

Such food of the gods deserves a cook's finest efforts. How can you make chocolate coatings as shiny and smooth as possible? How can you get the most intense chocolate flavor? How can you get the creamiest texture? How can you make chocolate behave properly when melting and blending with other ingredients?

Real chocolate and compound chocolate

Real chocolate contains fine cocoa particles (cocoa powder) and cocoa butter. Cocoa powder is the fine powder that you use to make hot chocolate, while cocoa butter is a rich chocolate-tasting fat. Compound chocolate (also called *summer coating* or *compound coating*) contains cocoa particles, but soybean oil, cottonseed oil, or palm kernel oil replaces the cocoa butter. This difference in fat creates a very different product.

Real chocolate

The cocoa butter in real chocolate causes problems in tempering (page 468), but it is a source of joy and one of the true characteristics of chocolate. Cocoa butter has not only irreplaceable flavor but also a sharp melting point. One split second it is a firm solid, then just a few degrees warmer it is a creamy, sensuous liquid. This melting point happens to be right at body temperature. Real chocolate literally melts in your mouth.

Cocoa particles have that dark, rich color and intense taste, and the cocoa butter has flavor and creaminess and gives you that incredible sensation as you feel it go from an aromatic solid to a satiny rich liquid in your mouth. For real chocolate, these two components are irreplaceable.

Natural cocoa beans are fermented, dried, stripped of their hulls, and roasted. Roasting develops flavor and aroma and loosens the bean shells so they can be removed more easily. The beans are crushed lightly and the shells and germ are removed, leaving broken particles of the main portion of the beans, or *nibs* as they are called. When the nibs are crushed, this mixture of unrefined cocoa particles in cocoa butter is called *chocolate liquor.*

In processing fine chocolate, the cocoa particles and cocoa butter are separated, refined, and recombined. The cocoa particles are ground very fine and smooth, an emulsifier is added, and the refined cocoa butter is added back. This mixture is then "conched" (kneaded) for up to 72 hours to further smooth and blend all particles, creating the creaminess of fine chocolates.

The flavor of a chocolate is influenced not only by the variety of the tree but also by the particular processes of fermenting and roasting. The same cocoa beans roasted in different ways produce differences in taste.

Types of real chocolate   There are five kinds of real chocolate and two kinds of cocoa on the market.

- Unsweetened chocolate (bitter or baking chocolate). In refining, the cocoa particles are ground very fine and smooth, and an emulsifier like lecithin may be added along with vanilla or artificial vanillin but no sugar.
- Semisweet (bittersweet) chocolate. The refined chocolate has sugar added along with emulsifiers

and vanilla or vanillin. The amount of sugar added varies from one processor to another.

- Couverture. This is very-high-quality chocolate that has a larger amount of cocoa butter than semisweet so that it is more free-flowing when melted. It is perfect for coating candy centers.
- Milk chocolate. In addition to sugar, this chocolate contains cream or dairy products.
- White chocolate. White chocolate, which until recently was not allowed to be called *chocolate* in the United States because it contains no chocolate liquor, contains cocoa butter, sugar, dairy products, and flavorings. Brand names include Alpine White and Narcissus. It is sometimes called *white confectionery.*

  Almond bark and many other white coatings that can now be called *white chocolate* do not even contain cocoa butter. They are made with palm kernel oil (lauric acid type) and are flavored to taste something like chocolate. They contain neither cocoa particles nor cocoa butter.
- Cocoa. Chocolate drinks were originally made by boiling chocolate nibs. Since these nibs are about 50 percent cocoa butter, the beverage had a very objectionable thick layer of fat floating on top. In 1828, Conrad van Houten developed hydraulic pressing of the nibs to remove most of the cocoa butter. This pressed cake of cocoa particles with 10 to 35 percent of cocoa butter is finely ground and sieved to produce cocoa powder. Natural cocoa is slightly acidic with a pH of about 5.5.
- Dutch process cocoa. About the same time that van Houten developed his process for making cocoa powder, he also began treating the beans, nibs, liquor, or powder with an alkaline solution. This improves color and flavor and produces a slight physical swelling of the cocoa particles as well as neutralizing free acids. Dutch process cocoa is neutral or slightly alkaline with a pH of about 7 or 8. The darker color, the milder flavor, and the lower acidity make Dutch process cocoa a favorite with many pastry makers.

Compound chocolate

When chocolates are made with fats other than cocoa butter, they look a lot like chocolate, their taste is similar to chocolate, but they don't necessarily feel or act like chocolate. They may not have the shine of real chocolate or melt in your mouth. You may have to chew them.

Major chocolate companies produce full lines of these compound chocolates, which are called *compound coatings* or *summer coatings*. Many are of excellent quality and often contain some cocoa butter. They are simpler to use than real chocolate, and some cooks feel that these easier-to-use chocolates have brought candy making to the public. They are like real chocolate in that you must melt them with tender, loving care and not get them over 120°F (49°C), and you must keep them dry. However, instead of having to temper them as you do real chocolate, you can use them as a coating by simply melting.

There are two major types of these coating compounds: those that are compatible with cocoa butter and real chocolate and those that are not. So when using these coatings, you need to read the labels care-

---

COCOA AND INTENSE FLAVOR

Many chocolate experts feel that you get a more concentrated chocolate flavor by using cocoa instead of or in addition to chocolate. You may notice the use of boiling water in some recipes with cocoa. Putting boiling water on cocoa releases an even more intense chocolate taste. You can convert recipes to this procedure by adding a little boiling water to cocoa before combining it with other ingredients. Reduce the amount of other liquid in the recipe by the amount of the water.

fully. Elaine González explains in her book *Chocolate Artistry* that one type of coating is made by substituting soybean or cottonseed oil for cocoa butter. These coatings are compatible with cocoa butter. In fact, many of them are flavored with chocolate liquor, which contains cocoa butter. These coatings are well suited for coating (enrobing) but are not as good for molding.

The other type of coating uses palm kernel oil (which contains fats with lauric acid) to replace cocoa butter. You cannot combine these lauric acid fats with cocoa butter, or you will get graying—called *bloom*—on the palm kernel chocolate as soon as it comes into direct contact with real chocolate.

If palm kernel oil is an ingredient, you know that you will have trouble using this with real chocolate. González does offer a solution, however: To pipe a coating containing palm kernel oil onto real chocolate, bring the coating to be piped down to the cool end of the working range (90°F/32°C) so that it will set right away. This will help prevent bloom on the palm kernel oil coating. Of course, you want to avoid situations like this, but sometimes you may be faced with them.

How chocolates differ

The same type of chocolate, such as semisweet, can taste different depending on the brand. There is also a big difference in types even when they sound similar. Semisweet and bittersweet chocolates can be quite distinct, depending on how much sugar they contain.

Lindt Courante contains 2 teaspoons of sugar per ounce of chocolate, while Tobler Bittersweet, Lindt Surfin, and most American semisweet chocolates contain $3^{1}/_{2}$ teaspoons of sugar per ounce of chocolate. In addition to differences in sugar content, which can make a considerable difference in taste, semisweet or bittersweet chocolates with the same sugar content can still taste very different from each other depending on the type of cocoa bean used and the degree of roasting.

If you're a true chocolate lover, you'll want to identify the very best there is. A fun way to do so is to hold a blind chocolate tasting like the one I participated in, where we compared a dozen different semisweet and a dozen different milk chocolates. My guess is you'll find plenty of willing friends to join in. See the box below for details.

SEARCHING FOR THE BEST: HOLDING A CHOCOLATE TASTING

Collect samples of different chocolates, then break off and number a sample of each chocolate for each participant. The sample piece has to be fairly large so your testers can observe how smoothly each chocolate breaks. Give each person a copy of the form on page 460 for grading each sample.

Tell your testers to start filling out the form by commenting on the appearance and aroma of a sample. Then have them break the piece to note whether it broke smoothly. Next have them place a piece in their mouth and let it melt. Did it melt smoothly? Does it have an oily,

creamy, or waxy feel? Finally, have everyone comment on taste and aftertaste.

A blind tasting is a good way to decide for yourselves which is your favorite without being influenced by label and price. You might be surprised by your choice. A good chocolate should be shiny, smell very chocolaty, break smoothly, melt smoothly, and have a rich chocolate taste and aftertaste. Fine chocolates end up scoring well across the whole row of characteristics—and the finest are not always the highest priced.

FORM FOR CHOCOLATE TASTING					
Number	Appearance *(Shiny, Dull)*	Aroma *(Chocolate, Orange)*	Break *(Smooth, Crumbly)*	Melt *(Even, Smooth, No grain, Waxy)*	Taste and Aftertaste *(Chocolate, Vanilla)*
1					
2					
3					
4					
•					
•					
•					

Substituting chocolates in recipes

If you can change the taste of a chocolate dish simply by changing the brand of chocolate, certainly you will get differences by substituting one type of chocolate for another. Still it can be done. Here are some general guidelines.

Irma Rombauer and Marion Becker in *The Joy of Cooking* give several substitutions for chocolates. Rose Levy Beranbaum's suggestions are more detailed. She uses cocoa powder in many cakes and believes that cocoa powder gives a richer, stronger chocolate flavor than chocolate.

CHOCOLATE SUBSTITUTIONS	
To Substitute for 1 Ounce	Use
Unsweetened (Bitter or Baking)	Rombauer—3 tablespoons cocoa and 1 tablespoon fat (butter or margarine). Beranbaum—1 ounce semisweet, and remove 1 tablespoon sugar and $1/3$ teaspoon butter from the recipe.
Semisweet (Bittersweet)	Rombauer—$3/5$ ounce unsweetened and $2^{1}/3$ teaspoons sugar. Beranbaum—1 tablespoon plus $1^{3}/4$ teaspoons cocoa, 1 tablespoon plus $1/2$ teaspoon sugar, and $1^{1}/2$ teaspoons unsalted butter.
Couverture	Beranbaum—1 ounce semisweet and $1/2$ teaspoon cocoa butter.

Working with chocolate

The two great problems when working with chocolate are moisture and excess heat.

Moisture—seizing

The number-one problem when working with chocolate is seizing. This may have happened to you. One moment, you have creamy melted chocolate, and the next you have a solid grainy mass that will not melt. What happened?

Dr. Richard A. Schwartz at the Wilbur Chocolate Company describes it as "the sugar bowl effect." If you dip the damp spoon you just stirred your coffee with into the sugar bowl, the moisture on the spoon creates little hard, grainy lumps. This happens because the sugar is dry, and when you add a few drops of moisture, the moisture causes the sugar crystals to glue together. On the other hand, if you pour a cup of boiling water into the sugar bowl, it dissolves the sugar—no lumps.

Seizing is similar to the damp spoon in the sugar bowl. In chocolate you have very fine, very dry particles in a rich fat, cocoa butter. The tiniest amount of moisture causes these dry particles to stick together. It does no good at all to heat it. The mass will not melt.

What can you do? Remember the sugar bowl. If you add enough water to wet all the particles, they will no longer stick together. This is the exact solution. This may seem all wrong, since it was water that got you into this mess in the first place. But it was not water as such; it was the small amount of water on dry particles that made them stick together. More water will moisten all of the particles, and they will come unstuck.

The best way to become a master of seizing is to do it on purpose and see for yourself. In a small saucepan, melt an ounce or two of real chocolate over very low heat, stirring constantly. When the chocolate has melted, remove it from the heat. Spoon a dab onto a small plate to compare it with later samples. Now stir $1/4$ teaspoon water into the melted chocolate. As you stir, you can see the chocolate become dull and tighten, turning into a solid, dull, grainy mass.

Spoon a little of this onto your sample plate. Next add 2 tablespoons of warm water to the seized chocolate and mash and whisk it in until the chocolate becomes smooth enough to whisk well. Shiny melted chocolate again! Spoon a little of this onto your sample plate. You will see that the original melted chocolate and the last sample with plenty of water are both smooth and shiny, although the water-added one is thinner and has a slightly different color.

What does this mean in cooking? Unfortunately, seizing has not been understood, and many chocolate recipes do not have enough liquid in them to prevent seizing. This is especially true of some truffle recipes. The rule of thumb that I have always used is that you must have a minimum of 1 tablespoon of liquid for every 2 ounces of chocolate to prevent seizing. This is a bare minimum, and some chocolates require a little more.

> *Rule of thumb:* Use at least 1 tablespoon liquid to 2 ounces chocolate to prevent seizing.

Adding any liquid to melted chocolate is risky. For a second, you have the dangerous situation of a little liquid and a lot of dry chocolate. It is much safer to whisk melted chocolate into other ingredients. This works well if there is a fair amount of liquid in the recipe and the ingredients that the melted chocolate is whisked into are not too cold.

Adding melted chocolate to ice-cold ingredients creates a new problem. The cold makes the fat in chocolate, the cocoa butter, harden instantly. For example, if cool melted chocolate is drizzled into ice cold whipped cream, tiny flecks of solid chocolate may form (a chocolate chip mousse!) rather than an even blend of chocolate and cream.

The fail-safe way to combine chocolate with other recipe ingredients is to melt the chocolate with any liquid or butter in the recipe. As long as you have at least 1 tablespoon of liquid for every 2 ounces of chocolate, you avoid all risk of seizing.

> **The fail-safe way to combine chocolate with other ingredients is to melt the chocolate with liquid or butter in the recipe (at least 1 tablespoon liquid to 2 ounces chocolate).**

If you notice, this is how many chocolate experts do it. To get a glassy smooth chocolate icing, Alice Medrich melts chocolate, butter (which contains about 18 percent water), and corn syrup together. For some brownies, Maida Heatter melts chocolate and butter together before combining them with other ingredients. For a chocolate cream glaze, Rose Levy Beranbaum pours hot cream over grated chocolate, lets it stand, then stirs, effectively melting the chocolate with the cream.

Temperature

Cocoa butter melts in the low 90s°F (30s°C). Chocolate separates into melted cocoa butter, a pale golden liquid, and burned, blackened cocoa particles somewhere below 130°F (54°C). This is an irreversible situation. You can drain off and save the cocoa butter, but the chocolate itself is gone forever.

As you can see, melting chocolate is a delicate procedure. Chocolate contains not only cocoa butter but also trace amounts of a number of other natural fats that melt at different temperatures. This does not cause a problem since even those with the highest melting points melt by about 118°F (48°C), which is not very hot. Remember, your body is at 98.6°F (37°C), so chocolate's final melting point is only a little warmer than you are. To completely melt chocolate, you need to get it to the 118°-to-120°F (48°-to-49°C) range, but no higher than 120°F (49°C). Just above that temperature it starts to separate.

Chocolate must be melted at a low temperature with constant stirring to keep the temperature even throughout. If there are big pieces, the chocolate that's already melted may get too hot and burn before the other pieces melt. For smooth, even melting of chocolate, break it into 1-inch or smaller pieces and process in a food processor with the steel knife until finely chopped.

Because chocolate melts at a low temperature, many recipes suggest melting chocolate in a double boiler. Since chocolate's worst enemy is a tiny bit of water, even steam, I really don't like to use a double boiler. There is too much risk that a little steam will get to the chocolate. You can melt chocolate over hot water, not touching the water and never boiling, and still the escaping steam can ruin the chocolate. Or you can melt it in a metal bowl placed in a smaller bowl of warm (120°F/49°C) water, taking great care that not one drop of water gets to the chocolate.

Pastry chefs and candy makers, who have to melt large quantities of chocolate, use many techniques. Some have a gas oven with a pilot light that heats the turned-off oven to an ideal temperature for melting chocolate. They place finely chopped chocolate in a metal bowl in the oven for several hours, stirring occasionally. Some use the lowest setting on a hot tray, electric skillet, or even a heating pad. If you want to use this method, you should check by placing a bowl of water on the tray or skillet first and measuring the temperature. Some are too hot even on the lowest setting. In that case you need to put a towel or other insulator between your bowl of chocolate and the heat source.

Many experts like to melt chocolate in the microwave. Alice Medrich, author of *Cocolat*, recommends melting semisweet and bittersweet chocolate at 50 percent or Medium power, milk and white chocolate at 30 percent. Microwave ovens vary, but 6 ounces of chopped semisweet chocolate generally take about 3 minutes to melt at 50 percent power.

When some chocolates melt in the microwave, they hold their form even though they are melted. For example, you can heat chocolate morsels or chips in the microwave above their melting temperature, and they will remain intact, so they don't appear to be melted. When you touch them, though, it turns out that they are completely melted and the chocolate may be burned. Since you can't tell by looking, to prevent burning be sure to stir several times while they are heating.

at a glance
Chocolate

What to Do	Why
Chop chocolate into small pieces before melting it.	For quicker and more even melting since chocolate may burn before large lumps melt.
Melt chocolate over very low heat, warm, not boiling, water, or in the microwave.	If chocolate is heated beyond 120°F (49°C), it separates and burns.
Avoid getting even a drop of water or steam into melted chocolate.	A small amount of water causes dry particles in chocolate to stick together and seize.
Stir chocolate as it melts.	To keep the temperature even.
When possible, melt chocolate with other liquid ingredients in a recipe.	Adding adequate liquid at the beginning of melting prevents seizing.
Use at least 1 tablespoon of liquid for every 2 ounces of chocolate.	The minimum of liquid necessary to prevent the dry particles from sticking together (seizing).
Add water to seized chocolate to bring it back to a liquid state.	Sufficient liquid wets all the particles so they no longer stick together.
Do not mix cooled melted chocolate into ice-cold ingredients.	Cold ingredients cause the cocoa butter to harden immediately.

In my Smoothest-Ever Truffles (page 464) and The Secret Marquise (page 467), I melted the chocolate with the cream and butter to avoid having to combine melted chocolate with a small amount of other ingredients. Remember, you can't melt a large amount of chocolate with a few tablespoons of butter—you need to have sufficient liquid to prevent seizing. Count on butter's being about 18 percent water and stick close to the rule of thumb of 1 tablespoon water-type liquid for each 2 ounces of chocolate. With the truffle recipe, I am right on the edge. For 16 to 17 ounces of chocolate, I have 4 tablespoons of cream (supplying 3 tablespoons of water), 4 tablespoons of butter (about 1 tablespoon of water), 4 tablespoons of Grand Marnier (about 2 tablespoons of water), and 5 egg yolks (about 3 tablespoons of water). This is a total of about 9 tablespoons of water.

Smoothest-Ever Truffles

MAKES ABOUT THIRTY 1-INCH TRUFFLES

When you bite into one of these, it is so smooth that your tooth marks on the truffle are shiny, satiny smooth like dark glass.

Roasting enhances the flavor of the nuts.

Egg yolks, nature's great emulsifiers, create sensual smoothness.

Diluting the egg yolks with cream and carefully heating them kills salmonella.

This recipe has the 9 tablespoons of water-type liquid (in cream, butter, egg yolks, and Grand Marnier) to prevent 17 ounces of chocolate from seizing.

Finely chopping the chocolate makes for smooth, even melting.

2 cups pecan pieces

3 tablespoons *and* 4 tablespoons butter
(7 tablespoons *total*)

¼ teaspoon salt

10 to 11 ounces good-quality semisweet chocolate, such as Lindt or Tobler, or 4 bars (2.6 ounces each) Hershey's King Size Special Dark, broken coarsely into 1-inch pieces

6 ounces milk chocolate, broken coarsely into 1-inch pieces

5 large egg yolks (see Note)

¼ cup heavy cream

¼ cup liqueur (Chambord, amaretto, or Grand Marnier)

1. Preheat the oven to 350°F (177°C).

2. Spread the pecans on a large baking sheet and roast until lightly browned, about 10 to 12 minutes. Remove from the oven and stir in 3 tablespoons butter and the salt while pecans are hot. Let stand until cool to the touch. Chop the pecans in 2 batches in the food processor with the steel knife with quick on/off pulses until finely chopped. Set aside.

3. Combine the semisweet chocolate and milk chocolate in a food processor with the steel knife and process until finely chopped. Set aside.

4. Heat the egg yolks and cream in an 8-inch heavy skillet over low heat, stirring constantly with a fork or spatula flat against the bottom of the pan. The split second that you feel thickening, remove the skillet from the heat and keep stirring. Add 4 tablespoons butter and stir in well. Add the chocolate and liqueur. Stir constantly until chocolate just melts. Continue to stir for 1 minute.

5. Place the skillet in the refrigerator to set. When the chocolate is partly firm, spoon up 1-inch balls and roll each in chopped pecans. Keep the truffles covered and refrigerated. Serve in gold fluted candy cups.

Note: *Because of the threat of salmonella, the egg yolks and cream are heated together in this recipe to kill possible bacteria. (See page 195 for more details.) If you use pasteurized yolks or are confident about the safety of your eggs, you can mix together the yolks and cream and heat until just hot enough to melt the chocolate as in Step 4.*

Melting chocolate with liquid even helps with the risky procedure of combining melted chocolate and cold whipped cream. I learned this from Jim Stacy, founder of Tarts Bakery in San Francisco. When you melt chocolate with water, the chocolate is thinned down and can be whisked into the cream much faster. Also, there are emulsifiers in the chocolate, so the cocoa butter may be slightly combined (emulsified) with the water, which makes it easier for the fat to blend in with the cream.

I took the safe way out with the white chocolate and whipped cream in the White Chocolate Mousse with Raspberry Sauce. I melted the white chocolate carefully with water to prevent seizing, cooled it a little until it was about body temperature, neither warm nor cool to the touch, then whisked it together with the cold whipped cream. You need to move fast once you've whisked in the chocolate. Get the mousse immediately into its mold. The cocoa butter and fats in the chocolate will harden in the cold mousse. (Chocolate and whipped cream are also combined in the mousse on page 471.)

White Chocolate Mousse with Raspberry Sauce

MAKES 10 TO 12 SERVINGS

This creamy white mousse in a pool of vivid red raspberry sauce is a beautiful holiday dessert, and the taste of white chocolate and whipped cream with raspberries and Chambord is hard to beat. I usually make a few dark chocolate leaves to garnish the top.

what this recipe shows

A small amount of gelatin ensures a good set to the mousse without any rubberiness.

Melting the white chocolate with water and rum prevents its seizing and makes it easier to mix with the whipped cream.

Cooling the white chocolate–gelatin mixture to about body temperature before folding it into the whipped cream allows it to blend well without deflating the cold whipped cream.

Nonstick cooking spray

1½ pounds good-quality white chocolate, finely chopped

¾ cup hot water

1 teaspoon good-quality dark Jamaican rum, preferably Myers's

1 envelope unflavored gelatin

¼ cup *and* 2¼ cups heavy or whipping cream (2½ cups *total*)

Raspberry Sauce (page 466)

3 dark chocolate leaves (page 471)

1 cup fresh raspberries

1. Spray a 6-cup brioche mold with nonstick cooking spray and line with parchment paper (see Note). Set aside.

continued

2. Pour 1½ inches of water into a skillet and bring to a boil. Turn off heat. Stir together the white chocolate, hot water, and rum in a medium-size heavy saucepan. Place the saucepan in the skillet of hot water and stir chocolate constantly until it melts. When melted, place the pan in a cool place to cool down to body temperature. Chocolate is ready to be folded into the whipped cream when it no longer feels warm to touch.

3. Sprinkle gelatin over ¼ cup cream in a very small saucepan or a large metal measuring cup. Let soften several minutes. Heat over very low heat just to melt gelatin. Do not boil.

4. Place a mixer bowl, beaters, and cream in the freezer for 5 minutes. Whip 2¼ cups cream to medium-firm peaks. Stir the dissolved gelatin into the melted chocolate, then fold chocolate and whipped cream together and spoon into prepared mold to chill. Chill several hours or overnight.

5. Unmold the mousse onto a pedestal cake dish with a rim. Carefully spoon the Raspberry Sauce onto the dish around the mousse. Garnish the top with three dark chocolate leaves and three perfect raspberries to resemble holly leaves with berries in the center. Sprinkle the remaining raspberries into the sauce at the base of the mousse.

Note: *Molded desserts are easier to unmold if you spray the mold with nonstick cooking spray. Cut a narrow strip of parchment paper (wax paper is not strong enough) about 1 inch wide and long enough to go down one side of the mold, across the bottom, and up the other side with a couple of inches to spare on each side. Put it in the mold. Cut a circle of parchment the size of the bottom of the mold, place it over the strip, and fill the mold. When ready to unmold, turn the mold upside down and pull gently on both ends of the strip. The mold comes right out. Remove and discard the parchment circle and strip.*

Raspberry Sauce

2 packages (10 ounces each) frozen
  raspberries

2 tablespoons arrowroot

3 tablespoons Chambord or other
  raspberry liqueur

Defrost the raspberries in a strainer over a medium saucepan, reserving all the juice. Stir the arrowroot into the juice and heat, stirring constantly, until it thickens. Remove from the heat and stir in the raspberries and Chambord. Refrigerate to chill.

My Secret Marquise applies the same principles as my truffle recipe—egg yolks for a smooth creamy texture and enough liquid to prevent seizing. But the marquise has a food-sleuth story attached to it. On an eat-our-way-through-France trip, Rose Beranbaum and I had dinner at Alain Chapel's restaurant in Mionnay and went wild over his chocolate marquise. It was a small wedge of sensationally smooth, dark, rich chocolate in a small, snow-white puddle of sweetened cream, whipped only to the consistency of a sauce.

Rose was writing an article for *The New York Times* and hoped that the chef might let her add his recipe with a story about his restaurant. Unfortunately, Chapel was out of town. The next day we told the charming lady who served us coffee how we had loved the marquise and hoped that the chef might share his recipe. She exclaimed, "C'est à moi!" ("It is mine!") She went on to say that it had been in the Rothschild family for years and that one of the Rothschilds had given it to her grandmother.

Now we were sure we would get the recipe, but we were disappointed. Since the chef was using the recipe the lady could not give it to anyone. All that she would divulge was that it was simple and did not require baking.

Rose later wrote to Chef Chapel requesting the recipe. The chef wrote back that it had been a secret recipe in *his* family for generations and he could not, of course, give it out!

Whose secret family recipe it is may be in question, but that Chapel would never give it out was a certainty. I learned that many had tried unsuccessfully to pry the recipe for the "secret marquise" from Chapel. Therefore, I was left to create my own. The texture of Chapel's was so smooth and so creamy, I knew that it had to contain egg yolks. Adding the clue that it was simple and not baked, I came up with my own Secret Marquise.

The Secret Marquise

MAKES 6 TO 8 SERVINGS

This is my family's "secret" marquise recipe.

what this recipe shows

Egg yolks, nature's great emulsifiers, create sensual smoothness.

Diluting egg yolks with cream and carefully heating them kills salmonella.

Finely chopping the chocolate makes for smooth, even melting.

This recipe has over the 6 tablespoons of water-type liquid required to prevent 12 ounces of chocolate from seizing.

Chilling cream, bowl, and beaters aids in whipping cream well.

7 to 8 ounces good-quality semisweet chocolate, such as Lindt or Tobler, or 3 bars (2.6 ounces each) Hershey's King Size Special Dark, broken coarsely into 1-inch pieces

4 ounces milk chocolate, broken coarsely into 1-inch pieces

5 large egg yolks (see Note)

1/3 cup heavy cream

5 tablespoons butter

Nonstick cooking spray

1 cup heavy or whipping cream

2 tablespoons sugar

1. Combine the semisweet chocolate and milk chocolate in a food processor with the steel knife and process until finely chopped. Set aside.

2. Heat the egg yolks and 1/3 cup heavy cream in a 7- to 8-inch heavy skillet over low heat, stirring constantly with a fork or spatula flat against the bottom of the pan. The split second that you feel thickening, remove the skillet from the heat and keep stirring. Add the butter and stir in well. Add the chocolate and stir constantly until chocolate just melts. Continue to stir for 1 minute.

continued

3. Remove the bottom from a 7- or 8-inch springform pan; you won't need it. Place the ring of the springform on a platter and spray the inside of the ring and the platter well with nonstick cooking spray. Spoon in the chocolate mixture and place in the refrigerator to chill and set firm (about an hour).

4. Place a medium mixing bowl and beaters in the freezer for 5 minutes to chill well. Whip 1 cup cream, not stiff but only to thicken slightly—the consistency of a thick cream sauce—add sugar, and stir in. Cover individual small serving plates with a thin layer of this cream.

5. Wet a towel with hot water and squeeze dry. Wrap warm towel around the springform ring for 20 seconds, then open and remove ring. Heat a knife under hot running water and wipe dry. To serve, cut a small wedge of the marquise while it is quite cold and place in the cream on the serving plate. Run the knife under hot water and dry before making each cut. May be served immediately after removing from refrigerator or cut and held at room temperature up to 30 minutes before serving.

Note: *Because of the threat of salmonella, the egg yolks and cream are heated together in this recipe to kill possible bacteria. (See page 195 for more details.) If you use pasteurized yolks or are confident about the safety of your eggs, you can mix together the yolks and cream and heat until just hot enough to melt the chocolate as in Step 2.*

Tempering chocolate

For most chocolate desserts, all you need to do is melt and blend the chocolate with other ingredients; you saw this technique in both the White Chocolate Mousse (page 465) and The Secret Marquise (page 467). To dip or mold chocolate or make fancy chocolate decorations, like curls and fans, the chocolate should be tempered.

Tempering chocolate is the process of melting and cooling chocolate in such a manner that exactly the right type of cocoa butter crystals form. When melted chocolate cools, the cocoa butter can harden (crystallize) into any one of four different types of crystals. Only one of these—beta crystals—yields a chocolate with a shiny, hard surface. Any of the other three types—gamma, alpha, or beta prime—makes the chocolate dull, streaked, or soft and gooey.

The chocolate that you buy has been tempered, but when you melt it the crystals melt. The right crystals are lost, and the chocolate has to be retempered. The chocolate must first be heated enough to completely melt all the fats: 118° to 120°F (48° to 49°C) for semisweet or bittersweet chocolate, 2°F (1°C) lower for milk chocolate and white chocolate. The

chocolate is cooled down to 80°F (27°C) by stirring constantly and adding unmelted (tempered) chocolate. Finally, the melted chocolate is reheated to 91°F (33°C) for semisweet or bittersweet, 85° to 87°F (29° to 30°C) for milk chocolate and white chocolate. This reheating remelts any of the unwanted types of crystals that formed in cooling. Alpha crystals melt at 70° to 75°F (21° to 24°C), gamma crystals at 63°F (17°C), and beta prime crystals at 81° to 84°F (27° to 29°C). The stable beta crystals do not melt until 95°F (35°C).

Tempering chocolate is not easy. To get a shiny, hard chocolate that does not become gooey at cool room temperature requires patience, attention to detail, and a good thermometer. Everything has to be kept bone dry (it's best to work in a cool, dry room); the chocolate has to be melted and cooled slowly and carefully; it has to be stirred well while cooling; and meticulous attention has to be paid to correct temperature.

There are many different techniques for tempering. I am a big believer in doing whatever works for you. You must get the chocolate to about 118°F (48°C) to be sure that all the fats are melted, but you cannot get it over 120°F (49°C), or you run the risk of having the chocolate separate. In the same way, when you cool the chocolate to about 80° to 81°F (about 27°C) to start crystallization of the beta crystals, be aware that you cannot get it below 77°F (25°C), or you will have to heat it back up to 118°F (48°C) and start over again.

To cool chocolate nearly 40°F (21°C) from 118° to 80°F (48° to 27°C) takes much longer than one might think. Some experts like to keep some of the melted and cooled chocolate at 90° to 92°F (32° to 33°C) and pour the rest on a cool marble slab, where they work it back and forth with a spatula to cool it down to 80°F (27°C). The cooled chocolate is then stirred back into some of the 90°F (32°C) chocolate to reheat it.

I like the technique of adding back finely chopped unmelted chocolate, which is still tempered, up to a fifth the weight of the melted chocolate. This tempered chocolate acts as seed crystals and encourages the right kind of beta crystals to form. Also, the melting of this unmelted chocolate shortens the cooling time. You just have to be careful not to add so much chopped chocolate that it does not all melt.

Some chefs add a lump of unmelted chocolate so that they can remove the lump when the chocolate has cooled enough. You can do both—add some chopped chocolate for fast cooling, then, when that is completely melted, add a lump for the final cooling. Go with whatever method you prefer (working on marble, adding unmelted tempered chocolate, simply stirring, or using a combination of these methods) to cool down the melted chocolate. Whatever procedure you use, be sure to stir constantly.

This is precise temperature work. Ordinary kitchen thermometers are as much as 10°F (5.5°C) off, which will not do at all for tempering. If you do a lot of chocolate tempering, you may want to invest in a very accurate digital thermometer from a laboratory supply company (see Sources—Thermometers, page 477). You can get a mercury thermometer designed specifically for chocolate work with a range of 70° to 125°F (21° to 52°C). These are available from Williams-Sonoma.

If you don't have a very accurate or chocolate thermometer, you may be better off using common sense and your own body temperature, which is about 98.6°F (37°C). A temperature of 118°F (48°C) feels fairly warm but not really hot. The area just above your upper lip is quite sensitive to temperature, so do as professional chocolatiers do: Put a dab of chocolate above your upper lip to gauge the temperature. At 80°F (27°C), the chocolate will feel cool and at 90°F (32°C) barely cool.

To test that you have a good temper, spoon a small dab of tempered chocolate on a plate and refrigerate for 3 minutes. It should be shiny and set. If not, reheat to 118°F (48°C) and start over.

While working with tempered chocolate, you must not let it cool too much, and you should stir it every few minutes. Some people like to keep the chocolate warm with a heat lamp, heating pad, or hot tray. I prefer these dry heat sources to a water bath. Just be sure that the heat source is not too hot. Semisweet and bittersweet chocolate should be kept at 90° to 91°F (about 32°C), milk and white chocolate at 85° to 87°F (about 30°C).

High drama with chocolate curls, fans, leaves, and ribbons Chocolate curls, fans, leaves, and ribbons make showstopping chocolate desserts. These techniques are not difficult to master and give you a lot of mileage for little effort. Some of them require tempered chocolate, but not all.

Chocolate curls, pencil curls, fans, and ruffles Chocolate curls can be simply shaved from a block of chocolate with a vegetable peeler or a melon baller. With this technique, temperature is the key factor. If the chocolate is cold, you will get flakes and small curls. For larger curls, the chocolate should be just a little warm on the surface. Alice Medrich recommends working under a gooseneck lamp with a 60-watt bulb placed about 12 inches above the chocolate to warm the surface evenly without melting the chocolate. Rose Levy Beranbaum suggests using a lamp or leaving the chocolate in an 80°F (27°C) room for a couple of hours or using 3-second bursts in the microwave on the High setting.

To keep the heat of your hand from melting the chocolate as you work, hold the chocolate in a piece of paper towel. With a long piece of chocolate you can even make 1- to 2-inch-long curls with a regular grater using the coarse grate and grating along the long side of the chocolate.

Larger curls, long, thin pencil curls, or wide curls can be made by spreading melted or tempered chocolate on a marble slab or the back of an upside-down baking sheet. Bruce Healy, author of *Mastering the Art of French Pastry* and *The French Cookie Book*, heats an upside-down baking sheet in a preheated 300°F (149°C) oven, then rubs a piece of chocolate back and forth across the hot surface to form a thin layer. He then lets this cool and makes curls, pencils, or fans. This produces shiny chocolate products just as if the chocolate had been tempered.

To make large curls, long, thin pencil curls, or wide curls, let the chocolate cool, then scrape a knife, putty knife, or metal spatula against the chocolate. Use a long straight-edged knife and pull the blade evenly across the chocolate to form long pencil curls. A spoon is good to use to make wide curls. Pull the bowl of the spoon toward you with a firm stroke. Again, the tem-perature of the chocolate is important. If the chocolate is slightly warm, it will roll up beautifully. Many recipes for curls or fans call for melted chocolate. For shinier, firmer curls, use Bruce's quick temper just described or temper the chocolate in the regular way.

Fans and ruffles need all the attention to detail that curls require and then some. For shiny fans and ruffles, tempered chocolate or Bruce's quick temper method is necessary, and the chocolate on the back of the pan needs to be cooled but still slightly warm. For fans, use a small metal spatula, holding it with both hands, one at each end of the blade. Move one hand toward you just a little while pulling the other hand toward you in a wide arc.

Long pencil curls can be piled high, radiating out from the center on the top of a cake. Curls made with the coarse grater are perfect to press around the sides of a cake. Wide curls made with a spoon can be piled high on top or pressed around the sides or both. Curls can be made more dramatic by sprinkling with confectioners' sugar.

Even simple angular chocolate slabs make impressive desserts as in Chocolate Stonehenge Slabs with Cappuccino Mousse.

Chocolate Slabs

Heat a large baking sheet for 10 minutes in a preheated 300°F (149°C) oven. Turn the sheet upside down and with a hot pad hold a large piece of parchment paper against the hot pan. Rub a large lump of chocolate over the parchment to form an even coating as in Bruce Healy's quick temper (above). Let cool in a cool, dry room or place in the refrigerator for a quick set. With a pizza cutter or sharp knife, cut out irregular shapes about 2 inches at the base, 1 inch at the top, and about 3 inches tall.

Chocolate Stonehenge Slabs with Cappuccino Mousse

MAKES ABOUT 10 SERVINGS

This is an easy and very impressive dessert. You have only to make a quick mousse, spoon up individual servings, and press jagged chocolate slabs against the sides of each.

> ### what this recipe shows
>
> Melting the chocolate with an adequate amount of water prevents seizing and makes it easier to blend with the whipped cream.
>
> Cooling the chocolate to body temperature before mixing it with whipped cream allows it to blend well without deflating the cold whipped cream.

$^{1}/_{2}$ cup hot water

2 tablespoons instant cappuccino powder

1 pound milk chocolate, broken or chopped into small pieces

2 cups heavy cream

Chocolate Slabs (page 470)

1. Place a skillet containing $1^{1}/_{2}$ inches of water on the heat and heat water to boiling for a water bath. Turn off heat.

2. Stir together a generous $^{1}/_{2}$ cup hot water and instant cappuccino in a medium-size heavy saucepan. Add milk chocolate and stir. Place saucepan with chocolate in the hot water bath and stir chocolate constantly until melted. The minute the chocolate melts, place it in a cool place to cool down to body temperature. Chocolate is ready to fold into whipped cream when it no longer feels warm to touch.

3. Place bowl, beaters, and cream in the freezer for 5 minutes to chill well. Whip cream in a cool room to medium-firm peaks. By hand, fold together melted chocolate and whipped cream.

4. Spoon about $^{1}/_{2}$ cup of mousse into the center of each serving plate. Press three or four Chocolate Slabs against the sides of the serving of mousse so that slabs are closer together at top and wider apart at base. Refrigerate until ready to serve.

Chocolate leaves You can make chocolate leaves from nearly any natural nonpoisonous leaf, but the prettiest ones are made from natural leaves that have a prominent vein pattern on the underside. The leaf must be perfect with no holes in it, or the chocolate will get down in the hole and you will be unable to separate the chocolate leaf from the real leaf.

Carefully melt the chocolate (page 462) or temper it (page 468) for very shiny leaves. Using a child's paintbrush, paint a fairly thick layer of chocolate on the underside of the leaf. Take care not to go over the edge (I do not go all the way to the edge) with the chocolate, or you will not be able to peel the leaf away from the chocolate leaf. Let the chocolate-coated leaves stand in a cool room or refrigerate until the chocolate is well set. Very carefully peel the natural leaves from the chocolate leaves.

I find it much easier to make a fairly thick chocolate leaf, which is less fragile and therefore easier to remove from the natural leaf. If the coated leaves stand overnight in a cool, dry room, the natural leaf dries out somewhat and is much easier to remove from the chocolate leaf.

You can be bold and coat large leaves like cabbage or collard leaves for high-drama desserts.

Modeling chocolate Chocolate ribbons can give you incredible effects with not too much effort. They look gorgeous and are not hard to make. For ribbons, you need to make what is called *modeling chocolate* or *modeling dough.*

> **For bold dessert decorations, make chocolate leaves using leaves like cabbage or collard.**

Modeling Chocolate

Chop 10 ounces good semisweet chocolate and melt it as described on page 462. When chocolate is melted, stir in $1/3$ cup corn syrup. Chocolate will tighten but still be a soft dough. Wrap tightly in plastic wrap and let stand refrigerated overnight. The modeling chocolate is now ready to shape into forms or ribbons.

Chocolate ribbons Prepare one recipe of Modeling Chocolate (above) and let stand refrigerated overnight. Chocolate will be rock-hard but will become pliable when it warms up. Warm under a gooseneck lamp or by working with your fingers. If it is very hard, soften it with a 10-second burst in the microwave on Low (30 percent). If the dough gets too soft, let it stand to harden. When the dough is pliable, roll a 3-inch-long cylinder between your hand and the counter. Flatten into a thick ribbon. Roll this thick ribbon thin between sheets of wax paper with a rolling pin or, without the wax paper, in a pasta machine. Start with widest setting on the pasta machine and run the strip through progressively thinner settings until you get the thickness you want. Use the ribbon to create bows, ruffles, and other elaborate decorations. When left uncovered, the ruffles or bows harden into firm shapes that hold well.

The Chocolate Walnut Ruffle Cake that follows is iced with a chocolate icing that is allowed to firm; then the remaining icing is reheated, thinned, and strained. This is poured over the cake to create a glossy, smooth, professional-looking chocolate cake. It is ideal for garnishing with chopped walnuts or chocolate curls pressed into the sides, with chocolate curls on top, or with chocolate ruffles as described in the recipe.

Chocolate Walnut Ruffle Cake

MAKES 1 ROUND 9-INCH LAYER

Intense, moist chocolate cake layers are glazed with shiny chocolate-cream glaze. The layers are assembled with a roasted walnut butter filling, reglazed, and topped with chocolate ruffles for high drama.

what this recipe shows -

Brown sugar imparts a fudgy taste to chocolate.

Using cake flour gives the cake a fine texture.

Using both cocoa powder and chocolate gives the cake an intense chocolate taste.

The double icing technique gives a glass-smooth professional finish.

- -

Nonstick cooking spray with flour, such as Baker's Joy, or 1 tablespoon shortening and 1 tablespoon flour, to grease pan

5 tablespoons butter

1 cup packed light brown sugar

1/4 cup vegetable oil

1/4 cup light corn syrup

6 large egg yolks

1 1/4 cups cake flour

2 tablespoons cocoa powder, preferably Dutch process

1 teaspoon baking powder

1/2 teaspoon salt

1 carton (8 ounces) sour cream

7 ounces good-quality semisweet chocolate, such as Hershey's King Size Special Dark, Lindt, or Tobler, cut into 1/2-inch pieces

1/4 cup light cream or half-and-half

2 teaspoons pure vanilla extract

Ganache Glaze (page 474)

Roasted Walnut Butter Filling (page 475)

Chocolate Ribbons (page 472)

1. Preheat the oven to 350°F (177°C). Line a 9 × 2-inch round cake pan with a parchment circle and spray with nonstick spray with flour, or grease by rubbing the pan with shortening, shaking the pan with 1 tablespoon flour, and shaking out the excess.

2. Cream butter and sugar with whisk beater on low to medium speed for 5 minutes in mixer bowl, scraping down the sides at least once. Beat in the oil and corn syrup. Beat in egg yolks. Scrape down the sides.

3. Using a sifter or two large bowls and a large strainer, sift flour, cocoa, baking powder, and salt three times to mix thoroughly.

4. Add half of dry mixture to mixer bowl and beat on low for 20 seconds. Add sour cream and beat to blend in well. Add the rest of the dry mixture and beat to blend in well, about 10 seconds.

5. Heat 1 1/2 inches of water to a boil in a large skillet, then remove from the heat and let stand 3 to 4 minutes. Place chocolate and cream in a metal bowl and place the bowl in skillet of hot water. Stir constantly with a plastic spatula and remove the bowl from the water when most of the chocolate is melted. When the chocolate is completely melted, beat it and the vanilla into the batter until well mixed, 5 to 10 seconds, scraping down the sides once. Pour the batter into the pan, level with a spatula, and drop the pan from 2 to 3 inches above the counter onto the counter to level and remove bubbles from the batter.

6. Place on a shelf in the middle of the oven and bake 30 to 35 minutes, until a toothpick inserted an inch from the center comes out clean. Place on a cooling rack and cool in the pan for 5 minutes, then invert onto the cooling rack. Cool completely.

7. Slice cake horizontally into three equal layers. Place the former top layer, cut side up, on a cake cardboard (or a 9-inch cardboard circle covered with foil). Reserve 1/2 cup ganache in a measuring cup and set aside. Spoon and spread with about 1/4 cup ganache. Let it sit 1 to 2 minutes to soak in a little. Spoon on half of the nut butter in tablespoon-size dabs dotted all over the layer. With a spatula that was dipped in hot water and dried, spread dabs onto the layer. Top with middle layer. Spread it with ganache and nut butter just as before, then top with bottom layer, cut side down. Glaze top and sides with ganache, spreading as needed with a large cake spatula.

8. Let the cake sit for 30 minutes in a cool place to set the glaze. Heat reserved ganache in the microwave for 30 seconds or a small saucepan over low heat just to thin. Strain ganache through a medium strainer. Now pour strained ganache on top of cake and tilt to spread. Do not touch with a spatula.

9. Flute a circle of chocolate ribbon into a ruffle circle with the outer edge slightly larger than the cake and arrange on the cake. Make two or three progressively smaller ruffle circles and arrange on the cake. Serve at room temperature.

Ganache Glaze

1 pound good-quality semisweet chocolate, such as Hershey's King Size Special Dark, Lindt, or Tobler, cut into 1/2-inch pieces

2 cups heavy cream
2 tablespoons corn syrup

1. Process chocolate in the food processor with the steel knife to chop fine.

2. In a heavy saucepan, carefully bring cream to a boil. Watch constantly. Let simmer for 1 minute. Pour the hot cream into a medium mixing bowl that has a wide surface. Let cool about 4 minutes. Sprinkle chopped chocolate over the entire surface. Start stirring in the middle, blending the melted chocolate and cream together. Stir steadily until all the chocolate is melted and blended. Stir in the corn syrup.

Roasted Walnut Butter Filling

I fell in love with nut butter fillings in cakes created by Marcel Desaulniers, cookbook author and chef-owner of The Trellis restaurant in Colonial Williamsburg.

1 pound walnuts, chopped

3 tablespoons *and* 4 tablespoons butter
  (7 tablespoons *total*), cut into
  ¹/₂-inch pieces

Pinch salt

1. Preheat the oven to 350°F (177°C).

2. Spread the walnuts on a large baking sheet and roast until lightly browned, about 10 to 12 minutes. Remove from the oven and stir in 3 tablespoons butter and the salt while walnuts are hot. Let stand until no longer hot to the touch. Process the walnuts and 4 tablespoons butter in the food processor with the steel knife with quick on/off pulses until the nuts are just short of a butter.

Chocolate storage

Chocolate should be stored in a cool, dry place. Both temperature and humidity have major effects on chocolate. Either can cause the familiar graying on the surface of chocolate.

Sugar bloom Chocolate is delicate. It melts at body temperature and can melt in your hand or in your pocket. I have plenty of childhood memories of getting into trouble because of this. To keep chocolate in top shape, you must watch both the temperature and the humidity carefully.

Suppose that you have a box of great chocolates and you live in a fairly cool climate. If it is a warm, moist day, then cold at night, the moisture in the air will condense on everything—your chocolate included. When it gets warm again the next day, the moisture on your chocolate evaporates. Is everything all right? Unfortunately, not exactly. When the moisture is on the chocolate, it dissolves some of the sugar out of the chocolate. When the moisture evaporates again, it leaves a fine gray film of superfine sugar crystals on the surface. Soon your chocolate has a gray coating. This "sugar bloom," as it is called, is caused by changes in temperature and humidity.

Sugar bloom can happen in many ways. If the weather has been very hot and you stored your chocolate in the refrigerator, when you bring it out into a warm moist room, moisture will condense on it. You can see that to prevent sugar bloom you must keep your chocolate tightly wrapped so that even the moisture in the air can't get to it. Chocolate is usually wrapped in foil, then covered with a paper label. This is not adequate protection for storage. You can seal tightly in an additional foil wrap or in a freezer-type zip-top plastic bag, squeezing out all the air before sealing.

You should try to avoid major temperature changes. You can store chocolate in the freezer for long storage. The chocolate should be well wrapped and sealed as just described. You should remove it from the freezer to the refrigerator and let it stay refrigerated a day or two. Then, ideally, remove it from the refrigerator to a cool room to come to room temperature. Anything that you can do to avoid sudden temperature changes will help. And, of course, always make sure that your chocolate is well sealed until it reaches room temperature.

Fat bloom   Unfortunately, sugar bloom is not all that you have to worry about. Suppose that, even if you have your chocolate tightly wrapped and avoid sudden temperature changes, the chocolate gets warm—not warm enough to melt, just warm enough so that some of the low-temperature melting fats in it start to melt. When the chocolate is stored for several months at a moderately warm temperature (in the high 70s°F/low to mid-20s°C), these melted fats float to the top, so to speak, and the surface of the chocolate becomes gray. This is a cloudy, oily gray called *fat bloom,* as opposed to the fine, grainy gray that you get with sugar bloom.

These blooms do not damage your chocolates. They are still perfectly edible and delicious. The chocolate can look bad, but there is nothing really wrong with it. A chocolate that has suffered sugar bloom, however, may seize when you melt it because of its exposure to moisture (page 461).

Sources

Equipment

Bake-Even Strips

Wilton Enterprises
2240 West 75th Street
Woodbridge, IL 60517-0750
Telephone: 630-963-7100
FAX: 630-963-7299

Available in stores handling Wilton products or from its catalog. (Catalog costs $6.99.)

Nonstick sheets, flexible

Von Snedaker's Magic Baking Sheets
12021 Wilshire Boulevard, Suite 231
Los Angeles, CA 90025
Telephone: 310-395-6365

Reusable Teflon-coated thin sheets that can be used instead of parchment paper to prevent sticking in baking.

Pie pans with perforated bottoms

Colonial Garden Kitchens
P. O. Box 66
Hanover, PA 17333-0066
Telephone: 800-752-5552
FAX: 800-757-9997

Has a free catalog of kitchen products.

Maid of Scandinavia Division
Sweet Celebrations Inc.
P. O. Box 39426
Edina, MN 55439-0426
Telephone: 800-328-6722
FAX: 612-943-1688

Has a free catalog.

Pie tape (for crusts)

Maid of Scandinavia Division
Sweet Celebrations Inc.
P. O. Box 39426
Edina, MN 55439-0426
Telephone: 800-328-6722
FAX: 612-943-1688

Has a free catalog.

Thermometers, digital (and digital scales)

Whatman LabSales
P. O. Box 1359
Hillsboro, OR 97123
Telephone: 800-942-8626 [WHATMAN]
FAX: 800-858-2243
Website: *http://www.whatman.com/labsales/*

Has a free catalog of laboratory-quality equipment.

Cooper Instruments Corp.
33 Reeds Gap Road
Middlefield, CT 06455
Telephone: 800-835-5011
FAX: 860-347-5735

Be sure to ask about laboratory-quality equipment since this company also sells restaurant thermometers that are not of laboratory quality. Has a free catalog.

Ingredients

B-V "The Beefer-Upper" Concentrate

Major Products Company, Inc.
Little Ferry, NJ 07643
Telephone: 201-641-5555

May be able to identify a local source, but also ships twelve 3-ounce bottles direct for $18.00, shipping included.

Cheese

The Mozzarella Company
2944 Elm Street
Dallas, TX 75226
Telephone: 800-798-2954
FAX: 214-741-4076
e-mail: mozzco@aol.com
Website: *http://www.foodwine.com/* and click on Mozzarella

Ships direct to consumer a wide variety of handmade specialty cheeses, including fresh mozzarella, goat's milk montasio, and mascarpone. Has a free catalog.

Flour—regular and specialty

The King Arthur Flour Baker's Catalogue
P. O. Box 876
Norwich, VT 05055-0876
Telephone: 800-827-6836
FAX: 800-343-3002
e-mail: info@King ArthurFlour.com
Website: *http://www.kingarthurflour.com/*

King Arthur flours are milled by Sands, Taylor & Wood and are distributed in stores and by free catalog. Soybean and chickpea flours, high-gluten flours, and others are available.

Flour—low-protein southern

The White Lily Foods Company
P. O. Box 871
Knoxville, TN 37901
Telephone: 423-546-5511
FAX: 423-521-7725 for brochure
Website: *http://www.whitelily.com/*

Will ship low-protein and other flours and mixes directly to consumer. Also sold in stores.

Williams-Sonoma, Inc.
Mail Order Department
P.O. Box 7456
San Francisco, CA 94120-7456
Telephone: 800-541-2233
FAX: 415-421-5153

Sells White Lily flours in most stores and through its free catalog.

Fondant, rolled

Maid of Scandinavia Division
Sweet Celebrations Inc.
P. O. Box 39426
Edina, MN 55439-0426
Telephone: 800-328-6722
FAX: 612-943-1688

Has a free catalog.

Wilton Enterprises
2240 West 75th Street
Woodbridge, IL 60517-0750
Telephone: 630-963-7100
FAX: 630-963-7299

Has a catalog. (Catalog costs $6.99.)

Gluten-free products

The Gluten-Free Pantry, Inc.
P. O. Box 881
Glastonbury, CT 06033
Telephone: 203-633-3826
FAX: 203-633-6853
Website: *http://www.glutenfree.com/*

The free mail-order catalog distributes the company's own mixes as well as products from other gluten-free sources.

Pomegranate molasses (concentrated juice)

Available from some Middle Eastern grocery stores or can be shipped from the following stores across the United States:

Kalustyan's
123 Lexington Avenue
New York, NY 10016
Telephone: 212-685-3451
FAX: 212-683-8458

Shiraz Food Market
9630 SW 77th Avenue
Miami, FL 33156
Telephone: 305-273-8888

Alvand Market
3033 South Briston, Suite G
Costa Mesa, CA 92626
Telephone: 714-545-7177
FAX: 714-751-2848

International Food Bazaar
915 SW 9th Avenue
Portland, OR 97205
Telephone: 503-228-1960

Ripe produce shippers

Harry and David
2518 S. Pacific Highway
Medford, OR 97501
Telephone: 800-547-3033
FAX: 800-648-6640
e-mail: service@harryanddavid.com
Website: *http://www.harryanddavid.com/*

The original seller of fruit by mail. Ships high-quality pears, apples, and other foods. Has a free catalog.

Gracewood Groves
9075 17th Place
Vero Beach, FL 32966
Telephone: 800-678-1154
Website: Indirectly in other sites (search "Gracewood Groves" or "Gracewood Fruit")

Ships Indian River citrus, mangoes, vine-ripe tomatoes (mid-November to March), and other foods. Has a free catalog.

Sourdough starters

Sourdoughs International
P. O. Box 670
Cascade, ID 83611
Telephone: 800-888-9567 for brochure
FAX: 208-382-3129

Dr. Ed Wood is a pathologist who has collected sourdough starters from around the world and sells them in dried form. They are from 10 regions: the Yukon, San Francisco, Finland, France, Austria, Egypt (Red Sea and Giza), Bahrain, Saudi Arabia, and a very active Russian starter. He is the author of *World Sourdoughs from Antiquity* (Ten Speed Press, 1996).

The King Arthur Flour Baker's Catalogue
P. O. Box 876
Norwich, VT 05055-0876
Telephone: 800-827-6836
FAX: 800-343-3002
e-mail: info@KingArthurFlour.com
Website: *http://www. kingarthurflour.com*

Has a free catalog, which includes sourdough starters and other baking supplies.

Information

FDA Food Code

National Technical Information Service (NTIS)
Department of Commerce
5285 Port Royal Road
Springfield, VA 22161
Telephone: 703-487-4650

The 1995 FDA Food Code is available from NTIS in hard copy as PB95-265492 or as a WordPerfect 6.1 diskette as PB96-500491 (also available on the World Wide Web at *http://www.agen.ufl.edu/˜foodsaf/fchome.html*). See especially Section 3-4.

Food allergies and sensitivities

The Food Allergy Network
10400 Eaton Place, Suite 107
Fairfax, VA 22030-5647
Telephone: 703-691-3179
Orders: 800-929-4040
FAX: 703-691-3179
e-mail: fan@worldwide.net
Website: *http://www.foodallergy.org/*

A national nonprofit organization established to help families living with food allergies and to increase public awareness about food allergies and anaphylaxis. Publishes a newsletter and has other publications and resources.

Food and Nutrition Information Center (FNIC)

National Agricultural Library (NAL)
Agricultural Research Service (ARS)
U.S. Department of Agriculture (USDA)
Website: *http://www.nal.usda.gov/fnic/*

Entry point for on-line USDA information, including the Index of Food and Nutrition Internet Resources (*http://www.nal.usda.fnic/fnic-etexts.html*), which links to over 500 U.S. and international sources. Many of these links are outside the domain of the USDA, which does not assume responsibility for their accuracy.

National Food Safety Database

University of Florida
Gainesville, FL
Website: *http://www.agen.ufl.edu/˜foodsaf/*

Has comprehensive food safety information from many sources in cooperation with the Food and Drug Administration, including the FDA Food Code.

Meat and Poultry Hotline

U.S. Department of Agriculture (USDA)
Washington, DC
Telephone: 800-535-4555

Seafood Hotline

U.S. Food and Drug Administration (FDA)
Washington, DC
Telephone: 800-332-4010 [FDA-4010]

White wheat information

American White Wheat Producers Association
P. O. Box 326
Atchison, KS 66002
Telephone: 913-367-4422

References and Bibliography

Following are the most frequently used and specific sources for some of the information in this book. The "General" section gives sources that were valuable as background for many parts of the book. Food-related technical and specialized books are listed first, followed by articles in similar areas from technical journals, food magazines, and popular sources. Government documents are then identified, and a short list of computerized documents completes this section.

Each chapter has lists of books and articles that pertain specifically to the topics covered in that chapter. The same source is listed in more than one chapter if material from that source is identified in the text of several chapters.

General

Books

Belitz, Hans-Dieter, and Werner Grosch. *Food Chemistry,* trans. D. Hadziyev from 2nd German ed. Berlin: Springer Verlag, 1986.

Bown, Deni. *Encyclopedia of Herbs & Their Uses,* 1st American ed. New York: Dorling Kindersley Publishing Inc., 1995.

Charley, Helen. *Food Science,* 1st ed. New York: The Ronald Press Company, 1970.

————. *Food Science,* 2nd ed. New York: John Wiley & Sons, 1982.

Child, Julia. *Cooking with Master Chefs.* New York: Alfred A. Knopf, Inc., 1993.

————. *The French Chef Cookbook.* New York: Alfred A. Knopf, Inc., 1968.

————. *From Julia Child's Kitchen.* New York: Alfred A. Knopf, Inc., 1975.

————. *Julia Child & Company.* New York: Alfred A. Knopf, Inc., 1978.

————. *Julia Child & More Company.* New York: Alfred A. Knopf, Inc., 1979.

————. *The Way to Cook.* New York: Alfred A. Knopf, Inc., 1989.

Child, Julia, Louisette Bertholle, and Simone Beck. *Mastering the Art of French Cooking,* Vol. I. New York: Alfred A. Knopf, Inc., 1961.

Child, Julia, and Simone Beck. *Mastering the Art of French Cooking,* Vol. II. New York: Alfred A. Knopf, Inc., 1970.

Coultate, Tom P. *Food: The Chemistry of Its Components,* 2nd ed. London: Royal Society of Chemistry, 1989.

DeMan, John M. *Principles of Food Chemistry,* rev. 3rd printing. Westport, Conn.: AVI Publishing Co., Inc., 1980.

Ensminger, Audrey H., et al. *Foods & Nutrition Encyclopedia,* 2nd ed. Boca Raton, Fla.: CRC Press, 1994. 2 Vols.

Eskin, N. A. Michael, ed. *Biochemistry of Foods,* 2nd ed. San Diego, Calif.: Academic Press, Inc., 1990.

Fennema, Owen R., ed. *Food Chemistry,* rev. and expanded 2nd ed. New York: Marcel Dekker, Inc., 1985.

FitzGibbon, Theodora. *The Food of the Western World: An Encyclopedia of Food from North America and Europe.* New York: Quadrangle/The New York Times Book Co., 1976.

Freydberg, Nicholas, and Willis A. Gortner. *The Food Additives Book.* New York: Bantam Books, 1982.

Fulton, Margaret. *Encyclopedia of Food and Cookery.* New York: W. H. Smith Publishers, Inc., 1986.

Herbst, Sharon Tyler. *The New Food Lover's Companion: Comprehensive Definitions of Over 4000 Food, Wine and Culinary Terms,* 2nd ed. Hauppauge, N.Y.: Barron's Educational Services, Inc., 1995.

Hughes, Osee, and Marion Bennion. *Introductory Foods,* 5th ed. New York: The Macmillan Company, 1970.

Kamman, Madeleine. *The Making of a Cook.* New York: Weathervane Books, 1971.

Lang, Jenifer Harvey, ed. *Larousse Gastronomique,* New American ed. New York: Crown Publishers, Inc., 1988.

McGee, Harold. *On Food and Cooking: The Science and Lore of the Kitchen.* New York: Charles Scribner's Sons, 1984.

————. *The Curious Cook: More Kitchen Science and Lore.* San Francisco: North Point Press, 1990.

Meyer, Lillian Hoagland. *Food Chemistry.* Westport, Conn.: AVI Publishing Co., Inc., 1960.

Montagné, Prosper. *Larousse Gastronomique,* 1st American ed. Charlotte Turgeon and Nina Froud, ed. New York: Crown Publishers, Inc., 1961.

Morris, Christopher, ed. *Academic Press Dictionary of Science and Technology.* San Diego, Calif.: Academic Press, Inc., 1992.

Penfield, Marjorie P., and Ada Marie Campbell. *Experimental Food Science,* 3rd ed. San Diego, Calif.: Academic Press, Inc., 1990.

Pennington, Jean A. T. *Bowles & Church's Food Values of Portions Commonly Used,* 16th ed. Philadelphia: J. B. Lippincott Company, 1994.

Pomeranz, Yeshajahu. *Functional Properties of Food Components.* New York: Academic Press, Inc., 1985.

Rombauer, Irma S., and Marion Rombauer Becker. *The Joy of Cooking.* New York: Bobbs-Merrill Company, Inc., 1975.

Whitman, Joan, and Dolores Simon. *Recipes into Type: A Handbook for Cookbook Writers and Editors.* New York: HarperCollins Publishers, Inc., 1993.

Government documents

Adams, Catherine F. *Nutritive Values of American Foods in Common Units.* U.S. Department of Agriculture, Agriculture Handbook No. 456. Washington, D.C.: U.S. Government Printing Office, November 1975.

Gebhardt, Susan E., Rene Cutrufelli, and Ruth H. Matthews (Principal Investigators). *Composition of Foods: Fruits and Fruit Juices—Raw, Processed, Prepared.* U.S. Department of Agriculture, Agriculture Handbook No. 8–9. Washington, D.C.: U.S. Government Printing Office, rev. August 1982.

Haytowitz, David B., and Ruth H. Matthews (Principal Investigators). *Composition of Foods: Vegetables and Vegetable Products—Raw, Processed, Prepared.* U.S. Department of Agriculture, Agriculture Handbook No. 8–11. Washington, D.C.: U.S. Government Printing Office, rev. August 1984.

Reeves, James B., III, and John L. Weihrauch (Principal Investigators). *Composition of Foods: Fats and Oils—Raw, Processed, Prepared.* U.S. Department of Agriculture, Agriculture Handbook No. 8-4. Washington, D.C.: U.S. Government Printing Office, rev. June 1979.

Watt, Bernice K., and Annabel L. Merill. *Composition of Foods: Raw, Processed, Prepared.* U.S. Department of Agriculture, Agriculture Handbook No. 8. Washington, D.C.: U.S. Government Printing Office, December 1963, rev. December 1975.

Computerized documents

Compton's Multimedia Encyclopedia (Special Edition), Macintosh Edition, Compton's New Media, Inc., 1992.

Encarta '95: The Complete Interactive Multimedia Encyclopedia, Microsoft Corporation, 1994.

NutriForm, Version 2.07d (Nutrition Database), Nutrition & Food Associates, Plymouth, Minn., 1993.

Nutritionist IV, Food Labeling Module Version 3.5.1 (Nutrition Database), N-Squared Computing, First DataBank Division, The Hearst Corporation, San Bruno, Calif., 1994.

Chapter one
the wonders of risen bread

Books

Amendola, Joseph, and Donald E. Lundberg. *Understanding Baking.* Boston: CBI Publishing Company, Inc., 1970.

Bilheux, Roland, et al. *Special and Decorative Breads,* trans. Rhona Poritzky-Lauvand and James Peterson. New York: Van Nostrand Reinhold, 1989.

Blanshard, J. M. V., P. J. Frazier, and T. Galliard, ed. *Chemistry and Physics of Baking: Materials, Processes, and Products.* London: Royal Society of Chemistry, 1986.

Campbell, Ada Marie. "Flour," *Food Theory and Applications,* Pauline C. Paul and H. H. Palmer, ed. Chapter 11. New York: John Wiley & Sons, 1972.

————. "Flour Mixtures," *Food Theory and Applications,* Pauline C. Paul and H. H. Palmer, ed. Chapter 12. New York: John Wiley & Sons, 1972.

Clayton, Bernard, Jr. *The Complete Book of Breads.* New York: Simon and Schuster, 1973.

Crane, Eva, ed. *Honey: A Comprehensive Survey.* New York: Crane, Russak & Company, Inc., 1975.

Cunningham, Marion. *The Breakfast Book.* New York: Alfred A. Knopf, 1988.

David, Elizabeth. *English Bread and Yeast Cookery,* American ed. New York: The Viking Press, 1980.

Editors of Time-Life Books. *Breads.* The Good Cook Techniques & Recipes Series. Alexandria, Va.: Time-Life Books, Inc., 1981.

Field, Carol. *The Italian Baker.* New York: Harper & Row, 1985.

Fletcher, Helen S. *The New Pastry Cook.* New York: William Morrow and Company, Inc., 1986.

German, Donna Rathmell, and Ed Wood. *Worldwide Sourdoughs from Your Bread Machine,* a Nitty Gritty Cookbook. San Leandro, Calif.: Bristol Publishing Enterprises, Inc., 1994.

Greenstein, George. *Secrets of a Jewish Baker: Authentic Jewish Rye and Other Breads.* Freedom, Calif.: The Crossing Press, 1993.

Guinard, J. Y., and P. Lesjean. *Le Pain Retrouvé: 30 Pains et Leurs Recettes.* Paris: L. T. Editions Jacques Lanore, 1982.

Kamman, Madeleine. *The Making of a Cook.* New York: Weathervane Books, 1971.

MacRitchie, F. "Baking Quality of Wheat Flours," *Advances in Food Research.* C. O. Chichester, E. M. Mrak, and B. S. Schweigert, ed., Vol. 29, pp. 201–277. New York: Academic Press, Inc., 1984.

McGee, Harold. *On Food and Cooking: The Science and Lore of the Kitchen.* New York: Charles Scribner's Sons, 1984.

Ortiz, Joe. *The Village Baker: European Regional Bread Recipes for the American Home Baker.* Berkeley, Calif.: Ten Speed Press, 1992.

Pomeranz, Yeshajahu. "Relationship Between Chemical Composition and Breadmaking Potentialities of Wheat Flour." *Advances in Food Research.* Vol. 16, pp. 335–455. New York: Academic Press, Inc., 1968.

———. ed. *Wheat: Chemistry and Technology,* 2nd ed. St. Paul, Minn.: American Association of Cereal Chemists, 1971.

———, *Wheat: Chemistry and Technology,* 3rd ed. St. Paul, Minn.: American Association of Cereal Chemists, 1988. 2 Vols.

———, and J. A. Shallenberger. *Bread Science and Technology.* Westport, Conn.: AVI Publishing Co., Inc., 1971.

Pyler, Ernst John. *Baking Science and Technology.* Chicago: Siebel Publishing Company, 1952. 2 Vols.

Reinhart, Peter. *Brother Juniper's Bread Book. Slow Rise as Method and Metaphor.* Reading, Mass.: Addison-Wesley Publishing Company, Inc., 1991.

Schünemann, Claus, and Günter Treu. *Baking: The Art and Science,* 1st English Edition. Calgary, Alberta, Canada: Baker Tech Inc., 1988.

Silverton, Nancy. *Nancy Silverton's Breads from the La Brea Bakery.* New York: Random House, Inc., 1996.

Sultan, William J. *Practical Baking,* 4th ed. New York: Van Nostrand Reinhold Company, 1986.

Weschberg, Joseph. *The Cooking of Vienna's Empire.* Foods of the World Series. Alexandria, Va.: Time-Life Books, 1974.

Wood, Ed. *World Sourdoughs from Antiquity.* Berkeley, Calif.: Ten Speed Press, 1996.

Articles

Barnes, Faye G., David R. Shenkenberg, and Eugene J. Guy. "Factors Affecting the Mixing Requirements of a Sourdough Bread Made with Acid Whey. I. Effects of Lactose, pH and Salt as Measured by the Farinograph." *Bakers Digest* (June 1973), pp. 16–18.

———. "Factors Affecting the Mixing Requirements of a Sourdough Bread Made with Acid Whey. II. Bread Mixing Studies." *Bakers Digest* (June 1973), pp. 19, 22–23, 63.

Baxter, Elma J., and E. Elizabeth Hester. "The Effect of Sucrose on Gluten Development and the Solubility of the Proteins of a Soft Wheat Flour." *Cereal Chemistry,* Vol. 35 (September 1958), pp. 366–374.

Biberoglu, Sevinc, Kadir Biberoglu, and Baki Komsuoglu. "Mad Honey." *J. American Medical Association,* Vol. 259, No. 13 (April 1, 1988), p. 1943.

Bietz, J. A., F. R. Huebner, and J. S. Wall. "Glutenin: The Strength Protein of Wheat Flour." *Bakers Digest* (February 1973), pp. 26–31, 34–35, 67.

Bruinsma, B. L., and K. F. Finney. "Functional (Bread-Making) Properties of a New Dry Yeast." *Cereal Chemistry,* Vol. 58, No. 5 (1981), pp. 477–480.

Bullerman, L. B., F. Y. Lieu, and S. A. Seier. "Inhibition of Growth and Aflatoxin Production by Cinnamon and Clove Oils, Cinnamic Aldehyde and Eugenol." *J. Food Science,* Vol. 42, No. 4 (1977), pp. 1107–1109, 1116. Abstract only.

Bushuk, W., and I. Hlynka. "The Bromate Reaction in Dough. III. Effect of Continuous Mixing and Flour Particle Size." *Cereal Chemistry,* Vol. 38 (March 1961), pp. 178–186.

———. "The Bromate Reaction in Dough. IV. Effect of Reducing Agents." *Cereal Chemistry,* Vol. 38, No. 4 (July 1961), pp. 309–316.

————. "The Bromate Reaction in Dough. V. Effect of Flour Components and Some Related Compounds." *Cereal Chemistry,* Vol. 38, No. 4 (July 1961), pp. 316–324.

Cole, Morton S. "An Overview of Modern Dough Conditioners." *Bakers Digest* (December 1973), pp. 21–23, 64.

Conner, D. E., et al. "Effects of Essential Oils and Oleoresins of Plants on Ethanol Production, Respiration and Sporulation of Yeasts." *International J. Food Microbiology,* Vol. 1, No. 2 (1984), pp. 63–74. Abstract only.

Cooper, Elmer J., and Gerald Reed. "Yeast Fermentation: Effects of Temperature, pH, Ethanol, Sugars, Salt, and Osmotic Pressure." *Bakers Digest,* Vol. 42, No. 6 (December 1968), pp. 22–24, 26, 28–29, 63.

Elkassabany, M., R. C. Hoseney, and P. A. Seib. "Ascorbic Acid as an Oxidant in Wheat Flour Dough. I. Conversion to Dehydroascorbic Acid." *Cereal Chemistry,* Vol. 57, No. 2 (1980), pp. 85–87.

Elkassabany, M., and R. C. Hoseney. "Ascorbic Acid as an Oxidant in Wheat Flour Dough. II. Rheological Effects." *Cereal Chemistry,* Vol. 57, No. 2 (1980), pp. 88–91.

Forsythe, Richard H., and Dwight H. Bergquist. "Functionality of Eggs in the Modern Bakery." *Bakers Digest* (October 1973), pp. 84–87, 90, 131.

Glezer, Maggie. "Salt in Bread Dough." *The Bread Bakers Guild of America* [Newsletter], Vol. 3, No. 3 (Summer 1995), p. 5.

————. "The Timing of the Salt Addition During Mixing." *The Bread Bakers Guild of America* [Newsletter], Vol. 3, No. 4 (Fall 1995), pp. 4–5.

————. "Q&A: Why Cinnamon Breads Don't Rise as High." *Fine Cooking.* Issue 15 (June/July 1996), p. 22.

Grant, D. R., and V. K. Sood. "Studies of the Role of Ascorbic Acid in Chemical Dough Development. II. Partial Purification and Characterization of an Enzyme Oxidizing Ascorbate in Flour." *Cereal Chemistry,* Vol. 57, No. 1 (1980), pp. 46–49.

Grosskreutz, J. C. "A Lipoprotein Model of Wheat Gluten Structure." *Cereal Chemistry,* Vol. 38, No. 4 (July 1961), pp. 336–349.

Johnson, John A., and Carlos R. S. Sanchez. "The Nature of Bread Flavor." *Bakers Digest* (October 1973), pp. 48 ff.

Kim, S. K., and B. L. D'Appolonia. "Bread Staling Studies. II. Effect of Protein Content and Storage Temperature on the Role of Starch." *Cereal Chemistry,* Vol. 54 (March–April 1977), pp. 216–224.

————. "Bread Staling Studies. III. Effect of Pentosans on Dough, Bread, and Bread Staling Rate." *Cereal Chemistry,* Vol. 54 (March–April 1977), pp. 225–229.

Kummer, Corby. "Holiday-Spanning Bread." *The Atlantic Monthly* (March 1989), pp. 84, 86–88.

Lampe, Kenneth F. "Rhododendrons, Mountain Laurel, and Mad Honey [editorial]." *J. American Medical Association,* Vol. 259, No. 13 (April 1, 1988), p. 2009.

Lorenz, Klaus. "Sourdough Processes—Methodology and Biochemistry." *Bakers Digest* (February 1981), pp. 32–36.

MacRitchie, F. "Differences in Baking Quality Between Wheat Flours." *J. Food Technology,* Vol. 13 (1978), pp. 187–194.

Mecham, D. K. "Wheat and Flour Proteins. Recent Research." *Bakers Digest* (October 1973), pp. 24–26, 28, 30, 32, 128.

Mertens, Michael H. "Functions of Dairy Ingredients in Baking." *Bakers Digest* (December 1969), pp. 57–60.

Pisesookbunterng, W., and B. L. D'Appolonia. "Bread Staling Studies. I. Effect of Surfactants on Moisture Migration from Crumb to Crust and Firmness Values of Bread Crumb." *Cereal Chemistry,* Vol. 60, No. 4 (1983), pp. 298–300.

Pisesookbunterng, W., B. L. D'Appolonia, and K. Kulp. "Bread Staling Studies. II. The Role of Refreshing." *Cereal Chemistry,* Vol. 60, No. 4 (1983), pp. 301–305.

Richard-Molard, D., M. C. Nago, and R. Drapron. "Influence of the Breadmaking Method on French Bread Flavor." *Bakers Digest* (June 1979), pp. 34–37.

Short, A. L., and E. A. Roberts. "Pattern of Firmness Within a Loaf." *J. Science of Food and Agriculture.* Vol. 22 (1971), pp. 470–472. Abstract only.

Sluimer, P. "Principles of Dough Retarding." *Bakers Digest* (August 1981), pp. 6–8, 10.

Stafford, H. R. "Biochemical Aspects of Breadmaking." *Process Biochemistry,* Vol. 2, No. 4 (April 1967), pp. 18–20.

Sternberg, George. "Practical Gluten Structure Control." *Bakers Digest* (April 1973), pp. 34–37.

Tipples, K. H., and R. H. Kilborn. "'Unmixing'—The Disorientation of Developed Bread Doughs by Slow Speed Mixing." *Cereal Chemistry,* Vol. 52 (March–April 1975), pp. 248–262.

Volpe, T., and M. E. Zabik. "A Whey Protein Contributing to Loaf Volume Depression." *Cereal Chemistry,* Vol. 52 (March–April 1975), pp. 188–197.

Wood, Ed. "Bake Like an Egyptian." *Modern Maturity,* Vol. 39, No. 5 (September–October 1996), pp. 66–67.

Wright, Wilma J., C. W. Bice, and J. M. Fogelberg. "The Effect of Spices on Yeast Fermentation." *Cereal Chemistry,* Vol. 31 (March 1954), pp. 100–112.

Chapter two
how rich it is!

Books

Amendola, Joseph. *The Bakers' Manual for Quantity Baking and Pastry Making,* rev. 3rd ed. Rochelle Park, N. J.: Hayden Book Company, Inc., 1972.

Babayan, V. K. "Medium Chain Triglycerides," *Dietary Fat Requirements in Health and Development.* J. Beare-Rogers, ed. Champaign, Ill.: American Oil Chemists' Society, 1988, p. 73.

Beranbaum, Rose Levy. *The Cake Bible.* New York: William Morrow and Company, Inc., 1988.

Braker, Flo. *The Simple Art of Perfect Baking.* New York: William Morrow and Company, Inc., 1985.

Carroll, K. K., ed. *Diet, Nutrition, and Health.* Montreal: McGill–Queen's University Press, 1990.

Chu, Grace Zia. *The Pleasures of Chinese Cooking.* New York: Simon & Schuster, 1962.

Dodge, Jim, with Elaine Ratner. *Baking with Jim Dodge.* New York: Simon & Schuster, 1991.

Dupree, Nathalie. *New Southern Cooking.* New York: Alfred A. Knopf, 1986.

Fletcher, Helen S. *The New Pastry Cook.* New York: William Morrow and Company, Inc., 1986.

Gottenbos, J. J. "Nutritional Evaluation of n-6 and n-3 Polyunsaturated Fatty Acids." *Dietary Fat Requirements in Health and Development.* J. Beare-Rogers, ed. Champaign, Ill.: American Oil Chemists' Society, 1988, p. 107.

Healy, Bruce, and Paul Bugat. *Mastering the Art of French Pastry.* Woodbury, N.Y.: Barron's Educational Series, Inc., 1984.

Jordan, Michele Anna. *The Good Cook's Book of Oil & Vinegar.* Reading, Mass.: Addison-Wesley Publishing Company, 1992.

Lake, Mark, and Judy Ridgway. *The Simon & Schuster Pocket Guide to Oils, Vinegars and Seasonings.* New York: A Fireside Book by Simon & Schuster, 1989.

Malgieri, Nick. *Nick Malgieri's Perfect Pastry.* New York: Macmillan Publishing USA, 1989.

Medrich, Alice. *Cocolat: Extraordinary Chocolate Desserts.* New York: Warner Books, Inc., 1990.

———. *Chocolate and the Art of Low-Fat Desserts.* New York: Warner Books, Inc., 1994.

Purdy, Susan G. *Have Your Cake and Eat It, Too.* New York: William Morrow and Company, Inc., 1993.

Pyler, Ernst John. *Baking Science and Technology.* Chicago: Siebel Publishing Company, 1952. 2 Vols.

Rombauer, Irma S., and Marion Rombauer Becker. *The Joy of Cooking.* New York: Bobbs-Merrill Company, Inc., 1975.

Scicolone, Michele. *La Dolce Vita.* New York: William Morrow and Company, Inc., 1993.

Strause, Monroe Boston. *Pie Marches On.* New York: Ahrens Publishing Company, Inc., 1951.

Swern, Daniel, ed., *Bailey's Industrial Oil and Fat Products,* 4th ed. New York: John Wiley and Sons, 1979. Reference only.

Vitale, J. J. "Lipids, Host Defenses and Immune Function," *Dietary Fat Requirements in Health and Development.* J. Beare-Rogers, ed., Champaign, Ill.: American Oil Chemists' Society, 1988, p. 137.

Wood, Randall. "Biological Effects of Palm Oil in Humans." *Fatty Acids in Foods and Their Health Implications.* Ching Kuang Chow, ed. New York: Marcel Dekker, Inc., 1992.

Articles

Addis, Paul B. "Further Research on the Physiological Consequences of Fat." Institute of Food Technologists Annual Meeting, IFT 88, New Orleans, June 19–22, 1988, Paper 366.

Anonymous. "P&G Sells Caprenin to Mars, Achieving Products' First Sale." *The Wall Street Journal* (January 20, 1992), p. B7.

Anonymous [review]. "Hydrogenated Vegetable Fats Shown to Increase Serum Cholesterol." *Nutrition Close-up,* Vol. 7, No. 3 (November 1990), pp. 1–2.

Anonymous [review]. "Fried Foods Stabilizer Keeps Moisture In, Fat Out." *Prepared Foods* (March 1991), p. 61.

Ash, David J., and John C. Colmey. "The Role of pH in Cake Baking." *Bakers Digest* (February 1973), pp. 36–39, 42, 68.

Babayan, V. K., and John R. Rosenau. "Medium-Chain-Triglyceride Cheese." *Food Technology* (February 1991), pp. 111–114.

Bell, A. V., et al. "A Study of the Micro-Baking of Sponges and Cakes Using Cine and Television Microscopy." *J. Food Technology,* Vol. 10, No. 2 (1975), pp. 147–156.

Best, Daniel. "Processors Target Diet and Cancer Connections." *Prepared Foods* (March 1991), pp. 40–41.

———. "Technology Presses Forward on Designer Fats and Oils." *Prepared Foods* (March 1991), pp. 47–48.

Blumenthal, Michael M. "A New Look at the Chemistry and Physics of Deep-Fat Frying." *Food Technology* (February 1991), pp. 68–71, 94.

Bonanome, A., and S. M. Grundy. "Effect of Dietary Stearic Acid on Plasma Cholesterol and Lipoprotein Levels." *New England J. Medicine,* Vol. 318, No. 19 (May 12, 1988), pp. 1244–1248.

Burton, A. F. "Oncolytic Effects of Fatty Acids in Mice and Rats." *American J. Clinical Nutrition,* Vol. 53 (1991), pp. 1082S–1086S.

Carr, Roy A. "Development of Deep-Frying Fats." *Food Technology* (February 1991), pp. 95–96.

Charley, Helen. "Effects of the Size and Shape of the Baking Pan on the Quality of Shortened Cakes." *J. Home Economics,* Vol. 44, No. 2 (February 1952), pp. 115–118.

Christakis, George. "The Effect of Dietary Cholesterol on Serum Cholesterol: An Interpretive Review." AEB/UEP Egg Nutrition Center, undated (ca. 1989).

Clark, Walter L., and George W. Serbia. "Safety Aspects of Frying Fats and Oils." *Food Technology* (February 1991), pp. 84–86, 88–89, 94.

Cohen, L. A., et al. "Influence of Dietary Medium Chain Triglycerides on the Development of N-methylnitrosourea-induced Rat Mammary Tumors." *Cancer Research,* Vol. 44 (1984), p. 5023.

Corriher, Shirley O. "Fat Expertise for the Cook." *CAREF: The Research Report,* International Association of Culinary Professionals, Vol. 7, No. 1 (February 1992), 7 pp.

Daly, J. M., et al. "Enteral Nutrition with Supplemental Arginine, RNA and Omega-3 Fatty Acids: A Prospective Clinical Trial." *J. of Parenteral and Enteral Nutrition,* Vol. 15, No. 1: Supplement (1991).

Denton, Minna C., Edith Wenoel, and Louise Pritchett. "Absorption of Fat by Fried Batters and Doughs and Causes of Variation." *J. Home Economics* (March 1920), pp. 111–127.

Deveny, Kathleen. "Unilever Unit Serves Up Fat Substitutes." *The Wall Street Journal* (Eastern Edition) (January 8, 1992), p. B1.

Edington, J. D., et al. "Serum Lipid Response to Dietary Cholesterol in Subjects Fed a Low-Fat, High-Fiber Diet." *American J. Clinical Nutrition,* Vol. 50, No. 1 (July 1989), pp. 58–62.

Giese, James. "Fats, Oils, and Fat Replacers." *Food Technology,* Vol. 50, No. 4 (April 1996), pp. 77–83.

———. "Fats and Fat Replacers: Balancing the Health Benefits." *Food Technology,* Vol. 50, No. 9 (September 1996), pp. 76, 78.

Glicksman, Martin. "Hydrocolloids and the Search for an 'Oily Grail.'" *Food Technology,* Vol. 45, No. 10 (October 1991), pp. 96–103.

Grundy, S. M. "Reevaluation of the Role of Fatty Acids in the American Diet." Institute of Food Technologists Annual Meeting, IFT 91, Dallas, June 1–5, 1991.

Handleman, Avrom R., James F. Conn, and John W. Lyons. "Bubble Mechanics in Thick Cake Foams and Their Effects on Cake Quality." *Cereal Chemistry,* Vol. 38 (May 1961), pp. 294–305.

Hansen, R. Garuth. "Why Calories Count: Communicating Moderation and a Balanced Diet." *Food Technology* (October 1991), pp. 86, 88, 90–93.

Howard, N. B. "The Role of Some Essential Ingredients in the Formation of Layer Cake Structures." *Bakers Digest* (October 1972), pp. 28–30, 32, 34, 36–37, 64.

———, D. H. Hughes, and R.G.K. Strobel. "Function of the Starch Granule in the Formation of Layer Cake Structure." *Cereal Chemistry,* Vol. 45 (July 1968), pp. 329–338.

Jackel, Simon S. "Bread Crumbs, Croutons, Breadings, Stuffings, and Textural Analysis." *Cereal Foods World,* Vol. 38, No. 9 (September 1993), pp. 704–705.

Jacobson, Glen A. "Quality Control in Deep-Fat Frying Operations." *Food Technology* (February 1991), pp. 72–74.

Jooste, Martha E., and Andrea Overman Mackey. "Cake Structure and Palatability as Affected by Emulsifying Agents and Baking Temperatures." *Food Research,* Vol. 17 (1952), pp. 185–196.

Kennedy, Joanne P. "Structured Lipids: Fats of the Future." *Food Technology,* Vol. 45, No. 11 (November 1991), pp. 76, 78, 80, 83.

Kissell, L. T., J. R. Donelson, and R. L. Clements. "Functionality in White Layer Cake of Lipids from Untreated and Chlorinated Patent Flours. I. Effects of Free Lipids." *Cereal Chemistry,* Vol. 56, No. 1 (1979), pp. 11–14.

Kuntz, Lynn A. "Selecting a Frying Fat." *Food Product Design* (July 1994), pp. 41, 44, 46–49.

Lowe, R. "Role of Ingredients in Batter Systems." *Cereal Foods World,* Vol. 38, No. 9 (September 1993), pp. 673–677.

Mayer, Caroline E. [*The Washington Post*]. "Milky Way II [caprenin]." *The Huntsville [Ala.] Times* (February 5, 1992), p. E2.

McBean, Lois D. "Health Effects of Dietary Fatty Acids." *Dairy Council Digest,* Vol. 63, No. 3 (May/June 1992), pp. 13–18.

Medrich, Alice. "Chocolate Delights." *House and Garden,* Vol. 152 (December 1980), pp. 128 ff.

Megremis, Cheryl J. "Medium-Chain Triglycerides: A Nonconventional Fat." *Food Technology* (February 1991), pp. 108, 110, 114.

Miller, Byron S., and Henry B. Trimbo. "Gelatinization of Starch and White Layer Cake Quality." *Food Technology* (April 1965), pp. 208–216.

Moore, Thomas J. "The Cholesterol Myth." *The Atlantic Monthly* (September 1989), pp. 37–41, 42–44, 46–52, 54, 56–57, 60.

Olewnik, M., and K. Kulp. "Factors Influencing Wheat Flour Performance in Batter System." *Cereal Foods World*, Vol. 38, No. 9 (September 1993), pp. 679–684.

Pinthus, E. J., Pnina Weinberg, and Israel Sam Saguy. "Criterion for Oil Uptake During Deep-fat Frying." *J. Food Science*, Vol. 58, No. 1 (1993), pp. 204–205, 222.

Pinthus, Eli J., and Israel Sam Saguy. "Initial Interfacial Tension and Oil Uptake by Deep-fat Fried Foods." *J. Food Science*, Vol. 59, No. 4 (1994), pp. 804–807, 823.

———, Pnina Weinberg, and I. S. Saguy. "Oil Uptake in Deep Fat Frying as Affected by Porosity." *J. Food Science*, Vol. 60, No. 4 (1995), pp. 767–769.

Procter & Gamble. "Background & Facts about Caprenin" (August 1, 1991). Leaflet.

Pszczola, Donald E. "Oatrim Finds Application in Fat-Free Cholesterol-Free Milk." *Food Technology*, Vol. 50, No. 9 (September 1996), pp. 80–81.

Rao, V. N. Mohan, and Rory A. M. Delaney. "An Engineering Perspective on Deep-Fat Frying of Breaded Chicken Pieces." *Food Technology* (April 1995), pp. 138–141.

Sabaté, Joan, et al. "Effects of Walnuts on Serum Lipid Levels and Blood Pressure in Normal Men." *New England J. Medicine*, Vol. 328 (March 4, 1993), pp. 603–607.

Saguy, Israel Sam, and Eli J. Pinthus. "Oil Uptake During Deep-Fat Frying: Factors and Mechanism." *Food Technology* (April 1995), pp. 142–145, 152.

Shukla, Triveni P. "Batters and Breadings for Traditional and Microwavable Foods." *Cereal Foods World*, Vol. 38, No. 9 (September 1993), pp. 701–702.

Singh, R. Paul. "Heat and Mass Transfer in Foods During Deep-Fat Frying." *Food Technology* (April 1995), pp. 134–137.

Swackhamer, Robert. "Responding to Customer Requirements for Improved Frying System Performance." *Food Technology* (April 1995), pp. 151–152.

Trimbo, Henry B., and Byron S. Miller. "The Development of Tunnels in Cakes." *Bakers Digest* (August 1973), pp. 24–26, 71.

Wilson, J. T., and D. H. Donelson. "Studies on the Dynamics of Cake-Baking. I. The Role of Water in Formation of Cake Layer Structure." *Cereal Chemistry*, Vol. 40 (September 1963), pp. 466–481.

Chapter three
eggs unscrambled

Books

Beranbaum, Rose Levy. *The Cake Bible*. New York: William Morrow and Company, Inc., 1988.

Grausman, Richard. *At Home with French Classics*. New York: Workman Publishing, 1988.

Kennedy, Diana. *The Cuisines of Mexico*. New York: Harper & Row, Publishers, 1972.

Medrich, Alice. *Chocolate and the Art of Low-Fat Desserts*. New York: Warner Books, Inc., 1994.

Purdy, Susan G. *Have Your Cake and Eat It, Too*. New York: William Morrow and Company, Inc., 1993.

Roux, Albert, and Michel Roux. *The New Classic Cuisine*. Woodbury, N. Y.: Barron's Educational Series, Inc., 1984.

Stadelman, William J., and Owen J. Cotterill, ed. *Egg Science and Technology*, 3rd ed. Westport, Conn.: AVI Publishing Company, Inc., 1986.

Wolfert, Paula. *The Cooking of South-West France*. Garden City, N. Y.: The Dial Press/Doubleday & Company, Inc., 1983.

Articles

American Egg Board. *Eggcyclopedia*. Park Ridge, Ill., 1989. Booklet.

American Egg Board. *The Incredible Edible Egg*. Park Ridge, Ill., 1989. Booklet.

Andross, M. "Effect of Cooking on Eggs." *Chemistry and Industry*, Vol. 59 (1940), pp. 449–454.

Bailey, M. Irene. "Foaming of Egg White." *Industrial and Engineering Chemistry*, Vol. 27, No. 8 (August 1935), pp. 973–976.

Baker, R. C., technical advisor. *A Scientist Speaks About: Salmonellae*. Park Ridge, Ill.: American Egg Board, 1987. Leaflet.

———, and Charlotte Bruce. *A Scientist Speaks About: The Microbiology of Eggs*. Park Ridge, Ill.: American Egg Board, 1987. Leaflet.

Barmore, Mark A. "The Influence of Chemical and Physical Factors on Egg-White Foams." The Colorado Agricultural College, Colorado [Agricultural] Experiment Station, Technical Bulletin No. 9, 1934, pp. 3–57.

————. "Baking Angel Food Cake at Any Altitude." The Colorado Agricultural College, Colorado [Agricultural] Experiment Station, Technical Bulletin No. 13, ca. 1935, pp. 4–15.

Carr, Ruth E., and G. M. Trout. "Some Cooking Qualities of Homogenized Milk. I. Baked and Soft Custard." *Food Research,* Vol. 7 (1942), pp. 360–369.

Clinger, Caroline, Arlene Young, Inez Prudent, and A. R. Winter. "The Influence of Pasteurization, Freezing, and Storage on the Functional Properties of Egg White." *Food Technology* (April 1951), pp. 166–170.

Cotterill, Owen J., and William J. Stadelman (revised by). *A Scientist Speaks About: Egg Products.* Park Ridge, Ill.: American Egg Board, 1990. Leaflet.

Dawson, E. H., C. Miller, and R. A. Redstrom. "Cooking Quality and Flavor of Eggs as Related to Candled Quality, Storage Conditions and Other Factors." U. S. Department of Agriculture, Agriculture Information Bulletin No. 164, 1956, 44 pages.

Forsythe, Richard H., and Dwight H. Bergquist. "The Effect of Physical Treatments on Some Properties of Egg White." *Poultry Science,* Vol. 30 (1951), pp. 302–311.

Gillis, Jean Neill, and Natalie K. Fitch. "Leakage of Baked Soft-Meringue Topping." *J. Home Economics,* Vol. 48, No. 9 (November 1956), pp. 703–707.

Grosser, Arthur E. "The Culinary Alchemy of Eggs." *The Exploratorium,* Vol. 7, Issue 4 (Winter 1983/1984), pp. 11–14.

Grunden, L. P., E. J. Mulnix, J. M. Darfler, and R. C. Baker. "Yolk Position in Hard Cooked Eggs as Related to Heredity, Age and Cooking Position." *Poultry Science,* Vol. 54 (1975), pp. 546–552.

Hasiak, R. J., D. V. Vadehra, R. C. Baker, and L. Hood. "Effect of Certain Physical and Chemical Treatments on the Microstructure of Egg Yolk." *J. Food Science,* Vol. 37 (1972), pp. 913–917.

Henry, W. C., and A. D. Barbour. "Beating Properties of Egg White." *Industrial and Engineering Chemistry,* Vol. 25, No. 9 (September 1933), pp. 1054–1058.

Hester, E. Elizabeth, and Catherine J. Personius. "Factors Affecting the Beading and Leakage of Soft Meringues." *Food Technology,* Vol. 3 (July 1949), pp. 236–240.

MacDonnell, L. R., et al. "The Functional Properties of the Egg White Proteins." *Food Technology,* Vol. 9 (February 1955), pp. 49–53.

Murthy, G. K. "Thermal Inactivation of Alpha-Amylase in Various Liquid Egg Products." *J. Food Science,* Vol. 35, No. 4 (July/August 1970), pp. 352–356.

Nielson, Hester J., Jean D. Hewitt, and Natalie K. Fitch. "Factors Influencing Consistency of a Lemon-Pie Filling." *J. Home Economics,* Vol. 44, No. 10 (December 1952), pp. 782–785.

Pyler, E. J. "Basic Factors in the Production of Angel Food Cake." *Bakers Digest,* Vol. 25 (April 1951), pp. 35–37, 39.

Reinke, William C., John V. Spencer, and Lydia J. Tryhnew. "The Effect of Storage Upon the Chemical, Physical and Functional Properties of Chicken Eggs." *Poultry Science,* Vol. 52 (1973), pp. 692–702.

Seidman, W. E., O. J. Cotterill, and E. M. Funk. "Factors Affecting Heat Coagulation of Egg White." *Poultry Science,* Vol. 42 (1963), pp. 406–417.

Swanson, M. H. "Some Observations on the Peeling Problem of Fresh and Shell Treated Eggs When Cooked." *Poultry Science,* Vol. 38 (1959), pp. 1253–1254.

Upchurch, R., and R. E. Baldwin. "Guar Gum and Triacetin in Meringues and a Meringue Product Cooked by Microwaves." *Food Technology,* Vol. 22 (October 1968), pp. 107–108.

Wang, Anne C., Kaye Funk, and Mary E. Zabik. "Effect of Sucrose on the Quality Characteristics of Baked Custards." *Poultry Science,* Vol. 53, No. 2 (March 1974), pp. 807–813.

Chapter four
sauce sense

Books

Beard, James A. *James Beard's Theory and Practice of Good Cooking.* New York: Alfred A. Knopf, 1977.

Beranbaum, Rose Levy. *The Cake Bible.* New York: William Morrow and Company, Inc., 1988.

Charley, Helen. *Food Science,* 1st ed. New York: The Ronald Press Company, 1970.

————. *Food Science,* 2nd ed. New York: John Wiley & Sons, 1982.

Child, Julia. *The Way to Cook.* New York: Alfred A. Knopf, Inc., 1989.

Corran, J. W. "Some Observations on a Typical Food Emulsion." *Emulsion Technology: Theoretical and Applied,* 2nd ed. Brooklyn: Chemical Publishing Company, 1946, pp. 176–192.

Friberg, S. "Emulsion Stability," *Food Emulsions.* S. Friberg, ed. New York: Marcel Dekker, Inc., 1976, pp. 1–37.

Krog, N., and J. B. Lauridsen. "Food Emulsifiers and Their Association with Water," *Food Emulsions.* S. Friberg, ed. New York: Marcel Dekker, Inc., 1976, pp. 67–139.

McGee, Harold. *The Curious Cook: More Kitchen Science and Lore.* San Francisco: North Point Press, 1990.

Osman, Elizabeth. "Starch and Other Polysaccharides." *Food Theory and Applications,* P. C. Paul and H. H. Palmer, eds. New York: John Wiley & Sons, Inc., 1972, pp. 151–212.

Paul, Pauline C., and Helen H. Palmer. "Colloidal Systems and Emulsions." *Food Theory and Applications,* P. C. Paul and H. H. Palmer, ed. New York: John Wiley & Sons, Inc., 1972, pp. 77–114.

Peterson, James. *Sauces: Classical and Contemporary Sauce Making.* New York: Van Nostrand Reinhold, 1991.

Prudhomme, Paul. *Chef Paul Prudhomme's Louisiana Kitchen.* New York: William Morrow and Company, Inc., 1983.

Rydhag, Lisbeth. "The Effect of Temperature and Time on Emulsion Stability." *Physical, Chemical and Biological Changes in Food Caused by Thermal Processing,* Tore Hoyem and Oskar Kvale, ed. International Union of Food Science and Technology. London: Applied Science Publishers, 1977, pp. 224–238.

Sokolov, Raymond. *The Saucier's Apprentice: A Modern Guide to Classic French Sauces for the Home.* New York: Alfred A. Knopf, 1976.

Articles

Campbell, Ada Marie, and Alice M. Briant. "Wheat Starch Pastes and Gels Containing Citric Acid and Sucrose." *Food Research,* Vol. 22 (1957), pp. 358–366.

Lauridsen, Jens Birk. "Food Emulsifiers: Surface Activity, Edibility, Manufacture, Composition, and Application." *J. American Oil Chemists' Society,* Vol. 53 (June 1976), pp. 400–407.

Leach, Harry W., and Thomas J. Schoch. "Structure of the Starch Granule: II. Action of Various Amylases on Granular Starches." *Cereal Chemistry,* Vol. 38 (January 1961), pp. 34–46.

Miller, B. S., R. I. Derby, and H. B. Trimbo. "A Pictorial Explanation for the Increase in Viscosity of a Heated Wheat Starch-Water Suspension." *Cereal Chemistry,* Vol. 50, No. 3 (1973), pp. 271–280.

Petrowski, Gary E. "Emulsion Stability and Its Relation to Foods." *Advances in Food Research,* Vol. 22 (1976), pp. 309–359.

Small, D. M., and Michael Bernstein. "Doctor in the Kitchen: Experiments on Sauce Béarnaise." *New England J. Medicine,* Vol. 300, No. 14 (April 5, 1979), pp. 801–802.

Walker, Jearl. "The Physics and Chemistry of a Failed Sauce Béarnaise." *Scientific American,* Vol. 241 (December 1979), pp. 178 ff.

———. "The Amateur Scientist [Béarnaise]." *Scientific American* (January 1981), pp. 168–169.

Chapter five
treasures of the earth

Books

Colbin, Annemarie. *The Book of Whole Meals: A Seasonal Guide to Assembling Balanced Vegetarian Breakfasts, Lunches and Dinners.* New York: Ballantine Books, 1985.

———. *The Natural Gourmet: Delicious Recipes for Healthy, Balanced Eating.* New York: Ballantine Books, 1989.

Goode, John, and Carol Willson. *Fruit and Vegetables of the World.* Melbourne: Lothian Publishing Company, 1987.

Hall, E. G., *Mixed Storage of Foodstuffs.* Sidney: CSIRO: Food Research Circular No. 9, 1973. Reference only.

National Research Council, Food Protection Committee. *Toxicants Occurring Naturally in Foods,* 2nd ed. Washington, D.C.: National Academy of Sciences, 1973.

Odigboh, E. U. "Cassava: Production, Processing, and Utilization." *Handbook of Tropical Foods.* Harvey T. Chan, Jr., ed. New York: Marcel Dekker, Inc., 1983.

Peirce, Lincoln C. *Vegetables: Characteristics, Production, and Marketing.* New York: John Wiley and Sons, 1987.

Schmutz, Ervin M., and Lucretia Breazeale Hamilton. *Plants That Poison.* Flagstaff, Ariz.: Northland Publishing, 1979.

Schneider, Elizabeth. *Uncommon Fruits and Vegetables: A Commonsense Guide.* New York: Harper & Row, 1986.

Shewfelt, Robert L. "Flavor and Color of Fruits as Affected by Processing." *Commercial Fruit Processing,* 2nd ed. Chapter 11. Westport, Conn.: AVI Publishing Co., Inc., 1986.

———. "Food Crops: Postharvest Deterioration." *Encyclopedia of Food Science and Technology.* New York: John Wiley & Sons, Inc., 1991. pp. 1019–1023. 4 Vols.

———. "Food Crops: Sensory Evaluation." *Encyclopedia of Food Science and Technology.* New York: John Wiley & Sons, Inc., 1991. pp. 1023–1026. 4 Vols.

———. "Food Crops: Varietal Differences, Maturation, Ripening, and Senescence." *Encyclopedia of Food Science and Technology.* New York: John Wiley & Sons, Inc., 1991. pp. 1029–1032. 4 Vols.

———, and Stanley E. Prussia, ed. *Postharvest Handling: A Systems Approach.* San Diego: Academic Press, Inc. (Harcourt Brace Jovanovich, Publishers), 1993.

Tropp, Barbara. *China Moon Cookbook.* New York: Workman Publishing Company, Inc., 1992.

Waldron, Maggie. *Cold Spaghetti at Midnight: Feel-Good Foods to Heal Your Body and Soothe Your Soul.* New York: William Morrow and Company, Inc., 1992.

Wills, R.B.H., et al. *Postharvest: An Introduction to the Physiology and Handling of Fruit and Vegetables.* New York: Van Nostrand Reinhold (An AVI Book), 1989.

Articles

Altman, Daryl, and Steve Taylor. "Understanding Food Allergies." Workshop No. 17, Seventeenth Annual Conference, International Association of Culinary Professionals, San Antonio, Texas, April 5–9, 1995.

Anonymous. "Preservation of Vegetables in Oil and Vinegar." Fact Sheet, Commonwealth Scientific and Research Organisation [CSIRO-Australia], Division of Food Science and Technology, April 1994.

Anonymous. "Herbs or Garlic in Oil: Dangerous Combos." *Environmental Nutrition,* Vol. 18, No. 8 (1995), p. 3. Abstract only.

Bushway, R. J., A. Yang, and A. M. Yamani. "Comparison of Alpha- and Beta-carotene Content of Supermarket versus Roadside Stand Produce." *J. Food Quality,* Vol. 9, No. 6 (1987), pp. 437–443.

Bushway, R. J., et al. "Comparison of Ascorbic Acid Content of Supermarket versus Roadside Stand Produce." *J. Food Quality,* Vol. 12, No. 2 (1989), pp. 99–105.

Doyle, Michael P. "Should Regulatory Agencies Reconsider the Policy of Zero-Tolerance of *Listeria monocytogenes* in all Ready-to-Eat Foods?" *Food Safety Notebook,* Vol. 2, Nos. 10/11 (October/November 1991), pp. 89–91.

Eitenmiller, Ronald, et al. "Nutrient Composition of Red Delicious Apples, Peaches, Honeydew Melons, Florida Pink and Texas Ruby Red Grapefruit, and Florida Oranges," Research Report 526, University of Georgia, Agricultural Experiment Stations, December 1987, 21 pp.

Fennema, O. "Loss of Vitamins in Fresh and Frozen Food." *Food Technology,* Vol. 31 (December 1977), pp. 32–35, 38.

Hefle, Susan L. "The Chemistry and Biology of Food Allergens." *Food Technology.* Vol. 50, No. 3 (March 1996), pp. 88–92.

Ko, N. P., et al. "Storage of Spinach Under Low Oxygen Atmosphere Above the Extinction Point." *J. Food Science,* Vol. 61, No. 2 (1996), pp. 398–400, 406.

Maduagwu, E. N., and D.H.E. Oben, "Effects of Processing of Grated Cassava Roots by the 'Screw Press' and by Traditional Fermentation Methods on the Cyanide Content of Gari," *J. Food Technology,* Vol. 16 (1981), pp. 299–302.

Sapers, Gerald M., and Robert L. Miller. "Heated Ascorbic/Citric Acid Solution as Browning Inhibitor for Pre-Peeled Potatoes." *J. Food Science,* Vol. 60, No. 4 (1995), pp. 762–766, 776.

Shewfelt, R. L. "Postharvest Treatment for Extending the Shelf Life of Fruits and Vegetables." *Food Technology,* Vol. 40, No. 5 (May 1986), pp. 70–72, 74, 76–78, 80, 89.

Speer, F. "Food Allergy: The 10 Common Offenders." *American Family Physician,* Vol. 13, No. 2 (1976), pp. 106–112. Abstract only.

Steingarten, Jeffrey. "Do the Mash [potatoes]." *HG* (January 1989), pp. 42, 44.

———. "Ripe Now [fruits]." *Vogue* (August 1992), pp. 164, 166, 168, 196.

Van Buren, J. P., et al. "Effects of Salts and pH on Heating-Related Softening of Snap Beans." *J. Food Science,* Vol. 55, No. 5 (September–October 1990), pp. 1312–1314, 1330.

Chapter six
fine fare from land, sea, and air

Books

Bugialli, Giuliano. *The Fine Art of Italian Cooking.* New York: Times Books, 1977.

Davidson, Alan. *Mediterranean Seafood.* Baton Rouge, La.: Louisiana State University Press, 1981.

———. *North Atlantic Seafood,* reprint ed. New York: HarperCollins Publishers, Inc., 1989.

Editors of Time-Life Books. *Beef and Veal.* The Good Cook Techniques & Recipes Series. Alexandria, Va.: Time-Life Books, Inc., 1978.

Ellis, Merle. *Cutting-up in the Kitchen: The Butcher's Guide to Saving Money on Meat and Poultry.* San Francisco: Chronicle Books, 1975.

Famularo, Joe. *The Joy of Grilling,* rev. ed. Hauppauge, N.Y.: Barron's Educational Series, Inc., 1988.

Fletcher, Helen S. *The New Pastry Cook.* New York: William Morrow and Company, Inc., 1986.

Hultin, Herbert O. "Characteristics of Muscle Tissue." *Principles of Food Science. Part 1: Food Chemistry,* Owen Fennema, ed. Chapter 13. Food Science Series, Vol. 4. New York: Marcel Dekker, 1976.

Kafka, Barbara. *Roasting.* New York: William Morrow and Company, Inc., 1995.

Kamman, Madeleine. *The Making of a Cook.* New York: Weathervane Books, 1971.

Lee, Frank A. "Proteins." *Basic Food Chemistry,* Chapter 3. Westport, Conn.: AVI Publishing Co., Inc., 1975.

Levie, Albert. *Meat Handbook,* 4th ed. Westport, Conn.: AVI Publishing Co., Inc., 1979.

McGee, Harold. *The Curious Cook: More Kitchen Science and Lore.* San Francisco: North Point Press, 1990.

Montagné, Prosper. *Larousse Gastronomique,* 1st American ed. Charlotte Turgeon and Nina Froud, ed. New York: Crown Publishers, Inc., 1961.

National Live Stock and Meat Board. *The Meat Board's Lessons on Meat.* Chicago: National Live Stock and Meat Board, 1991.

Price, James F., and Bernard S. Schweigert. *The Science of Meat and Meat Products,* 3rd ed. Westport, Conn.: Food & Nutrition Press, Inc., 1987.

Schlesinger, Chris, and John Willoughby. *The Thrill of the Grill.* New York: William Morrow and Company, Inc., 1990.

Schulz, Phillip S. *Cooking with Fire and Smoke.* New York: Simon & Schuster, 1986.

Sinnes, A. Cort. *The Grilling Encyclopedia: An A-to-Z Compendium of How to Grill Almost Anything.* New York: Atlantic Monthly Press, 1992.

Articles

Anderson, Pam, with Karen Tack. "The Holiday Turkey Perfected." *Cook's Illustrated,* Vol. 1, No. 5 (November/December 1993), pp. 18–21.

———. "Perfect Prime Rib." *Cook's Illustrated,* No. 17 (November/December 1995), pp. 12–14.

———. "Roast Turkey Revisited." *Cook's Illustrated,* No. 17 (November/December 1995), p. 9.

Bendall, J. R., and D. J. Restall. "The Cooking of Single Myofibres, Small Myofibre Bundles and Muscle Strips from Beef. *M. psoas* and *M. sternomandibularis* Muscles at Varying Heating Rates and Temperatures." *Meat Science,* Vol. 8 (1983), pp. 93–117.

Gilbert, J., and M. E. Knowles. "The Chemistry of Smoked Foods: A Review." *J. Food Technology,* Vol. 10 (1975), pp. 245–261.

McGee, Harold. "The Way to Roast a Chicken." *Cook's Illustrated,* Charter Issue (1992), pp. 26–27.

Schwartz, David M. "Play with Your Food More!" *Smithsonian,* Vol. 23, No. 9 (December 1992), pp. 110–112, 114, 116, 118–119.

Steingarten, Jeffrey. "Fish Out of Water." *Vogue* (September 1992), pp. 572–575, 623–624.

Ubaldi, Jack, and Elizabeth Crossman. "Beef Basics—All the Cuts and How to Cook Them." *Food & Wine* (January 1993), pp. 85–86, 89.

Chapter seven
sweet thoughts
and chocolate dreams

Books

Arbuckle, W. S. *Ice Cream,* 3rd ed. Westport, Conn.: AVI Publishing Company, Inc., 1977.

Charley, Helen. *Food Science,* 1st ed. New York: The Ronald Press Company, 1970.

———. *Food Science,* 2nd ed. New York: John Wiley & Sons, 1982.

Editors of Time-Life Books. *Candy.* The Good Cook Techniques & Recipes Series. Alexandria, Va.: Time-Life Books, Inc., 1981.

Fletcher, Helen S. *The New Pastry Cook.* New York: William Morrow and Company, Inc., 1986.

González, Elaine. *Chocolate Artistry: Techniques for Molding, Decorating, and Designing with Chocolate.* Chicago: Contemporary Books, Inc., 1983.

Healy, Bruce, and Paul Bugat. *Mastering the Art of French Pastry.* Woodbury, N.Y.: Barron's Educational Series, Inc., 1984.

Healy, Bruce, with Paul Bugat. *The French Cookie Book: Classic and Contemporary Recipes for Easy and Elegant Cookies.* New York: William Morrow and Company, Inc., 1994.

Heatter, Maida. *Maida Heatter's Book of Great Chocolate Desserts.* New York: Alfred A. Knopf, 1984.

Hoffman, Mable, and Gar Hoffman. *Ice Cream.* Tucson, Ariz.: H.P. Books, 1981.

McGee, Harold. *The Curious Cook: More Kitchen Science and Lore.* San Francisco: North Point Press, 1990.

Medrich, Alice. *Cocolat: Extraordinary Chocolate Desserts.* New York: Warner Books, Inc., 1990.

————. *Chocolate and the Art of Low-Fat Desserts.* New York: Warner Books, Inc., 1994.

Minifie, Bernard W. *Chocolate, Cocoa and Confectionery: Science and Technology,* 2nd ed. Westport, Conn.: AVI Publishing Company, Inc., 1980.

Prichard, Anita. *Anita Prichard's Complete Candy Cookbook.* New York: Harmony Books, 1978.

Rombauer, Irma S., and Marion Rombauer Becker. *The Joy of Cooking.* New York: Bobbs-Merrill Company, Inc., 1975.

Articles

Anonymous. "Accompaniments [granitas]." *Cooking,* Issue No. 3 (May 1978) pp. 26–28.

Anonymous. "An Ice Cream that Comes from Afar [gelato]." *Cucina,* Vol. 4, No. 3 (June–July 1994) pp. 6–15.

Beranbaum, Rose Levy. "All You Need to Know About Cooking with Chocolate." *Food & Wine* (February 1984), pp. 34, 36, 38, 72.

Beard, James. "Sorbets." *Cooking,* Issue No. 3 (May 1978) pp. 22–25.

Berger, K. G., and G. W. White. "An Electron Microscopical Investigation of Fat Destabilization in Ice Cream." *J. Food Technology,* Vol. 6 (1971) pp. 285–294.

Bryant, A. P. "Why Candy Sweats." *Candy Manufacturer* (1922), 1 p.

Duck, William. "A Study of the Consistency of Caramel." *Manufacturing Confectioner,* Vol. 39 (June 1959), pp. 29–31.

Lachmann, Alfred, and Henry Voll. "Structure and Behavior of Icings." *Bakers Digest* (April 1969), pp. 40–41, 44–45, 73.

Lin, Po-Min, and J. G. Leeder. "Mechanism of Elmusifier Action in an Ice Cream System." *J. Food Science,* Vol. 39 (1974), pp. 108–111.

Mariani, John. "Sensuous Sorbets." *Harper's Bazaar* (April 1991) pp. 173, 182.

Mastro, Diane Posner. "Granitas." *Fine Cooking,* Issue No. 2 (April/May 1994), pp. 53–57.

Nielsen, Chat, Jr. "The Story of Vanilla." Lake Forest, Ill.: Nielsen-Massey Vanillas, Inc., October 1981. Booklet.

O'Neill, Molly. "A Jolt of Cool [ice cream]." *The New York Times Magazine* (June 28, 1992) pp. 41–42.

Schultz, Dodi. "All About Chocolate." *FDA Consumer* (July–August 1994), pp. 24–25.

————. "Candy: How Sweet It Is." *FDA Consumer* (July–August 1994), pp. 21–23.

Schwain, Frank R. "Icing Pointers." *Bakers Digest* (August 1951), pp. 30–32.

Sherman, P. "The Texture of Ice Cream." *J. Food Science,* Vol. 30 (1965), pp. 201–211.

Steingarten, Jeffrey. "The Big Chill [ice cream]." *HG,* Vol. 160 (August 1988), pp. 132–133, 166–167.

Toll, Jack. "'Fats, Oils and Emulsifiers' . . . Used in Confectionery." *Manufacturing Confectioner,* Vol. 37, No. 4 (April 1957), pp. 19–21, 23–24, 26.

Welch, R. C. "Milk and Milk Chocolate." *Manufacturing Confectioner,* Vol. 38, No. 11 (November 1958), pp. 17, 19, 21, 46.

Index